Spectroscopic Probes of Quantum Matter

Spectroscopic Probes of Quantum Matter

Christophe Berthod
University of Geneva, Switzerland

IOP Publishing, Bristol, UK

ISBN 978-0-7503-1741-2 (ebook)
ISBN 978-0-7503-1739-9 (print)
ISBN 978-0-7503-1740-5 (mobi)

DOI 10.1088/978-0-7503-1741-2

Version: 20181101

IOP Expanding Physics
ISSN 2053-2563 (online)
ISSN 2054-7315 (print)

British Library Cataloguing-in-Publication Data: A catalogue record for this book is available from the British Library.

Published by IOP Publishing, wholly owned by The Institute of Physics, London

IOP Publishing, Temple Circus, Temple Way, Bristol, BS1 6HG, UK

US Office: IOP Publishing, Inc., 190 North Independence Mall West, Suite 601, Philadelphia, PA 19106, USA

To my beloved large family, which is also great.

‘Science is what we understand well enough to explain to a computer. Art is everything else we do.’

Donald Knuth

Contents

Preface

Students entering condensed-matter physics are exposed to a remarkable diversity of physical systems. No less remarkable is the diversity of experimental and theoretical tools that are used to study them. Behind this profusion lies the unifying principle that experiments probe correlations in quantum systems. What kind of correlations are probed by specific experiments and how to address these correlations theoretically is the common theme of this textbook, which focuses on spectroscopic probes.

Many excellent books and reviews describe a particular experimental technique and, along the way, digress to launch general notions of many-body quantum theory. The present textbook uses a different strategy. It grew out of a course I first taught in 2008 at the University of Geneva. The course was mostly devoted to students starting their graduate studies in experimental condensed-matter physics. As the intent was to describe several techniques using a common formalism and a unified notation, the approach has been to present the required elements of many-body quantum theory in a first part and use this common language in a second part dedicated to spectroscopies.

With few exceptions, the presentation is self-contained and goes into detail through the maths. I made every effort to separate steps that are conceptually distinct: general principles, physically founded assumptions, and mere mathematical developments are marked as such. As the latter would easily burden the text, they are moved to separate sections at the end of each chapter. This is complemented by a number of practical applications, often accompanied by a little computer program, that bridge the gap between abstract formula and concrete numbers.

Although the core material displayed in this textbook is familiar to experts, several developments are not so easily found in the literature. I assume responsibility for what is shown and look forward to learning from readers about possible mistakes. The bibliography is cursory. My goal was not to provide an extensive review of published works and I apologize to all those who deserve to be cited and are not.

It is a pleasure to thank the students for their involvement during the course and many colleagues for numerous discussions and continued support, especially Bernard Giovannini, Thierry Giamarchi, Antoine Georges, Dirk van der Marel, Thomas Jarlborg and Shintaro Takayoshi.

Author biography

Christophe Berthod

Christophe Berthod is senior lecturer in the Department of Quantum Matter Physics at the University of Geneva, Switzerland. He obtained his physics degree in 1994 for work on ballistic electron transport and his PhD in 1998 for electronic-structure calculations of metal-semiconductor interfaces. He joined University of Geneva in 1999 and since then conducts research on superconductivity, correlated electrons, and computational physics.

List of symbols

$\langle \cdots \rangle$	thermal average, equation (2.2)
$\langle \cdots \rangle_0$	thermal average with respect to a quadratic Hamiltonian
$\langle \cdots \rangle_{\mathrm{imp}}$	impurity average
$\langle \cdots \rangle_V$	thermal average in the presence of an external perturbation V
$[\cdot, \cdot]_\pm$	commutator or anti-commutator, equation (4)
$\hat{X}(t)$	operator X in the interaction picture, equation (2.19*a*)
0^+	positive infinitesimal
$\lvert \varnothing \rangle$	vacuum state
$\mathbb{1}$	identity operator
(1)	abbreviated set of coordinates: $(1) \equiv (\boldsymbol{r}_1\sigma_1)$
$\int d1$	short notation for $\int d\boldsymbol{r}_1 \sum_{\sigma_1}$
$*$	general scalar product: $A*B = \sum_\mu \int d\boldsymbol{r} A_\mu(\boldsymbol{r}) B_\mu(\boldsymbol{r})$
A	generic operator
$\boldsymbol{A}(\boldsymbol{r}, t)$	vector potential
$A(\boldsymbol{k}, \varepsilon)$	one-electron spectral function in momentum space
$\lvert a \rangle$	many-particle state, eigenstate of H
a_α	generic annihilation operator for single-particle state α
$a_\alpha^\dagger$	generic creation operator for single-particle state α
a_0	Bohr radius, $a_0 = 4\pi\epsilon_0\hbar^2/(me^2)$.
α	general single-particle state index
B	generic operator
$B_{\boldsymbol{q}\lambda}^\dagger$	phonon operator, equation (5.49)
$\lvert b \rangle$	many-particle state, eigenstate of H
$b_\alpha^\dagger$	creation operator for single-particle bosonic state α
$b_{\boldsymbol{q}\lambda}^\dagger$	creation operator for phonon of momentum $\boldsymbol{q}$ and polarization λ
$b(\xi)$	Bose–Einstein distribution function, $b(\xi) = (e^{\beta\xi} - 1)^{-1}$
b	nuclear scattering length (chapter 10)
β	general single-particle state index
β	inverse temperature, $\beta = (k_{\mathrm{B}}T)^{-1}$
$C_{AB}(t)$	time-ordered correlation function of the operators A and B, equation (3.1)
$C_{AB}^R(t)$	retarded correlation function of the operators A and B, equation (3.4)
$C_{AB}^A(t)$	advanced correlation function of the operators A and B, equation (3.5)
$C_{AB}^>(t)$	greater correlation function of the operators A and B, equation (3.6)
$C_{AB}^<(t)$	lesser correlation function of the operators A and B, equation (3.7)
$\mathscr{C}_{AB}(\tau)$	imaginary-time correlation function of the operators A and B, equation (4.5)
$\mathscr{C}_{\alpha\beta\gamma\delta}(\tau)$	particle–hole propagator, equation (5.59)
$C_{\mathscr{V}}$	specific heat at constant volume, equation (2.9)
$C_{\mathscr{V}}^{\mathrm{el}}$	electronic specific heat
$C_{\mathscr{V}}^{\mathrm{ph}}$	phononic specific heat
$c_\alpha^\dagger$	creation operator for single-particle fermionic state α
$\chi_{AB}(t)$	linear susceptibility, equation (6.8)

$\chi^{(2)}_{AB}(t, t')$	second-order susceptibility, equation (6.10)
$\chi^{\mu\nu}_{jj}(\boldsymbol{q}, \varepsilon)$	current–current correlation function
$\chi_{nn}(\boldsymbol{q}, \varepsilon)$	density–density correlation function, or charge susceptibility
$\chi_{ss}(\boldsymbol{q}, \varepsilon)$	spin–spin correlation function, or spin susceptibility
$\chi_{\mathcal{M}\mathcal{M}}(\boldsymbol{q}, \varepsilon)$	magnetization–magnetization correlation function
$\boldsymbol{D}(\boldsymbol{r}, t)$	electric displacement field
$D(\boldsymbol{k})$	dynamical matrix
$\mathscr{D}^0_\lambda(\boldsymbol{q}, i\Omega_n)$	free phonon propagator, equation (5.51)
$d_\pm(\xi)$	Bose–Einstein and Fermi–Dirac distribution functions, equation (6)
$d\sigma/d\Omega$	scattering cross section
∂_t	derivative with respect to t
Δ	superconducting order parameter
$\delta_{\alpha\beta}$	Kronecker symbol: $\delta_{\alpha\beta} = 1$ if $\alpha = \beta$, $\delta_{\alpha\beta} = 0$ otherwise
$\delta(x)$	Dirac delta function
$\boldsymbol{E}(\boldsymbol{r}, t)$	electric field
E_a	energy of the many-body state $\lvert a\rangle$
$E_{\boldsymbol{k}}$	energy of the Bogoliubov excitations, $E_{\boldsymbol{k}} = \sqrt{\xi_{\boldsymbol{k}}^2 + \lvert\Delta_{\boldsymbol{k}}\rvert^2}$; quasi-particle energy
e	electron charge, $e = -\lvert e\rvert$.
ϵ_0	vacuum permittivity
$\epsilon_\parallel(\boldsymbol{q}, \omega)$	longitudinal dielectric function, equation (5.91)
ε	energy variable
ε^+	short notation for $\varepsilon + i0^+$
ε_α	single-particle energy level
ε_{F}	Fermi energy
$\varepsilon_{\boldsymbol{k}n}$	electronic energy level (band structure)
$\epsilon_{\alpha\beta\gamma}$	Levi-Civita symbol, equation (23)
$\boldsymbol{\epsilon}_{\boldsymbol{q}\lambda}$	vector giving the displacement of atoms for the phonon $(\boldsymbol{q}, \lambda)$
$F_\lambda(\boldsymbol{q})$	phonon form factor, equation (10.20)
$\mathscr{F}(\boldsymbol{q})$	magnetic form factor, equation (10.35)
$\mathscr{F}^\dagger_{\alpha\beta}(\tau)$	imaginary-time anomalous Green's function, equation (5.116)
$f(\xi)$	Fermi–Dirac distribution function, $f(\xi) = (e^{\beta\xi} + 1)^{-1}$
ϕ	work function
$\phi(\boldsymbol{r}, t)$	electric potential
$\varphi_\alpha(1)$	generic one-particle basis function
$\boldsymbol{G}$	reciprocal-lattice vector, equation (2.51)
$\mathscr{G}_{\alpha\beta}(\tau)$	imaginary-time one-particle Green's function, equation (5.6)
$G(\boldsymbol{k}, \varepsilon)$	one-particle Green's function in momentum space
$g_{\boldsymbol{q}\lambda}$	electron–phonon coupling vertex, equation (2.74)
$\boldsymbol{\Gamma}(K, Q)$	renormalized current vertex
$\Gamma_c(K, Q)$	renormalized density vertex
$\Gamma_{a\to b}$	transition rate between states $\lvert a\rangle$ and $\lvert b\rangle$
$\gamma^\dagger_{\alpha\sigma}$	creation operator for the single-electron state φ_α
H	Hamiltonian

$H_{\text{el-ph}}$	electron–phonon Hamiltonian, equation (2.73)
H_{A}	Anderson Hamiltonian, equation (8.23)
H_{sd}	Kondo Hamiltonian, equation (8.24)
$\boldsymbol{H}(\boldsymbol{r}, t)$	applied magnetic field
$h(\boldsymbol{k})$	band Hamiltonian, equation (2.52)
η	specifies the particle statistics: $\eta = +1$ for bosons, $\eta = -1$ for fermions
I_s	single-particle tunneling current
$I_J(t)$	Josephson tunneling current
Im	imaginary part
i	$\sqrt{-1}$
$i\nu_n$	generic Matsubara frequency $i\Omega_n$ or $i\omega_n$
$i\bar{\nu}_j$	short notation for $i\nu_n^{(j)}$; see remark at the end of Doc A5.3
$i\Omega_n$	bosonic Matsubara frequency, $\Omega_n = 2n\pi k_{\text{B}}T$
$i\omega_n$	fermionic Matsubara frequency, $\omega_n = (2n + 1)\pi k_{\text{B}}T$
J	exchange coupling (Section 8.5.1)
$J_n(x)$	Bessel function of the first kind
$\boldsymbol{j}(\boldsymbol{r})$	current operator, equation (2.49)
$\boldsymbol{j}^p(\boldsymbol{r})$	paramagnetic current operator, equation (2.49)
$\boldsymbol{j}^d(\boldsymbol{r})$	diamagnetic current operator, equation (2.49)
K	grand Hamiltonian, $K = H - \mu N$
K	short 'four-vector' notation for $(\boldsymbol{k}, \sigma, i\omega_n)$
K_0	one-body part of K, equation (2.45)
K_a	eigenvalue of K for eigenstate $\vert a\rangle$: $K_a = E_a - \mu N_a$
k	norm of the vector $\boldsymbol{k}$
$\boldsymbol{k}$	vector in reciprocal space
k_{B}	Boltzmann constant
k_{F}	Fermi wave vector
k_{TF}	Thomas–Fermi wave vector, $k_{\text{TF}}^2 = (e^2/\epsilon_0)N_0^{\text{el}}(0)/\mathscr{V}$
κ_T	isothermal compressibility, equation (2.11)
$\Lambda(K, K', Q)$	vertex function, equation (8.37)
λ	dispersion renormalization, coupling strength, equation (7.22)
M_ν	mass of atom ν in a crystal
$\boldsymbol{\mathcal{M}}(\boldsymbol{q})$	dimensionless total magnetization (section 10.3)
m	electron mass
m^*	dynamical effective mass
m_b	band mass
m_n	neutron mass
$\mu,\ \mu_i$	generic index
μ	chemical potential
N	number of particles, particle-number operator
N_a	number of atoms in the unit cell of a crystal
N_a	number of particles in the many-body state $\vert a\rangle$
N_i	number of impurities
$\mathscr{N}$	number of lattice sites

$N(\varepsilon)$	density of states (DOS)
$N(\boldsymbol{r}, \varepsilon)$	local density of states (LDOS)
$N^{\text{el}}(0)$	Fermi-level density of states; depending upon the context, it should be understood as the total DOS or the DOS per unit volume
$\tilde{N}^{\text{el}}(0)$	Fermi-level density of states *per spin*, $\tilde{N}^{\text{el}}(0) = \frac{1}{2}N^{\text{el}}(0)$
$N^{\text{ph}}(\varepsilon)$	phonon density of states
n	particle number density
$\boldsymbol{n}$	unit vector
$n(\boldsymbol{r})$	local particle number density
$n(\boldsymbol{q})$	Fourier transform of $n(\boldsymbol{r})$
n_α	particle number operator for single-particle state α
n_i	impurity density, $n_i = N_i/\mathscr{V}$
$\nu, \quad \nu_i$	generic index
$\mathscr{P}$	Cauchy principal value, equation (9)
$\mathscr{P}$	permulation
p	pressure, equation (2.8)
$\Pi_{\alpha\beta\gamma\delta}(\tau)$	polarization, equation (5.68)
$\Pi(\boldsymbol{q}, \varepsilon)$	polarization in momentum space, equation (5.68)
$\boldsymbol{Q}$	special $\boldsymbol{q}$ vector; nesting vector
Q	short 'four-vector' notation for $(\boldsymbol{q}, i\Omega_n)$
q	norm of the vector $\boldsymbol{q}$
$\boldsymbol{q}$	vector in reciprocal space
Re	real part
$\boldsymbol{R}_n$	node of a real-space lattice
r	norm of the vector $\boldsymbol{r}$
$\boldsymbol{r}$	vector in real space
r_e	classical electron radius, $r_e = \mu_0 e^2/(4\pi m)$
r_s	dimensionless density parameter, $(4/3)\pi(r_s a_0)^3 = 1/n$
ρ	statistical density matrix, equation (2.3)
$\rho(T)$	temperature-dependent resistivity
$\rho^{>}_{AB}(\varepsilon)$	greater spectral function for operators A and B, equation (3.12)
$\rho^{<}_{AB}(\varepsilon)$	lesser spectral function for operators A and B, equation (3.14)
$\rho_{AB}(\varepsilon)$	total spectral function for operators A and B, $\rho_{AB} = \rho^{>}_{AB} + \rho^{<}_{AB}$
S	entropy, equation (2.7)
$\boldsymbol{S}$	spin operator
S^z	spin component along quantization axis, equation (2.48)
$S^{\pm}$	spin raising and lowering operators, equation (2.48)
$S(\boldsymbol{q}, \varepsilon)$	scattering function (chapter 10)
$\boldsymbol{\mathcal{S}}$	dimensionless spin, $\boldsymbol{\mathcal{S}} = \boldsymbol{S}/\hbar$
$\Sigma(\boldsymbol{k}, \varepsilon)$	self-energy in momentum space
σ, σ_i	spin index
$\sigma(\mathscr{P})$	signature of the permutation $\mathscr{P}$
$\sigma_{\mu\nu}, \bar{\sigma}_{\mu\nu}$	conductivity tensor and spatial average (for broken translational symmetry)
σ_{dc}	dc conductivity

$\sigma(V)$	differential tunneling conductance, $\sigma(V) = dI_s/dV$.
T	temperature
T_η	time-ordering operator, equation (5)
T_τ	imaginary-time-ordering operator, equation (4.6)
T_{D}	Debye temperature, $k_{\mathrm{B}}T_{\mathrm{D}} = \hbar\omega_{\mathrm{D}}$
$T(\varepsilon)$	t-matrix, equation (2.26)
$T_{\alpha\beta}$	matrix element of generic one-body operator, equation (2.43)
$T_{\lambda\rho}$	tunneling matrix element, equation (9.11)
$T(\boldsymbol{l}, \boldsymbol{r})$	tunneling matrix element in real space
Tr	matrix trace, sum of the diagonal matrix elements
t	time
τ	imaginary-time variable
τ_{tr}	transport life-time, equation (8.11)
$\boldsymbol{\tau}$	vector of Pauli matrices with components (τ^x, τ^y, τ^z), equation (22)
$\boldsymbol{\tau}_\nu$	position of the atom ν in the elementary cell of a crystal
$\tau(\boldsymbol{k}, \varepsilon)$	quasi-particle life-time, equation (7.16*c*)
$\theta(x)$	Heaviside theta function (step function): $\theta(x < 0) = 0$, $\theta(x > 0) = 1$
ϑ	angle variable
U	elastic energy (section 2.5.1)
$U(t)$	evolution operator
$U(\tau)$	imaginary-time evolution operator
$\boldsymbol{u}_\nu(\boldsymbol{R}_n)$	displacement of atom ν in the cell located at $\boldsymbol{R}_n$
$u_{\boldsymbol{k}}$	BCS electron coherence factor
V	general interaction or perturbation operator
$V(\boldsymbol{r})$	local potential
V_{H}	Hartree potential
V_{x}	exchange potential
$V_\nu(\boldsymbol{r})$	potential generated by the atom ν in the elementary cell of a crystal
$V_{\alpha\beta}$	matrix element of generic one-body potential, equation (5.9)
$V_{\alpha\beta\gamma\delta}$	matrix element of generic two-body potential, equation (2.44)
V_{Cb}	Coulomb interaction, equation (5.55)
$V_{\mathrm{el-el}}^{\mathrm{ph}}$	phonon-mediated electron–electron interaction, equation (5.54)
$\mathscr{V}$	volume of the system, subject to periodic boundary conditions
$\mathscr{V}_{\mathrm{cell}}$	lattice unit-cell volume, $\mathscr{V}_{\mathrm{cell}} = \mathscr{V}/\mathscr{N}$
v_{F}	Fermi velocity $v_{\mathrm{F}} = \hbar k_{\mathrm{F}}/m$
$v_{\boldsymbol{k}}$	BCS hole coherence factor
$\boldsymbol{v}_{\boldsymbol{k}n}$	group velocity, $\boldsymbol{v}_{\boldsymbol{k}n} = \frac{1}{\hbar}\nabla\varepsilon_{\boldsymbol{k}n}$
$W(\boldsymbol{q}, i\Omega_n)$	screened Coulomb interaction, equation (5.90)
W	band width
Ω	grand potential, equation (2.5)
ω	frequency
$\omega_{\boldsymbol{q}\lambda}$	frequency of phonon with wave vector $\boldsymbol{q}$ and polarization λ
ω_{D}	Debye frequency, largest phonon frequency
ω_p	plasma frequency, equation (5.84)

ϖ	free energy density, $\varpi = \Omega/\mathscr{V}$
x	variable, usually meant as real
ξ_α	single-particle excitation energy, $\xi_\alpha = \varepsilon_\alpha - \mu$
$\xi_{\alpha\beta}$	matrix element of a general one-particle Hamiltonian K_0, equation (2.45)
$\|\Psi_0\rangle$	ground state
Z	partition function, equation (2.4)
Z	atomic valence
$Z(\boldsymbol{k}_\mathrm{F})$	quasi-particle residue, equation (7.16*a*)
z	arbitrary complex number

Conventions and notations

Grand Hamiltonian Throughout this book, we will consider many-particle systems at finite temperature in the grand-canonical ensemble. The relevant operator in this case is the *grand Hamiltonian*, denoted K and defined as

$$K = H - \mu N. \tag{1}$$

H is the Hamiltonian, μ the chemical potential, and N the particle-number operator. The thermal average of operator A will be denoted $\langle A \rangle$ and is defined in equation (2.2).

Fourier transforms For the space–time Fourier transforms, we use the convention:

$$f(\boldsymbol{r}, t) = \frac{1}{\mathcal{V}} \sum_{\boldsymbol{k}} \int_{-\infty}^{\infty} \frac{d\omega}{2\pi} f(\boldsymbol{k}, \omega) e^{i(\boldsymbol{k}\cdot\boldsymbol{r} - \omega t)} \tag{2a}$$

$$f(\boldsymbol{k}, \omega) = \int d\boldsymbol{r} \int_{-\infty}^{\infty} dt\, f(\boldsymbol{r}, t) e^{-i(\boldsymbol{k}\cdot\boldsymbol{r} - \omega t)}. \tag{2b}$$

The real space is continuous and confined to a normalization volume $\mathcal{V}$. The notation $\int d\boldsymbol{r}$ represents integration over the volume $\mathcal{V}$. The reciprocal space is discrete with $\boldsymbol{k}$ points defined such that the plane waves satisfy periodic boundary conditions on the volume $\mathcal{V}$. This convention translates into the following closure relations:

$$\int d\boldsymbol{r} e^{i(\boldsymbol{k}-\boldsymbol{k}')\cdot\boldsymbol{r}} = \mathcal{V} \delta_{\boldsymbol{k}\boldsymbol{k}'}, \qquad \frac{1}{\mathcal{V}} \sum_{\boldsymbol{k}} e^{i\boldsymbol{k}\cdot(\boldsymbol{r}-\boldsymbol{r}')} = \delta(\boldsymbol{r} - \boldsymbol{r}') \tag{3a}$$

$$\int_{-\infty}^{\infty} dt e^{i(\omega-\omega')t} = 2\pi\delta(\omega - \omega'), \qquad \int_{-\infty}^{\infty} \frac{d\omega}{2\pi} e^{i\omega(t-t')} = \delta(t - t'). \tag{3b}$$

In cases where the real space is discrete, such as lattice models, the space integration $\int d\boldsymbol{r}$ is replaced by $\sum_{\boldsymbol{R}_n}$ with $\boldsymbol{R}_n$ the lattice sites, and in reciprocal space $\mathcal{V}^{-1} \sum_{\boldsymbol{k}}$ is replaced by $\mathcal{N}^{-1} \sum_{\boldsymbol{k}}$ with $\mathcal{N}$ the number of elementary cells. The unit-cell volume $\mathcal{V}/\mathcal{N}$ is denoted $\mathcal{V}_{\text{cell}}$.

Bosons and fermions Whenever possible, the formulas are written in a form valid for bosons and fermions. We use the symbol η to distinguish the two cases: $\eta = +1$ for bosons and $\eta = -1$ for fermions. For instance, the commutator and anti-commutator are written at once as

$$[A, B]_{-\eta} = AB - \eta BA \tag{4}$$

(when we write $[A, B]$, we mean $[A, B]_{-}$) and the time-ordering operators are

$$T_\eta\{A(t)B(t')\} = \theta(t - t')A(t)B(t') + \eta\theta(t' - t)B(t')A(t). \tag{5}$$

A generic creation operator for a boson or a fermion in state α is denoted $a_\alpha^\dagger$. When we refer specifically to bosons, we use instead the notation $b_\alpha^\dagger$ and for fermions we use $c_\alpha^\dagger$. Likewise, the Bose–Einstein and Fermi–Dirac distributions are generically written as

$$d_{-\eta}(\varepsilon) \equiv \frac{1}{e^{\beta\varepsilon} - \eta}, \tag{6}$$

while the notations $b(\varepsilon) \equiv d_{-1}(\varepsilon)$ and $f(\varepsilon) \equiv d_{+1}(\varepsilon)$ are used when appropriate.

Scalars and operators Ideally, a notation should be light and exact, which are antithetical requirements. In order to lighten the notation by avoiding redundancy, in this book we generally rely on function's arguments to help identify the function. The example of the particle density is eloquent:

- n without argument is the average particle-number density, $n = N/\mathscr{V}$;
- n_α is the occupation number for the one-particle state α;
- $n(\boldsymbol{r})$ is the local particle number density;
- $n(\boldsymbol{q})$ is the Fourier transform of $n(\boldsymbol{r})$.

Furthermore, n, n_α, $n(\boldsymbol{r})$, and $n(\boldsymbol{q})$ can be scalars or operators depending upon the context. In this way, we avoid fat notations like $\langle\langle\widehat{\tilde{n}}(\boldsymbol{q})\rangle\rangle$ for the thermodynamic average of the Fourier transform of the density operator.

Units We use international units and, errors excepted, write all $\hbar$'s explicitly. This is a little cumbersome in a few places but, on the other hand, a global and coherent '$\hbar = 1$' convention also has drawbacks.

For convenience, we use energy rather than frequency variables in the definition of certain functions. In general, the Fourier transform of the time-dependent correlation function $C_{AB}(t)$ is the frequency-dependent $C_{AB}(\omega)$. This convention is followed in particular for the dielectric function $\epsilon(\boldsymbol{q}, \omega)$ and the conductivity tensor $\sigma_{\mu\nu}(\boldsymbol{q}, \omega)$. The quantities defined as functions of the energy $\varepsilon = \hbar\omega$ include the single-particle Green's function $G_{\alpha\beta}(\varepsilon)$, all susceptibilities $\chi_{AB}(\varepsilon)$ and all spectral functions $\rho_{AB}(\varepsilon)$, as well as the Matsubara function $\mathscr{C}_{AB}(z)$ where $z \in \mathbb{C}$ has the unit of energy.

Useful mathematical formula

Mathematical identities

- *Residue theorem*

$$\oint_{\circlearrowleft} dz\, F(z) = \sum_{z_0} \frac{2\pi i}{(n-1)!} \lim_{z\to z_0} \frac{d^{n-1}}{dz^{n-1}}[(z-z_0)^n F(z)]. \tag{7}$$

z_0 are the poles of $F(z)$ and n is the order of each pole. As such, the formula applies only if the orientation of the closed contour is counter-clockwise. If the contour is followed in the clockwise direction, a minus sign must be inserted after the '=' sign.

- *Fourier transform of the Heaviside function*

$$\int_{-\infty}^{\infty} dt\, e^{i\omega t}\theta(\pm t) = \frac{\pm i}{\omega \pm i0^+}. \tag{8}$$

For the proof, see Doc A1.

- *Cauchy principal value and Dirac delta function*

$$\frac{1}{x \pm i0^+} = \mathscr{P}\frac{1}{x} \mp i\pi\delta(x). \tag{9}$$

The principal value $\mathscr{P}(1/x)$ is defined as $\mathscr{P}(1/x) = 1/x$ for $x \neq 0$ and $\mathscr{P}(1/x) = 0$ for $x = 0$. Equation (9) is best visualized as the limit of the function $1/(x \pm i\epsilon) = (x \mp i\epsilon)/(x^2 + \epsilon^2)$ as $\epsilon \to 0^+$.

- *A very useful variant*

$$-\frac{1}{\pi}\,\mathrm{Im}\left(\frac{1}{x \pm i0^+}\right) = \pm\delta(x). \tag{10}$$

- *Composition of the Dirac delta function*

$$\delta[f(x)] = \sum_{x_0} \frac{\delta(x-x_0)}{|f'(x_0)|}, \qquad f(x_0) = 0. \tag{11}$$

The sum extends over all solutions x_0 of the equation $f(x) = 0$.

- *Laplacian and Dirac delta function in three dimensions*

$$\nabla^2 \frac{1}{|\boldsymbol{x}|} = -4\pi\delta(\boldsymbol{x}). \tag{12}$$

This relation is just the Poisson equation for a point-charge at the origin.

- *Plane waves and Bessel functions*

$$e^{ix\cos\vartheta} = \sum_{n=-\infty}^{+\infty} i^n J_n(x) e^{in\vartheta}. \tag{13}$$

$x \in \mathbb{R}$ and $J_n(x)$ is the Bessel function of the first kind.

- *Fourier transform of the Yukawa potential in three dimensions*

$$\int \frac{d^3q}{(2\pi)^3} \frac{e^{i\mathbf{q}\cdot\mathbf{r}}}{q^2 + k_0^2} = \frac{e^{-k_0 r}}{4\pi r}. \tag{14}$$

- *Fourier transform of the square of the Yukawa potential*

$$\int \frac{d^3q}{(2\pi)^3} \frac{e^{i\mathbf{q}\cdot\mathbf{r}}}{(q^2 + k_0^2)^2} = \frac{e^{-k_0 r}}{8\pi k_0}. \tag{15}$$

For the proof, see Doc A2.

- *Sums over Matsubara frequencies*

$$\frac{1}{\beta}\sum_{i\nu_n} F(i\nu_n) = \text{sum of the residues of } -\eta d_{-\eta}(z)F(z) \text{ at the poles of } F(z). \tag{16}$$

This formula applies provided that $F(z)$ is analytic and that the function $d_{-\eta}(z)F(z)$ vanishes for $|z| \to \infty$. The proof is presented in Doc A3.

$$\frac{1}{\beta}\sum_{i\nu_n} \frac{e^{-i\nu_n\tau}}{i\nu_n - \varepsilon} = -[\theta(\tau) + \eta d_{-\eta}(\varepsilon)]e^{-\varepsilon\tau}. \tag{17}$$

This is the analogous of equation (8) for imaginary time. It seems to contradict equation (16), but does not. For the proof see Doc A4.

- *Three useful integrals* ($a \in \mathbb{R}$)

$$\int_0^\pi \frac{\sin\vartheta \, d\vartheta}{a - \cos\vartheta \pm i0^+} = \ln\left|\frac{1+a}{1-a}\right| \mp i\pi\theta(1 - |a|) \tag{18}$$

$$\int_0^1 du\, u \ln\left|\frac{u+a}{u-a}\right| = a + \frac{1}{2}(1 - a^2)\ln\left|\frac{1+a}{1-a}\right| \tag{19}$$

$$\lim_{a\to\infty}\int_{-a}^{a} du \frac{\tanh(u/2)}{2u} = \ln\left(\frac{2e^\gamma}{\pi}a\right) \approx \ln(1.134\, a). \tag{20}$$

$\gamma \approx 0.577$ is the Euler constant.

Quantum mechanics

- *Closure relation*

$$\sum_a |a\rangle\langle a| = \mathbb{1}. \tag{21}$$

The set of states $|a\rangle$ must be complete, such as for instance the set of the eigenstates of some Hamiltonian.

- *Pauli matrices*

$$\tau^x = \begin{pmatrix} 0 & 1 \\ 1 & 0 \end{pmatrix}, \qquad \tau^y = \begin{pmatrix} 0 & -i \\ i & 0 \end{pmatrix}, \qquad \tau^z = \begin{pmatrix} 1 & 0 \\ 0 & -1 \end{pmatrix}. \tag{22}$$

Together with the identity, these three matrices form a 'basis' on which any 2×2 complex matrix can be expanded.

- *Traces of Pauli matrices and their products*

$$\mathrm{Tr}\,\tau^{\mu} = 0, \qquad \mathrm{Tr}\,\tau^{\mu_1}\tau^{\mu_2} = 2\delta_{\mu_1\mu_2}, \qquad \mathrm{Tr}\,\tau^{\mu_1}\tau^{\mu_2}\tau^{\mu_3} = 2i\epsilon_{\mu_1\mu_2\mu_3}. \tag{23}$$

$\epsilon_{\mu_1\mu_2\mu_3}$ is the Levi-Civita symbol, which is $(-1)^{\sigma_{\{\mu_1,\mu_2,\mu_3\}}}$ if $\mu_1 \neq \mu_2 \neq \mu_3$ and 0 otherwise, $\sigma_{\{\mu_1,\mu_2,\mu_3\}}$ being the signature of the permutation.

- *Other relations involving Pauli matrices*

$$\sum_{\mu_1\mu_2} \tau^{\mu_1}\tau^{\mu_2}\delta_{\mu_1\mu_2} = 3 \times \mathbb{1}, \qquad \sum_{\mu_1\mu_2\mu_3} \tau^{\mu_1}\tau^{\mu_2}\tau^{\mu_3}\epsilon_{\mu_1\mu_2\mu_3} = 6i \times \mathbb{1}. \tag{24}$$

Here, $\mathbb{1}$ is the 2×2 identity matrix, sometimes denoted τ^0.

Second quantization

- *Commutators of creation and annihilation operators with the Hamiltonian*

$$[a_\alpha, K] = \sum_\beta \xi_{\alpha\beta} a_\beta + \sum_{\beta\gamma\delta} V_{\alpha\beta\gamma\delta} a^\dagger_\beta a_\delta a_\gamma \tag{25}$$

$$[a^\dagger_\alpha, K] = -[a_\alpha, K]^\dagger = -\sum_\beta \xi^*_{\alpha\beta} a^\dagger_\beta - \sum_{\beta\gamma\delta} V^*_{\alpha\beta\gamma\delta} a^\dagger_\gamma a^\dagger_\delta a_\beta \tag{26}$$

$$[a^\dagger_\alpha a_\beta, K] = a^\dagger_\alpha [a_\beta, K] + [a^\dagger_\alpha, K] a_\beta. \tag{27}$$

K is defined in equation (2.45). For the proof, see Doc A5.

- *Commutators of creation and annihilation operators with the number operator*

$$[N, a^\dagger_\alpha] = a^\dagger_\alpha, \qquad [N, a_\alpha] = -a_\alpha. \tag{28}$$

- *A variant*:

$$e^{zN} a^\dagger_\alpha = a^\dagger_\alpha e^{z(N+1)}, \qquad e^{zN} a_\alpha = a_\alpha e^{z(N-1)}, \tag{29}$$

for any complex number z. For the proof, see Doc A5.

- *Average of the product of two exponentials*

Let the operators A and B be linear combinations of boson creation and annihilation operators. If the Hamiltonian is quadratic in these creation and annihilation operators, the following holds:

$$\langle e^{iA} e^{iB} \rangle = e^{-\frac{1}{2}\langle A^2 + 2AB + B^2 \rangle}. \tag{30}$$

For the proof, see Doc A6.

Doc A1 Fourier transform of the Heaviside function

Equation (8) can be established by using the residue theorem, equation (7). It is more convenient to prove the inverse-Fourier transform relation, namely

$$\int_{-\infty}^{\infty} \frac{d\omega}{2\pi} e^{-i\omega t} \frac{\pm i}{\omega \pm i0^+} = \theta(\pm t).$$

Let us consider the + sign first. The integrand has a pole at $\omega = -i0^+$ with residue $i/(2\pi)$. If $t < 0$, we must close the contour in the upper half of the complex plane, so that $e^{-i\omega t}$ is zero on that part of the contour; there is no pole inside the contour, so the integral vanishes. If $t > 0$, we must close the contour in the lower half of the complex plane, thus enclosing the pole, so that the integral equals $-2\pi i \times i/(2\pi) = 1$. Thus we get $\theta(t)$. For the minus sign, the pole is at $\omega = +i0^+$ with residue $-i/(2\pi)$, so that we get 1 for $t < 0$ and 0 for $t > 0$, i.e. $\theta(-t)$.

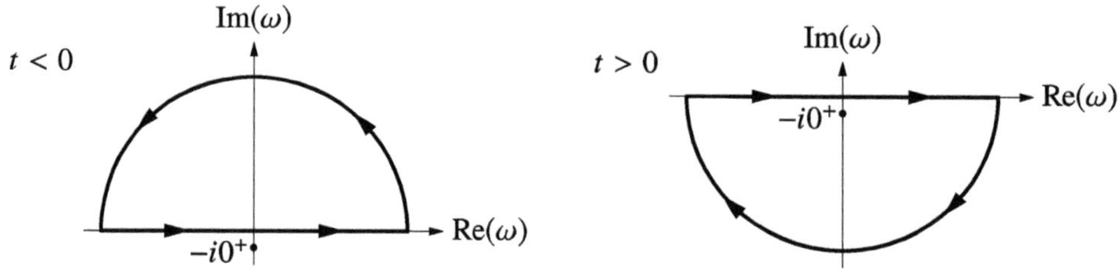

Doc A2 Powers of the Yukawa potential and their Fourier transform

We study the Fourier transform of the function $1/(q^2 + k_0^2)^n$ in the limit of a continuous reciprocal space, i.e. $\mathcal{V}^{-1} \sum_{\boldsymbol{q}} \to (2\pi)^{-3} \int d^3q$. The integral to evaluate is

$$I = \int \frac{d^3q}{(2\pi)^3} \frac{e^{i\boldsymbol{q}\cdot\boldsymbol{r}}}{(q^2 + k_0^2)^n}.$$

We choose spherical coordinates with the z axis parallel to $\boldsymbol{r}$, so that $\boldsymbol{q} \cdot \boldsymbol{r} = qr \cos\vartheta$.

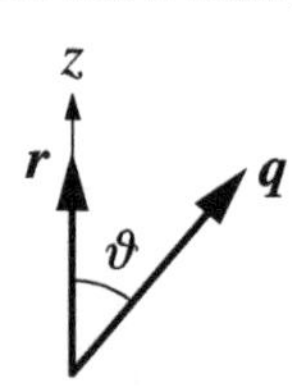

There is no dependence on the angular variable φ:

$$I = \frac{1}{(2\pi)^2} \int_0^{\infty} dq \int_0^{\pi} d\vartheta \frac{q^2 \sin\vartheta \, e^{iqr\cos\vartheta}}{(q^2 + k_0^2)^n}.$$

The ϑ integration poses no problem, $\int_0^\pi d\vartheta \sin\vartheta \, e^{iqr\cos\vartheta} = \frac{2\sin qr}{qr}$, so that we get

$$I = \frac{1}{2\pi^2 r}\int_0^\infty dq \frac{q \sin qr}{(q^2 + k_0^2)^n} = \frac{1}{4\pi^2 r}\frac{1}{i}\int_{-\infty}^{+\infty} dq \frac{q\, e^{iqr}}{(q^2 + k_0^2)^n}.$$

Since the integrand is even, we could extend the domain of integration to the whole real axis and divide by two. Furthermore, we could replace $\sin qr$ by e^{iqr}/i because the added $\cos qr$ term gives a vanishing contribution to the integral. We are now in the position to use the residue theorem, equation (7), by continuing q into the complex plane. The integrand has a pole of order n at $q = ik_0$. Thanks to the e^{iqr} numerator, we can close the integration contour in the upper half of the complex plane, thus enclosing the pole. Equation (7) gives:

$$I = \frac{1}{4\pi^2 r}\frac{2\pi}{(n-1)!}\lim_{q\to ik_0}\frac{d^{n-1}}{dq^{n-1}}\underbrace{\left[(q - ik_0)^n \frac{q\, e^{iqr}}{(q^2 + k_0^2)^n}\right]}_{\dfrac{q\, e^{iqr}}{(q + ik_0)^n}}.$$

Equations (14) and (15) result by setting $n = 1$ and $n = 2$, respectively.

Doc A3 Sums over Matsubara frequencies

The key observation that allows one to evaluate the sums over Matsubara frequencies is that the distribution function $d_{-\eta}(z)$, once analytically continued to complex variables z, has poles on the imaginary axis precisely at the Matsubara frequencies: $1/d_{-\eta}(i\nu_n) = e^{i\nu_n\beta} - \eta = 0$ (see equation (4.9) and figure 4.2). Expanding $1/d_{-\eta}(z)$ around $z = i\nu_n$ gives $1/d_{-\eta}(z) = \beta e^{i\nu_n\beta}(z - i\nu_n) + \mathcal{O}[(z - i\nu_n)^2] = \beta\eta(z - i\nu_n) + \mathcal{O}[(z - i\nu_n)^2]$. Therefore, the residue of $d_{-\eta}(z)$ at the Matsubara frequencies is simply $1/(\beta\eta)$. The idea, then, is to consider the function $G(z) = -\eta d_{-\eta}(z)F(z)$. Let us assume that $G(|z| \to \infty) = 0$: since $d_{-\eta}(z)$ is regular at infinity, this amounts to assuming that $F(z)$ decreases at infinity. The contour integral of $G(z)$ on a circle of infinite diameter in the complex plane therefore vanishes. By the residue theorem, this integral must equal the sum of residues of $G(z)$ at its poles. $G(z)$ has two types of poles: the poles at the Matsubara frequencies originating from $d_{-\eta}(z)$, with residues $-\eta/(\beta\eta)F(i\nu_n)$, and the poles originating from $F(z)$. Hence

$$\oint_{\circlearrowleft} dz\, G(z) = 0 = -\frac{1}{\beta}\sum_{i\nu_n} F(i\nu_n) + \sum_{z_0}(-\eta)d_{-\eta}(z_0)\mathrm{Res}_{z_0}F(z),$$

where z_0 are the poles of $F(z)$. This is equation (16). If the function $F(z)$ is not analytic, for instance if it has branch cuts, one must deform the contour in the complex plane such as to avoid these branch cuts. Equation (16) gets additional terms depending on the discontinuity across the branch cuts.

Doc A4 Non-interacting Green's function in imaginary time

Equation (17) is just the Fourier transform of the free imaginary-time Green's function $\mathscr{G}_0(\varepsilon, i\nu_n) = 1/(i\nu_n - \varepsilon)$. The Matsubara sum in equation (17) cannot be performed for $\tau > 0$ using equation (16), because the integrand $d_{-\eta}(z)e^{-z\tau}/(z - \varepsilon)$ diverges as $|z| \to \infty$ for $\tau > 0$. For $\tau < 0$, the integrand vanishes at infinity and equation (16) can therefore be directly used. In order the establish equation (17) for arbitrary τ, it is simplest to check the inverse imaginary-time Fourier transform as defined in equation (4.10):

$$\int_0^\beta d\tau\left\{-[\theta(\tau) + \eta d_{-\eta}(\varepsilon)]e^{-\varepsilon\tau}\right\}e^{i\nu_n\tau} = -[1 + \eta d_{-\eta}(\varepsilon)]\underbrace{\int_0^\beta d\tau e^{(i\nu_n-\varepsilon)\tau}}_{\left.\frac{e^{(i\nu_n-\varepsilon)\tau}}{i\nu_n-\varepsilon}\right|_0^\beta = \frac{\eta e^{-\beta\varepsilon}-1}{i\nu_n-\varepsilon}} = \frac{1}{i\nu_n - \varepsilon},$$

where we have used $e^{i\nu_n\beta} = \eta$ (equation (4.9)) and $-[1 + \eta d_{-\eta}(\varepsilon)](\eta e^{-\beta\varepsilon} - 1) = 1$.

Doc A5 Useful commutators

The first term in the right-hand side of equation (25) reads

$$[a_\alpha, K_0] = \sum_{\gamma\beta}\xi_{\gamma\beta}[a_\alpha, a_\gamma^\dagger a_\beta].$$

The commutator $[a_\alpha, a_\gamma^\dagger a_\beta]$ contains two terms and is evaluated by making use of the commutation relations, equation (2.41):

$$[a_\alpha, a_\gamma^\dagger a_\beta] = \underbrace{a_\alpha a_\gamma^\dagger}_{\delta_{\alpha\gamma} + \eta a_\gamma^\dagger a_\alpha} a_\beta - a_\gamma^\dagger a_\beta a_\alpha = \delta_{\alpha\gamma}a_\beta + \eta a_\gamma^\dagger \underbrace{a_\alpha a_\beta}_{\eta a_\beta a_\alpha} - a_\gamma^\dagger a_\beta a_\alpha = \delta_{\alpha\gamma}a_\beta.$$

Hence we obtain $[a_\alpha, K_0] = \sum_\beta \xi_{\alpha\beta}a_\beta$ as shown in equation (25). The second term in equation (25) reads

$$[a_\alpha, V] = \frac{1}{2}\sum_{\mu\beta\gamma\delta} V_{\mu\beta\gamma\delta}[a_\alpha, a_\mu^\dagger a_\beta^\dagger a_\delta a_\gamma].$$

The commutator is evaluated in the same way as the previous one,

$$[a_\alpha, a_\mu^\dagger a_\beta^\dagger a_\delta a_\gamma] = a_\alpha a_\mu^\dagger a_\beta^\dagger a_\delta a_\gamma - a_\mu^\dagger a_\beta^\dagger a_\delta a_\gamma a_\alpha,$$

by using the commutation rules to transform the first term on the right until it cancels the second, thereby generating additional terms. Consider the first term:

$$
\underbrace{a_\alpha a_\mu^\dagger}_{\delta_{\alpha\mu}+\eta a_\mu^\dagger a_\alpha} a_\beta^\dagger a_\delta a_\gamma = \delta_{\alpha\mu} a_\beta^\dagger a_\delta a_\gamma + \eta a_\mu^\dagger \underbrace{a_\alpha a_\beta^\dagger}_{\delta_{\alpha\beta}+\eta a_\beta^\dagger a_\alpha} a_\delta a_\gamma = \delta_{\alpha\mu} a_\beta^\dagger a_\delta a_\gamma + \eta\delta_{\alpha\beta} a_\mu^\dagger a_\delta a_\gamma
$$

$$
+ a_\mu^\dagger a_\beta^\dagger \underbrace{a_\alpha a_\delta}_{\eta a_\delta a_\alpha} a_\gamma = \delta_{\alpha\mu} a_\beta^\dagger a_\delta a_\gamma + \eta\delta_{\alpha\beta} a_\mu^\dagger a_\delta a_\gamma + \eta a_\mu^\dagger a_\beta^\dagger a_\delta \underbrace{a_\alpha a_\gamma}_{\eta a_\gamma a_\alpha}
$$

$$
= \delta_{\alpha\mu} a_\beta^\dagger a_\delta a_\gamma + \eta\delta_{\alpha\beta} a_\mu^\dagger a_\delta a_\gamma + a_\mu^\dagger a_\beta^\dagger a_\delta a_\gamma a_\alpha.
$$

We see that the last term cancels the second term in the commutator, while two additional terms have been generated:

$$
[a_\alpha,\, a_\mu^\dagger a_\beta^\dagger a_\delta a_\gamma] = \delta_{\alpha\mu} a_\beta^\dagger a_\delta a_\gamma + \eta\delta_{\alpha\beta} a_\mu^\dagger a_\delta a_\gamma.
$$

Inserting this into the expression of $[a_\alpha,\, V]$ we get two terms which can be recast as one single term by using the symmetry property of the interaction, namely $V_{\alpha\beta\gamma\delta} = V_{\beta\alpha\delta\gamma}$ (or, in real space, $V(\boldsymbol{r}_1,\, \boldsymbol{r}_2) = V(\boldsymbol{r}_2,\, \boldsymbol{r}_1)$):

$$
\begin{aligned}
[a_\alpha,\, V] &= \frac{1}{2}\sum_{\mu\beta\gamma\delta} V_{\mu\beta\gamma\delta}\left[\delta_{\alpha\mu} a_\beta^\dagger a_\delta a_\gamma + \eta\delta_{\alpha\beta} a_\mu^\dagger a_\delta a_\gamma\right] \\
&= \frac{1}{2}\sum_{\beta\gamma\delta} V_{\alpha\beta\gamma\delta} a_\beta^\dagger a_\delta a_\gamma + \frac{\eta}{2}\sum_{\mu\gamma\delta} \underbrace{V_{\mu\alpha\gamma\delta}}_{V_{\alpha\mu\delta\gamma}} a_\mu^\dagger a_\delta a_\gamma \\
&= \frac{1}{2}\sum_{\beta\gamma\delta} V_{\alpha\beta\gamma\delta} a_\beta^\dagger a_\delta a_\gamma + \frac{\eta}{2}\sum_{\beta\gamma\delta} V_{\alpha\beta\delta\gamma} a_\beta^\dagger a_\delta a_\gamma.
\end{aligned}
$$

In the last term, we have renamed the dummy index μ as β. To finish, we exchange the dummy indices γ and δ, $V_{\alpha\beta\delta\gamma} a_\beta^\dagger a_\delta a_\gamma \rightarrow V_{\alpha\beta\gamma\delta} a_\beta^\dagger a_\gamma a_\delta = \eta V_{\alpha\beta\gamma\delta} a_\beta^\dagger a_\delta a_\gamma$, and we obtain the second term in the right-hand side of equation (25).

Writing the particle number $N = \sum_\beta a_\beta^\dagger a_\beta$ and using the commutation rules (2.41), we find

$$
N a_\alpha^\dagger = \sum_\beta a_\beta^\dagger \underbrace{a_\beta a_\alpha^\dagger}_{\delta_{\alpha\beta}+\eta a_\alpha^\dagger a_\beta} = a_\alpha^\dagger + \eta\sum_\beta \underbrace{a_\beta^\dagger a_\alpha^\dagger}_{\eta a_\alpha^\dagger a_\beta^\dagger} a_\beta = a_\alpha^\dagger + a_\alpha^\dagger N,
$$

which proves the first of equation (28). Taking the conjugate we have $a_\alpha N = a_\alpha + N a_\alpha$, which proves the second. Equation (29) can be derived by using equation (28) and the definition of the exponential:

$$
e^{zN} a_\alpha^\dagger = \sum_{n=0}^{\infty} \frac{(zN)^n}{n!} a_\alpha^\dagger = \sum_{n=0}^{\infty} \frac{z^n}{n!} a_\alpha^\dagger (N+1)^n = a_\alpha^\dagger e^{z(N+1)}.
$$

The conjugate gives $a_\alpha e^{z^*N} = e^{z^*(N+1)} a_\alpha$. Since this is valid for any z we can replace z^* by z and, after multiplying by the *number* e^{-z}, we find $a_\alpha e^{z(N-1)} = e^{zN} a_\alpha$.

Doc A6 Average of the product of two exponentials

Various ways of proving equation (30) can be found in the literature, sometimes giving the false impression that the result is peculiar to harmonic oscillators. We propose a proof relying only on elementary calculus, namely, Taylor expansion of the exponentials, and making use of Wick's theorem with some tricky combinatorics. A quadratic Hamiltonian such as equation (2.65) conserves particle number, such that the average value of any operator that does not conserve particle number vanishes. As a result, if A and B are linear combinations of creation and annihilation operators, $\langle A\rangle = \langle B\rangle = 0$ and likewise for any combination A^nB^m if $n + m$ is odd: such a combination is a sum of products of an odd number of creation and/or annihilation operators, hence does not conserve particle number and therefore vanishes on average. Another property of quadratic Hamiltonians is that Wick's theorem applies, which is equally important in order to prove equation (30) than particle-number conservation. Wick's theorem allows one to evaluate averages like $\langle A^nB^m\rangle$ as a sum of all possible products of pairs. For fermionic operators, there would be sign changes for certain of these products, while for bosonic operators there is none. For instance, we have $\langle A^4\rangle = 3\langle A^2\rangle^2$ because there are three pairwise contractions of $AAAA$. Expanding the exponentials on the left-hand side of equation (30) and eliminating all terms that do not conserve particle number, we can write

$$\langle e^{iA}e^{iB}\rangle = \sum_{n,m=0}^{\infty}\frac{i^{n+m}}{n!m!}\langle A^nB^m\rangle = \sum_{k=0}^{\infty}\frac{(-1/2)^k}{k!}\sum_{m=0}^{2k}\frac{2^kk!}{(2k-m)!m!}\langle A^{2k-m}B^m\rangle.$$

For the second equality, we have imposed $n + m = 2k$. It turns out that, when Wick's theorem applies,

$$\sum_{m=0}^{2k}\frac{2^kk!}{(2k-m)!m!}\langle A^{2k-m}B^m\rangle = \langle A^2 + 2AB + B^2\rangle^k.$$

By substituting this into the previous relation, we recognize the Taylor expansion of $e^{-\frac{1}{2}\langle A^2+2AB+B^2\rangle}$ and obtain equation (30). To convince oneself of the above relation, one can go on and check order by order. The relation holds trivially for $k = 0$ and $k = 1$; for $k = 2$, one must use $\langle A^4\rangle = 3\langle A^2\rangle^2$, $\langle A^3B\rangle = 3\langle A^2\rangle\langle AB\rangle$, $\langle A^2B^2\rangle = \langle A^2\rangle\langle B^2\rangle + 2\langle AB\rangle^2$, $\langle AB^3\rangle = 3\langle AB\rangle\langle B\rangle^2$, and $\langle B^4\rangle = 3\langle B^2\rangle^2$. One can then set up a systematic proof by recurrence. Assume that the relation holds for k and consider the left-hand side for $k + 1$. It involves the average $\langle A^{2k+2-m}B^m\rangle$, which we partly evaluate as the sum of all possible products singling out only one pair of operators:

$$\begin{aligned}
\langle A^{2k+2-m}B^m\rangle = \langle A^2\rangle\langle A^{2k-m}B^m\rangle \times \frac{(2k+2-m)(2k+1-m)}{2(k+1)} \qquad & (m \leqslant 2k) \\
+ \langle AB\rangle\langle A^{2k+1-m}B^{m-1}\rangle \times \frac{(2k+2-m)m}{k+1} \qquad & (1 \leqslant m \leqslant 2k+1) \\
+ \langle B^2\rangle\langle A^{2k+2-m}B^{m-2}\rangle \times \frac{m(m-1)}{2(k+1)} \qquad & (m \geqslant 2).
\end{aligned}$$

Each term occurs only for a limited range of m values, as indicated on the right. Getting the correct combinatorial factors here is the only nontrivial step in this proof. They can be checked by expanding completely both sides of the equation, distinguishing even and odd values of m, with the help of the general expressions

$$\langle A^{2n}B^{2m}\rangle = \sum_{p=0}^{\min(n,m)} \frac{(2n)!(2m)!}{(2p)!(2n-2p)!!(2m-2p)!!}\langle A^2\rangle^{n-p}\langle B^2\rangle^{m-p}\langle AB\rangle^{2p}$$

$$\begin{aligned}
\langle A^{2n-1}B^{2m+1}\rangle = & \sum_{p=0}^{\min(n-1,m)} \frac{(2n-1)!(2m+1)!}{(2p+1)!(2n-2p-2)!!(2m-2p)!!} \\
& \times \langle A^2\rangle^{n-p-1}\langle B^2\rangle^{m-p}\langle AB\rangle^{2p+1}.
\end{aligned}$$

It is now easy to proceed. The first term in the expansion of $\langle A^{2k+2-m}B^m\rangle$ yields a sum that is cut at $m = 2k$ and where the combinatorial factor is just what is needed to recover the sum at order k, where the relation to be proven holds by assumption:

$$\begin{aligned}
& \sum_{m=0}^{2k} \frac{2^{k+1}(k+1)!}{(2k+2-m)!m!}\langle A^2\rangle\langle A^{2k-m}B^m\rangle\frac{(2k+2-m)(2k+1-m)}{2(k+1)} \\
& = \langle A^2\rangle \sum_{m=0}^{2k} \frac{2^k k!}{(2k-m)!m!}\langle A^{2k-m}B^m\rangle = \langle A^2\rangle\langle A^2 + 2AB + B^2\rangle^k.
\end{aligned}$$

The second term yields a sum running from $m = 1$ to $m = 2k + 1$, which we can recast as a sum from $m = 0$ to $m = 2k$:

$$\begin{aligned}
& \sum_{m=1}^{2k+1} \frac{2^{k+1}(k+1)!}{(2k+2-m)!m!}\langle AB\rangle\langle A^{2k+1-m}B^{m-1}\rangle\frac{(2k+2-m)m}{k+1} \\
& = 2\langle AB\rangle \sum_{m=1}^{2k+1} \frac{2^k k!}{(2k+1-m)!(m-1)!}\langle A^{2k+1-m}B^{m-1}\rangle \\
& = 2\langle AB\rangle \sum_{m=0}^{2k} \frac{2^k k!}{(2k-m)!m!}\langle A^{2k-m}B^m\rangle = 2\langle AB\rangle\langle A^2 + 2AB + B^2\rangle^k.
\end{aligned}$$

Finally, the last term yields a sum running from $m = 2$ to $m = 2k + 2$, which can also be recast as a sum from $m = 0$ to $m = 2k$:

$$
\begin{aligned}
&\sum_{m=2}^{2k+2} \frac{2^{k+1}(k+1)!}{(2k+2-m)!m!} \langle B^2 \rangle \langle A^{2k+2-m} B^{m-2} \rangle \frac{m(m-1)}{2(k+1)} \\
&= \langle B^2 \rangle \sum_{m=2}^{2k+2} \frac{2^k k!}{(2k+2-m)!(m-2)!} \langle A^{2k+2-m} B^{m-2} \rangle \\
&= \langle B^2 \rangle \sum_{m=0}^{2k} \frac{2^k k!}{(2k-m)!m!} \langle A^{2k-m} B^m \rangle = \langle B^2 \rangle \langle A^2 + 2AB + B^2 \rangle^k .
\end{aligned}
$$

Thus we have found that the sum at order $k + 1$ indeed yields $(\langle A^2 \rangle + 2\langle AB \rangle + \langle B^2 \rangle)\langle A^2 + 2AB + B^2 \rangle^k = \langle A^2 + 2AB + B^2 \rangle^{k+1}$.

IOP Publishing

Chapter 1

Introduction

The concept of correlation function is at the heart of the modern quantum theory of matter. It is also central in our understanding of the techniques that we use to scrutinize this matter. In most—if not all—instances, the measurements performed in condensed-matter physics are more or less directly probing one of the system's correlation functions. Understanding this connection for each type of experiment is an indispensable first step before we can start thinking about what the experimental results actually tell us. This whole book is about correlation functions and how they are revealed by various experiments. The focus is on spectroscopic techniques, which either use particles to probe samples in a scattering geometry or fields for measuring their response to a stimulation. This introductory chapter picks one example of each kind, neutron scattering and linear-response, and briefly explains the key role played by correlation functions in each case. These two subjects are revisited in depth in chapters 6 and 10. The thermodynamic techniques are another source of information that is not considered in this book. The third section of this introduction highlights a remarkable result—which, admittedly, has more academic than practical interest—showing that the thermodynamic functions too are tightly bound to correlation functions.

1.1 Nuclear scattering and density–density correlation function

Consider a neutron with momentum $\hbar \boldsymbol{k}$ and energy $E_{\boldsymbol{k}}$ falling on a sample cut from an elementary material (composed of identical atoms). Let us ignore the fact that the neutron owns a magnetic moment and assume that it only 'feels' the target when hitting a nucleus. We can model this interaction with the contact potential

$$V(\boldsymbol{r}) = \sum_{\ell} V_0\, \delta(\boldsymbol{r} - \boldsymbol{r}_{\ell}) = V_0\, n(\boldsymbol{r}), \tag{1.1}$$

doi:10.1088/978-0-7503-1741-2ch1

where $n(\boldsymbol{r})$ is the operator giving the density of nuclei in the target. The neutron is scattered to a new state with momentum $\hbar\boldsymbol{k}'$ and energy $E_{k'}$. Let us call $\hbar\boldsymbol{q}$ and $\hbar\omega$ the momentum and energy transferred from the neutron to the target:

$$\boldsymbol{q} \equiv \boldsymbol{k} - \boldsymbol{k}', \qquad \hbar\omega \equiv E_k - E_{k'}. \tag{1.2}$$

In the process, the target has evolved from an initial state $|a\rangle$ with energy E_a to a final state $|b\rangle$ with energy E_b. We do not know the initial state of the target, but we may describe it as a mixed thermal state characterized by the Boltzmann probability distribution $\rho_a = e^{-\beta E_a}/(\sum_a e^{-\beta E_a})$. For the neutron, the transition rate from the state $|\boldsymbol{k}\rangle$ to the state $|\boldsymbol{k}'\rangle$ is therefore given at lowest order, according to the Fermi golden rule, by

$$\Gamma_{|k\rangle\to|k'\rangle} = \frac{2\pi}{\hbar}\sum_{ab}\rho_a \mid \langle \boldsymbol{k}, a|V|\boldsymbol{k}', b\rangle \mid^2 \delta(E_k + E_a - E_{k'} - E_b). \tag{1.3}$$

The a sum accounts for the thermodynamic average over the possible initial states of the target, while the b sum, together with the delta function, selects all possible final states of the target such that the energy gain $E_b - E_a$ is equal to the energy $\hbar\omega = E_k - E_{k'}$ provided by the neutron. We may assume that the states $|\boldsymbol{k}, a\rangle$ and $|\boldsymbol{k}', b\rangle$ are actually product states $|\boldsymbol{k}\rangle|a\rangle$ and $|\boldsymbol{k}'\rangle|b\rangle$ since the probing particle and the target are well separated before and after the scattering event. We can therefore rewrite the transition rate as (see Doc A1.1):

$$\Gamma_{|k\rangle\to|k'\rangle} = \frac{V_0^2}{\hbar^2}\int_{-\infty}^{\infty} dt\, e^{i\omega t}\langle n(\boldsymbol{q}, t)n(-\boldsymbol{q}, 0)\rangle. \tag{1.4}$$

Hence the measured scattering cross section, which is proportional to the transition rate, is just the time Fourier transform of the density–density correlation function $\langle n(\boldsymbol{q}, t)n(-\boldsymbol{q}, 0)\rangle$. This function involves two kinds of terms: the 'diagonal' ones giving the temporal correlations in the motion of a given atom, $\langle e^{-i\boldsymbol{q}\cdot[\boldsymbol{r}_\ell(t)-\boldsymbol{r}_\ell(0)]}\rangle$, and the 'off-diagonal' ones representing the correlated motion of a pair of atoms, $\langle e^{-i\boldsymbol{q}\cdot[\boldsymbol{r}_\ell(t)-\boldsymbol{r}_{\ell'}(0)]}\rangle$. If the atomic motion is periodic in time, as occurs when a phonon is excited, these correlation functions are also periodic functions of time and their Fourier transform has peaks at the corresponding frequencies. The neutron enters in resonance with these oscillations and the scattering is strong. This is how nuclear neutron scattering can measure the dispersion relation of phonons. We shall return to this in chapter 10.

Of course, the simple result equation (1.4) echoes the simple form of the scattering potential equation (1.1). We will see in the next section that this kind of correspondence is not limited to nuclear scattering, but is the rule rather than the exception.

1.2 Linear response and retarded correlation functions

As a second example showing the importance of correlation functions, we consider the linear response of a quantum system to an applied external field. Imagine that a space- and time-dependent field $\boldsymbol{F}(\boldsymbol{r}, t)$ is turned on at time $t = 0$. Most often, this field is coupled to some observable $\boldsymbol{A}(\boldsymbol{r})$ of the system through a term

$$V = \int d\boldsymbol{r}\, \boldsymbol{A}(\boldsymbol{r}) \cdot \boldsymbol{F}(\boldsymbol{r}, t) \tag{1.5}$$

in the Hamiltonian. As a result, the average value of the observable $\boldsymbol{A}$ will change at times $t > 0$. This change can be expanded in powers of the applied field: the zeroth order term is just $\langle \boldsymbol{A}(\boldsymbol{r}, t)\rangle^{(0)}$, i.e. the value of the observable in the absence of external field, and the first-order term can be written in the form

$$\big\langle A_\mu(\boldsymbol{r}, t)\big\rangle^{(1)} = \sum_\nu \int d\boldsymbol{r}' \int_{-\infty}^{\infty} dt'\, \chi_{\mu\nu}(\boldsymbol{r}, t; \boldsymbol{r}', t') F_\nu(\boldsymbol{r}', t'). \tag{1.6}$$

This is nothing but the most general linear relationship between the two vector fields $\langle \boldsymbol{A}(\boldsymbol{r}, t)\rangle$ and $\boldsymbol{F}(\boldsymbol{r}, t)$. The tensor field χ is called a *susceptibility*—some authors call it *generalized susceptibility*—since it tells how 'susceptible' the observable $\boldsymbol{A}$ is to the external field. The linear-response theory ought to derive an explicit expression for the susceptibility. We shall give this derivation in chapter 6 when we have all relevant tools at hand, but for the time being we just quote the result:

$$\chi_{\mu\nu}(\boldsymbol{r}, t; \boldsymbol{r}', t') = -\frac{i}{\hbar}\theta(t - t')\big\langle\big[A_\mu(\boldsymbol{r}, t), A_\nu(\boldsymbol{r}', t')\big]\big\rangle. \tag{1.7}$$

This is called the retarded correlation function of the operators $A_\mu(\boldsymbol{r})$ and $A_\nu(\boldsymbol{r}')$, which will be introduced in chapter 3. Hence the linear response of a condensed-matter system to external perturbations can be entirely described in terms of correlation functions. The most important examples are the spin–spin correlation function—i.e. the magnetic susceptibility—which gives the linear response to an external magnetic field and the density–density correlation function—or charge susceptibility—for the linear response to an electric field.

1.3 Thermodynamic properties and Green's function

A large body of experiments in condensed-matter physics address thermodynamic quantities, such as entropy, specific heat, compressibility, magnetization, etc. We have already seen how some of these quantities can be computed, to linear order, using the theory of linear response. Here, we show how these quantities can in principle be obtained exactly from a particular correlation function.

Since we will always be considering the grand-canonical ensemble at finite temperature in this book, the appropriate thermodynamic potential for us is the grand potential Ω. All thermodynamic quantities can be deduced from Ω: entropy, pressure, specific heat, compressibility and density can all be obtained from particular derivatives of Ω, as will be recalled in section 2.1.2. Now, assume that the system's Hamiltonian is

$$H = H_0 + \lambda V, \tag{1.8}$$

where H_0 is something we can solve and V is some difficult two-body interaction. In this case, as we will see in section 5.2.2.2, the grand potential can be related to the correlation function of the creation and annihilation operators according to

$$\Omega = \Omega_0 - \frac{\eta}{2}\int_0^1 \frac{d\lambda}{\lambda} \sum_\alpha \left[\sum_\beta \left(-\partial_\tau \delta_{\alpha\beta} - \xi_{\alpha\beta}\right)\mathscr{C}^{\lambda}_{a_\beta a_\alpha^\dagger}(\tau) - \delta(\tau)\right]_{\tau=0^-}. \tag{1.9}$$

Here Ω_0 is the grand potential for $\lambda = 0$ and $\xi_{\alpha\beta}$ are matrix elements of H_0. The meaning of this expression will become clear later, but the point here is that, if we know the correlation function $\mathscr{C}^{\lambda}_{a_\beta a_\alpha^\dagger}(\tau)$—which is in fact the imaginary-time one-particle Green's function—for all values of the coupling constant λ, then we can in principle deduce from it all thermodynamic properties.

Appendix A1

Doc A1.1 Nuclear scattering rate

The matrix element entering the transition rate in equation (1.3) is

$$|\langle \boldsymbol{k}, a|V|\boldsymbol{k}', b\rangle|^2 = \langle a|\langle \boldsymbol{k}|V|\boldsymbol{k}'\rangle|b\rangle\langle b|\langle \boldsymbol{k}'|V^\dagger|\boldsymbol{k}\rangle|a\rangle$$

with

$$\begin{aligned}\langle \boldsymbol{k}|V|\boldsymbol{k}'\rangle &= V_0 \int d\boldsymbol{r} e^{-i\boldsymbol{k}\cdot\boldsymbol{r}} n(\boldsymbol{r}) e^{i\boldsymbol{k}'\cdot\boldsymbol{r}} = V_0 \int d\boldsymbol{r} e^{-i(\boldsymbol{k}-\boldsymbol{k}')\cdot\boldsymbol{r}} n(\boldsymbol{r})\\ &= V_0\, n(\boldsymbol{k} - \boldsymbol{k}') = V_0\, n(\boldsymbol{q})\\ \langle \boldsymbol{k}'|V^\dagger|\boldsymbol{k}\rangle &= V_0\, n(\boldsymbol{k}' - \boldsymbol{k}) = V_0\, n(-\boldsymbol{q}),\end{aligned}$$

where we have used equation (2*b*) and ignored the normalization of the plane waves, which is inessential for the present discussion (all factors are properly taken into account in chapter 10). On the other hand, we can rewrite the delta function using equation (3*b*) as

$$2\pi\delta(E_{\boldsymbol{k}} + E_a - E_{\boldsymbol{k}'} - E_b) = \frac{1}{\hbar}\int_{-\infty}^{\infty} dt\ e^{i(\hbar\omega + E_a - E_b)t/\hbar}$$

and it follows that

$$\begin{aligned}\Gamma_{|\boldsymbol{k}\rangle\to|\boldsymbol{k}'\rangle} &= \frac{V_0^2}{\hbar^2}\int_{-\infty}^{\infty} dt\ e^{i\omega t} \sum_{ab} \rho_a \underbrace{\langle a|e^{iE_a t/\hbar}}_{\langle a|e^{iHt/\hbar}} n(\boldsymbol{q}) \underbrace{e^{-iE_b t/\hbar}|b\rangle}_{e^{-iHt/\hbar}|b\rangle} \langle b|n(-\boldsymbol{q})|a\rangle\\ &= \frac{V_0^2}{\hbar^2}\int_{-\infty}^{\infty} dt\ e^{i\omega t} \sum_{ab} \rho_a \langle a| \underbrace{e^{iHt/\hbar} n(\boldsymbol{q}) e^{-iHt/\hbar}}_{n(\boldsymbol{q},t)} |b\rangle\langle b|n(-\boldsymbol{q})|a\rangle\\ &= \frac{V_0^2}{\hbar^2}\int_{-\infty}^{\infty} dt\ e^{i\omega t} \sum_{a} \rho_a \langle a|n(\boldsymbol{q}, t) n(-\boldsymbol{q}, 0)|a\rangle\\ &= \frac{V_0^2}{\hbar^2}\int_{-\infty}^{\infty} dt\ e^{i\omega t}\, \mathrm{Tr}\, \rho\, n(\boldsymbol{q}, t) n(-\boldsymbol{q}, 0)\\ &= \frac{V_0^2}{\hbar^2}\int_{-\infty}^{\infty} dt\ e^{i\omega t} \langle n(\boldsymbol{q}, t) n(-\boldsymbol{q}, 0)\rangle.\end{aligned}$$

We have used the fact that the many-particle states $|a\rangle$ are by assumption eigenstates of the target's Hamiltonian H, and at the third line we used the closure relation $\sum_b |b\rangle\langle b| = \mathbb{1}$. The other ingredients of this calculation are reviewed in chapter 2.

Part I

Digest of many-body theory

Chapter 2

Elements of quantum mechanics

The theoretical tools pervading this whole book are the equilibrium time-dependent correlation functions at finite temperature. The formalism describing these objects rests on four pillars: quantum statistical mechanics, time-dependent quantum mechanics, second quantization, and complex analysis. In this chapter, we give a brief overview of the first three subjects. This is not meant to be comprehensive: we introduce all the ingredients that are used later in the book, but refrain from presenting elements that are not. We first describe the thermodynamics of quantum systems and recall the basic concepts and relations. We then discuss the question of time dependence, especially the interaction picture and the time-dependent perturbation theory; as an illustration we obtain the famous Fermi golden rule. Finally, we review the second quantization formalism as well as the basic properties of independent electrons and phonons. After deriving the electron–phonon coupling, we finish this quick tour with an account of the magnetic interactions and the theory of magnons. The fourth pillar, complex analysis, is assumed known and will be copiously illustrated in the subsequent chapters.

2.1 Thermodynamics of quantum systems

2.1.1 Thermodynamic average

The fundamental postulate of quantum thermodynamics is that, for a quantum system in equilibrium with a bath at a given temperature T, the average value $\langle A\rangle$ of a property A is a weighted average of the expectation values $\langle a|A|a\rangle$ of this property in each of the system's eigenstates $|a\rangle$. The weight of each eigenstate in this average is given by the Boltzmann factor

$$\rho_a = \frac{e^{-\beta(E_a-\mu N_a)}}{\sum_a e^{-\beta(E_a-\mu N_a)}} \tag{2.1}$$

doi:10.1088/978-0-7503-1741-2ch2

with $\beta = (k_BT)^{-1}$. The average value of an observable is therefore defined as

$$\boxed{\langle A\rangle = \sum_a \rho_a \langle a|A|a\rangle = \mathrm{Tr}\,\rho A.} \tag{2.2}$$

We have introduced the so-called *statistical density matrix*

$$\boxed{\rho = \frac{1}{Z}e^{-\beta K} = \sum_a \rho_a |a\rangle\langle a|,} \tag{2.3}$$

with $K = H - \mu N$, and the *partition function*

$$\boxed{Z = \mathrm{Tr}e^{-\beta K}.} \tag{2.4}$$

Equation (2.2) shows that a quantum system is entirely characterized by its density matrix ρ: knowing ρ, we can compute the expectation value of any observable. For a quantum system in a *pure state* $|\Psi\rangle$—any state in the Hilbert space, not necessarily an eigenstate of H—the expectation value is $\langle A\rangle = \langle\Psi|A|\Psi\rangle$ and the wave function contains all the relevant information. In thermal equilibrium, however, the system is not in a pure state but in a so-called *mixed state* specified by the density matrix ρ, which replaces the wave function $|\Psi\rangle$. We also see in equation (2.2) that the end result for $\langle A\rangle$ is a trace. This is very satisfactory because the trace of an operator does not depend upon the choice of the basis. Therefore, although we have formulated the problem using the eigenstates $|a\rangle$ of H as a basis, the result is nevertheless general. It is also seen that $\langle A\rangle$ naturally reduces to the conventional ground-state average $\langle\Psi_0|A|\Psi_0\rangle$ at zero temperature, since all weights $e^{-\beta(E_a-\mu N_a)}/Z$ become negligible relative to the weight of the ground state which approaches unity in the limit $\beta \to \infty$.

2.1.2 Grand potential and thermodynamic properties

The thermodynamic potential—the grand potential Ω in our case—can be deduced from the partition function using

$$\boxed{\Omega = -k_BT\ln Z.} \tag{2.5}$$

Ω is a natural function of the variables $\mathscr{V}$, T, and μ: $d\Omega = -pd\mathscr{V} - SdT - Nd\mu$. We use the symbol $\mathscr{V}$ for the volume in order to avoid confusion with the interaction or perturbation V. Ω, moreover, is, in the thermodynamic limit, an extensive homogeneous function which can be written as

$$\Omega(\mathscr{V}, T, \mu) = \mathscr{V}\varpi(T, \mu). \tag{2.6}$$

For future reference, we provide here the expression of the main thermodynamic quantities in terms of Ω or ϖ. The entropy is given by

$$S = -\left.\frac{d\Omega}{dT}\right)_{\mathscr{V},\mu} = -\mathscr{V}\frac{\partial\varpi}{\partial T} \tag{2.7}$$

while the pressure is just

$$p = -\left.\frac{d\Omega}{d\mathcal{V}}\right)_{T,\,\mu} = -\varpi. \tag{2.8}$$

Note that equations (2.8) and (2.5) lead to the *equation of state*: $p/(k_{\rm B}T) = \frac{\partial}{\partial\mathcal{V}}\ln Z$. From the entropy, we deduce the specific heat

$$C_{\mathcal{V}} = T\left.\frac{dS}{dT}\right)_{\mathcal{V}} = -\mathcal{V}T\left(\frac{\partial^2\varpi}{\partial T^2} + \frac{\partial^2\varpi}{\partial T\partial\mu}\frac{d\mu}{dT}\right) \tag{2.9}$$

and, with the particle-number density

$$n = -\frac{1}{\mathcal{V}}\left.\frac{d\Omega}{d\mu}\right)_{\mathcal{V},\,T} = -\frac{\partial\varpi}{\partial\mu}, \tag{2.10}$$

we also obtain the isothermal compressibility (see Doc A2.1):

$$\kappa_T = -\frac{1}{\mathcal{V}}\left.\frac{d\mathcal{V}}{dp}\right)_T = \frac{1}{n^2}\left.\frac{dn}{d\mu}\right)_T = -\frac{1}{n^2}\frac{\partial^2\varpi}{\partial\mu^2}. \tag{2.11}$$

An illustration of these various relations is provided in Bonus B2.1, where the cases of free fermions and free bosons in dimensions $d = 1$, 2, and 3 are treated.

2.1.3 Independent particles

For independent particles (bosons or fermions), the partition function takes the form $Z = \prod_\alpha z_\alpha$, where the index α numbers the excitation energies $\xi_\alpha = \varepsilon_\alpha - \mu$ of the system and (see Doc A2.2)

$$z_\alpha = (1 - \eta e^{-\beta\xi_\alpha})^{-\eta} = \begin{cases} \dfrac{1}{1 - e^{-\beta\xi_\alpha}} & \text{Bosons} \\ 1 + e^{-\beta\xi_\alpha} & \text{Fermions.} \end{cases} \tag{2.12}$$

Inserting equation (2.12) into equation (2.5) and performing the derivatives in equation (2.9), we obtain the specific heat

$$C_{\mathcal{V}} = \frac{k_{\rm B}}{2}\sum_\alpha \frac{(\beta\xi_\alpha)^2 - \beta\xi_\alpha\dfrac{1}{k_{\rm B}}\dfrac{d\xi_\alpha}{dT}}{\cosh(\beta\xi_\alpha) - \eta}. \tag{2.13}$$

This expression assumes that the energies ε_α are independent of T; the term involving $d\xi_\alpha/dT$ accounts for the temperature dependence of the chemical potential. Using equation (2.10), we also easily recover the well-known expression of the density in terms of the Bose–Einstein and Fermi–Dirac distribution functions:

$$n = \frac{1}{\mathcal{V}}\sum_\alpha \frac{1}{e^{\beta\xi_\alpha} - \eta} = \frac{1}{\mathcal{V}}\sum_\alpha d_{-\eta}(\xi_\alpha). \tag{2.14}$$

We will use these expressions later, when reviewing the properties of independent electrons and phonons. In practical calculations, the variable n is often used instead of μ: equation (2.14) must then be inverted to yield μ as a function of n. The thermodynamic properties of free fermions and bosons are discussed in Bonus B2.1.

2.2 Time dependence

2.2.1 Schrödinger, Heisenberg, and interaction pictures

In the *Schrödinger picture*, the time dependence in a quantum system is carried by the wave function according to $i\hbar\partial_t\Psi(t) = K\Psi(t)$. It is customary to introduce the *evolution operator* $U(t)$ defined in such a way that $\Psi(t) \equiv U(t)\Psi(0)$. Substituting in the Schrödinger equation, we see that the evolution operator satisfies the equation of motion $i\hbar\partial_t U(t) = KU(t)$. If the Hamiltonian is time-independent, the solution with initial value $U(0) = \mathbb{1}$ is obvious: $U(t) = e^{-iKt/\hbar}$. In the *Heisenberg picture*, on the contrary, the wave function is time-independent and the time dependence is carried by the observables according to

$$A(t) = U^{\dagger}(t)AU(t) = e^{iKt/\hbar}Ae^{-iKt/\hbar}, \tag{2.15}$$

with the last equality holding only for a time-independent K. The equivalence of both pictures is apparent if one considers the average values of observables: $\langle\Psi(t)|A|\Psi(t)\rangle = \langle\Psi(0)U^{\dagger}(t)|A|U(t)\Psi(0)\rangle = \langle\Psi(0)|A(t)|\Psi(0)\rangle$. These descriptions are convenient when the Hamiltonian is time-independent, but are advantageously replaced by the *interaction picture* in the opposite case, where the evolution operator cannot be solved as $e^{-iKt/\hbar}$. In the interaction picture, we write $K = K_0 + V_t$ where V_t is some time-dependent contribution[1], and we define $U = U_0\hat{U}$ where, by assumption, $i\hbar\partial_t U_0(t) = K_0U_0(t)$ with the solution $U_0(t) = e^{-iK_0t/\hbar}$. Then, the equation of motion of the evolution operator gives

$$KU(t) = i\hbar\partial_t U_0(t)\hat{U}(t) = K_0U(t) + U_0(t)i\hbar\partial_t\hat{U}(t) \tag{2.16}$$

which implies that

$$U_0(t)i\hbar\partial_t\hat{U}(t) = (K - K_0)U(t) = V_tU(t). \tag{2.17}$$

Multiplying on the left by $U_0^{-1}(t)$, we find

$$i\hbar\partial_t\hat{U}(t) = U_0^{-1}(t)V_tU_0(t)\hat{U}(t) \equiv \hat{V}_t(t)\hat{U}(t). \tag{2.18}$$

[1] It is necessary to distinguish the 'external' time dependence—due, e.g. to the fantasy of the experimenter—from the 'internal' time dependence governed by the quantum dynamics of the system. In this book, the external time dependence is denoted by the index t (often omitted) while the internal time dependence is indicated in parentheses. Most textbooks refer to 'explicit' and 'implicit' time dependencies, but we find this terminology misleading.

We will use the notation $\hat{A}(t)$ to denote the time evolution of A in the interaction picture, while $A(t)$ means the Heisenberg picture. The main relations are collected below:

$$\boxed{\hat{A}(t) = e^{iK_0t/\hbar}Ae^{-iK_0t/\hbar}, \quad K_0 = H_0 - \mu N} \tag{2.19a}$$

$$\boxed{U(t) = e^{-iK_0t/\hbar}\hat{U}(t)} \tag{2.19b}$$

$$\boxed{i\hbar\partial_t\hat{U}(t) = \hat{V}(t)\hat{U}(t).} \tag{2.19c}$$

In summary, in the interaction picture the time evolution of the operators is governed by K_0, equation (2.19a), while the equation of motion of the evolution operator only depends on the interaction V, equation (2.19c). One additional useful equation is the one relating the operators in the Heisenberg and interaction pictures:

$$A(t) = U^{\dagger}(t)AU(t) = \hat{U}^{\dagger}(t)e^{iK_0t/\hbar}Ae^{-iK_0t/\hbar}\hat{U}(t) = \hat{U}^{\dagger}(t)\hat{A}(t)\hat{U}(t). \tag{2.20}$$

2.2.2 Feynman–Dyson expansion of the evolution operator

Let us assume that the time-dependent term V_t is switched on adiabatically starting from $t = -\infty$, i.e. $V_{-\infty} = 0$. In the interaction picture, the evolution operator obeys the equation of motion equation (2.19c) with the boundary condition $\hat{U}(-\infty) = \mathbb{1}$. This differential equation can be solved formally by writing

$$\hat{U}(t) = \mathbb{1} - \frac{i}{\hbar}\int_{-\infty}^{t} dt_1\ \hat{V}(t_1)\hat{U}(t_1). \tag{2.21}$$

The same expression can then be used to substitute $\hat{U}(t_1)$ and the process iterated in order to generate an infinite series of terms with increasing powers of V. This is the Feynman–Dyson expansion. The series reads

$$\hat{U}(t) = \mathbb{1} + \sum_{n=1}^{\infty}\left(-\frac{i}{\hbar}\right)^n \int_{-\infty}^{t} dt_1 \cdots \int_{-\infty}^{t_{n-1}} dt_n\ \hat{V}(t_1)\cdots\hat{V}(t_n). \tag{2.22}$$

Each term in the series can be rewritten in a more symmetric and convenient way by means of the time-ordering operator T_+ (see Doc A2.3):

$$\boxed{\hat{U}(t) = \sum_{n=0}^{\infty}\frac{(-i/\hbar)^n}{n!}\int_{-\infty}^{t} dt_1\cdots dt_n T_+\{\hat{V}(t_1)\cdots\hat{V}(t_n)\}.} \tag{2.23}$$

This is one of the most beautiful formulas in the quantum theory of many-particle systems. It can be expressed symbolically by the elegant and suggestive notation

$$\hat{U}(t) = T_+\exp\left\{-\frac{i}{\hbar}\int_{-\infty}^{t} dt'\ \hat{V}(t')\right\}. \tag{2.24}$$

The Feynman–Dyson expansion lays the foundations of time-dependent perturbation theory, including the Fermi golden rule, all diagrammatic techniques as well as the theory of the response. We will therefore heavily rely on equation (2.23).

2.2.3 Generalized Fermi golden rule

As an illustration of how to use of equation (2.22), we calculate the rate of transition between an initial state and a final state induced by a perturbation V at all orders in V. The first-order result is the well-known Fermi golden rule.

Consider a system characterized by the Hamiltonian K. This can be a fully interacting many-particle system just as well as a single particle. The system is initially in the stationary state $|a\rangle$, an eigenstate of K with energy $K_a = E_a - \mu N_a$. A perturbation V is adiabatically turned on at time $t = -\infty$ and the state of the system starts to evolve in time according to $|\Psi(t)\rangle = U(t)|a\rangle$. The probability that the system is found in another eigenstate $|b\rangle$ of K after some time t is $P_{a\to b}(t) = |\langle b|\Psi(t)\rangle|^2$. Therefore, the rate of transition from state $|a\rangle$ to state $|b\rangle$ is $\Gamma_{a\to b}(t) = dP_{a\to b}(t)/dt$. Using equations (2.22) and (2.19) and performing all time integrations, we find (see Doc A2.4)

$$\langle b|\Psi(t)\rangle = \langle b|T(K_a)|a\rangle \frac{e^{-i(K_a+i0^+)t/\hbar}}{K_a - K_b + i0^+}, \tag{2.25}$$

where the matrix T is the so-called t-matrix:

$$\begin{aligned} T(\varepsilon) &= V + V(\varepsilon + i0^+ - K)^{-1}T(\varepsilon) \\ &= [V^{-1} - (\varepsilon + i0^+ - K)^{-1}]^{-1}. \end{aligned} \tag{2.26}$$

The infinitesimal shift $i0^+$ reflects the adiabatic switching of the interaction. Finally, taking the time derivative we find (see Doc A2.4)

$$\boxed{\Gamma_{a\to b} = \frac{2\pi}{\hbar}|\langle b|T(K_a)|a\rangle|^2 \delta(K_a - K_b).} \tag{2.27}$$

At lowest order in V, we have $T(\varepsilon) = V$ and the Fermi golden rule is recovered. We have considered a perturbation V that, apart from the adiabatic switching, does not depend on time. Another important case is when the perturbation oscillates in time such as $2\cos(\omega t)$. One can perform the calculation of Doc A2.4 in that case as well, but it is significantly more complicated because we have two terms $e^{i\omega t_k} + e^{-i\omega t_k}$ associated with each intermediate time t_k. Instead of just one term at each order, there are now 2^n terms that correspond physically to the absorption or emission of one or several quanta of energy $\hbar\omega$. The calculation is simple at first order, where instead of a single delta function like in equation (2.27) we get the sum $\delta(K_a - K_b - \hbar\omega) + \delta(K_a - K_b + \hbar\omega)$ as well as time-dependent terms that oscillate with the frequency 2ω and vanish upon averaging over a period.

2.3 Second quantization

2.3.1 Occupation-number representation

The second quantization is a convenient way of dealing with quantum systems containing many particles. An N-particle wave function in the real-space representation is a complex function of the N coordinates of the particles:

$$\Psi(\boldsymbol{r}_1\sigma_1, \ldots, \boldsymbol{r}_N\sigma_N) \equiv \Psi(1, \ldots, N) \tag{2.28}$$

where '1' is a short-hand notation for '$(\boldsymbol{r}_1\sigma_1)$'. More generally, if the number of particles is not fixed (as, for example, in a metal connected to electrical contacts), the state of the system is a linear superposition of many-particle states with different numbers of particles.

In order to construct a basis on which we can expand such a many-particle state, we start by choosing a complete basis for the one-particle problem: $\varphi_\alpha(1)$. The Greek indices can represent any relevant quantum number(s), such as momentum, spin, position in real space, etc. In a solid, for instance, a convenient basis is formed by the plane waves

$$\varphi_{\boldsymbol{k}\sigma}(1) \equiv \frac{\delta_{\sigma\sigma_1}}{\sqrt{\mathcal{V}}} e^{i\boldsymbol{k}\cdot\boldsymbol{r}_1}, \tag{2.29}$$

which satisfy the relations $\langle \boldsymbol{k}\sigma|\boldsymbol{k}'\sigma'\rangle = \delta_{\sigma\sigma'}\delta_{\boldsymbol{k}\boldsymbol{k}'}$ and $\sum_{\boldsymbol{k}\sigma}|\boldsymbol{k}\sigma\rangle\langle\boldsymbol{k}\sigma| = \mathbb{1}$ as required for a complete basis (see Doc A2.5). Dual to this momentum-space basis, there is the real-space basis

$$\varphi_{\boldsymbol{r}\sigma}(1) \equiv \delta_{\sigma\sigma_1}\delta(\boldsymbol{r} - \boldsymbol{r}_1). \tag{2.30}$$

Other basis sets are, of course, also possible: the general second-quantization formalism does not depend upon this arbitrary choice.

The second step is to construct basis functions for systems with N particles. These are the so-called Slater determinants for fermions and sign-less determinants (or 'permanents') for bosons. For two particles they read

$$\varphi_{\alpha,\beta}(1, 2) = \langle 1, 2|\alpha, \beta\rangle = \frac{1}{\sqrt{2}}\left[\varphi_\alpha(1)\varphi_\beta(2) + \eta\varphi_\alpha(2)\varphi_\beta(1)\right]. \tag{2.31}$$

Throughout this book, we shall use η to distinguish bosons ($\eta = +1$) from fermions ($\eta = -1$). The two-particle wave function in equation (2.31) is either invariant (if $\eta = +1$), or changes sign (if $\eta = -1$) under the exchange of the coordinates 1 and 2, as required for indistinguishable bosons and fermions, respectively. In particular, it is immediately seen that $\varphi_{\alpha,\alpha} \equiv 0$ for fermions, which encodes the Pauli exclusion principle. The ensemble of wave functions obtained by taking all possible pairs (α, β) of one-particle states in equation (2.31) constitutes a complete orthonormal basis for the two-particle problem[2]. The procedure is readily generalized to N particles:

[2] The indices α, β in the pair (α, β) must be ordered in some way such that only one of the pairs (α, β) and (β, α), which represent the same two-particle state, appears in the basis.

$$\varphi_{\alpha_1,\,\ldots,\,\alpha_N}(1,\,\ldots\,,\,N) = \frac{1}{\sqrt{N!}}\sum_{\mathscr{P}}\eta^{\sigma(\mathscr{P})}\varphi_{\alpha_1}(\mathscr{P}_1)\cdots\varphi_{\alpha_N}(\mathscr{P}_N). \tag{2.32}$$

Here $\mathscr{P}$ represents a permutation of the set $\{1,\,\ldots\,,\,N\}$, $\mathscr{P}_i$ is the ith element of that permutation, and $\sigma(\mathscr{P})$ is the number of transpositions in $\mathscr{P}$. Fortunately, it is the first and last time that we write a wave function in this form! One can indeed formulate the many-body problem in this 'conventional' representation, but the cost is a deluge of indices and tricky combinatorics.

The solution is to adopt a more compact (and abstract) representation called the *occupation-number representation*. In order to specify a Slater determinant (or a permanent), rather than giving the list of one-particle wave functions φ_α from which it is constructed, one tells how many times each of the φ_α's is used, by means of the following notation:

$$|n_1,\,n_2,\,\ldots\,,\,n_\alpha,\,\ldots\rangle. \tag{2.33}$$

For example, the one-particle state $|\alpha\rangle$ is denoted $|0,\,\ldots\,,\,0,\,n_\alpha = 1,\,0,\,\ldots\rangle$. Figure 2.1 illustrates the procedure with more examples. Obviously, for fermions the *occupation numbers* n_α are either 0 or 1, while for bosons they can be any non-negative integer. In this collection of states, we find one that is of particular importance, namely the vacuum state for which all occupation numbers are zero:

$$|0,\,0,\,\ldots\rangle \equiv |\varnothing\rangle. \tag{2.34}$$

Let us define a *creation operator* $a_\alpha^\dagger$ for the state $|\alpha\rangle$, in such a way that

$$\langle 1|a_\alpha^\dagger|\varnothing\rangle = \varphi_\alpha(1). \tag{2.35}$$

This is not enough to completely specify $a_\alpha^\dagger$. For this, we must indicate how $a_\alpha^\dagger$ acts on an arbitrary state $|n_1,\,n_2,\,\ldots\,,\,n_\alpha,\,\ldots\rangle$. We therefore move one step forward and require that

$$\langle 1,\,2|a_\beta^\dagger a_\alpha^\dagger|\varnothing\rangle = \varphi_{\alpha,\,\beta}(1,\,2). \tag{2.36}$$

Exchanging α and β we obtain, using equation (2.31),

$$\langle 1,\,2|a_\alpha^\dagger a_\beta^\dagger|\varnothing\rangle = \varphi_{\beta,\,\alpha}(1,\,2) = \varphi_{\alpha,\,\beta}(2,\,1). \tag{2.37}$$

	Slater determinants		
	Usual representation		Occupation-number representation
1-particle	$|\mathbf{1}\rangle, |\mathbf{2}\rangle, |\mathbf{3}\rangle, \ldots$	$\longleftrightarrow$	$|1,0,0,0,\cdots\rangle, |0,1,0,0,\cdots\rangle, |0,0,1,0,\cdots\rangle, \ldots$
2-particles	$|\mathbf{1},\mathbf{2}\rangle, |\mathbf{1},\mathbf{3}\rangle, |\mathbf{2},\mathbf{3}\rangle, \ldots$	$\longleftrightarrow$	$|1,1,0,0,\cdots\rangle, |1,0,1,0,\cdots\rangle, |0,1,1,0,\cdots\rangle, \ldots$
3-particles	$|\mathbf{1},\mathbf{2},\mathbf{3}\rangle, |\mathbf{1},\mathbf{2},\mathbf{4}\rangle, |\mathbf{2},\mathbf{3},\mathbf{4}\rangle, \ldots$	$\longleftrightarrow$	$|1,1,1,0,\cdots\rangle, |1,1,0,1,\cdots\rangle, |0,1,1,1,\cdots\rangle, \ldots$
N-particles	$|\alpha_1,\cdots,\alpha_N\rangle$	$\longleftrightarrow$	$|n_1,n_2,\cdots,n_\alpha,\cdots\rangle, \quad \sum_\alpha n_\alpha = N$

Figure 2.1. Correspondence between Slater determinants in the usual and occupation-number representations for fermions. The one-particle basis states φ_α are denoted here with bold letters which can take the values $\alpha \in \{\mathbf{1},\,\mathbf{2},\,\mathbf{3},\,\cdots\}$. For bosons we would have additional states, for example $|\mathbf{1},\,\mathbf{1},\,\mathbf{3}\rangle$, which would be denoted $|2,\,0,\,1,\,0,\,\cdots\rangle$ in the occupation-number representation.

In order to fulfill the requirement that $\varphi_{\alpha,\beta}(2, 1) = \eta\varphi_{\alpha,\beta}(1, 2)$, we must require that $a_\alpha^\dagger a_\beta^\dagger = \eta a_\beta^\dagger a_\alpha^\dagger$, in other words that $[a_\alpha^\dagger, a_\beta^\dagger]_{-\eta} = 0$. Therefore the creation operators must commute for bosons and anti-commute for fermions. Likewise, the Hermitian conjugated operators $a_\alpha = (a_\alpha^\dagger)^\dagger$ (anti-)commute. From the definition, it is clear that the only nonzero matrix elements of the creation operator $a_\alpha^\dagger$ are between states with all occupation numbers identical except for the one-particle state α, for which the occupation numbers must differ by one:

$$\langle \dots, n_\alpha + 1, \dots | a_\alpha^\dagger | \dots, n_\alpha, \dots \rangle \neq 0. \tag{2.38}$$

Taking the complex conjugate, we see that the nonzero matrix elements of a_α are of the form

$$\langle \dots, n_\alpha - 1, \dots | a_\alpha | \dots, n_\alpha, \dots \rangle \neq 0. \tag{2.39}$$

This shows that a_α removes a particle in state α, and is therefore called an *annihilation operator*. If $\alpha \neq \beta$, the creation and annihilation operators a_α and $a_\beta^\dagger$ also (anti-)commute, reflecting the (anti-)symmetry of the wave functions. However, for $\alpha = \beta$ we must verify the two properties

$$a_\alpha a_\alpha^\dagger |\varnothing\rangle = |\varnothing\rangle \qquad \text{and} \qquad a_\alpha^\dagger a_\alpha |\varnothing\rangle = 0. \tag{2.40}$$

The first has an obvious meaning, while the second holds because it is not possible to remove a particle from the vacuum. These relations imply that the creation and annihilation operators for the same state do not (anti)-commute but satisfy the commutation rules $[a_\alpha, a_\alpha^\dagger]_{-\eta} = 1$.

In summary, if the creation and annihilation operators are to be consistent with the symmetry properties of the bosonic and fermionic wave functions, they must obey the following commutation rules:

$$[a_\alpha, a_\beta^\dagger]_{-\eta} = \delta_{\alpha\beta}, \qquad [a_\alpha^\dagger, a_\beta^\dagger]_{-\eta} = [a_\alpha, a_\beta]_{-\eta} = 0. \tag{2.41}$$

These relations are central and must be remembered: they fully take care of the problem of symmetrizing or anti-symmetrizing the wave functions of bosons and fermions. We are now in the position to specify the action of a_α and $a_\alpha^\dagger$ on an arbitrary state in the occupation-number representation:

$$a_\alpha^\dagger | \dots, n_\alpha, \dots \rangle = \eta^{(\sum_{i<\alpha} n_i)} \sqrt{1 + \eta n_\alpha} | \dots, n_\alpha + 1, \dots \rangle \tag{2.42a}$$

$$a_\alpha | \dots, n_\alpha, \dots \rangle = \eta^{(\sum_{i<\alpha} n_i)} \sqrt{n_\alpha} | \dots, n_\alpha - 1, \dots \rangle. \tag{2.42b}$$

These definitions fulfill the required commutation rules, as well as all needed properties (see Doc A2.6). Note that the states given in equation (2.33) are eigenstates of the *operator* $n_\alpha = a_\alpha^\dagger a_\alpha$ and that the corresponding eigenvalue is the *number* n_α. Therefore, $a_\alpha^\dagger a_\alpha$ is the operator measuring the occupation number for the one-particle state φ_α.

2.3.2 Operators in second-quantized form

At this point, the benefit of introducing creation and annihilation operators does not appear clearly. But the benefit becomes obvious when we realize that all other operators can be expressed in terms of a_α and $a_\alpha^\dagger$ through simple and intuitive formulas. For any one-body operator T, like for instance the kinetic energy, this formula is

$$\boxed{T = \sum_{\alpha\beta} T_{\alpha\beta}\, a_\alpha^\dagger a_\beta, \qquad T_{\alpha\beta} = \int d1\; \varphi_\alpha^*(1) T(1) \varphi_\beta(1).} \tag{2.43}$$

The proof is given in many textbooks and will not be reproduced here; see for example Bruus & Flensberg (2004, p 14). Hence, the operator is represented by a matrix whose matrix elements $T_{\alpha\beta}$ are the conventional matrix elements evaluated in the one-particle basis φ_α. If the basis φ_α diagonalizes the operator T, like the plane waves equation (2.29) do for the kinetic energy, then the matrix representing T is also diagonal and the expression of the operator simplifies accordingly to $T = \sum_\alpha T_\alpha n_\alpha$. For any two-body operator V, like the Coulomb interaction, we have

$$\boxed{\begin{aligned} V &= \frac{1}{2} \sum_{\alpha\beta\gamma\delta} V_{\alpha\beta\gamma\delta}\, a_\alpha^\dagger a_\beta^\dagger a_\delta a_\gamma, \\ V_{\alpha\beta\gamma\delta} &= \int d1 d2\; \varphi_\alpha^*(1) \varphi_\beta^*(2) V(1, 2) \varphi_\gamma(1) \varphi_\delta(2). \end{aligned}} \tag{2.44}$$

Note the interchange in the order of indices for the two annihilation operators. This time, the operator V is represented by a rank-4 tensor, whose matrix elements are again standard matrix elements in the basis φ_α.

One sees at this stage that a considerable simplification has been achieved. Not only do we have a compact formalism to treat quantum systems with an arbitrary number of particles—the symmetry properties related to the particle statistics being encoded in the commutation rules of the creation and annihilation operators—but, in addition, this whole formalism rests on one single set of operators, a_α. It must be kept in mind that these operators are not defined once and for all, but relative to a choice of the one-particle basis. We shall assume that the index is sufficient in order to identify the basis to which the operators refer: for instance $a_{k\sigma}$ refers to the plane-wave basis (2.29) while $a_{r\sigma}$ refers to the real-space basis (2.30)[3]. For future reference, we reproduce here the expression of some frequently used operators.

[3] The creation operator in the real-space representation is sometimes denoted $\psi^\dagger(\boldsymbol{r})$ in the literature. In terms of the operators $\psi^\dagger(\boldsymbol{r})$ and $\psi(\boldsymbol{r})$, the kinetic energy operator has the form

$$T = -\frac{\hbar^2}{2m} \int d\boldsymbol{r}\; \psi^\dagger(\boldsymbol{r}) \nabla^2 \psi(\boldsymbol{r}),$$

as can be shown from equation (2.43). This expression looks like the average value of the one-particle kinetic energy $-\hbar^2/(2m)\nabla^2$ for the one-particle *wave function* $\psi(\boldsymbol{r})$. The terminology of 'second quantization' finds an illustration here: it seems that we obtain the second-quantized formulas by replacing the wave functions by operators, like in the 'first quantization' we replace the dynamical variables by operators.

Hamiltonian In this book, we will be interested in interacting many-particle systems. The generic Hamiltonian for such systems involves one-body terms (like the kinetic energy and the Hartree and exchange potentials) as well as two-body terms (like the part of the Coulomb interaction that is not contained in the Hartree and exchange potentials). Hence the generic (grand-)Hamiltonian K contains a one-body part, which we denote K_0, and a two-body part, which we denote V:

$$\boxed{K = \sum_{\alpha\beta} \xi_{\alpha\beta} a_\alpha^\dagger a_\beta + \frac{1}{2} \sum_{\alpha\beta\gamma\delta} V_{\alpha\beta\gamma\delta} a_\alpha^\dagger a_\beta^\dagger a_\delta a_\gamma \equiv K_0 + V.} \tag{2.45}$$

We have chosen to write $\xi_{\alpha\beta}$ rather than $[K_0]_{\alpha\beta}$ for the matrix elements of K_0, because it is customary to denote as $\xi_\alpha \equiv \varepsilon_\alpha - \mu$ the excitation energies, i.e. the one-particle energies measured relative to the chemical potential. In the basis where K_0 is diagonal, we recover the usual form $K_0 = \sum_\alpha \xi_\alpha a_\alpha^\dagger a_\alpha$.

Particle density The particle-number operator $n_\alpha = a_\alpha^\dagger a_\alpha$ allows one to define the particle density operator in real space, $n(\boldsymbol{r}) = \sum_\sigma a_{\boldsymbol{r}\sigma}^\dagger a_{\boldsymbol{r}\sigma}$. Here σ is most often meant as a spin, but more generally it could be any quantum number needed in order to specify the state of the particle, in addition to its position. It is seen that the operator $n(\boldsymbol{r})$ is diagonal in the real-space representation. In order to perform the Fourier transform and obtain $n(\boldsymbol{q})$ in terms of the plane-wave operators $a_{\boldsymbol{k}\sigma}$, we need the relation between $a_{\boldsymbol{r}\sigma}$ and $a_{\boldsymbol{k}\sigma}$. This relation is *not* what one would guess by looking at our convention for the Fourier transforms, equation (2). The correct relation is derived in Doc A2.7. We then obtain in the plane-wave basis

$$\boxed{n(\boldsymbol{q}) = \sum_{\boldsymbol{k}\sigma} a_{\boldsymbol{k}\sigma}^\dagger a_{\boldsymbol{k}+\boldsymbol{q}\sigma}.} \tag{2.46}$$

This is the Fourier transform of the density—which, in a translation-invariant system, is proportional to $\delta_{\boldsymbol{q},0}$—and should not be confused with the occupation number of the state $|\boldsymbol{k}\sigma\rangle$, $n_{\boldsymbol{k}\sigma} \equiv \langle a_{\boldsymbol{k}\sigma}^\dagger a_{\boldsymbol{k}\sigma}\rangle$, also known as the momentum distribution function. Equation (2.46) has this very simple form because the Fourier transform is defined over plane waves. On a generic basis, $n(\boldsymbol{q})$ depends on the functions φ_α.

Spin density A spin 1/2 is represented by the three operators $S^\mu = (\hbar/2)\tau^\mu$, where τ^μ are the Pauli matrices given in equation (22). This is consistent with equation (2.43) if we define creation operators that act only in the 2×2 spin space and if the matrix elements are $\langle\sigma|S^\mu|\sigma'\rangle = (\hbar/2)\tau_{\sigma\sigma'}^\mu$. For a distribution of spins-1/2 electrons, we must consider the operators $c_{\boldsymbol{r}\sigma}$ and we can define the local spin density and its Fourier transform as

$$\boldsymbol{S}(\boldsymbol{r}) = \frac{\hbar}{2} \sum_{\sigma\sigma'} \boldsymbol{\tau}_{\sigma\sigma'} c_{\boldsymbol{r}\sigma}^\dagger c_{\boldsymbol{r}\sigma'}, \qquad \boldsymbol{S}(\boldsymbol{q}) = \frac{\hbar}{2} \sum_{\boldsymbol{k}\sigma\sigma'} \boldsymbol{\tau}_{\sigma\sigma'} c_{\boldsymbol{k}\sigma}^\dagger c_{\boldsymbol{k}+\boldsymbol{q}\sigma'}. \tag{2.47}$$

This gives for S^z and $S^\pm \equiv S^x \pm iS^y$:

$$S^z(\boldsymbol{r}) = \frac{\hbar}{2}\left(c^\dagger_{\boldsymbol{r}\uparrow}c_{\boldsymbol{r}\uparrow} - c^\dagger_{\boldsymbol{r}\downarrow}c_{\boldsymbol{r}\downarrow}\right), \qquad S^+(\boldsymbol{r}) = \hbar c^\dagger_{\boldsymbol{r}\uparrow}c_{\boldsymbol{r}\downarrow}, \qquad S^-(\boldsymbol{r}) = \hbar c^\dagger_{\boldsymbol{r}\downarrow}c_{\boldsymbol{r}\uparrow}. \tag{2.48}$$

S^+ and S^- are called spin raising and spin lowering operators, respectively, since they increase and decrease the value of the spin at point $\boldsymbol{r}$ by one.

Current density The expression of the current operator for particles of charge e and mass m is derived in Doc A2.8. The current has two components, called the paramagnetic current $\boldsymbol{j}^p$ and the diamagnetic current $\boldsymbol{j}^d$:

$$\boxed{\boldsymbol{j}(\boldsymbol{r}) = \boldsymbol{j}^p(\boldsymbol{r}) + \boldsymbol{j}^d(\boldsymbol{r})} \tag{2.49a}$$

$$\boxed{\boldsymbol{j}^p(\boldsymbol{r}) = \frac{i\hbar}{2m}\sum_\sigma\left[\left(\nabla_{\boldsymbol{r}}a^\dagger_{\boldsymbol{r}\sigma}\right)a_{\boldsymbol{r}\sigma} - a^\dagger_{\boldsymbol{r}\sigma}\left(\nabla_{\boldsymbol{r}}a_{\boldsymbol{r}\sigma}\right)\right]} \tag{2.49b}$$

$$\boxed{\boldsymbol{j}^d(\boldsymbol{r}) = -\frac{e}{m}\sum_\sigma \boldsymbol{A}(\boldsymbol{r})a^\dagger_{\boldsymbol{r}\sigma}a_{\boldsymbol{r}\sigma}.} \tag{2.49c}$$

This is a *particle* current, not an electric current. More precisely, it is a particle-current density which has the units of velocity per volume. The paramagnetic current assumes a simpler form in the plane-wave basis because the gradients of the field operators can be evaluated explicitly (see Doc A2.8):

$$\boldsymbol{j}^p(\boldsymbol{q}) = \frac{\hbar}{m}\sum_{\boldsymbol{k}\sigma}\left(\boldsymbol{k} + \frac{\boldsymbol{q}}{2}\right)a^\dagger_{\boldsymbol{k}\sigma}a_{\boldsymbol{k}+\boldsymbol{q}\sigma}. \tag{2.50}$$

For $\boldsymbol{q} = 0$, we recover that the total current is proportional to the total momentum, $\boldsymbol{J} = (\hbar/m)\sum_{\boldsymbol{k}\sigma}\boldsymbol{k}n_{\boldsymbol{k}\sigma}$.

It is important to realize that, although equations (2.46) and (2.50) express the density and the current in the plane-wave basis, they *do not* assume translation invariance of the physical system under study: these definitions are completely general and apply to gases, liquids, and solids, as long as their one-particle wave functions can be normalized in the volume $\mathscr{V}$ and therefore represented in the plane-wave basis. When the plane waves are not the best choice, a change of basis like in Doc A2.7 allows one to obtain the most appropriate representation of the operators. This representation depends on the wave functions φ_α, leading to expressions more complicated than equations (2.46) and (2.50). We will stick to the plane waves for the general considerations and move to other basis sets when appropriate, as for instance in Doc A3.4 for the density operator and in Bonus B8.1 for the current operator.

2.4 Independent electrons

2.4.1 Electrons in a periodic potential

A seminal problem in condensed-matter physics is that of independent electrons living in a periodic atomic lattice. The lattice acts on the electrons via a periodic potential

$$V(\boldsymbol{r}) = V(\boldsymbol{r} + \boldsymbol{R}_n) = \sum_{\boldsymbol{G}} V(\boldsymbol{G}) e^{i\boldsymbol{G}\cdot\boldsymbol{r}}, \tag{2.51}$$

where $\boldsymbol{R}_n$ are the lattice sites and $\boldsymbol{G}$ are the vectors of the reciprocal lattice defined by the condition $e^{i\boldsymbol{G}\cdot\boldsymbol{R}_n} = 1$. The electrons are not independent in nature because the Coulomb interaction is not a small interaction compared with typical electron energies. Nevertheless, in many circumstances, it turns out that replacing the Coulomb interaction by an effective one-body potential is not a bad approximation. Depending upon the level of sophistication, this one-body potential can take into account only the classical part of the Coulomb interaction (Hartree approximation), add to this the effect of the exchange interaction (Hartree–Fock approximation), or build in approximate correlation effects (density-functional theory and Kohn–Sham approximation). But, in the end, the structure of the problem remains that of independent electrons moving in a periodic potential such as the one in equation (2.51).

In this problem, it is a good idea to use the plane-wave basis set (2.29) and decompose the wave vector $\boldsymbol{k}$ as $\boldsymbol{k} + \boldsymbol{G}$, where in the latter expression $\boldsymbol{k}$ lies in the first Brillouin zone. This decomposition is unique such that, if $\boldsymbol{k} + \boldsymbol{G} = \boldsymbol{k}' + \boldsymbol{G}'$, then $\boldsymbol{k} = \boldsymbol{k}'$ *and* $\boldsymbol{G} = \boldsymbol{G}'$. Using this property, we can write the Hamiltonian of the system as (see Doc A2.9)

$$H_0 = \sum_{\boldsymbol{k}} \sum_{\boldsymbol{G}\boldsymbol{G}'} \left[\frac{\hbar^2}{2m} (\boldsymbol{k} + \boldsymbol{G})^2 \delta_{\boldsymbol{G}\boldsymbol{G}'} + V(\boldsymbol{G} - \boldsymbol{G}') \right] c^{\dagger}_{\boldsymbol{k}+\boldsymbol{G}} c_{\boldsymbol{k}+\boldsymbol{G}'}. \tag{2.52}$$

We have omitted the spin indices for simplicity. This is just another way of stating the Bloch theorem: in the plane-wave representation, the Hamiltonian of electrons in a periodic potential is diagonal in the index $\boldsymbol{k}$. Therefore, solving the problem reduces to diagonalizing the matrix $h_{\boldsymbol{G}\boldsymbol{G}'}(\boldsymbol{k}) = \frac{\hbar^2}{2m}(\boldsymbol{k} + \boldsymbol{G})^2 \delta_{\boldsymbol{G}\boldsymbol{G}'} + V(\boldsymbol{G} - \boldsymbol{G}')$ independently for each $\boldsymbol{k}$ in the first Brillouin zone. One thus obtains series of energy levels $\varepsilon_{\boldsymbol{k}n}$ known as electron bands. Working on the basis which makes the matrix $h_{\boldsymbol{G}\boldsymbol{G}'}(\boldsymbol{k})$ diagonal, we can express the grand-Hamiltonian in the simple form

$$K_0 = \sum_{\boldsymbol{k} \in 1^{\text{st}}\,\text{BZ}} \sum_n \xi_{\boldsymbol{k}n} c^{\dagger}_{\boldsymbol{k}n} c_{\boldsymbol{k}n} \tag{2.53}$$

with the excitation energies $\xi_{\boldsymbol{k}n} \equiv \varepsilon_{\boldsymbol{k}n} - \mu$. Note that $c^{\dagger}_{\boldsymbol{k}n}$ in this expression is the operator creating an electron in the Bloch state with quantum numbers $(\boldsymbol{k}, n)$ and energy $\varepsilon_{\boldsymbol{k}n}$, not to be confused with $c^{\dagger}_{\boldsymbol{k}}$ in equation (2.52), which creates an electron in a plane-wave state.

The document Bonus B2.2 provides a nice practical implementation of equation (2.52) for calculating the band structure of fourteen semiconductors with just a few lines of code. To calculate energy bands in solids is a discipline per se and a great deal of ingeniousness has been put into finding efficient and accurate methods for this task. We will not touch this vast subject, which has gained a new impetus recently with burgeoning research on topological materials[4]; however, we shall often refer to one-particle excitation energies ξ_α, as we already have in section 2.1.3. In doing so, we generally have in mind the energy levels which would result from some independent-electron treatment of the problem at hand. This serves as a basis for including the effect of additional interactions not taken into account in the independent-electron approximation.

2.4.2 Specific heat of independent electrons

We close this section by a problem of great historical importance, namely calculating the temperature dependence of the electronic specific heat in a system of independent electrons. In general, one has to inject the one-electron excitation energies ξ_α into equation (2.13) and perform the summation numerically. An approximate analytical result is obtained by recognizing that the temperature dependence stems mostly from the occupation numbers while contributions due to the energy dependence of the spectrum and temperature dependence of the chemical potential are comparatively much smaller[5]. Neglecting the latter, we write

$$C_{\mathscr{V}}^{\text{el}} = \frac{k_{\text{B}}}{2}\sum_\alpha \frac{(\beta\xi_\alpha)^2}{\cosh(\beta\xi_\alpha)+1} = \frac{k_{\text{B}}}{2\beta}\int_{-\infty}^{\infty} dx \frac{x^2 N^{\text{el}}(x/\beta)}{\cosh x + 1}. \tag{2.54}$$

We have introduced the density of states,

$$N^{\text{el}}(\varepsilon) = \sum_\alpha \delta(\varepsilon - \xi_\alpha), \tag{2.55}$$

and changed the integration variable from ε to $x = \beta\varepsilon$. The function $x^2/(\cosh x + 1)$ is of order one near $|x| = 2$ and becomes negligible for $|x| \gtrsim 10$, in other words $|\xi| \gtrsim 10k_{\text{B}}T$. Therefore, if the density of states can be considered constant in the interval $-10k_{\text{B}}T < \xi < 10k_{\text{B}}T$ we can take it out of the integral and arrive at the well-known result

[4] See, e.g. Bansil A, Lin H and Das T 2016 *Rev. Mod. Phys.* **88** 021004.

[5] The chemical potential generally depends on temperature such as to keep the density fixed as the temperature changes. From the condition $dn/dT = 0$ and the expression (2.14) giving the density of independent fermions, we find

$$\frac{d\mu}{dT} = -\frac{1}{T}\frac{\sum_\alpha \xi_\alpha f(\xi_\alpha)f(-\xi_\alpha)}{\sum_\alpha f(\xi_\alpha)f(-\xi_\alpha)}.$$

The product $f(\xi_\alpha)f(-\xi_\alpha)$ is sensibly different from zero in an energy range of the order $\pm 10k_{\text{B}}T$ around the Fermi energy. The DOS can often be considered constant in this energy range, such that the numerator vanishes because the integrand is odd. This explains why the temperature variation of the chemical potential can often be neglected at low temperature. For a particle–hole symmetric system, the chemical potential is exactly temperature independent.

$$C_{\mathcal{V}}^{\text{el}} = \frac{\pi^2}{3} k_{\text{B}}^2 N^{\text{el}}(0) T. \tag{2.56}$$

This formula provides a route to the experimental determination of the Fermi-level density of states—a quantity of primary importance for solids—by measuring the slope of the linear term in the specific heat at very low temperatures. An exact calculation of the specific heat for free electrons, including the temperature dependence of the chemical potential, is presented in Bonus B2.1.

2.5 Phonons

Phonons are the quanta of lattice vibrations in solids. In order to obtain a quantum description of the phonons, one first writes the Hamiltonian of classical lattice vibrations. At second order in the atomic displacements, this Hamiltonian describes a superposition of independent harmonic oscillators. The classical oscillators are then quantized in the usual way and their quanta are the phonons.

2.5.1 Classical lattice vibrations

Let us consider a lattice with unit cells at positions $\boldsymbol{R}_n$ and let us denote by $u_{\nu i}(\boldsymbol{R}_n)$ the displacement in the Cartesian direction i of the atom ν in the cell located at $\boldsymbol{R}_n$. If these displacements are measured relative to the equilibrium positions, the whole lattice is stationary with respect to these displacements, which means that $\partial U / \partial u_{\nu i}(\boldsymbol{R}_n) = 0$, where U is the elastic energy. Hence the expansion of U starts at second order in the displacements:

$$U = U_0 + \frac{1}{2} \sum_{nm} \sum_{\nu i \mu j} \left. \frac{\partial^2 U}{\partial u_{\nu i}(\boldsymbol{R}_n) \partial u_{\mu j}(\boldsymbol{R}_m)} \right|_0 u_{\nu i}(\boldsymbol{R}_n) u_{\mu j}(\boldsymbol{R}_m) + \cdots \tag{2.57}$$

In the harmonic approximation, one neglects higher-order terms (anharmonic terms) and one searches the eigenmodes of the harmonic Hamiltonian. In a perfect crystal, the matrix of *force constants*

$$W_{\nu i \mu j}(\boldsymbol{R}_n, \boldsymbol{R}_m) \equiv \frac{1}{\sqrt{M_\nu M_\mu}} \left. \frac{\partial^2 U}{\partial u_{\nu i}(\boldsymbol{R}_n) \partial u_{\mu j}(\boldsymbol{R}_m)} \right|_0, \tag{2.58}$$

where M_ν is the mass of atom ν, is invariant upon translation by a lattice vector, i.e. $W_{\nu i \mu j}(\boldsymbol{R}_n, \boldsymbol{R}_m) = W_{\nu i \mu j}(\boldsymbol{R}_n - \boldsymbol{R}_m)$. Its Fourier transform is called the *dynamical matrix*:

$$D(\boldsymbol{k}) \equiv \sum_n W(\boldsymbol{R}_n) e^{-i\boldsymbol{k}\cdot\boldsymbol{R}_n}. \tag{2.59}$$

Diagonalizing this matrix (which has dimension s^{-2}), we obtain a set of eigenvalues and eigenvectors:

$$D(\boldsymbol{k}) \boldsymbol{\epsilon}_{\boldsymbol{k}\lambda} = \omega_{\boldsymbol{k}\lambda}^2 \boldsymbol{\epsilon}_{\boldsymbol{k}\lambda}. \tag{2.60}$$

As the matrix W is real and positive-definite, the eigenvalues of D are real and positive such that we could write them as $\omega_{k\lambda}^2$ without loss of generality. The size of the dynamical matrix is $N_a d \times N_a d$, where N_a is the number of atoms per cell and d is the spatial dimension. Hence there are $N_a d$ eigenvalues for each wave vector $\boldsymbol{k}$. d of those are acoustic modes such that $\omega_{k\lambda} \propto k$ as $\boldsymbol{k} \to 0$. The others are optical modes with a finite frequency at $\boldsymbol{k} = 0$. We introduce the *normal coordinates* $q_{k\lambda}$, which represent the displacements $u_{\nu i}(\boldsymbol{R}_n)$ in the basis formed by the eigenvectors $\epsilon_{k\lambda}$ of the dynamical matrix:

$$u_{\nu i}(\boldsymbol{R}_n) = \frac{1}{\sqrt{M_\nu}} \frac{1}{\sqrt{\mathcal{N}}} \sum_{k\lambda} q_{k\lambda} [\epsilon_{k\lambda}]_{\nu i} e^{i\boldsymbol{k}\cdot\boldsymbol{R}_n}. \tag{2.61}$$

$\mathcal{N}$ is the number of elementary cells in the system. In terms of these new variables, the harmonic potential becomes simply (see Doc A2.10)

$$U = U_0 + \frac{1}{2} \sum_{k\lambda} |q_{k\lambda}|^2 \omega_{k\lambda}^2. \tag{2.62}$$

Adding to this harmonic potential the kinetic energy (see Doc A2.10)

$$\frac{1}{2} \sum_{n\nu i} M_\nu \dot{u}_{\nu i}^2(\boldsymbol{R}_n) = \frac{1}{2} \sum_{k\lambda} |\dot{q}_{k\lambda}|^2 \equiv \frac{1}{2} \sum_{k\lambda} |p_{k\lambda}|^2, \tag{2.63}$$

we obtain the Hamiltonian describing classical lattice vibrations:

$$H_{\text{lattice}} = \sum_{k\lambda} \frac{1}{2} \left(|p_{k\lambda}|^2 + \omega_{k\lambda}^2 |q_{k\lambda}|^2 \right). \tag{2.64}$$

This is a collection of independent classical harmonic oscillators[6]. Like for the one-electron bands ε_{kn}, to calculate the phonon frequencies $\omega_{k\lambda}$ is a whole field of research that we will not touch here. For a qualitative understanding of the various vibrational modes, one can use spring models as described in Bonus B2.3.

2.5.2 Hamiltonian of the phonons

The classical oscillators of equation (2.64) are quantized in the usual way by introducing the boson creation and annihilation operators $b_{k\lambda}^\dagger$ and $b_{k\lambda}$ for the oscillator's quanta. The quantized Hamiltonian takes the well-known form

$$H_{\text{ph}} = \sum_{k\lambda} \hbar\omega_{k\lambda} \left(b_{k\lambda}^\dagger b_{k\lambda} + \frac{1}{2} \right), \tag{2.65}$$

[6] Quick reminder: a classical body of mass M in a harmonic potential centered at $x = 0$ has an energy $H = \frac{1}{2}M\dot{x}^2 + \frac{1}{2}Kx^2$. It oscillates with a frequency $\omega = \sqrt{K/M}$. In terms of the conjugated variables $q \equiv \sqrt{M}x$ and $p \equiv \sqrt{M}\dot{x}$, the Hamiltonian is $H = \frac{1}{2}(p^2 + \omega^2 q^2)$. The canonical equations of motion are $\dot{q} = \partial H/\partial p = p$, and $\dot{p} = -\partial H/\partial q = -\omega^2 q$.

and the displacement operators in real space read

$$u_{\nu i}(\boldsymbol{R}_n) = \frac{1}{\sqrt{\mathcal{N}}} \sum_{k\lambda} [u_{k\lambda}]_{\nu i} e^{i\boldsymbol{k}\cdot\boldsymbol{R}_n} \tag{2.66a}$$

with

$$[u_{k\lambda}]_{\nu i} = \sqrt{\frac{\hbar}{2M_\nu \omega_{k\lambda}}} [\epsilon_{k\lambda}]_{\nu i} \left(b_{k\lambda} + b^\dagger_{-k\lambda}\right). \tag{2.66b}$$

Note that the $\boldsymbol{k}$ sums in equations (2.61)–(2.66) span the first Brillouin zone.

2.5.3 Einstein and Debye models for the phonon specific heat

The phonons provide a large contribution to the specific heat of solids, even the dominant contribution at ambient temperature. In order to estimate this contribution, we again use the general expression, equation (2.13), specialized to the case of bosons with temperature-independent frequencies $\omega_{k\lambda}$:

$$C_{\mathcal{V}}^{\text{ph}} = \frac{k_{\text{B}}}{2} \sum_{k\lambda} \frac{(\beta\omega_{k\lambda})^2}{\cosh(\beta\omega_{k\lambda}) - 1} = \frac{k_{\text{B}}}{2\beta} \int_{-\infty}^{\infty} dx \frac{x^2 N^{\text{ph}}(x/\beta)}{\cosh x - 1}. \tag{2.67}$$

$N^{\text{ph}}(\varepsilon) = \sum_{k\lambda} \delta(\varepsilon - \hbar\omega_{k\lambda})$ is the phonon density of states. The Einstein and Debye models for the phonon specific heat result from two different crude models for the phonon density of states. Einstein's model simply assumes that all phonons have the same frequency ω_{E}. This frequency plays the role of an average phonon frequency, related to a characteristic temperature $k_{\text{B}} T_{\text{E}} = \hbar\omega_{\text{E}}$. The resulting phonon DOS is $N_{\text{E}}^{\text{ph}}(\varepsilon) = N_{\text{modes}} \delta(\varepsilon - \hbar\omega_{\text{E}})$. In the Debye model, it is assumed that the dispersion relation of the phonons is exactly linear and cut at some frequency ω_{D}, which therefore plays the role of the largest phonon frequency in the system. The resulting phonon DOS is quadratic up to ω_{D} and reads (see Doc A2.11)

$$N_{\text{D}}^{\text{ph}}(\varepsilon) = 3N_{\text{modes}} \frac{\varepsilon^2}{(\hbar\omega_{\text{D}})^3} \theta(\hbar\omega_{\text{D}} - \varepsilon). \tag{2.68}$$

Both models satisfy the sum rule $\int_0^\infty d\varepsilon\, N^{\text{ph}}(\varepsilon) = N_{\text{modes}}$, where N_{modes} is the total number of phonon modes in the system. We obtain

$$C_{\mathcal{V},\,\text{E}}^{\text{ph}} = k_{\text{B}} N_{\text{modes}} \frac{1}{2} \frac{(T_{\text{E}}/T)^2}{\cosh(T_{\text{E}}/T) - 1} \tag{2.69}$$

$$C_{\mathcal{V},\,\text{D}}^{\text{ph}} = k_{\text{B}} N_{\text{modes}} \frac{3}{2} \left(\frac{T}{T_{\text{D}}}\right)^3 \int_0^{T_{\text{D}}/T} \frac{x^4\, dx}{\cosh x - 1}. \tag{2.70}$$

Both models agree on giving $C_{\mathcal{V}}^{\text{ph}} = k_{\text{B}} N_{\text{modes}}$ at high temperature, which is the law of Dulong and Petit. To see this in the Debye model, replace the integrand by its expansion near $x = 0$, namely $2x^2$, as is justified if $T \gg T_{\text{D}}$. At low temperature,

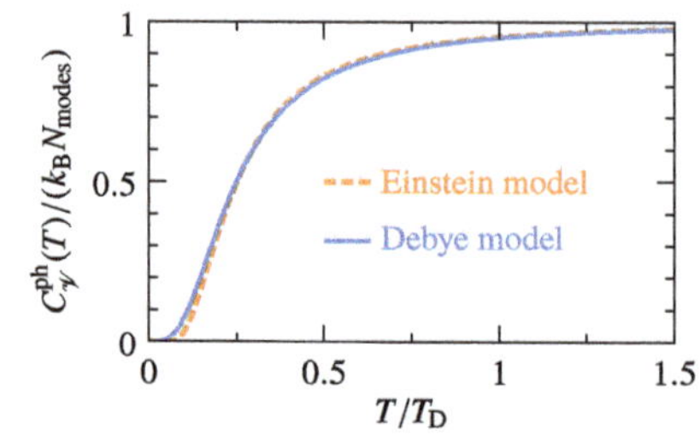

Figure 2.2. Temperature dependence of the phonon specific heat in the Einstein and Debye models. For a meaningful comparison, ω_{E} is taken as the 'center of gravity' of the Debye density of states: $\hbar\omega_{\mathrm{E}} = \int_0^{\infty} d\varepsilon\, \varepsilon N_{\mathrm{D}}^{\mathrm{ph}}(\varepsilon) \Big/ \int_0^{\infty} d\varepsilon\, N_{\mathrm{D}}^{\mathrm{ph}}(\varepsilon) = \frac{3}{4}\hbar\omega_{\mathrm{D}}$, so that $T_{\mathrm{E}} = \frac{3}{4}T_{\mathrm{D}}$.

$C_{\mathscr{V}}^{\mathrm{ph}}$ increases exponentially in the Einstein model as $(T_{\mathrm{E}}/T)^2 e^{-T_{\mathrm{E}}/T}$. In the Debye model, we may extend the upper bound of the integral to $+\infty$, and get a T^3 behavior: $C_{\mathscr{V},\,\mathrm{D}}^{\mathrm{ph}}(T \ll T_{\mathrm{D}}) = k_{\mathrm{B}}N_{\mathrm{modes}}(4\pi^4/5)(T/T_{\mathrm{D}})^3$. The two models are compared in figure 2.2.

2.5.4 Electron–phonon coupling

The electrons interact with the ionic lattice through the one-body Hamiltonian

$$H_{\text{el-ion}} = \int d\boldsymbol{r}(-e)n(\boldsymbol{r}) \sum_{n\nu} V_{\nu}(\boldsymbol{r} - \boldsymbol{R}_n - \boldsymbol{\tau}_{\nu}), \tag{2.71}$$

where $n(\boldsymbol{r})$ is the electron density, $V_{\nu}(\boldsymbol{r})$ is the potential of atom ν and $\boldsymbol{R}_n + \boldsymbol{\tau}_{\nu}$ is the position of the atom ν in the cell located at $\boldsymbol{R}_n$. The vector $\boldsymbol{\tau}_{\nu}$ is decomposed as $\boldsymbol{\tau}_{\nu} = \boldsymbol{\tau}_{\nu}^0 + \boldsymbol{u}_{\nu}(\boldsymbol{R}_n)$ with $\boldsymbol{u}_{\nu}(\boldsymbol{R}_n)$ the displacement from the equilibrium position. For small displacements, the potential can be expanded:

$$V_{\nu}(\boldsymbol{r} - \boldsymbol{R}_n - \boldsymbol{\tau}_{\nu}) = V_{\nu}(\boldsymbol{r} - \boldsymbol{R}_n - \boldsymbol{\tau}_{\nu}^0) - \nabla V_{\nu}(\boldsymbol{r} - \boldsymbol{R}_n - \boldsymbol{\tau}_{\nu}^0) \cdot \boldsymbol{u}_{\nu}(\boldsymbol{R}_n) + \cdots \tag{2.72}$$

The first term gives the interaction of the electrons with the static undeformed lattice in the absence of phonons, while the second term gives the electron–phonon coupling in the harmonic approximation. Using the known expressions for the phonons, this term can be rewritten as (see Doc A2.12):

$$\boxed{H_{\text{el-ph}} = \sum_{\boldsymbol{k}\sigma} \sum_{\boldsymbol{q}\lambda} g_{\boldsymbol{q}\lambda} c_{\boldsymbol{k}+\boldsymbol{q}\sigma}^{\dagger} c_{\boldsymbol{k}\sigma} \left(b_{\boldsymbol{q}\lambda} + b_{-\boldsymbol{q}\lambda}^{\dagger}\right),} \tag{2.73}$$

with the electron–phonon coupling vertex given by

$$g_{\boldsymbol{q}\lambda} = \frac{ie}{\mathscr{V}_{\text{cell}}} \sum_{\nu} \sqrt{\frac{\hbar}{2M_{\nu}\omega_{\boldsymbol{q}\lambda}\mathscr{N}}} \left([\boldsymbol{\epsilon}_{\boldsymbol{q}\lambda}]_{\nu} \cdot \boldsymbol{q}\right) V_{\nu}(\boldsymbol{q}) e^{-i\boldsymbol{q}\cdot\boldsymbol{\tau}_{\nu}^0}. \tag{2.74}$$

In an isotropic system, the polarization vectors are either parallel or perpendicular to the phonon propagation vector $\boldsymbol{q}$. Only the longitudinal phonons couple to the electrons in this case. Note that the electron–phonon coupling has the dimension of energy but contains a factor $1/\sqrt{\mathcal{N}}$. This is consistent with the requirement that the Hamiltonian (2.73) be extensive, i.e. $\propto \mathcal{N}$: the $\boldsymbol{k}$ sum scales like a density $n(\boldsymbol{q})$ (see equation (2.46)) which goes like $\mathcal{N}$, while equation (2.66) shows that, without $g_{q\lambda}$, the $\boldsymbol{q}$ sum scales like $\sqrt{\mathcal{N}}$. Note also that the coupling vertex $g_{q\lambda}$ does not depend on the electron wave vector $\boldsymbol{k}$ because equation (2.73) is expressed in the plane-wave basis using equation (2.46); on a different basis, for instance for Bloch states, the coupling vertex depends on the electronic quantum numbers.

2.6 Magnons

Magnons are wavelike elementary magnetic excitations of a solid. Ideally, in the ordered state of a ferromagnet or anti-ferromagnet, all spins have a well-defined orientation in order to minimize the magnetic energy. If all spins coherently deviate slightly from this equilibrium value, the energy cost is low. Therefore, the low-energy excitations take the form of a coherent precession of all spins around their equilibrium orientation. In this section, we recall the origin of magnetic forces and the theory of magnon excitations.

2.6.1 Quantum mechanical origin of the magnetic interactions

The magnetic interactions can be triggered by a variety of mechanisms in solids. In systems where the electrons are delocalized, the minimization of the Coulomb repulsion goes by adding nodes to the wave function such as to reduce the electronic overlap, which in turn favors anti-symmetry in the orbital sector and thus parallel alignment of the spins. Such systems are more prone to ferromagnetism. If the electrons are localized, on the other hand, minimization of the kinetic energy favors symmetric combinations of the orbitals, which reduces the wave function gradients and thereby promotes anti-ferromagnetism. We recall here the most basic mechanisms of magnetic interaction, namely Coulomb and kinetic exchange.

In a general basis, the Coulomb interaction has the form given by equation (2.44). To describe magnetism, we need to pull out the spin degrees of freedom from the generic Greek indices. We therefore consider basis functions of the kind $\varphi_{\alpha\sigma_\alpha}(1) \equiv \varphi_{\alpha\sigma_\alpha}(\boldsymbol{r}_1, \sigma_1) = \phi_\alpha(\boldsymbol{r}_1)\delta_{\sigma_\alpha\sigma_1}$. The matrix element $V_{\alpha\beta\gamma\delta}$ contains an orbital part given by

$$U_{\alpha\beta\gamma\delta} = \int d\boldsymbol{r}d\boldsymbol{r}'\phi^*_\alpha(\boldsymbol{r})\phi^*_\beta(\boldsymbol{r}')V(|\boldsymbol{r} - \boldsymbol{r}'|)\phi_\gamma(\boldsymbol{r})\phi_\delta(\boldsymbol{r}') \tag{2.75}$$

as well as a spin part, which is simply $\delta_{\sigma_\alpha\sigma_\gamma}\delta_{\sigma_\beta\sigma_\delta}$ because the Coulomb interaction conserves the spin; the Coulomb energy then reads

$$V_{\text{Cb}} = \frac{1}{2}\sum_{\alpha\beta\gamma\delta}\sum_{\sigma\sigma'} U_{\alpha\beta\gamma\delta}c^\dagger_{\alpha\sigma}c^\dagger_{\beta\sigma'}c_{\delta\sigma'}c_{\gamma\sigma}. \tag{2.76}$$

We can always use a localized basis, even if the electrons are delocalized. For instance, the Wannier functions of a Bloch crystal provide a localized basis on which the Hamiltonian can be projected. For a localized basis, the matrix element

(2.75) is largest if ϕ_γ has a large overlap with ϕ_α and ϕ_δ with ϕ_β. The optimal situation is clearly that of the *direct* terms $U_{\alpha\beta\alpha\beta}$ with $\gamma = \alpha$ and $\delta = \beta$. These terms contribute even if the overlap of the functions ϕ_α and ϕ_β is negligible. If ϕ_α and ϕ_β have non-negligible overlap, as may happen if they are centered on the same atom, the *exchange* terms $U_{\alpha\beta\beta\alpha}$ are of the same order of magnitude as the direct ones. It is shown in Doc A2.13 that if all matrix elements are neglected but the direct and exchange ones, the Coulomb interaction can be recast in the form

$$V_{\text{Cb}} \approx \sum_{\alpha} U_{\alpha\alpha\alpha\alpha} n_{\alpha\uparrow} n_{\alpha\downarrow} + \frac{1}{2} \sum_{\alpha\neq\beta} \left(U_{\alpha\beta\alpha\beta} - \frac{1}{2} U_{\alpha\beta\beta\alpha} \right) n_\alpha n_\beta - \frac{1}{\hbar^2} \sum_{\alpha\neq\beta} U_{\alpha\beta\beta\alpha} \boldsymbol{S}_\alpha \cdot \boldsymbol{S}_\beta. \quad (2.77)$$

The first term is the intra-orbital direct repulsion between two electrons of opposite spins. The second term corresponds to the classical (Hartree) repulsion (because $U_{\alpha\beta\alpha\beta} > 0$) between the electron densities in the orbitals α and β, which is reduced by the exchange process. One can indeed see that $U_{\alpha\beta\beta\alpha}$ is also positive:

$$\begin{aligned} U_{\alpha\beta\beta\alpha} &= \int d\boldsymbol{r} d\boldsymbol{r}' \, \phi_\alpha^*(\boldsymbol{r}) \phi_\beta^*(\boldsymbol{r}') \frac{1}{\mathcal{V}} \sum_{\boldsymbol{q}} V(\boldsymbol{q}) e^{i\boldsymbol{q}\cdot(\boldsymbol{r}-\boldsymbol{r}')} \phi_\beta(\boldsymbol{r}) \phi_\alpha(\boldsymbol{r}') \\ &= \frac{1}{\mathcal{V}} \sum_{\boldsymbol{q}} V(\boldsymbol{q}) \left| \int d\boldsymbol{r} \, \phi_\alpha^*(\boldsymbol{r}) e^{i\boldsymbol{q}\cdot\boldsymbol{r}} \phi_\beta(\boldsymbol{r}) \right|^2 > 0. \end{aligned}$$

The last term in equation (2.77) corresponds to a Heisenberg model

$$H = \frac{1}{2} \sum_{\alpha\neq\beta} J_{\alpha\beta} \boldsymbol{S}_\alpha \cdot \boldsymbol{S}_\beta \quad (2.78)$$

where the magnetic coupling $J_{\alpha\beta} = -2U_{\alpha\beta\beta\alpha}/\hbar^2 < 0$ is ferromagnetic, favoring parallel alignment of the spins. In atoms with partially-filled shells, this is the origin of the first Hund's rule stating that the total spin tends to be maximal.

Of course, let aside the case of atoms, the kinetic energy is in general not diagonal in the localized basis. Whether or not the spins actually align in the ground state also depends on whether the kinetic energy favors or disfavors this alignment. As it turns out, the kinetic energy may favor anti-ferromagnetic ordering, such that a competition can take place. The mechanism by which the kinetic energy produces anti-ferromagnetic interactions is known as kinetic exchange. To illustrate it, we consider a case where the localized orbitals ϕ_α are well separated from one another in space, such that the Coulomb repulsion is dominated by the intra-orbital terms. We disregard all terms but these and come to a Hamiltonian known as the Hubbard Hamiltonian:

$$H = \sum_{\alpha\beta\sigma} \xi_{\alpha\beta} c_{\alpha\sigma}^\dagger c_{\beta\sigma} + U \sum_{\alpha} n_{\alpha\uparrow} n_{\alpha\downarrow}. \quad (2.79)$$

The intra-orbital repulsion $U = U_{\alpha\alpha\alpha\alpha}$ takes by assumption the same value for all orbitals. In the simplest setup, the kinetic term is given by $\xi_{\alpha\alpha} = -\mu$ and $\xi_{\alpha\beta} = 0$ for all pairs of orbitals except the nearest-neighbor ones, for which $\xi_{\alpha\beta} = t$. Solving equation (2.79) remains a key challenge for the condensed-matter community. While powerful methods exist in one dimension and mean-field approximations perform reasonably well in three dimensions and higher, the two-dimensional case remains unsolved. The kinetic exchange mechanism shows up already in the most elementary version of this model, which has only one pair of orbitals occupied by two electrons. For such a 'dimer', it is shown in Doc A2.14 that, in the limit $U \gg t$, the ground state is a singlet and the ground-state projected Hamiltonian is

$$H_{\mathrm{GS}}^{\alpha\beta} = \frac{4t^2}{U}\left(\frac{1}{\hbar^2}\boldsymbol{S}_\alpha \cdot \boldsymbol{S}_\beta - \frac{n_\alpha n_\beta}{4}\right). \tag{2.80}$$

It is seen that the minimization of the kinetic energy produces an effective anti-ferromagnetic coupling of strength $4t^2/U$ between the spins. A generalization of these considerations to more than two orbitals leads to the t–J model, which has anti-ferromagnetic coupling between the nearest-neighbor spins and a modified kinetic energy without the hopping processes that produce doubly-occupied orbitals.

2.6.2 Spin waves of a ferromagnet

We consider a periodic lattice defined by the vectors $\boldsymbol{R}_n$ and we attach to each lattice site a local spin operator $\boldsymbol{S}_n = \hbar\boldsymbol{\mathcal{S}}_n$. It will prove convenient to work with the dimensionless operators $\boldsymbol{\mathcal{S}}_n$. All spins have the same value S. This is an idealized picture of a magnetic material. In general, the spin density is a function that varies continuously throughout space, following the peculiarities of the electronic wave functions. If the unpaired electrons giving rise to magnetism occupy sufficiently localized orbitals, one can think of their spins as being bound to one atomic shell and pretend that the whole spin density is concentrated at this atomic position. We furthermore, for simplicity, assume that these localized spins interact pairwise via forces that only depend on the relative position of the two spins. In doing so, we disregard a possible anisotropy of the matrix element, e.g. $U_{\alpha\beta\beta\alpha}$ in equation (2.77), with respect to the various directions in the lattice. These ideas lead us to the Heisenberg Hamiltonian for describing the magnetic energy:

$$H_{\mathrm{FM}} = -\frac{1}{2}\sum_{nm} J(\boldsymbol{R}_n - \boldsymbol{R}_m)\,\boldsymbol{\mathcal{S}}_n \cdot \boldsymbol{\mathcal{S}}_m - g\mu_{\mathrm{B}}B\sum_n \mathcal{S}_n^z. \tag{2.81}$$

Having a ferromagnet in mind, we pull out a minus sign and will consider hereafter that $J(\boldsymbol{R}_n - \boldsymbol{R}_m)$ is (mostly) positive. The second term accounts for the energy of the spins in the magnetic field B oriented along the spin-quantization axis, which result from the external applied field and the system's magnetization. It is also an idealization to take this field as constant in intensity and direction throughout space. If the spins are completely decoupled from the other degrees of freedom, e.g. phonons and electrons, we can study equation (2.81) on its own to find the ground state and excitations of H_{FM}. This is in general a difficult problem. The linear spin-wave theory simplifies this problem by considering that the deviations of the spin projections from full saturation along the z axis are small, as is justified if the ground state and the low-lying excitations are strongly polarized. There are various ways of linearizing equation (2.81). We shall follow the bosonization technique here, as it naturally leads to the notion of spin-wave excitations being bosons analogous to phonons. The Holstein–Primakoff transformation detailed in Doc A2.15 maps a spin $\boldsymbol{S}$ onto boson creation and annihilation operators $b^\dagger$ and b, while preserving the commutation relations of the spin components. The operator $b^\dagger$ acts like $S^-$and reduces the value of S^z by one unit. The complete mapping reads

$$S_n^z = S - b_n^\dagger b_n \tag{2.82a}$$

$$\boldsymbol{S}_n \cdot \boldsymbol{S}_m = S^2 - S(b_n^\dagger b_n + b_m^\dagger b_m) + S(b_n b_m^\dagger + b_n^\dagger b_m) + \cdots, \tag{2.82b}$$

where the ellipsis stands for interaction terms involving four or more boson operators. The mapping is exact if all interaction terms are kept. Neglecting those terms, which correspond to large deviations of S_n^z from full saturation, we are lead to a quadratic model of bosons whose solutions are the spin waves. The solution is presented in Doc A2.16: a Fourier transform is sufficient to bring the Hamiltonian to diagonal form:

$$H_{\mathrm{FM}}^{(1)} = \mathcal{N}\varepsilon_0 + \sum_{\boldsymbol{k}} \hbar\omega_{\boldsymbol{k}}\left(b_{\boldsymbol{k}}^\dagger b_{\boldsymbol{k}} + \frac{1}{2}\right). \tag{2.83a}$$

The first term is the ground-state energy, with $\mathcal{N}$ the number of spins and ε_0 the energy per spin, while the second term is a collection of independent harmonic oscillators, like the phonons in equation (2.65). The parameters are

$$\varepsilon_0 = -J(\boldsymbol{0})\frac{S(S+1)}{2} - g\mu_{\mathrm{B}}B\left(S + \frac{1}{2}\right) \tag{2.83b}$$

$$\hbar\omega_{\boldsymbol{k}} = [J(\boldsymbol{0}) - J(\boldsymbol{k})]S + g\mu_{\mathrm{B}}B \sim J_1 S(ka)^2 + g\mu_{\mathrm{B}}B, \tag{2.83c}$$

where $J(\boldsymbol{0})$ means $J(\boldsymbol{k} = \boldsymbol{0})$. The last expression holds at low k, for instance on a square lattice in dimension d, where $J(\boldsymbol{k}) = 2J_1\sum_{i=1}^{d}\cos(k_i a)$ and in this case J_1 is the coupling of nearest-neighbor spins. Equation (2.83c) shows that the excitations are gapped by the field B if $B > 0$: the latter reinforces the stability of the ferromagnet, such that a finite energy is needed to create excitations. If $B < 0$, the appearance of negative excitation energies at low k represents the instability towards magnetization reversal, which is triggered by these long-wavelength excitations[7]. Above the gap, the dispersion is quadratic like for free bosons. This is different from phonons, which have linear dispersion for $k \to 0$ describing sound propagation. Since no velocity can be defined for ferromagnetic spin waves, one defines instead the spin-wave stiffness D such that $\hbar\omega_k = Dk^2$. The elementary excitations, called magnons, correspond to plane waves of spins precessing around the quantization axis. The easiest way to see

[7] If $B < 0$, the ground state of equation (2.81) has all spins polarized along $-\hat{z}$: the spin-wave theory must be redone using that state as a starting point.

this is to note that, if $|\Psi_0\rangle$ is the fully polarized ground state of energy $\mathcal{N}\varepsilon_0$, the state $\mathcal{S}_{\bar{k}}^{-}|\Psi_0\rangle$ is also an eigenstate with energy $\mathcal{N}\varepsilon_0 + \hbar\omega_k$ (see Doc A2.16). For a single such mode, we have in real space $\mathcal{S}_n^{-}(t) = [\mathcal{S}_{\bar{k}}^{-}(0)/\sqrt{\mathcal{N}}]e^{-i(k\cdot R_n-\omega_k t)} \equiv Ae^{-i(k\cdot R_n-\omega_k t)}$, which implies $\mathcal{S}_n^{x}(t) = A\cos(\boldsymbol{k}\cdot\boldsymbol{R}_n - \omega_k t)$, $\mathcal{S}_n^{y}(t) = A\sin(\boldsymbol{k}\cdot\boldsymbol{R}_n - \omega_k t)$, while the z component is by assumption $\mathcal{S}_n^{z}(t) \approx S$.

Very much like the harmonic approximation for phonons amounts to neglecting phonon–phonon interactions, the linear approximation here amounts to neglecting the interaction between spin waves. From this starting point, one can construct theories of interacting spin waves. The bosonization indicates a systematic route to this, as it provides interaction terms that can be included to improve the theory.

2.6.3 Spin waves of an anti-ferromagnet

In anti-ferromagnets, the spins order in the ground state in such a way that the magnetizations of two or more sublattices cancel on average and the magnetization remains zero. Spin waves can still appear with this type of ground state, but they are slightly different from the spin waves in ferromagnets, as they are mixtures of excitations that decrease the up spins and others that increase the down spins. These excitations are analogous to the coherent electron–hole mixtures that form the Bogoliubov excitations in a superconductor. As a result, the magnons in anti-ferromagnets can disperse linearly for $k \to 0$, as opposed to the quadratic dispersion of ferromagnetic magnons. We consider here two sublattices denoted '+' and '−', formed by spins $\mathcal{S}_{n\nu}$ located at positions $\boldsymbol{R}_n + \nu\boldsymbol{\tau}$ with $\nu = \pm 1$. Like for the ferromagnet, we take a coupling that depends only on the vector connecting the two spins and a uniform magnetic field applied along the z direction: our Hamiltonian is

$$H_{\mathrm{AF}} = \frac{1}{2}\sum_{nm}\sum_{\nu\mu=\pm1} J(\boldsymbol{R}_n + \nu\boldsymbol{\tau} - \boldsymbol{R}_m - \mu\boldsymbol{\tau})\,\boldsymbol{\mathcal{S}}_{n\nu}\cdot\boldsymbol{\mathcal{S}}_{m\mu} - g\mu_{\mathrm{B}}B\sum_{n\nu}\mathcal{S}_{n\nu}^{z}. \tag{2.84}$$

J is positive for the vectors connecting neighboring spins in different sublattices, i.e. $J(\pm 2\boldsymbol{\tau}) > 0$. Depending on the sign of the intra-sublattice coupling $J(\boldsymbol{R}_n)$, the anti-ferromagnet will be reinforced or destabilized. In particular, we expect to find negative-energy magnons if $J(\boldsymbol{R}_n)$ is positive and exceeds some critical value, indicating that a ferromagnetic solution becomes more stable. In order to construct a viable spin-wave theory, we must start from the classical ground state of the spins[8]. If $B = 0$, the Hamiltonian is invariant under simultaneous rotation of all spins and all directions of the sublattice magnetization are equivalent. The classical state can be chosen with half the spins pointing up and the other half pointing down. For $B > 0$, the symmetry between the z axis and the (x, y) plane is broken and the state with all spins aligned along z is not the classical ground state. As shown in Doc A2.17, the state minimizing the classical energy has the spins slightly tilted out of the (x, y) plane by an angle ϑ for one sublattice and $\pi - \vartheta$ for the other sublattice, with ϑ related to the applied field via

[8] For the ferromagnet model (2.81), the classical state with all spins up is an eigenstate and corresponds to the true ground state, but for the anti-ferromagnet, the true ground state differs from the classical one.

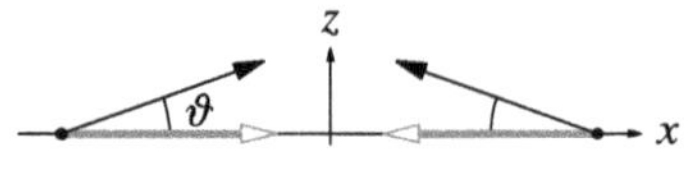

$$\sin\vartheta = \frac{g\mu_B B}{2SJ_1(\mathbf{0})}. \tag{2.85}$$

$J_1(\boldsymbol{k})$ is the Fourier transform of the inter-sublattice coupling. If $g\mu_B B > 2SJ_1(\mathbf{0})$, the system is fully polarized along z and $\sin\vartheta = 1$. As the Holstein–Primakoff bosons are designed to describe excitations of spins nearly saturated along the quantization axis, we need to rotate the quantization axis such that it coincides with the direction of the spins in the classical ground state. Let us call $\tilde{\boldsymbol{\mathcal{S}}}_{n\nu}$ the spins expressed in the new coordinate system, then the transformation reads

$$\boldsymbol{\mathcal{S}}_{n\nu} = \begin{pmatrix} \sin\vartheta & 0 & \nu\cos\vartheta \\ 0 & 1 & 0 \\ -\nu\cos\vartheta & 0 & \sin\vartheta \end{pmatrix} \tilde{\boldsymbol{\mathcal{S}}}_{n\nu}. \tag{2.86}$$

Now that all spins $\tilde{\boldsymbol{\mathcal{S}}}_{n\nu}$ are up, we can bosonize (see Doc A2.18). The occurrence of $b^\dagger b^\dagger$ and bb terms is the only difference with respect to the ferromagnetic case; due to these terms, the Hamiltonian is not diagonal in momentum space and a Bogoliubov transformation is needed to diagonalize it. The result is again a superposition of harmonic waves, which for a bipartite lattice takes the form

$$H_{\mathrm{AF}}^{(1)} = \mathcal{N}\varepsilon_0 + \sum_{\boldsymbol{k}} \hbar\omega_{\boldsymbol{k}}\left(\beta_{\boldsymbol{k}}^\dagger\beta_{\boldsymbol{k}} + \frac{1}{2}\right), \tag{2.87a}$$

however, with more complex ground-state energy and dispersion law with respect to the ferromagnet:

$$\frac{\varepsilon_0}{J_1(\mathbf{0})S} = -\left[1 + 2b^2 - \frac{J_2(\mathbf{0})}{J_1(\mathbf{0})}\right]S - \frac{1}{2\mathcal{N}}\sum_{\boldsymbol{k}} \left|\, 1 + b^2 j_1(\boldsymbol{k}) + j_2(\boldsymbol{k})\right| \tag{2.87b}$$

$$\frac{\hbar\omega_{\boldsymbol{k}}}{J_1(\mathbf{0})S} = \sqrt{\left[1 + b^2 j_1(\boldsymbol{k}) + j_2(\boldsymbol{k})\right]^2 - \left[(1 - b^2) j_1(\boldsymbol{k})\right]^2}. \tag{2.87c}$$

We have used $J_1(\mathbf{0})S$ as the natural unit of energy and introduced the dimensionless quantities

$$j_1(\boldsymbol{k}) = \mathrm{Re}\left[\frac{J_1(\boldsymbol{k})}{J_1(\mathbf{0})}\right], \quad j_2(\boldsymbol{k}) = \frac{J_2(\boldsymbol{k}) - J_2(\mathbf{0})}{J_1(\mathbf{0})}, \quad b = \sin\vartheta = \min\left(\frac{g\mu_B B}{2SJ_1(\mathbf{0})}, 1\right),$$

where $J_1(\boldsymbol{k})$ and $J_2(\boldsymbol{k})$ are the Fourier transforms of the inter- and intra-sublattice couplings, respectively (see Doc A2.17). The energy per spin ε_0 is lower than the

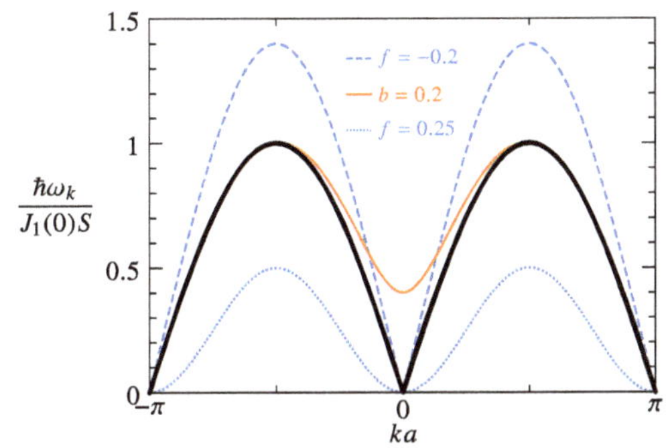

Figure 2.3. Typical examples of spin-wave dispersions in a one-dimensional chain with nearest-neighbor anti-ferromagnetic coupling $J_1(k) = J_1(0)\cos(ka)$ and next-nearest neighbor coupling $J_2(k) = J_2(0)\cos(2ka)$ with $J_2(0)/J_1(0) = f$. Black curve: $b = f = 0$, simplest case with linear dispersion. Orange curve: $b = 0.2, f = 0$, the size of the gap at $k = 0$ is $2b$. Blue dashed curve: $b = 0, f = -0.2$, reinforced anti-ferromagnetism by ferromagnetic next-nearest neighbor coupling. Dotted blue curve: $b = 0, f = 0.25$, weakened anti-ferromagnetism by anti-ferromagnetic next-nearest neighbor coupling; the anti-ferromagnet is unstable for $f > 0.25$.

energy of the classical state given by the first term in the right-hand side of equation (2.87*b*). In the simplest case, $b = 0$ and $j_2(\boldsymbol{k}) = 0$, the classical and quantum energies are $-J_1(\mathbf{0})S^2$ and $-J_1(\mathbf{0})S(S + 1/2)$, respectively, not counting the zero-point energy in equation (2.87*a*).

Consider for instance a one-dimensional spin-1/2 chain with nearest neighbor coupling $J' > 0$ and next-nearest neighbor coupling J''. In this case, we have $J_1(\mathbf{0})S = J'$, $j_1(k) = \cos ka$, and $j_2(k) = (J''/J')(\cos 2ka - 1)$. For $J'' = 0$, there is no instability. The dispersion is linear at $k = 0$ and $k = \pi/a$ in the absence of field, and becomes gapped at $k = 0$—but not at $k = \pi/a$—if $b > 0$ (see figure 2.3). If $J'' \neq 0$ and $b = 0$, the dispersion close to $k = 0$ behaves as $J'\sqrt{(1 - 4J''/J')(ka)^2}$. This indicates an instability towards a ferromagnetic solution if $J'' > J'/4$. If $J'' < 0$, on the other hand, the ferromagnetic next-nearest neighbor coupling reinforces the stability of the anti-ferromagnetic state and increases the magnon velocity.

Appendix A2

Doc A2.1 Isothermal compressibility

As $\Omega = \mathcal{V}\varpi$ and $\varpi = -p$, we have $\Omega = -\mathcal{V}p$ and therefore $d\Omega = -\mathcal{V}dp - pd\mathcal{V}$. Since, on the other hand, $d\Omega = -SdT - pd\mathcal{V} - Nd\mu$ we deduce $\mathcal{V}dp = SdT + Nd\mu$ or equivalently

$$dp = \frac{S}{\mathcal{V}}dT + \frac{N}{\mathcal{V}}d\mu.$$

It follows that

$$\left.\frac{d}{dp}\right)_T = \frac{1}{n}\left.\frac{d}{d\mu}\right)_T$$

where $n = N/\mathcal{V}$ and consequently

$$\kappa_T = -\frac{1}{\mathscr{V}}\frac{d\mathscr{V}}{dp}\bigg)_T = -\frac{1}{\mathscr{V}}\frac{1}{n}\frac{d(N/n)}{d\mu}\bigg)_T = -\frac{N}{\mathscr{V}}\frac{1}{n}\left(\frac{d}{d\mu}\frac{1}{n}\right)_T = \frac{1}{n^2}\frac{dn}{d\mu}\bigg)_T.$$

Doc A2.2 Partition function of independent fermions and bosons

A quantum system of independent particles in thermal equilibrium is entirely characterized by (i) its one-particle energy levels ε_α—or rather, in the grand-canonical ensemble, its excitation energies $\xi_\alpha = \varepsilon_\alpha - \mu$—and (ii) their occupation numbers n_α, which give the number of particles actually in state α. We can therefore specify any individual state of the system by the list $\{n_1, n_2, \ldots, n_\alpha, \ldots\}$ of occupation numbers and the corresponding energy $K = \sum_\alpha n_\alpha \xi_\alpha$. For fermions, n_α is either 0 or 1 due to the Pauli exclusion principle, while for bosons n_α is any non-negative integer. The indistinguishability of quantum particles is taken into account since we do not require to know which particle goes into which state, but only the number of them in each level. The partition function is obtained by summing over all possible states according to equation (2.4):

$$Z = \mathrm{Tr} e^{-\beta K} = \begin{cases} \displaystyle\sum_{\{n_\alpha = 0,\ldots,\infty\}} e^{-\beta \sum_\alpha n_\alpha \xi_\alpha} & \text{Bosons} \\ \displaystyle\sum_{\{n_\alpha = 0,1\}} e^{-\beta \sum_\alpha n_\alpha \xi_\alpha} & \text{Fermions.} \end{cases}$$

Let us start with the case of fermions, which is simpler. Assume first that there exists only one state α, say $\alpha = 1$. The system can accommodate either zero ($n_1 = 0$) or one ($n_1 = 1$) particle, and its energy is $K = 0$ in the first case and $K = \xi_1$ in the second case. Hence we have

$$Z = \underbrace{1}_{0\text{ fermion}} + \underbrace{e^{-\beta\xi_1}}_{1\text{ fermion}}.$$

If there are two levels, the system can accommodate zero, one, or two particles and the four possible states are $\{n_1, n_2\} = \{0,0\}$, $\{1,0\}$, $\{0,1\}$, and $\{1,1\}$:

$$Z = \underbrace{1}_{0\text{ fermion}} + \underbrace{e^{-\beta\xi_1} + e^{-\beta\xi_2}}_{1\text{ fermion}} + \underbrace{e^{-\beta\xi_1}e^{-\beta\xi_2}}_{2\text{ fermions}} = (1 + e^{-\beta\xi_1})(1 + e^{-\beta\xi_2}).$$

For three levels we have

$$\begin{aligned} Z = {} & \underbrace{1}_{0\text{ fermion}} + \underbrace{e^{-\beta\xi_1} + e^{-\beta\xi_2} + e^{-\beta\xi_3}}_{1\text{ fermion}} \\ & + \underbrace{e^{-\beta\xi_1}e^{-\beta\xi_2} + e^{-\beta\xi_1}e^{-\beta\xi_3} + e^{-\beta\xi_2}e^{-\beta\xi_3}}_{2\text{ fermions}} + \underbrace{e^{-\beta\xi_1}e^{-\beta\xi_2}e^{-\beta\xi_3}}_{3\text{ fermions}} \\ = {} & (1 + e^{-\beta\xi_1})(1 + e^{-\beta\xi_2})(1 + e^{-\beta\xi_3}). \end{aligned}$$

Generalizing to an arbitrary number of levels is obvious:

$$Z = \prod_{\alpha} (1 + e^{-\beta\xi_\alpha}) \qquad \text{(Fermions)}.$$

For bosons we proceed along the same lines, but in the case of only one level there are already infinitely many states:

$$\begin{aligned} Z &= \underbrace{1}_{\text{0 boson}} + \underbrace{e^{-\beta\xi_1}}_{\text{1 boson}} + \underbrace{e^{-2\beta\xi_1}}_{\text{2 bosons}} + \underbrace{e^{-3\beta\xi_1}}_{\text{3 bosons}} + \cdots \\ &= 1 + e^{-\beta\xi_1} + (e^{-\beta\xi_1})^2 + (e^{-\beta\xi_1})^3 + \cdots = \frac{1}{1 - e^{-\beta\xi_1}}. \end{aligned}$$

For two levels we have

$$\begin{aligned} Z &= \underbrace{1}_{\text{0 boson}} + \underbrace{e^{-\beta\xi_1} + e^{-\beta\xi_2}}_{\text{1 boson}} + \underbrace{e^{-2\beta\xi_1} + e^{-2\beta\xi_2} + e^{-\beta\xi_1}e^{-\beta\xi_2}}_{\text{2 bosons}} + \cdots \\ &= [1 + e^{-\beta\xi_1} + (e^{-\beta\xi_1})^2 + \cdots][1 + e^{-\beta\xi_2} + (e^{-\beta\xi_2})^2 + \cdots] \\ &= \frac{1}{1 - e^{-\beta\xi_1}} \frac{1}{1 - e^{-\beta\xi_2}}. \end{aligned}$$

The generalization is again obvious:

$$Z = \prod_{\alpha} \frac{1}{1 - e^{-\beta\xi_\alpha}} \qquad \text{(Bosons)}.$$

Thus we have established equation (2.12).

Doc A2.3 Evolution operator and time ordering

At orders zero and one, the two equations (2.22) and (2.23) give the same result. At order two, equation (2.22) gives

$$\hat{U}_2 = \left(-\frac{i}{\hbar}\right)^2 \int_{-\infty}^{t} dt_1 \int_{-\infty}^{t_1} dt_2\, \hat{V}(t_1)\hat{V}(t_2) = \left(-\frac{i}{\hbar}\right)^2 \int_{-\infty}^{t} dt_1 dt_2\, \theta(t_1 - t_2)\hat{V}(t_1)\hat{V}(t_2).$$

We have extended the upper bound of the second integral from t_1 to t and set the integrand to zero in this range with the Heaviside θ function. Exchanging the dummy variables t_1 and t_2 in the last expression, we also have

$$\hat{U}_2 = \left(-\frac{i}{\hbar}\right)^2 \int_{-\infty}^{t} dt_1 dt_2\, \theta(t_2 - t_1)\hat{V}(t_2)\hat{V}(t_1).$$

Finally, averaging the two equations, we obtain

$$\hat{U}_2 = \frac{(-i/\hbar)^2}{2!} \int_{-\infty}^{t} dt_1 dt_2 T_+\{\hat{V}(t_1)\hat{V}(t_2)\}$$

with T_+ defined in equation (5). This is the second-order term in equation (2.23). The trick is readily generalized. At order n we have from equation (2.22):

$$\hat{U}_n = \left(-\frac{i}{\hbar}\right)^n \int_{-\infty}^{t} dt_1 \cdots \int_{-\infty}^{t_{n-1}} dt_n\ \hat{V}(t_1)\cdots\hat{V}(t_n)$$
$$= \left(-\frac{i}{\hbar}\right)^n \int_{-\infty}^{t} dt_1 \cdots dt_n\ \theta(t_1 - t_2)\cdots\theta(t_{n-1} - t_n)\hat{V}(t_1)\cdots\hat{V}(t_n).$$

There are $n!$ ways of shuffling the dummy variables $\{t_1, \ldots, t_n\}$ in the last expression. In this way we generate the $n!$ terms of $T_+\{\hat{V}(t_1)\cdots\hat{V}(t_n)\}$.

Doc A2.4 Time evolution and the t-matrix

Using the expression (2.19b) of the evolution operator in the interaction picture, $U(t) = e^{-iKt/\hbar}\hat{U}(t)$, the expansion equation (2.22) of $\hat{U}(t)$, and the time evolution equation (2.19a), we obtain for the matrix element $\langle b|\Psi(t)\rangle = \langle b|U(t)|a\rangle$:

$$\langle b|\Psi(t)\rangle = \langle b|e^{-iKt/\hbar}\Bigg[\mathbb{1} + \sum_{n=1}^{\infty}\left(-\frac{i}{\hbar}\right)^n \int_{-\infty}^{t} dt_1 \cdots \int_{-\infty}^{t_{n-1}} dt_n$$
$$\times\, e^{iKt_1/\hbar}\, V e^{-iK(t_1-t_2)/\hbar} \ldots e^{-iK(t_{n-1}-t_n)/\hbar}\, V e^{-iKt_n/\hbar}\Bigg]|a\rangle.$$

Let us lighten the notation by temporarily introducing $\mathcal{K} = K/\hbar$. Since $|a\rangle$ and $|b\rangle$ are (orthogonal) eigenstates of $\mathcal{K}$ with eigenvalues $\mathcal{K}_a$ and $\mathcal{K}_b$, respectively, we have $e^{-i\mathcal{K}t_n}|a\rangle = e^{-i\mathcal{K}_a t_n}|a\rangle$ and $\langle b|e^{-i\mathcal{K}t} = e^{-i\mathcal{K}_b t}\langle b|$. The first term of the expansion thus reads $e^{-i\mathcal{K}_b t}\langle b|a\rangle = 0$. In the other terms, we insert the identity $\mathbb{1} = \sum_a |a\rangle\langle a|$ after each interaction V except the last:

$$\langle b|\Psi(t)\rangle = \sum_{n=1}^{\infty}\left(-\frac{i}{\hbar}\right)^n \sum_{a_1}\cdots\sum_{a_{n-1}} \int_{-\infty}^{t} dt_1 \cdots \int_{-\infty}^{t_{n-1}} dt_n$$
$$\times\, e^{-i\mathcal{K}_b(t-t_1)}\langle b|V|a_1\rangle\langle a_1|e^{-i\mathcal{K}(t_1-t_2)}\ldots|a_{n-1}\rangle\langle a_{n-1}|e^{-i\mathcal{K}(t_{n-1}-t_n)}V|a\rangle e^{-i\mathcal{K}_a t_n}.$$

Again, we can take advantage of the fact that the states $|a\rangle$ are eigenstates of $\mathcal{K}$:

$$\langle b|\Psi(t)\rangle = \sum_{n=1}^{\infty}\left(-\frac{i}{\hbar}\right)^n \sum_{a_1}\cdots\sum_{a_{n-1}} \langle b|V|a_1\rangle\langle a_1|V|a_2\rangle\cdots\langle a_{n-1}|V|a\rangle$$
$$\times \int_{-\infty}^{t} dt_1 \cdots \int_{-\infty}^{t_{n-1}} dt_n e^{-i\mathcal{K}_b(t-t_1)} e^{-i\mathcal{K}_{a_1}(t_1-t_2)} \ldots e^{-i\mathcal{K}_{a_{n-1}}(t_{n-1}-t_n)} e^{-i\mathcal{K}_a t_n} e^{0^+ t_n}.$$

At the very end, we have added a factor $e^{0^+ t_n}$ to take into account the adiabatic switching of the interaction. The justification is as follows. Looking at the domains of integration, we see that $-\infty < t_n < t_{n-1} < \cdots < t_1 < t$. Therefore, if the interaction is turned on at time t_n, it will be automatically on at all other times in the expansion.

The factor $e^{0^+ t_n}$ ensures that $V_{t_n \to -\infty} = 0$, such that the interaction is switched on exponentially from $t = -\infty$. The t_n integration can now be performed:

$$\int_{-\infty}^{t_{n-1}} dt_n e^{i(\mathcal{K}_{a_{n-1}} - \mathcal{K}_a - i0^+)t_n} = i \frac{e^{i(\mathcal{K}_{a_{n-1}} - \mathcal{K}_a - i0^+)t_{n-1}}}{\mathcal{K}_a - \mathcal{K}_{a_{n-1}} + i0^+}.$$

It is seen that the factor $e^{0^+ t_{n-1}}$ in the numerator carries the adiabatic switching factor over to the integration on t_{n-1}; furthermore, the factor $e^{i\mathcal{K}_{a_{n-1}} t_{n-1}}$ cancels the $e^{-i\mathcal{K}_{a_{n-1}} t_{n-1}}$ originating from $\hat{V}(t_{n-1})$ such that, in the end, the t_{n-1} integration takes exactly the same form as the above t_n integration; however, with n replaced by $n-1$ everywhere. Iterating the process from t_n to t_1, we get

$$\begin{aligned}
&\int_{-\infty}^{t} dt_1 \cdots \int_{-\infty}^{t_{n-1}} dt_n e^{-i\mathcal{K}_b(t-t_1)} e^{-i\mathcal{K}_{a_1}(t_1-t_2)} \ldots e^{-i\mathcal{K}_{a_{n-1}}(t_{n-1}-t_n)} e^{-i\mathcal{K}_a t_n} e^{0^+ t_n} \\
&= i^n \frac{e^{-i(\mathcal{K}_a + i0^+)t}}{\mathcal{K}_a - \mathcal{K}_b + i0^+} \frac{1}{\mathcal{K}_a - \mathcal{K}_{a_1} + i0^+} \cdots \frac{1}{\mathcal{K}_a - \mathcal{K}_{a_{n-1}} + i0^+} \\
&= (i\hbar)^n \frac{e^{-i(\mathcal{K}_a + i0^+)t}}{K_a - K_b + i0^+} \frac{1}{K_a - K_{a_1} + i0^+} \cdots \frac{1}{K_a - K_{a_{n-1}} + i0^+}.
\end{aligned}$$

Noting that $(-i)^n i^n = 1$, the matrix element becomes

$$\begin{aligned}
\langle b|\Psi(t)\rangle &= \frac{e^{-i(\mathcal{K}_a + i0^+)t}}{K_a - K_b + i0^+} \langle b| V \sum_{n=1}^{\infty} \underbrace{\sum_{a_1} \frac{|a_1\rangle\langle a_1|}{K_a - K_{a_1} + i0^+}}_{(K_a - K + i0^+)^{-1}} V \cdots \\
&\quad \times \underbrace{\sum_{a_{n-1}} \frac{|a_{n-1}\rangle\langle a_{n-1}|}{K_a - K_{a_{n-1}} + i0^+}}_{(K_a - K + i0^+)^{-1}} V|a\rangle \\
&= \frac{e^{-i(K_a + i0^+)t/\hbar}}{K_a - K_b + i0^+} \langle b| V \sum_{n=1}^{\infty} [(K_a - K + i0^+)^{-1} V]^{n-1} |a\rangle.
\end{aligned}$$

This is just equation (2.25) with the t-matrix defined as

$$\begin{aligned}
T(\varepsilon) &= V \sum_{n=1}^{\infty} [(\varepsilon + i0^+ - K)^{-1} V]^{n-1} \\
&= V + V(\varepsilon + i0^+ - K)^{-1} V \\
&\quad + V(\varepsilon + i0^+ - K)^{-1} V(\varepsilon + i0^+ - K)^{-1} V + \cdots \\
&= V + V(\varepsilon + i0^+ - K)^{-1} T(\varepsilon),
\end{aligned}$$

consistently with equation (2.26).

In order to evaluate the transition rate, we need the time derivative of $|\langle b|\Psi(t)\rangle|^2$. We temporarily replace 0^+ by ϵ and write

$$\frac{d}{dt} |\langle b|\Psi(t)\rangle|^2 = \frac{|\langle b|T(K_a)|a\rangle|^2}{(K_a - K_b)^2 + \epsilon^2} \frac{d}{dt} e^{2\epsilon t/\hbar} = \frac{2}{\hbar} |\langle b|T(K_a)|a\rangle|^2 \frac{e^{2\epsilon t/\hbar} \epsilon}{(K_a - K_b)^2 + \epsilon^2}.$$

We thus arrive at equation (2.27) by noting that the Lorentzian of width ϵ, $L_\epsilon(x) = (\epsilon/\pi)/(x^2 + \epsilon^2)$, becomes a delta function as $\epsilon \to 0^+$:

$$\Gamma_{a\to b} = \frac{2\pi}{\hbar}|\langle b|T(K_a)|a\rangle|^2 \underbrace{\lim_{\epsilon\to 0^+} e^{2\epsilon t/\hbar}\frac{\epsilon/\pi}{(K_a - K_b)^2 + \epsilon^2}}_{\delta(K_a - K_b)}.$$

Doc A2.5 The plane-wave basis set

The orthonormality of the plane waves in equation (2.29) is verified as follows:

$$\begin{aligned}\langle \boldsymbol{k}\sigma|\boldsymbol{k}'\sigma'\rangle &= \int d1\, \varphi^*_{\boldsymbol{k}\sigma}(1)\varphi_{\boldsymbol{k}'\sigma'}(1) = \sum_{\sigma_1}\int d\boldsymbol{r}_1 \frac{\delta_{\sigma\sigma_1}}{\sqrt{\mathcal{V}}}e^{-i\boldsymbol{k}\cdot\boldsymbol{r}_1}\frac{\delta_{\sigma'\sigma_1}}{\sqrt{\mathcal{V}}}e^{i\boldsymbol{k}'\cdot\boldsymbol{r}_1}\\ &= \delta_{\sigma\sigma'}\frac{1}{\mathcal{V}}\int d\boldsymbol{r}_1 e^{-i(\boldsymbol{k}-\boldsymbol{k}')\cdot\boldsymbol{r}_1} = \delta_{\sigma\sigma'}\delta_{\boldsymbol{k}\boldsymbol{k}'}.\end{aligned}$$

The completeness requires

$$\begin{aligned}\sum_{\boldsymbol{k}\sigma}|\boldsymbol{k}\sigma\rangle\langle\boldsymbol{k}\sigma| = \mathbb{1} &\Leftrightarrow \sum_{\boldsymbol{k}\sigma}\langle 1|\boldsymbol{k}\sigma\rangle\langle\boldsymbol{k}\sigma|2\rangle = \sum_{\boldsymbol{k}\sigma}\varphi_{\boldsymbol{k}\sigma}(1)\varphi^*_{\boldsymbol{k}\sigma}(2) = \delta(1-2)\\ &\Leftrightarrow \sum_{\boldsymbol{k}\sigma}\frac{\delta_{\sigma\sigma_1}}{\sqrt{\mathcal{V}}}e^{i\boldsymbol{k}\cdot\boldsymbol{r}_1}\frac{\delta_{\sigma\sigma_2}}{\sqrt{\mathcal{V}}}e^{-i\boldsymbol{k}\cdot\boldsymbol{r}_2} = \delta_{\sigma_1\sigma_2}\delta(\boldsymbol{r}_1 - \boldsymbol{r}_2)\\ &\Leftrightarrow \frac{1}{\mathcal{V}}\sum_{\boldsymbol{k}} e^{i\boldsymbol{k}\cdot(\boldsymbol{r}_1-\boldsymbol{r}_2)} = \delta(\boldsymbol{r}_1 - \boldsymbol{r}_2).\end{aligned}$$

This last condition is part of our conventions for the Fourier transforms (see equation (3*a*)).

Doc A2.6 Commutation rules of creation and annihilation operators

Using the definitions in equation (2.42), we calculate the action of $a_\alpha a^\dagger_\beta$ on an arbitrary state $|n_1, n_2, \ldots\rangle$. We must distinguish the three cases $\alpha < \beta$, $\alpha = \beta$, and $\alpha > \beta$:

$$\begin{aligned}\alpha < \beta: \quad & a_\alpha a^\dagger_\beta\, |\ldots, n_\alpha, \ldots, n_\beta, \ldots\rangle\\ &= \eta^{\left(\sum_{i=\alpha}^{\beta-1} n_i\right)}\sqrt{n_\alpha}\sqrt{1+\eta n_\beta}\,|\ldots, n_\alpha - 1, \ldots, n_\beta + 1, \ldots\rangle\\ \alpha = \beta: \quad & a_\alpha a^\dagger_\alpha\, |\ldots, n_\alpha, \ldots\rangle = \underbrace{\sqrt{n_\alpha + 1}\sqrt{1+\eta n_\alpha}}_{1+\eta n_\alpha}\,|\ldots, n_\alpha, \ldots\rangle\\ \alpha > \beta: \quad & a_\alpha a^\dagger_\beta\, |\ldots, n_\beta, \ldots, n_\alpha, \ldots\rangle\\ &= \eta^{\left(\sum_{i=\beta}^{\alpha-1} n_i\right)+1}\sqrt{n_\alpha}\sqrt{1+\eta n_\beta}\,|\ldots, n_\beta + 1, \ldots, n_\alpha - 1, \ldots\rangle.\end{aligned}$$

At the third line, we have replaced $\sqrt{n_\alpha + 1}\sqrt{1 + \eta n_\alpha}$ by $1 + \eta n_\alpha$: this is obvious for $\eta = +1$; for $\eta = -1$ it can also be verified case by case since $n_\alpha = 0$ or 1. At the fifth line, the additional factor η stems from the fact that a_α acts on a state in which the occupation number of state β is increased by one. We now calculate the action of $a_\beta^\dagger a_\alpha$:

$$\alpha < \beta: \quad a_\beta^\dagger a_\alpha \,|\ldots, n_\alpha, \ldots, n_\beta, \ldots\rangle = \eta^{\left(\sum_{i=\alpha}^{\beta-1} n_i\right)-1} \sqrt{1+\eta n_\beta}\sqrt{n_\alpha}\,|\ldots, n_\alpha - 1, \ldots, n_\beta + 1, \ldots\rangle$$

$$\alpha = \beta: \quad a_\alpha^\dagger a_\alpha \,|\ldots, n_\alpha, \ldots\rangle = \underbrace{\sqrt{1+\eta(n_\alpha - 1)}\sqrt{n_\alpha}}_{n_\alpha}\,|\ldots, n_\alpha, \ldots\rangle$$

$$\alpha > \beta: \quad a_\beta^\dagger a_\alpha \,|\ldots, n_\beta, \ldots, n_\alpha, \ldots\rangle = \eta^{\left(\sum_{i=\beta}^{\alpha-1} n_i\right)} \sqrt{1+\eta n_\beta}\sqrt{n_\alpha}\,|\ldots, n_\beta + 1, \ldots, n_\alpha - 1, \ldots\rangle.$$

Subtracting these two sets of equations, we obtain the action of $a_\alpha a_\beta^\dagger - \eta a_\beta^\dagger a_\alpha$:

$$a_\alpha a_\beta^\dagger - \eta a_\beta^\dagger a_\alpha = [a_\alpha, a_\beta^\dagger]_{-\eta} = \begin{cases} 0 & \alpha < \beta \\ \mathbb{1} & \alpha = \beta \\ 0 & \alpha > \beta. \end{cases}$$

This proves the first expression in equation (2.41). The other relations are easily established using the same method.

From the definitions, we readily check that $a_\alpha|\ldots, n_\alpha = 0, \ldots\rangle = 0$. Furthermore, for fermions, $c_\alpha^\dagger|\ldots, n_\alpha = 1, \ldots\rangle = 0$. Finally, we see above that the states in the occupation number representation are eigenstates of $a_\alpha^\dagger a_\alpha$ with eigenvalue n_α. Therefore we can define the *operator* $n_\alpha = a_\alpha^\dagger a_\alpha$ as the operator measuring the number of particles in state α.

Doc A2.7 Change of basis

We are looking for a relation between $a_{\boldsymbol{k}}^\dagger$ and $a_{\boldsymbol{r}}^\dagger$. We omit writing quantum numbers other than $\boldsymbol{k}$ and $\boldsymbol{r}$ for briefness. The desired relation takes the generic form

$$a_{\boldsymbol{k}}^\dagger = \int d\boldsymbol{r}\; U_{\boldsymbol{kr}} a_{\boldsymbol{r}}^\dagger.$$

The commutation rules require that the matrix U be unitary. Acting on the vacuum with both sides of the equation and multiplying on the left by $\langle \boldsymbol{r}_1|$, we find

$$\langle \boldsymbol{r}_1|a_{\boldsymbol{k}}^\dagger|\varnothing\rangle = \varphi_{\boldsymbol{k}}(\boldsymbol{r}_1) = \int d\boldsymbol{r}\; U_{\boldsymbol{kr}} \underbrace{\langle \boldsymbol{r}_1|a_{\boldsymbol{r}}^\dagger|\varnothing\rangle}_{\delta(\boldsymbol{r} - \boldsymbol{r}_1)} = U_{\boldsymbol{kr}_1},$$

which gives the expression of the matrix U and the relation between $a_{\boldsymbol{k}}^\dagger$ and $a_{\boldsymbol{r}}^\dagger$:

$$a_{\boldsymbol{k}}^\dagger = \frac{1}{\sqrt{\mathcal{V}}} \int d\boldsymbol{r} e^{i\boldsymbol{k}\cdot\boldsymbol{r}} a_{\boldsymbol{r}}^\dagger. \tag{2.88}$$

Note that this is different from our convention for Fourier transforms, equation (2*b*), both for the sign of the phase and for the normalization coefficient. Conjugating, we obtain

$$a_k = \frac{1}{\sqrt{\mathcal{V}}} \int d\boldsymbol{r} e^{-i\boldsymbol{k}\cdot\boldsymbol{r}} a_{\boldsymbol{r}}, \tag{2.89}$$

which shows that the creation and annihilation operators do not transform like ordinary scalar functions. The inverse relations are

$$a_{\boldsymbol{r}}^{\dagger} = \frac{1}{\sqrt{\mathcal{V}}} \sum_{\boldsymbol{k}} e^{-i\boldsymbol{k}\cdot\boldsymbol{r}} a_{\boldsymbol{k}}^{\dagger} \quad \text{and} \quad a_{\boldsymbol{r}} = \frac{1}{\sqrt{\mathcal{V}}} \sum_{\boldsymbol{k}} e^{i\boldsymbol{k}\cdot\boldsymbol{r}} a_{\boldsymbol{k}}. \tag{2.90}$$

The argument is readily extended to arbitrary one-particle bases and leads to:

$$a_{\boldsymbol{r}}^{\dagger} = \sum_{\alpha} \varphi_{\alpha}^{*}(\boldsymbol{r}) a_{\alpha}^{\dagger} \quad \text{and} \quad a_{\alpha}^{\dagger} = \int d\boldsymbol{r}\ \varphi_{\alpha}(\boldsymbol{r}) a_{\boldsymbol{r}}^{\dagger}. \tag{2.91}$$

With these transformation rules, we can evaluate the Fourier transform of the real-space density operator:

$$\begin{aligned} n(\boldsymbol{q}) &= \int d\boldsymbol{r}\ n(\boldsymbol{r}) e^{-i\boldsymbol{q}\cdot\boldsymbol{r}} = \int d\boldsymbol{r} \underbrace{a_{\boldsymbol{r}}^{\dagger} a_{\boldsymbol{r}}}_{\frac{1}{\mathcal{V}} \sum_{\boldsymbol{k}\boldsymbol{k}'} a_{\boldsymbol{k}}^{\dagger} a_{\boldsymbol{k}'} e^{-i(\boldsymbol{k}-\boldsymbol{k}')\cdot\boldsymbol{r}}} e^{-i\boldsymbol{q}\cdot\boldsymbol{r}} \\ &= \frac{1}{\mathcal{V}} \sum_{\boldsymbol{k}\boldsymbol{k}'} a_{\boldsymbol{k}}^{\dagger} a_{\boldsymbol{k}'} \underbrace{\int d\boldsymbol{r} e^{i(\boldsymbol{k}'-\boldsymbol{k}-\boldsymbol{q})\cdot\boldsymbol{r}}}_{\mathcal{V}\delta_{\boldsymbol{k}',\,\boldsymbol{k}+\boldsymbol{q}}} = \sum_{\boldsymbol{k}} a_{\boldsymbol{k}}^{\dagger} a_{\boldsymbol{k}+\boldsymbol{q}}. \end{aligned}$$

Doc A2.8 Current operator and continuity equation

As we know from analytical mechanics, the effect of an electromagnetic field on particles is to shift their momenta by the quantity $-e\boldsymbol{A}(\boldsymbol{r})$, where e is the charge of the particle and $\boldsymbol{A}(\boldsymbol{r})$ is the vector potential (assume the gauge is such that the scalar potential is zero). The resulting change in the particle energy is the product of its velocity (i.e. the current) by the change in momentum. For a collection of particles, the change in energy is therefore $\delta H' = \int d\boldsymbol{r}\ \boldsymbol{j}(\boldsymbol{r}) \cdot (-e)\delta\boldsymbol{A}(\boldsymbol{r})$ with H' the part of the Hamiltonian that does depend on $\boldsymbol{A}(\boldsymbol{r})$. In differential form, this becomes

$$\boldsymbol{j}(\boldsymbol{r}) = -\frac{1}{e}\frac{\delta H'}{\delta \boldsymbol{A}(\boldsymbol{r})}.$$

The current is the functional derivative of the Hamiltonian with respect to $-e\boldsymbol{A}(\boldsymbol{r})$. If we can express the Hamiltonian in terms of the creation and annihilation operators, the above relation can serve as a definition of the current operator. The effect of the electromagnetic field can be accounted for by replacing in the unperturbed Hamiltonian the momentum operator $\boldsymbol{p} = -i\hbar\nabla$ by $\boldsymbol{p} - e\boldsymbol{A}(\boldsymbol{r})$. Since, in a solid, the operator $\boldsymbol{p}$ appears only in the kinetic energy, the energy of interaction with the electromagnetic field is, in second-quantization notation,

$$H' = \int d\boldsymbol{r} \sum_{\sigma} a^{\dagger}_{\boldsymbol{r}\sigma} \frac{[\boldsymbol{p} - e\boldsymbol{A}(\boldsymbol{r})]^2}{2m} a_{\boldsymbol{r}\sigma} - \int d\boldsymbol{r} \sum_{\sigma} a^{\dagger}_{\boldsymbol{r}\sigma} \frac{\boldsymbol{p}^2}{2m} a_{\boldsymbol{r}\sigma}$$
$$= \frac{1}{2m} \int d\boldsymbol{r} \sum_{\sigma} a^{\dagger}_{\boldsymbol{r}\sigma} [ie\hbar \nabla_{\boldsymbol{r}} \cdot \boldsymbol{A}(\boldsymbol{r}) + ie\hbar \boldsymbol{A}(\boldsymbol{r}) \cdot \nabla_{\boldsymbol{r}} + e^2 \boldsymbol{A}^2(\boldsymbol{r})] a_{\boldsymbol{r}\sigma}.$$

We can rearrange this into a more convenient form with the help of the relation

$$\int d\boldsymbol{r}\; a^{\dagger}_{\boldsymbol{r}\sigma} \nabla_{\boldsymbol{r}} \cdot \boldsymbol{A}(\boldsymbol{r}) a_{\boldsymbol{r}\sigma} = -\int d\boldsymbol{r}\; \boldsymbol{A}(\boldsymbol{r}) \cdot \left(\nabla_{\boldsymbol{r}} a^{\dagger}_{\boldsymbol{r}\sigma}\right) a_{\boldsymbol{r}\sigma},$$

which is readily verified through integration by parts and neglect of the surface term. We then have

$$H' = -e \int d\boldsymbol{r} \left\{ \frac{i\hbar}{2m} \sum_{\sigma} \boldsymbol{A}(\boldsymbol{r}) \cdot \left[\left(\nabla_{\boldsymbol{r}} a^{\dagger}_{\boldsymbol{r}\sigma}\right) a_{\boldsymbol{r}\sigma} - a^{\dagger}_{\boldsymbol{r}\sigma} (\nabla_{\boldsymbol{r}} a_{\boldsymbol{r}\sigma}) \right] - \frac{e}{2m} \sum_{\sigma} \boldsymbol{A}^2(\boldsymbol{r}) a^{\dagger}_{\boldsymbol{r}\sigma} a_{\boldsymbol{r}\sigma} \right\}.$$

Evaluating the functional derivative, we deduce the formulas given in equation (2.49) for the paramagnetic and diamagnetic current densities. It will be proven below that this definition of the current satisfies the continuity equation.

The Fourier transform of $\boldsymbol{j}(\boldsymbol{r})$ is easily evaluated by using the expressions of the field operators in the plane-wave basis (see Doc A2.7):

$$\boldsymbol{j}^p(\boldsymbol{q}) = \int d\boldsymbol{r} e^{-i\boldsymbol{q}\cdot\boldsymbol{r}} \frac{i\hbar}{2m} \sum_{\sigma} \frac{1}{\mathscr{V}} \sum_{\boldsymbol{k}\boldsymbol{k}'} a^{\dagger}_{\boldsymbol{k}\sigma} a_{\boldsymbol{k}'\sigma} \left[\underbrace{\left(\nabla_{\boldsymbol{r}} e^{-i\boldsymbol{k}\cdot\boldsymbol{r}}\right) e^{i\boldsymbol{k}'\cdot\boldsymbol{r}}}_{-i\boldsymbol{k} e^{-i(\boldsymbol{k}-\boldsymbol{k}')\cdot\boldsymbol{r}}} - \underbrace{e^{-i\boldsymbol{k}\cdot\boldsymbol{r}} \left(\nabla_{\boldsymbol{r}} e^{i\boldsymbol{k}'\cdot\boldsymbol{r}}\right)}_{i\boldsymbol{k}' e^{-i(\boldsymbol{k}-\boldsymbol{k}')\cdot\boldsymbol{r}}} \right]$$
$$= \frac{\hbar}{2m} \sum_{\sigma} \frac{1}{\mathscr{V}} \sum_{\boldsymbol{k}\boldsymbol{k}'} (\boldsymbol{k} + \boldsymbol{k}') a^{\dagger}_{\boldsymbol{k}\sigma} a_{\boldsymbol{k}'\sigma} \underbrace{\int d\boldsymbol{r} e^{-i\boldsymbol{q}\cdot\boldsymbol{r}} e^{-i(\boldsymbol{k}-\boldsymbol{k}')\cdot\boldsymbol{r}}}_{\mathscr{V} \delta_{\boldsymbol{k}',\, \boldsymbol{k}+\boldsymbol{q}}}$$
$$= \frac{\hbar}{2m} \sum_{\boldsymbol{k}\sigma} (2\boldsymbol{k} + \boldsymbol{q}) a^{\dagger}_{\boldsymbol{k}\sigma} a_{\boldsymbol{k}+\boldsymbol{q}\sigma},$$

consistently with equation (2.50), and

$$\boldsymbol{j}^d(\boldsymbol{q}) = \int d\boldsymbol{r} e^{-i\boldsymbol{q}\cdot\boldsymbol{r}} \left(-\frac{e}{m}\right) \sum_{\sigma} \boldsymbol{A}(\boldsymbol{r}) \frac{1}{\mathscr{V}} \sum_{\boldsymbol{k}\boldsymbol{k}'} a^{\dagger}_{\boldsymbol{k}\sigma} a_{\boldsymbol{k}'\sigma} e^{-i\boldsymbol{k}\cdot\boldsymbol{r}} e^{i\boldsymbol{k}'\cdot\boldsymbol{r}}$$
$$= -\frac{e}{m} \frac{1}{\mathscr{V}} \sum_{\boldsymbol{k}\boldsymbol{k}'\sigma} a^{\dagger}_{\boldsymbol{k}\sigma} a_{\boldsymbol{k}'\sigma} \underbrace{\int d\boldsymbol{r}\; \boldsymbol{A}(\boldsymbol{r}) e^{-i(\boldsymbol{q}+\boldsymbol{k}-\boldsymbol{k}')\cdot\boldsymbol{r}}}_{\boldsymbol{A}(\boldsymbol{q}+\boldsymbol{k}-\boldsymbol{k}')}$$
$$= -\frac{e}{m} \frac{1}{\mathscr{V}} \sum_{\boldsymbol{k}\boldsymbol{k}'\sigma} \boldsymbol{A}(\boldsymbol{k} - \boldsymbol{k}') a^{\dagger}_{\boldsymbol{k}\sigma} a_{\boldsymbol{k}'+\boldsymbol{q}\sigma},$$

where we have introduced the Fourier transform of the vector potential.

We check now that the current operator, equation (2.49), satisfies the continuity equation $\nabla \cdot \boldsymbol{j}(\boldsymbol{r}) + \partial_t n(\boldsymbol{r}) = 0$ with $n(\boldsymbol{r}) = \sum_\sigma a^\dagger_{\boldsymbol{r}\sigma} a_{\boldsymbol{r}\sigma}$ the particle density. The time evolution of the density is given by equation (2.15), $n(t) = e^{iKt/\hbar} n e^{-iKt/\hbar}$, which yields $\partial_t n = \frac{i}{\hbar}[K, n]$. Furthermore, the only term in the Hamiltonian which does not commute with the density is the kinetic energy. Hence we must calculate $[H, n]$, where H is

$$\begin{aligned} H &= \frac{1}{2m} \int d\boldsymbol{r} \sum_\sigma a^\dagger_{\boldsymbol{r}\sigma} [\boldsymbol{p} - e\boldsymbol{A}(\boldsymbol{r})]^2 a_{\boldsymbol{r}\sigma} \\ &= \frac{1}{\mathcal{V}} \sum_{\boldsymbol{k}\boldsymbol{k}'\sigma} \frac{1}{2m} [\hbar^2 k^2 \mathcal{V} \delta_{\boldsymbol{k}\boldsymbol{k}'} - e\hbar(\boldsymbol{k} + \boldsymbol{k}') \cdot \boldsymbol{A}(\boldsymbol{k} - \boldsymbol{k}') + e^2 A^2(\boldsymbol{k} - \boldsymbol{k}')] a^\dagger_{\boldsymbol{k}\sigma} a_{\boldsymbol{k}'\sigma}. \end{aligned}$$

We have performed the change of basis $\boldsymbol{r} \to \boldsymbol{k}$. In the last expression, $A^2(\boldsymbol{q})$ should be understood as the Fourier transform of $A^2(\boldsymbol{r})$, not as the Fourier transform of $\boldsymbol{A}(\boldsymbol{r})$ squared. The calculations are more easily done in momentum space, where the continuity equation reads

$$i\boldsymbol{q} \cdot \boldsymbol{j}(\boldsymbol{q}) + \frac{i}{\hbar}[H, n(\boldsymbol{q})] = 0.$$

We now evaluate the commutator using equations (2.46) and (2.41):

$$\begin{aligned} [H, n(\boldsymbol{q})] &= \frac{1}{2m} \frac{1}{\mathcal{V}} \sum_{\boldsymbol{k}\boldsymbol{k}'\sigma} \sum_{\boldsymbol{k}''\sigma''} [\ldots] \underbrace{\left[a^\dagger_{\boldsymbol{k}\sigma} a_{\boldsymbol{k}'\sigma}, a^\dagger_{\boldsymbol{k}''\sigma''} a_{\boldsymbol{k}''+\boldsymbol{q}\sigma''}\right]}_{\delta_{\sigma\sigma''}\delta_{\boldsymbol{k}'\boldsymbol{k}''} a^\dagger_{\boldsymbol{k}\sigma} a_{\boldsymbol{k}'+\boldsymbol{q}\sigma} - \delta_{\sigma\sigma''}\delta_{\boldsymbol{k},\boldsymbol{k}''+\boldsymbol{q}} a^\dagger_{\boldsymbol{k}-\boldsymbol{q}\sigma} a_{\boldsymbol{k}'\sigma}} \\ &= \frac{1}{2m} \frac{1}{\mathcal{V}} \sum_{\boldsymbol{k}\boldsymbol{k}'\sigma} [\hbar^2 k^2 \mathcal{V} \delta_{\boldsymbol{k}\boldsymbol{k}'} - e\hbar(\boldsymbol{k} + \boldsymbol{k}') \cdot \boldsymbol{A}(\boldsymbol{k} - \boldsymbol{k}') \\ &\quad - \hbar^2 (\boldsymbol{k} + \boldsymbol{q})^2 \mathcal{V} \delta_{\boldsymbol{k}\boldsymbol{k}'} + e\hbar(\boldsymbol{k} + \boldsymbol{k}' + 2\boldsymbol{q}) \cdot \boldsymbol{A}(\boldsymbol{k} - \boldsymbol{k}')] a^\dagger_{\boldsymbol{k}\sigma} a_{\boldsymbol{k}'+\boldsymbol{q}\sigma}. \end{aligned}$$

In the second series of terms coming from the commutator, we have shifted both momenta $\boldsymbol{k}$ and $\boldsymbol{k}'$ by $\boldsymbol{q}$. Collecting the terms, we obtain

$$\begin{aligned} [H, n(\boldsymbol{q})] &= \frac{-\hbar^2}{2m} \sum_{\boldsymbol{k}\sigma} (2\boldsymbol{q} \cdot \boldsymbol{k} + q^2) a^\dagger_{\boldsymbol{k}\sigma} a_{\boldsymbol{k}+\boldsymbol{q}\sigma} + \frac{e\hbar}{m} \boldsymbol{q} \cdot \frac{1}{\mathcal{V}} \sum_{\boldsymbol{k}\boldsymbol{k}'\sigma} \boldsymbol{A}(\boldsymbol{k} - \boldsymbol{k}') a^\dagger_{\boldsymbol{k}\sigma} a_{\boldsymbol{k}'+\boldsymbol{q}\sigma} \\ &= -\hbar \boldsymbol{q} \cdot [\boldsymbol{j}^p(\boldsymbol{q}) + \boldsymbol{j}^d(\boldsymbol{q})], \end{aligned}$$

which proves that the current (2.49) satisfies the continuity equation.

Doc A2.9 Hamiltonian of Bloch electrons

The Hamiltonian describing independent electrons subject to the periodic potential (2.51) is a one-body Hamiltonian of the generic form (2.43). We work in the plane-wave basis and thus replace α and β by $(\boldsymbol{k}\sigma)$ and $(\boldsymbol{k}'\sigma')$. We furthermore omit the spin indices—the whole problem is diagonal in the spin variables—and we use the decomposition $\boldsymbol{k} \to \boldsymbol{k} + \boldsymbol{G}$ with $\boldsymbol{k}$ in the first Brillouin zone. The matrix element becomes

$$T_{\alpha\beta} \longrightarrow T_{\boldsymbol{k}+\boldsymbol{G},\, \boldsymbol{k}'+\boldsymbol{G}'} = \int d\boldsymbol{r}_1 \frac{e^{-i(\boldsymbol{k}+\boldsymbol{G})\cdot\boldsymbol{r}_1}}{\sqrt{\mathcal{V}}}\left[\frac{\boldsymbol{p}_1^2}{2m} + V(\boldsymbol{r}_1)\right]\frac{e^{i(\boldsymbol{k}'+\boldsymbol{G}')\cdot\boldsymbol{r}_1}}{\sqrt{\mathcal{V}}}$$

with $\boldsymbol{p}_1 = -i\hbar\nabla_{\boldsymbol{r}_1}$. Performing the integration and making use of the expansion (2.51), we get

$$T_{\boldsymbol{k}+\boldsymbol{G},\, \boldsymbol{k}'+\boldsymbol{G}'} = \frac{\hbar^2}{2m}(\boldsymbol{k}' + \boldsymbol{G}')^2 \delta_{\boldsymbol{k}+\boldsymbol{G},\, \boldsymbol{k}'+\boldsymbol{G}'} + \sum_{\boldsymbol{G}''} V(\boldsymbol{G}'')\delta_{\boldsymbol{k}+\boldsymbol{G},\, \boldsymbol{k}'+\boldsymbol{G}'+\boldsymbol{G}''}.$$

Due to the uniqueness of the decomposition $\boldsymbol{k} \to \boldsymbol{k} + \boldsymbol{G}$, we have $\delta_{\boldsymbol{k}+\boldsymbol{G},\, \boldsymbol{k}'+\boldsymbol{G}'} = \delta_{\boldsymbol{k}\boldsymbol{k}'}\delta_{\boldsymbol{G}\boldsymbol{G}'}$ and $\delta_{\boldsymbol{k}+\boldsymbol{G},\, \boldsymbol{k}'+\boldsymbol{G}'+\boldsymbol{G}''} = \delta_{\boldsymbol{k}\boldsymbol{k}'}\delta_{\boldsymbol{G},\, \boldsymbol{G}'+\boldsymbol{G}''}$. It follows that

$$T_{\boldsymbol{k}+\boldsymbol{G},\, \boldsymbol{k}'+\boldsymbol{G}'} = \delta_{\boldsymbol{k}\boldsymbol{k}'}\left[\frac{\hbar^2}{2m}(\boldsymbol{k} + \boldsymbol{G})^2\delta_{\boldsymbol{G}\boldsymbol{G}'} + V(\boldsymbol{G} - \boldsymbol{G}')\right],$$

which proves equation (2.52).

Doc A2.10 Normal coordinates

Using the definition (2.58), the second-order term in equation (2.57) becomes

$$U_2 = \frac{1}{2}\sum_{nm}\sum_{\nu i\mu j} u_{\nu i}(\boldsymbol{R}_n)\sqrt{M_\nu M_\mu}\; W_{\nu i\mu j}(\boldsymbol{R}_n - \boldsymbol{R}_m)u_{\mu j}(\boldsymbol{R}_m).$$

We inject the Fourier transform $W_{\nu i\mu j}(\boldsymbol{R}_n - \boldsymbol{R}_m) = \frac{1}{\mathcal{N}}\sum_{\boldsymbol{k}} D_{\nu i\mu j}(\boldsymbol{k})e^{i\boldsymbol{k}\cdot(\boldsymbol{R}_n-\boldsymbol{R}_m)}$ as well as the displacements, equation (2.61), and find

$$\begin{aligned}
U_2 &= \frac{1}{2}\frac{1}{\mathcal{N}^2}\sum_{\boldsymbol{k}\,\boldsymbol{k}_1\boldsymbol{k}_2}\sum_{\lambda_1\lambda_2}\sum_{\nu i\mu j} q_{\boldsymbol{k}_1\lambda_1}[\boldsymbol{\epsilon}_{\boldsymbol{k}_1\lambda_1}]_{\nu i} q_{\boldsymbol{k}_2\lambda_2}[\boldsymbol{\epsilon}_{\boldsymbol{k}_2\lambda_2}]_{\mu j} D_{\nu i\mu j}(\boldsymbol{k}) \\
&\quad \times \underbrace{\sum_{nm} e^{i\boldsymbol{k}_1\cdot\boldsymbol{R}_n}e^{i\boldsymbol{k}\cdot(\boldsymbol{R}_n-\boldsymbol{R}_m)}e^{i\boldsymbol{k}_2\cdot\boldsymbol{R}_m}}_{\mathcal{N}^2\delta_{\boldsymbol{k}_1,\,-\boldsymbol{k}}\delta_{\boldsymbol{k}_2\boldsymbol{k}}} \\
&= \frac{1}{2}\sum_{\boldsymbol{k}}\sum_{\lambda_1\lambda_2}\sum_{\nu i\mu j} q_{-\boldsymbol{k}\lambda_1}[\boldsymbol{\epsilon}_{-\boldsymbol{k}\lambda_1}]_{\nu i} q_{\boldsymbol{k}\lambda_2}[\boldsymbol{\epsilon}_{\boldsymbol{k}\lambda_2}]_{\mu j} D_{\nu i\mu j}(\boldsymbol{k}).
\end{aligned}$$

The matrix of force constants W being real, we have the property $D(-\boldsymbol{k}) = D^*(\boldsymbol{k})$ which also implies $\boldsymbol{\epsilon}_{-\boldsymbol{k}\lambda} = \boldsymbol{\epsilon}^*_{\boldsymbol{k}\lambda}$. Furthermore, the reality of the displacements $u_{\nu i}(\boldsymbol{R}_n)$ implies $q_{-\boldsymbol{k}\lambda} = q^*_{\boldsymbol{k}\lambda}$. Using these properties and the fact that the dynamical matrix is diagonal in the representation of its eigenstates,

$$D_{\nu i\mu j}(\boldsymbol{k}) = \sum_{\lambda}[\boldsymbol{\epsilon}_{\boldsymbol{k}\lambda}]_{\nu i}\omega^2_{\boldsymbol{k}\lambda}[\boldsymbol{\epsilon}^*_{\boldsymbol{k}\lambda}]_{\mu j},$$

we obtain

$$U_2 = \frac{1}{2}\sum_{k\lambda}\sum_{\lambda_1\lambda_2} q^*_{k\lambda_1} q_{k\lambda_2}\omega^2_{k\lambda} \underbrace{\sum_{\nu i}[\epsilon^*_{k\lambda_1}]_{\nu i}[\epsilon_{k\lambda}]_{\nu i}}_{\delta_{\lambda\lambda_1}} \underbrace{\sum_{\mu j}[\epsilon^*_{k\lambda}]_{\mu j}[\epsilon_{k\lambda_2}]_{\mu j}}_{\delta_{\lambda\lambda_2}}$$

$$= \frac{1}{2}\sum_{k\lambda}|q_{k\lambda}|^2\omega^2_{k\lambda}.$$

The kinetic energy of the lattice expressed in terms of the normal coordinates reads

$$\frac{1}{2}\sum_{n\nu i} M_\nu \dot{u}^2_{\nu i}(\boldsymbol{R}_n) = \frac{1}{2}\sum_{n\nu i}\frac{1}{\mathcal{N}}\sum_{k_1\lambda_1}\sum_{k_2\lambda_2}\dot{q}_{k_1\lambda_1}[\epsilon_{k_1\lambda_1}]_{\nu i} e^{ik_1\cdot\boldsymbol{R}_n}\dot{q}_{k_2\lambda_2}[\epsilon_{k_2\lambda_2}]_{\nu i} e^{ik_2\cdot\boldsymbol{R}_n}.$$

The $\boldsymbol{R}_n$ sum yields $\mathcal{N}\delta_{k_1,-k_2}$. Equation (2.63) follows:

$$\frac{1}{2}\sum_{n\nu i} M_\nu \dot{u}^2_{\nu i}(\boldsymbol{R}_n) = \frac{1}{2}\sum_{k}\sum_{\lambda_1\lambda_2}\dot{q}_{k\lambda_1}\dot{q}^*_{k\lambda_2}\underbrace{\sum_{\nu i}[\epsilon_{k\lambda_1}]_{\nu i}[\epsilon^*_{k\lambda_2}]_{\nu i}}_{\delta_{\lambda_1\lambda_2}} = \frac{1}{2}\sum_{k\lambda}|\dot{q}_{k\lambda}|^2.$$

Doc A2.11 Phonon density of states in the Debye model

In the model of Debye, the phonon spectrum is composed of only acoustic branches that remain linear up to the Brillouin-zone boundary, where they reach the maximum frequency ω_D, i.e. $\omega_{k\lambda} = ck$. The phonon density of states in this model is, in three dimensions,

$$N_D^{ph}(\varepsilon) = \sum_{k\lambda}\delta(\varepsilon - \hbar\omega_{k\lambda}) = \theta(\hbar\omega_D - \varepsilon)\frac{3N_{at}\mathcal{V}}{(2\pi)^3}\int d^3k\ \delta(\varepsilon - \hbar ck)$$

$$= \theta(\hbar\omega_D - \varepsilon)\frac{3N_{at}\mathcal{V}}{2\pi^2\hbar c}\int_0^\infty dk\ k^2\delta\left(\frac{\varepsilon}{\hbar c} - k\right) = \theta(\hbar\omega_D - \varepsilon)\frac{3N_{at}\mathcal{V}}{2\pi^2\hbar c}\left(\frac{\varepsilon}{\hbar c}\right)^2.$$

The various factors can be expressed in terms of ω_D thanks to the normalization:

$$N_{modes} = \int_0^\infty d\varepsilon\ N_D^{ph}(\varepsilon) = \frac{3N_{at}\mathcal{V}}{2\pi^2(\hbar c)^3}\frac{(\hbar\omega_D)^3}{3}.$$

In this way we get equation (2.68). The lowest-frequency phonons are always acoustic modes that disperse linearly for $k \to 0$. The ε^2 behavior of the phonon DOS for $\varepsilon \to 0$ is therefore expected to be a robust feature and should be controlled by the sound velocity. When relating the curvature of the phonon DOS to the sound velocity, one must be careful that the latter is not necessarily equal to c. For a cubic crystal, for instance, the sound velocity is $v_s = (v_x^2 + v_y^2 + v_z^2)^{1/2} = \sqrt{3}\,c$ because each of the three modes has a velocity $c = (d\omega/dk_x)_{k_y=k_z=0}$. On the other hand, we have $3N_{at}\mathcal{V} = 3N_{at}\mathcal{N}\mathcal{V}_{cell} = N_{modes}\mathcal{V}_{cell}$. It follows that

$$\frac{N^{ph}(\varepsilon \to 0)}{N_{modes}} = \frac{3^{\frac{3}{2}}\mathcal{V}_{cell}}{2\pi^2(\hbar v_s)^3}\varepsilon^2.$$

Doc A2.12 Hamiltonian for the electron–phonon interaction

The electron–phonon coupling Hamiltonian is obtained by inserting in the electron–ion Hamiltonian (2.71) the second term in the right-hand side of equation (2.72):

$$H_{\text{el-ph}} = \int d\boldsymbol{r}\; n(\boldsymbol{r}) \sum_{n\nu} e\, \boldsymbol{u}_\nu(\boldsymbol{R}_n) \cdot \nabla V_\nu(\boldsymbol{r} - \boldsymbol{R}_n - \boldsymbol{\tau}_\nu^0).$$

The displacements are given by equation (2.66),

$$\boldsymbol{u}_\nu(\boldsymbol{R}_n) = \sum_{\boldsymbol{q}\lambda} \sqrt{\frac{\hbar}{2M_\nu \omega_{\boldsymbol{q}\lambda}\mathcal{N}}}\, [\boldsymbol{\epsilon}_{\boldsymbol{q}\lambda}]_\nu \left(b_{\boldsymbol{q}\lambda} + b_{-\boldsymbol{q}\lambda}^\dagger\right) e^{i\boldsymbol{q}\cdot\boldsymbol{R}_n},$$

and the gradient of the lattice potential is most easily evaluated in momentum space:

$$\nabla V_\nu(\boldsymbol{r} - \boldsymbol{R}_n - \boldsymbol{\tau}_\nu^0) = \frac{1}{\mathcal{V}} \sum_{\boldsymbol{k}} V_\nu(\boldsymbol{k}) \nabla e^{i\boldsymbol{k}\cdot(\boldsymbol{r}-\boldsymbol{R}_n-\boldsymbol{\tau}_\nu^0)} = \frac{1}{\mathcal{V}} \sum_{\boldsymbol{k}} V_\nu(\boldsymbol{k})(i\boldsymbol{k}) e^{i\boldsymbol{k}\cdot(\boldsymbol{r}-\boldsymbol{R}_n-\boldsymbol{\tau}_\nu^0)}.$$

Inserting these two expressions in $H_{\text{el-ph}}$, we see that the $\boldsymbol{R}_n$ sum yields $\mathcal{N}\delta_{\boldsymbol{k}\boldsymbol{q}}$:

$$H_{\text{el-ph}} = \frac{1}{\mathcal{V}_{\text{cell}}} \sum_{\boldsymbol{q}\lambda\nu} e \sqrt{\frac{\hbar}{2M_\nu \omega_{\boldsymbol{q}\lambda}\mathcal{N}}}\, [\boldsymbol{\epsilon}_{\boldsymbol{q}\lambda}]_\nu \left(b_{\boldsymbol{q}\lambda} + b_{-\boldsymbol{q}\lambda}^\dagger\right) \cdot V_\nu(\boldsymbol{q})(i\boldsymbol{q}) e^{-i\boldsymbol{q}\cdot\boldsymbol{\tau}_\nu^0} \underbrace{\int d\boldsymbol{r}\; n(\boldsymbol{r}) e^{i\boldsymbol{q}\cdot\boldsymbol{r}}}_{n(-\boldsymbol{q})}.$$

The electron density operator is given by equation (2.46) as $n(-\boldsymbol{q}) = \sum_{\boldsymbol{k}\sigma} c_{\boldsymbol{k}\sigma}^\dagger c_{\boldsymbol{k}-\boldsymbol{q}\sigma} = \sum_{\boldsymbol{k}\sigma} c_{\boldsymbol{k}+\boldsymbol{q}\sigma}^\dagger c_{\boldsymbol{k}\sigma}$, which yields

$$H_{\text{el-ph}} = \sum_{\boldsymbol{k}\sigma} \sum_{\boldsymbol{q}\lambda} \left[\frac{ie}{\mathcal{V}_{\text{cell}}} \sum_\nu \sqrt{\frac{\hbar}{2M_\nu \omega_{\boldsymbol{q}\lambda}\mathcal{N}}} \left([\boldsymbol{\epsilon}_{\boldsymbol{q}\lambda}]_\nu \cdot \boldsymbol{q}\right) V_\nu(\boldsymbol{q}) e^{-i\boldsymbol{q}\cdot\boldsymbol{\tau}_\nu^0} \right] c_{\boldsymbol{k}+\boldsymbol{q}\sigma}^\dagger c_{\boldsymbol{k}\sigma} \left(b_{\boldsymbol{q}\lambda} + b_{-\boldsymbol{q}\lambda}^\dagger\right),$$

consistently with equations (2.73) and (2.74).

Doc A2.13 Direct and exchange terms of the Coulomb interaction

Ignoring all terms except the direct and exchange ones and using the commutation rules (2.41), the Coulomb interaction (2.76) becomes

$$\begin{aligned} V_{\text{Cb}} &\approx \frac{1}{2} \sum_{\alpha\beta\sigma\sigma'} (U_{\alpha\beta\alpha\beta} c_{\alpha\sigma}^\dagger \underbrace{c_{\beta\sigma'}^\dagger c_{\beta\sigma'} c_{\alpha\sigma}}_{-\delta_{\alpha\beta}\delta_{\sigma\sigma'} c_{\alpha\sigma} + c_{\alpha\sigma} c_{\beta\sigma'}^\dagger c_{\beta\sigma'}} + (1 - \delta_{\alpha\beta}) U_{\alpha\beta\beta\alpha} c_{\alpha\sigma}^\dagger \underbrace{c_{\beta\sigma'}^\dagger c_{\alpha\sigma'}}_{\delta_{\alpha\beta} - c_{\alpha\sigma'} c_{\beta\sigma'}^\dagger} c_{\beta\sigma}) \\ &= \sum_\alpha U_{\alpha\alpha\alpha\alpha} n_{\alpha\uparrow} n_{\alpha\downarrow} + \frac{1}{2} \sum_{\alpha\neq\beta} U_{\alpha\beta\alpha\beta} n_\alpha n_\beta - \frac{1}{2} \sum_{\alpha\neq\beta} \sum_{\sigma\sigma'} U_{\alpha\beta\beta\alpha} c_{\alpha\sigma}^\dagger c_{\alpha\sigma'} c_{\beta\sigma'}^\dagger c_{\beta\sigma}. \end{aligned}$$

On going from the first to the second line, we have taken advantage of the relation $n_\alpha^2 = n_\alpha + 2n_{\alpha\uparrow}n_{\alpha\downarrow}$. The spin sum in the last term can be expressed by means of the density and spin operators:

$$\begin{aligned}\sum_{\sigma\sigma'} c_{\alpha\sigma}^\dagger c_{\alpha\sigma'} c_{\beta\sigma'}^\dagger c_{\beta\sigma} &= \frac{1}{2}\left(c_{\alpha\uparrow}^\dagger c_{\alpha\uparrow} + c_{\alpha\downarrow}^\dagger c_{\alpha\downarrow}\right)\left(c_{\beta\uparrow}^\dagger c_{\beta\uparrow} + c_{\beta\downarrow}^\dagger c_{\beta\downarrow}\right)\\ &+ \frac{1}{2}\left(c_{\alpha\uparrow}^\dagger c_{\alpha\uparrow} - c_{\alpha\downarrow}^\dagger c_{\alpha\downarrow}\right)\left(c_{\beta\uparrow}^\dagger c_{\beta\uparrow} - c_{\beta\downarrow}^\dagger c_{\beta\downarrow}\right)\\ &+ c_{\alpha\uparrow}^\dagger c_{\alpha\downarrow} c_{\beta\downarrow}^\dagger c_{\beta\uparrow} + c_{\alpha\downarrow}^\dagger c_{\alpha\uparrow} c_{\beta\uparrow}^\dagger c_{\beta\downarrow}\\ &= \frac{1}{2} n_\alpha n_\beta + \frac{1}{2}\left(\frac{2}{\hbar}\right)^2 S_\alpha^z S_\beta^z + \left(\frac{1}{\hbar}\right)^2 \left(S_\alpha^+ S_\beta^- + S_\alpha^- S_\beta^+\right).\end{aligned}$$

We have made use of equation (2.48) and defined orbital-resolved spin operators. Finally, using the fact that $\boldsymbol{S}_\alpha \cdot \boldsymbol{S}_\beta = S_\alpha^z S_\beta^z + \frac{1}{2}(S_\alpha^+ S_\beta^- + S_\alpha^- S_\beta^+)$ we arrive at equation (2.77).

Doc A2.14 Kinetic exchange

We consider the two-orbitals Hubbard model

$$\begin{aligned}H^{\alpha\beta} &= -\mu(n_\alpha + n_\beta) + t\sum_\sigma\left(c_{\alpha\sigma}^\dagger c_{\beta\sigma} + c_{\beta\sigma}^\dagger c_{\alpha\sigma}\right) + U(n_{\alpha\uparrow}n_{\alpha\downarrow} + n_{\beta\uparrow}n_{\beta\downarrow})\\ &\equiv -\mu(n_\alpha + n_\beta) + H_0^{\alpha\beta} + H_U^{\alpha\beta}\end{aligned}$$

in the two-electrons subspace. This subspace has dimension 6 and is spanned by the state vectors $|\uparrow\uparrow\rangle, |\downarrow\downarrow\rangle, |\uparrow\downarrow\rangle, |\downarrow\uparrow\rangle, |\uparrow\downarrow,\cdot\rangle$, and $|\cdot,\uparrow\downarrow\rangle$, the last two being the double-occupancy states in which both electrons sit on the same orbital. For all these states, the operator $n_\alpha + n_\beta$ is equal to the number 2 and the term $-\mu(n_\alpha + n_\beta)$ amounts to a rigid shift of all energies by -2μ. We ignore this term which plays no role in our discussion. Two eigenstates of $H^{\alpha\beta}$ are immediately obvious: the two triplet states $|\uparrow\uparrow\rangle$ and $|\downarrow\downarrow\rangle$ are annihilated by $H_0^{\alpha\beta}$, because no hopping is allowed from one orbital to the other, and they are eigenstates of $H_U^{\alpha\beta}$ with zero eigenvalue. The third triplet state $(|\uparrow\downarrow\rangle + |\downarrow\uparrow\rangle)/\sqrt{2}$ is also a zero-energy eigenstate. Using the basic rules of equation (2.42), one indeed sees that $H_0^{\alpha\beta}|\uparrow\downarrow\rangle = t(|\uparrow\downarrow,\cdot\rangle + |\cdot,\uparrow\downarrow\rangle)$ and $H_0^{\alpha\beta}|\downarrow\uparrow\rangle = -t(|\uparrow\downarrow,\cdot\rangle + |\cdot,\uparrow\downarrow\rangle)$, such that the kinetic energy, like the interaction energy, is zero for this eigenstate. The non-trivial physics occurs in the subspace spanned by the remaining three linearly independent vectors, which are the singlet $|1\rangle \equiv (|\uparrow\downarrow\rangle - |\downarrow\uparrow\rangle)/\sqrt{2}$ and the two states with double occupancy, $|2\rangle \equiv |\uparrow\downarrow,\cdot\rangle$ and $|3\rangle \equiv |\cdot,\uparrow\downarrow\rangle$. The following table shows how the Hamiltonian transforms the three vectors, allowing us to build the 3×3 Hamiltonian matrix in this subspace:

	$\|1\rangle$	$\|2\rangle$	$\|3\rangle$
$H_0^{\alpha\beta}$	$\sqrt{2}t(\|2\rangle+\|3\rangle)$	$\sqrt{2}t\|1\rangle$	$\sqrt{2}t\|1\rangle$
$H_U^{\alpha\beta}$	0	$U\|2\rangle$	$U\|3\rangle$

$$\Rightarrow \quad H_0^{\alpha\beta} + H_U^{\alpha\beta} = \begin{pmatrix} 0 & \sqrt{2}t & \sqrt{2}t \\ \sqrt{2}t & U & 0 \\ \sqrt{2}t & 0 & U \end{pmatrix}.$$

The lowest-energy eigenstate has negative energy $[U - \sqrt{U^2 + (4t)^2}]/2$ and is a mixture of the singlet and the two double-occupancy states; this mixture becomes the pure singlet in the limit $U \gg t$, while the energy approaches $-4t^2/U$. The other two states have energies of order U. One can understand this by using standard perturbation theory in $H_0^{\alpha\beta}$. For $t = 0$, the singlet has energy $E_1^{(0)} = 0$ and the double-occupancy states have energy U. The second-order perturbation theory then gives

$$E_1^{(2)} = 0 + \langle 1|H_0^{\alpha\beta}|1\rangle + \frac{|\langle 2|H_0^{\alpha\beta}|1\rangle|^2}{0 - U} + \frac{|\langle 3|H_0^{\alpha\beta}|1\rangle|^2}{0 - U} = -\frac{4t^2}{U}.$$

In the large-U limit, the projection of the Hamiltonian on its ground state is

$$H_{\rm GS}^{\alpha\beta} = -\frac{4t^2}{U}|1\rangle\langle 1| = -\frac{4t^2}{U}\frac{1}{2}\left(c_{\beta\downarrow}^\dagger c_{\alpha\uparrow}^\dagger|\varnothing\rangle - c_{\beta\uparrow}^\dagger c_{\alpha\downarrow}^\dagger|\varnothing\rangle\right)\left(\langle\varnothing|c_{\alpha\uparrow}c_{\beta\downarrow} - \langle\varnothing|c_{\alpha\downarrow}c_{\beta\uparrow}\right).$$

One can see that, in the two-electrons subspace, the projector on the vacuum $|\varnothing\rangle\langle\varnothing|$ in this expression is equivalent to the identity. Because all operators anti-commute, we have

$$\begin{aligned}
H_{\rm GS}^{\alpha\beta} = &-\frac{4t^2}{U}\frac{1}{2}\left[\frac{1}{2}\left(c_{\alpha\uparrow}^\dagger c_{\alpha\uparrow} + c_{\alpha\downarrow}^\dagger c_{\alpha\downarrow}\right)\left(c_{\beta\uparrow}^\dagger c_{\beta\uparrow} + c_{\beta\downarrow}^\dagger c_{\beta\downarrow}\right)\right.\\
&\left.-\frac{1}{2}\left(c_{\alpha\uparrow}^\dagger c_{\alpha\uparrow} - c_{\alpha\downarrow}^\dagger c_{\alpha\downarrow}\right)\left(c_{\beta\uparrow}^\dagger c_{\beta\uparrow} - c_{\beta\downarrow}^\dagger c_{\beta\downarrow}\right) - c_{\alpha\uparrow}^\dagger c_{\alpha\downarrow} c_{\beta\downarrow}^\dagger c_{\beta\uparrow} - c_{\alpha\downarrow}^\dagger c_{\alpha\uparrow} c_{\beta\uparrow}^\dagger c_{\beta\downarrow}\right]\\
= &-\frac{4t^2}{U}\frac{1}{2}\left[\frac{1}{2}n_\alpha n_\beta - \frac{1}{2}\left(\frac{2}{\hbar}\right)^2 S_\alpha^z S_\beta^z - \left(\frac{1}{\hbar}\right)^2\left(S_\alpha^+ S_\beta^- + S_\alpha^- S_\beta^+\right)\right].
\end{aligned}$$

Proceeding like in Doc A2.13, this leads to equation (2.80).

Doc A2.15 Holstein–Primakoff transformation

The Holstein–Primakoff bozonisation[9] expresses a spin operator in terms of bosonic creation and annihilation operators. This is to be contrasted with equation (2.48), where the spin density is given in terms of fermionic operators. Consider a spin $\boldsymbol{\mathcal{S}}$ of length S and the basis of common eigenstates of $\mathcal{S}^2$ and $\mathcal{S}^z$ denoted $|S, m\rangle$, with $m = -S, \dots, +S$. The idea is to regard the fully up state $|S, m = +S\rangle$ as a vacuum and generate the other states by acting on this vacuum with some creation operator $b^\dagger$. Let us define $b^\dagger$ and its conjugate b by their action on the states $|S, m\rangle$:

$$\begin{aligned}
b^\dagger|S, m\rangle &= \sqrt{S - m + 1}\,|S, m - 1\rangle\\
b\,|S, m\rangle &= \sqrt{S - m}\,|S, m + 1\rangle.
\end{aligned}$$

$b^\dagger$, like the spin lowering operator $\mathcal{S}^- = \mathcal{S}^x - i\mathcal{S}^y$, reduces the projection of the spin along z by one unit: one should see this as creating an excitation; that this excitation

[9] Holstein T and Primakoff H 1940 *Phys. Rev.* **58** 1098.

is a boson will become clear below. Our definitions look different than equation (2.42) but this is because the ordering of the states is inverted. The operator $n = b^\dagger b$ should count the number of bosons, i.e. how far m is from the maximum value $m = S$. This measure is $S - m$ and indeed

$$\begin{aligned} b^\dagger b|S, m\rangle &= \sqrt{S-m}\, b^\dagger|S, m+1\rangle = \sqrt{S-m}\sqrt{S-(m+1)+1}\,|S, m\rangle \\ &= (S-m)|S, m\rangle. \end{aligned}$$

To see whether $b^\dagger$ is a boson or a fermion, we calculate $[b, b^\dagger]$. Using

$$\begin{aligned} bb^\dagger|S, m\rangle &= \sqrt{S-m+1}\, b|S, m-1\rangle = \sqrt{S-m+1}\sqrt{S-(m-1)}\,|S, m\rangle \\ &= (S-m+1)|S, m\rangle, \end{aligned}$$

we see that $[b, b^\dagger] = 1$, hence equation (2.41) tells us that this is a boson. The operator $S - n$ measures m, like $\mathcal{S}^z$:

$$(S - b^\dagger b)|S, m\rangle = [S - (S-m)]|S, m\rangle = m|S, m\rangle = \mathcal{S}^z|S, m\rangle.$$

This gives us the expression of the z component as $\mathcal{S}^z = S - n$. We recall that the actions of the spin-raising and spin-lowering operators on the basis states are

$$\mathcal{S}^\pm|S, m\rangle = \sqrt{S(S+1) - m(m\pm 1)}\,|S, m\pm 1\rangle$$

and we wish to reproduce this same behavior with the b and $b^\dagger$. $\mathcal{S}^+$ is proportional to b, with a proportionality factor $\sqrt{2S-n}$. Indeed, one checks

$$\begin{aligned} \sqrt{2S-n}\, b|S, m\rangle &= \sqrt{2S-n}\sqrt{S-m}\,|S, m+1\rangle \\ &= \sqrt{2S-[S-(m+1)]}\sqrt{S-m}\,|S, m+1\rangle \\ &= \sqrt{S(S+1)-m(m+1)}\,|S, m+1\rangle = \mathcal{S}^+|S, m\rangle. \end{aligned}$$

Similarly,

$$\begin{aligned} b^\dagger\sqrt{2S-n}\,|S, m\rangle &= b^\dagger\sqrt{2S-(S-m)}\,|S, m\rangle = \sqrt{S-m+1}\sqrt{S+m}\,|S, m-1\rangle \\ &= \sqrt{S(S+1)-m(m-1)}\,|S, m-1\rangle = \mathcal{S}^-|S, m\rangle. \end{aligned}$$

The complete mapping therefore reads

$$\mathcal{S}^z = S - n, \qquad \mathcal{S}^+ = \sqrt{2S-n}\, b, \qquad \mathcal{S}^- = b^\dagger\sqrt{2S-n}, \qquad n = b^\dagger b.$$

This mapping is exact as long as we stay in the space spanned by the states $|S, m\rangle$. If we forget the spins and work in a purely bosonic world, then unphysical states appear because $b^\dagger$ can generate infinitely many states, while, in the original space, it cannot reduce the projection below the value $-S$. This is not a concern per se, because the representation of $\mathcal{S}^-$, $b^\dagger\sqrt{2S-n}$ *does* annihilate the state $|S, -S\rangle$—in other words, a Hamiltonian containing only the operator $\boldsymbol{\mathcal{S}}$ will not visit the unphysical states—but it may become a concern if we adopt approximations such that the Hamiltonian we work with contains terms that cannot be expressed by means of $\boldsymbol{\mathcal{S}}$ only.

The mapping naturally implements the fact that different spin operators commute. In the boson language, the scalar product of two spins takes the form

$$\begin{aligned}
\mathcal{S}_\alpha \cdot \mathcal{S}_\beta &= S^z_\alpha S^z_\beta + \frac{1}{2}\left(S^+_\alpha S^-_\beta + S^-_\alpha S^+_\beta\right) \\
&= (S - n_\alpha)(S - n_\beta) \\
&\quad + \frac{1}{2}\left(\sqrt{2S - n_\alpha}\, b_\alpha b^\dagger_\beta \sqrt{2S - n_\beta} + b^\dagger_\alpha \sqrt{2S - n_\alpha}\sqrt{2S - n_\beta}\, b_\beta\right) \\
&= S^2\left(1 - \frac{n_\alpha}{S}\right)\left(1 - \frac{n_\beta}{S}\right) \\
&\quad + S\left(\sqrt{1 - \frac{n_\alpha}{2S}}\, b_\alpha b^\dagger_\beta \sqrt{1 - \frac{n_\beta}{2S}} + b^\dagger_\alpha \sqrt{1 - \frac{n_\alpha}{2S}}\sqrt{1 - \frac{n_\beta}{2S}}\, b_\beta\right).
\end{aligned}$$

The last writing emphasizes the role of n/S as an expansion parameter. If the spins are close to saturation, n/S can be considered as small, especially for large S, and the expression can be expanded. We see that this gives rise to quadratic terms in the bosons as well as higher-order terms corresponding to interactions:

$$\mathcal{S}_\alpha \cdot \mathcal{S}_\beta = S^2 - S(n_\alpha + n_\beta) + S\left(b_\alpha b^\dagger_\beta + b^\dagger_\alpha b_\beta\right) + \text{interaction terms.}$$

The linear spin-wave theory is built by ignoring the interaction terms.

Doc A2.16 Linearized ferromagnetic Hamiltonian

After bosonization with equation (2.82), the Hamiltonian (2.81) becomes

$$\begin{aligned}
H_{\mathrm{FM}} &= -\frac{1}{2}\sum_{nm} J(\boldsymbol{R}_n - \boldsymbol{R}_m)\left[S^2 - S\left(b^\dagger_n b_n + b^\dagger_m b_m\right) + S\left(b_n b^\dagger_m + b^\dagger_n b_m\right)\right] \\
&\quad - g\mu_{\mathrm{B}} B \sum_n \left(S - b^\dagger_n b_n\right) + \text{interaction terms}
\end{aligned}$$

$$\begin{aligned}
H^{(1)}_{\mathrm{FM}} &= -\frac{S^2}{2}\sum_{nm} J(\boldsymbol{R}_n - \boldsymbol{R}_m) + S\sum_{nm} J(\boldsymbol{R}_n - \boldsymbol{R}_m) b^\dagger_n b_n \\
&\quad - \frac{S}{2}\left[\sum_{nm} J(\boldsymbol{R}_n - \boldsymbol{R}_m) b^\dagger_n b_m + \text{h. c.}\right] - g\mu_{\mathrm{B}} B\left(\mathcal{N}S - \sum_n b^\dagger_n b_n\right).
\end{aligned}$$

For the second expression, we have discarded the interaction terms and used the properties $J(\boldsymbol{R}_n - \boldsymbol{R}_m) = J(\boldsymbol{R}_m - \boldsymbol{R}_n)$, $J(\boldsymbol{R}_n - \boldsymbol{R}_m) \in \mathbb{R}$, and $J(\boldsymbol{R} = \boldsymbol{0}) = 0$. We move now to momentum space using (see Doc A2.7 and equation (2*a*))

$$b^\dagger_n = \frac{1}{\sqrt{\mathcal{N}}}\sum_{\boldsymbol{k}} e^{-i\boldsymbol{k}\cdot\boldsymbol{R}_n} b^\dagger_{\boldsymbol{k}} \quad \text{and} \quad J(\boldsymbol{R}_n - \boldsymbol{R}_m) = \frac{1}{\mathcal{N}}\sum_{\boldsymbol{k}} J(\boldsymbol{k}) e^{i\boldsymbol{k}\cdot(\boldsymbol{R}_n - \boldsymbol{R}_m)}.$$

All calculations are elementary:

$$\sum_{nm} J(\boldsymbol{R}_n - \boldsymbol{R}_m) = \mathcal{N} \sum_{n} J(\boldsymbol{R}_n) = \mathcal{N} J(\boldsymbol{k} = \boldsymbol{0})$$

$$\sum_{nm} J(\boldsymbol{R}_n - \boldsymbol{R}_m) b_n^\dagger b_n = \frac{1}{\mathcal{N}} \sum_{\boldsymbol{k}} J(\boldsymbol{k}) \frac{1}{\mathcal{N}} \sum_{\boldsymbol{k}_1\boldsymbol{k}_2} b_{\boldsymbol{k}_1}^\dagger b_{\boldsymbol{k}_2} \underbrace{\sum_{nm} e^{i\boldsymbol{k}\cdot(\boldsymbol{R}_n - \boldsymbol{R}_m)} e^{-i\boldsymbol{k}_1\cdot\boldsymbol{R}_n} e^{i\boldsymbol{k}_2\cdot\boldsymbol{R}_n}}_{\mathcal{N}^2 \delta_{\boldsymbol{k}\boldsymbol{0}} \delta_{\boldsymbol{k}_1\boldsymbol{k}_2}}$$

$$= J(\boldsymbol{0}) \sum_{\boldsymbol{k}} b_{\boldsymbol{k}}^\dagger b_{\boldsymbol{k}}$$

$$\sum_{nm} J(\boldsymbol{R}_n - \boldsymbol{R}_m) b_n^\dagger b_m = \frac{1}{\mathcal{N}} \sum_{\boldsymbol{k}} J(\boldsymbol{k}) \frac{1}{\mathcal{N}} \sum_{\boldsymbol{k}_1\boldsymbol{k}_2} b_{\boldsymbol{k}_1}^\dagger b_{\boldsymbol{k}_2} \underbrace{\sum_{nm} e^{i\boldsymbol{k}\cdot(\boldsymbol{R}_n - \boldsymbol{R}_m)} e^{-i\boldsymbol{k}_1\cdot\boldsymbol{R}_n} e^{i\boldsymbol{k}_2\cdot\boldsymbol{R}_m}}_{\mathcal{N}^2 \delta_{\boldsymbol{k}\boldsymbol{k}_1} \delta_{\boldsymbol{k}\boldsymbol{k}_2}}$$

$$= \sum_{\boldsymbol{k}} J(\boldsymbol{k}) b_{\boldsymbol{k}}^\dagger b_{\boldsymbol{k}}$$

$$\sum_{n} b_n^\dagger b_n = \frac{1}{\mathcal{N}} \sum_{\boldsymbol{k}_1\boldsymbol{k}_2} b_{\boldsymbol{k}_1}^\dagger b_{\boldsymbol{k}_2} \underbrace{\sum_{n} e^{-i\boldsymbol{k}_1\cdot\boldsymbol{R}_n} e^{i\boldsymbol{k}_2\cdot\boldsymbol{R}_n}}_{\mathcal{N} \delta_{\boldsymbol{k}_1\boldsymbol{k}_2}} = \sum_{\boldsymbol{k}} b_{\boldsymbol{k}}^\dagger b_{\boldsymbol{k}}.$$

Collecting all terms and noting that $J(\boldsymbol{k})$ is real, we get

$$H_{\text{FM}}^{(1)} = \mathcal{N}\left[-J(\boldsymbol{0})\frac{S^2}{2} - g\mu_{\text{B}} BS\right] + \sum_{\boldsymbol{k}} \{[J(\boldsymbol{0}) - J(\boldsymbol{k})]S + g\mu_{\text{B}} B\} b_{\boldsymbol{k}}^\dagger b_{\boldsymbol{k}}.$$

Upon adding the zero-point energy of the oscillators in the second term and subtracting it from the first term, remarking that $\sum_{\boldsymbol{k}} J(\boldsymbol{k}) = J(\boldsymbol{R} = \boldsymbol{0}) = 0$, we arrive at equation (2.83).

We show now that the state $\mathcal{S}_{\boldsymbol{k}}^-|\Psi_0\rangle$, where $|\Psi_0\rangle$ is the ferromagnetic ground state, is an eigenstate of energy $\mathcal{N}\varepsilon_0 + \hbar\omega_{\boldsymbol{k}}$ and therefore corresponds to a magnon. To linear order, we have $\mathcal{S}_{\boldsymbol{k}}^- = \sqrt{2S} b_{\boldsymbol{k}}^\dagger$ (see Doc A2.15), which gives

$$H_{\text{FM}}^{(1)} \mathcal{S}_{\boldsymbol{k}}^- |\Psi_0\rangle = \underbrace{[H_{\text{FM}}^{(1)}, \mathcal{S}_{\boldsymbol{k}}^-]}_{\hbar\omega_{\boldsymbol{k}} \mathcal{S}_{\boldsymbol{k}}^-} |\Psi_0\rangle + \mathcal{S}_{\boldsymbol{k}}^- \underbrace{H_{\text{FM}}^{(1)} |\Psi_0\rangle}_{\mathcal{N}\varepsilon_0 |\Psi_0\rangle} = (\mathcal{N}\varepsilon_0 + \hbar\omega_{\boldsymbol{k}}) \mathcal{S}_{\boldsymbol{k}}^- |\Psi_0\rangle,$$

where we have used the relations (2.41) to calculate the commutator. We can deduce the equation of motion of the operator using equation (2.15):

$$i\hbar\partial_t \mathcal{S}_{\boldsymbol{k}}^- = [\mathcal{S}_{\boldsymbol{k}}^-, H_{\text{FM}}^{(1)}] = -\hbar\omega_{\boldsymbol{k}} \mathcal{S}_{\boldsymbol{k}}^- \Rightarrow \mathcal{S}_{\boldsymbol{k}}^-(t) = \mathcal{S}_{\boldsymbol{k}}^-(0) e^{i\omega_{\boldsymbol{k}} t}.$$

In real space, remembering that operators do not transform like functions (see Doc A2.7), this becomes $\mathcal{S}_n^-(t) = [\mathcal{S}_{\boldsymbol{k}}^-(0)/\sqrt{\mathcal{N}}] e^{-i(\boldsymbol{k}\cdot\boldsymbol{R}_n - \omega_{\boldsymbol{k}} t)}$.

Doc A2.17 Classical ground state of the anti-ferromagnet

If all spins $\boldsymbol{\mathcal{S}}_{n+}$ are parallel to $\hat{z}$ and all spins $\boldsymbol{\mathcal{S}}_{n-}$ are parallel to $-\hat{z}$, i.e. $\boldsymbol{\mathcal{S}}_{n\nu} = \hat{z} S\nu$, the classical energy of the system (2.84) is

$$E_{\uparrow\downarrow} = \frac{1}{2}\sum_{nm}\sum_{\nu\mu} J(\boldsymbol{R}_n + \nu\boldsymbol{\tau} - \boldsymbol{R}_m - \mu\boldsymbol{\tau})S^2\nu\mu - g\mu_{\rm B}B\sum_{n\nu} S\nu = \mathcal{N}[-J_1(\boldsymbol{0}) + J_2(\boldsymbol{0})]S^2.$$

We have introduced different functions for the Fourier transforms of the inter- and intra-sublattice couplings:

$$J_1(\boldsymbol{k}) = \sum_n J(\boldsymbol{R}_n + 2\boldsymbol{\tau})e^{-i\boldsymbol{k}\cdot(\boldsymbol{R}_n+2\boldsymbol{\tau})}, \qquad J_2(\boldsymbol{k}) = \sum_n J(\boldsymbol{R}_n)e^{-i\boldsymbol{k}\cdot\boldsymbol{R}_n}.$$

Note that $J_2(\boldsymbol{k})$ is real and $J_2(-\boldsymbol{k}) = J_2(\boldsymbol{k})$ but that, in general, $J_1(\boldsymbol{k})$ is complex and $J_1(-\boldsymbol{k}) = J_1^*(\boldsymbol{k})$. $J_1(\boldsymbol{0}) > 0$ is the strength of the main anti-ferromagnetic coupling. If the spins are mainly along $\hat{\boldsymbol{x}}$ with a tilt angle ϑ towards $\hat{\boldsymbol{z}}$, as expected in the presence of a field oriented along $\hat{\boldsymbol{z}}$, i.e. $\boldsymbol{S}_{n\nu} = \hat{\boldsymbol{x}}S\nu\cos\vartheta + \hat{\boldsymbol{z}}S\sin\vartheta$, the energy becomes

$$\begin{aligned}E_{\rightarrow\leftarrow} &= \frac{1}{2}\sum_{nm}\sum_{\nu\mu} J(\boldsymbol{R}_n + \nu\boldsymbol{\tau} - \boldsymbol{R}_m - \mu\boldsymbol{\tau})S^2(\nu\mu\cos^2\vartheta + \sin^2\vartheta) - g\mu_{\rm B}B\sum_{n\nu} S\sin\vartheta \\ &= \mathcal{N}\big\{[-J_1(\boldsymbol{0})(1 - 2\sin^2\vartheta) + J_2(\boldsymbol{0})]S^2 - 2g\mu_{\rm B}BS\sin\vartheta\big\}.\end{aligned}$$

This energy is minimal if $\sin\vartheta$ is given by equation (2.85). At the minimum, this state has lower energy than the up–down state:

$$E_{\rightarrow\leftarrow} - E_{\uparrow\downarrow} = \mathcal{N}\big[2J_1(\boldsymbol{0})S^2\sin^2\vartheta - 2g\mu_{\rm B}BS\sin\vartheta\big] = -\mathcal{N}\frac{(g\mu_{\rm B}B)^2}{2J_1(\boldsymbol{0})} < 0.$$

Doc A2.18 Linearized anti-ferromagnetic Hamiltonian

The substitution of equation (2.86) in equation (2.84) yields

$$\begin{aligned}H_{\rm AF} &= \frac{1}{2}\sum_{nm}\sum_{\nu\mu} J(\boldsymbol{R}_n + \nu\boldsymbol{\tau} - \boldsymbol{R}_m - \mu\boldsymbol{\tau})\Big[(\sin^2\vartheta + \nu\mu\cos^2\vartheta)\big(\tilde{S}^x_{n\nu}\tilde{S}^x_{m\mu} + \tilde{S}^z_{n\nu}\tilde{S}^z_{m\mu}\big) \\ &\quad + \tilde{S}^y_{n\nu}\tilde{S}^y_{m\mu} + (\nu - \mu)\sin\vartheta\cos\vartheta\big(\tilde{S}^z_{n\nu}\tilde{S}^x_{m\mu} - \tilde{S}^x_{n\nu}\tilde{S}^z_{m\mu}\big)\Big] \\ &\quad - g\mu_{\rm B}B\sum_{n\nu}(-\nu\cos\vartheta\tilde{S}^x_{n\nu} + \sin\vartheta\tilde{S}^z_{n\nu}).\end{aligned}$$

All spins $\tilde{\boldsymbol{S}}$ are close to their saturation along the quantization axis and can be expressed in terms of Holstein–Primakoff bosons. To linear order, the mapping is (see Doc A2.15)

$$\tilde{S}^x = \frac{1}{2}(\tilde{S}^+ + \tilde{S}^-) = \frac{\sqrt{2S}}{2}(b^\dagger + b), \quad \tilde{S}^y = \frac{\sqrt{2S}}{2i}(b^\dagger - b), \quad \tilde{S}^z = S - b^\dagger b.$$

The terms like $\tilde{S}^z_{n\nu}\tilde{S}^x_{m\mu}$ give one-boson contributions of the form $S(b^\dagger + b)$ as well as three-boson contributions that we neglect. Collecting all one-boson terms, we find that they vanish on account of equation (2.85). Indeed

$$\frac{1}{2}\sum_{nm}\sum_{\nu\mu}J(\boldsymbol{R}_n+\nu\boldsymbol{\tau}-\boldsymbol{R}_m-\mu\boldsymbol{\tau})(\nu-\mu)\sin\vartheta\cos\vartheta\left(S\tilde{\mathcal{S}}^x_{m\mu}-\tilde{\mathcal{S}}^x_{n\nu}S\right)$$
$$=S\sin\vartheta\cos\vartheta\sum_{nm}\sum_{\nu}\nu J(\boldsymbol{R}_n-\boldsymbol{R}_m+2\nu\boldsymbol{\tau})\left(\tilde{\mathcal{S}}^x_{m,-\nu}-\tilde{\mathcal{S}}^x_{n\nu}\right)$$
$$=-2SJ_1(\boldsymbol{0})\sin\vartheta\cos\vartheta\sum_{n\nu}\nu\tilde{\mathcal{S}}^x_{n\nu}=-g\mu_{\mathrm{B}}B\sum_{n\nu}\nu\cos\vartheta\tilde{\mathcal{S}}^x_{n\nu}.$$

For the last equality, we have used equation (2.85) and we find a result which exactly cancels the term in H_{AF} due to the field and involving $\tilde{\mathcal{S}}^x_{n\nu}$. The remaining terms of H_{AF}, after bosonization, neglect of the interaction terms, use of the properties $J(\boldsymbol{r})=J(-\boldsymbol{r})$ and $J(\boldsymbol{0})=0$, and equation (2.85) to replace $g\mu_{\mathrm{B}}B$ by $2SJ_1(\boldsymbol{0})\sin\vartheta$, give the linearized Hamiltonian

$$H^{(1)}_{\mathrm{AF}}=E_{\to\leftarrow}+[J_1(\boldsymbol{0})-J_2(\boldsymbol{0})]S\sum_{n\nu}b^\dagger_{n\nu}b_{n\nu}$$
$$+S\sum_{nm}\sum_{\nu\mu}J(\boldsymbol{R}_n+\nu\boldsymbol{\tau}-\boldsymbol{R}_m-\mu\boldsymbol{\tau})\left(\sin^2\vartheta+\delta_{\nu\mu}\cos^2\vartheta\right)b^\dagger_{n\nu}b_{m\mu}$$
$$-\frac{S}{2}\cos^2\vartheta\sum_{nm}\sum_{\nu}J(\boldsymbol{R}_n-\boldsymbol{R}_m+2\nu\boldsymbol{\tau})\left(b_{n\nu}b_{m,-\nu}+b^\dagger_{n\nu}b^\dagger_{m,-\nu}\right),$$

where $E_{\to\leftarrow}$ is the classical ground-state energy given in Doc A2.17. We move to reciprocal space using $b^\dagger_{n\nu}=(2\mathcal{N})^{-\frac{1}{2}}\sum_{\boldsymbol{k}}e^{-i\boldsymbol{k}\cdot(\boldsymbol{R}_n+\nu\boldsymbol{\tau})}b^\dagger_{\boldsymbol{k}}$[10]. Calculations are again elementary and give the following result:

$$H^{(1)}_{\mathrm{AF}}=E_{\to\leftarrow}+\sum_{\boldsymbol{k}}\xi_{\boldsymbol{k}}b^\dagger_{\boldsymbol{k}}b_{\boldsymbol{k}}+\frac{1}{2}\sum_{\boldsymbol{k}}\Delta_{\boldsymbol{k}}\left(b^\dagger_{\boldsymbol{k}}b^\dagger_{-\boldsymbol{k}}+\text{h. c.}\right)$$
$$\xi_{\boldsymbol{k}}=J_1(\boldsymbol{0})S+\mathrm{Re}[J_1(\boldsymbol{k})]S\sin^2\vartheta+[J_2(\boldsymbol{k})-J_2(\boldsymbol{0})]S$$
$$\Delta_{\boldsymbol{k}}=-\,\mathrm{Re}[J_1(\boldsymbol{k})]S\cos^2\vartheta.$$

In this form, the Hamiltonian resembles the BCS Hamiltonian for a superconductor, if not for the fact that the $b^\dagger_{\boldsymbol{k}}$ are spinless bosons rather than spinful fermions. The diagonalization proceeds via a Bogoliubov transformation in complete analogy with the superconducting case. One introduces a new boson that mixes the original ones,

$$\beta^\dagger_{\boldsymbol{k}}=u_{\boldsymbol{k}}b^\dagger_{\boldsymbol{k}}+v_{\boldsymbol{k}}b_{-\boldsymbol{k}},$$

where $u_{\boldsymbol{k}}$ and $v_{\boldsymbol{k}}$ are real numbers satisfying $u^2_{\boldsymbol{k}}-v^2_{\boldsymbol{k}}=1$ and $u_{\boldsymbol{k}}v_{-\boldsymbol{k}}=u_{-\boldsymbol{k}}v_{\boldsymbol{k}}$, two sufficient conditions to ensure that the $\beta_{\boldsymbol{k}}$ obey the commutation rules of bosons. Calculating $\sum_{\boldsymbol{k}}E_{\boldsymbol{k}}\beta^\dagger_{\boldsymbol{k}}\beta_{\boldsymbol{k}}$, we readily find that this is equal to $\sum_{\boldsymbol{k}}\xi_{\boldsymbol{k}}b^\dagger_{\boldsymbol{k}}b_{\boldsymbol{k}}+\frac{1}{2}\sum_{\boldsymbol{k}}\Delta_{\boldsymbol{k}}(b^\dagger_{\boldsymbol{k}}b^\dagger_{-\boldsymbol{k}}+\text{h. c.})+\sum_{\boldsymbol{k}}E_{\boldsymbol{k}}v^2_{\boldsymbol{k}}$ if the following equations are satisfied:

[10] Here, we limit ourselves for simplicity to a bipartite lattice. In the general case, one must introduce two excitations $b^\dagger_{\boldsymbol{k}\nu}$ for each wave vector $\boldsymbol{k}$.

$$E_k u_k^2 + E_{-k} v_{-k}^2 = \xi_k, \qquad E_k u_k v_k = \frac{1}{2}\Delta_k.$$

The second equation, together with the normalization $u_k^2 - v_k^2 = 1$, yields

$$u_k^2 = \frac{1}{2}\left(1 + \sqrt{1 + \Delta_k^2/E_k^2}\right).$$

Furthermore, by subtracting from the first equation itself (the first equation) with $\boldsymbol{k}$ changed to $-\boldsymbol{k}$ and using the normalization, we obtain the relation $E_k - E_{-k} = \xi_k - \xi_{-k} = 0$. This, together with the property $\Delta_k = \Delta_{-k}$, shows that the amplitudes u_k and u_k are even functions of $\boldsymbol{k}$. Forming the sum rather than the difference and inserting our expressions, we are led to a closed equation for E_k, namely $E_k(1 + \Delta_k^2/E_k^2)^{1/2} = \xi_k$. This equation admits no real and positive solution for E_k if ξ_k is negative or if $|\Delta_k| > |\xi_k|$. These conditions express the instability of the anti-ferromagnetic solution under a frustration induced by competing couplings. Whenever it exists, the solution is given by

$$E_k = \sqrt{\xi_k^2 - \Delta_k^2}\,.$$

It is convenient to introduce the dimensionless quantities

$$j_1(\boldsymbol{k}) = \mathrm{Re}\left[\frac{J_1(\boldsymbol{k})}{J_1(\boldsymbol{0})}\right], \qquad j_2(\boldsymbol{k}) = \frac{J_2(\boldsymbol{k}) - J_2(\boldsymbol{0})}{J_1(\boldsymbol{0})}, \qquad b = \frac{g\mu_{\mathrm{B}}B}{2SJ_1(\boldsymbol{0})} = \sin\vartheta,$$

in terms of which the dispersion can be recast as

$$\frac{E_k}{J_1(\boldsymbol{0})S} = \sqrt{\left[1 + b^2 j_1(\boldsymbol{k}) + j_2(\boldsymbol{k})\right]^2 - \left[(1 - b^2)j_1(\boldsymbol{k})\right]^2}\,.$$

In summary, we have found

$$\begin{aligned} H_{\mathrm{AF}}^{(1)} &= E_{\rightarrow\leftarrow} + \sum_k E_k \beta_k^\dagger \beta_k - \sum_k E_k v_k^2 \\ &= \mathcal{N}\left\{-[J_1(\boldsymbol{0})(1 + 2\sin^2\vartheta) + J_2(\boldsymbol{0})]S^2 - \frac{1}{\mathcal{N}}\sum_k E_k\left(v_k^2 + \frac{1}{2}\right)\right\} \\ &\quad + \sum_k E_k\left(\beta_k^\dagger\beta_k + \frac{1}{2}\right) \end{aligned}$$

which, upon using the normalization and the expression of u_k^2 and renaming E_k as $\hbar\omega_k$, gives equation (2.87).

While the operators β_k diagonalize $H_{\mathrm{AF}}^{(1)}$, the original spin operators are expressed in terms of the b_k. We need to invert the relation $\beta_k^\dagger = u_k b_k^\dagger + v_k b_{-k}$, which gives

$$b_k^\dagger = u_k \beta_k^\dagger - v_k \beta_{-k}.$$

With this, we can obtain the representation of the spin density $\boldsymbol{S}(\boldsymbol{q})$ in momentum space in terms of the operators β_k, at linear order. We must use equation (2.86) to transform back from the coordinate system of the $\tilde{\boldsymbol{S}}_{n\nu}$ to that of the $\boldsymbol{S}_{n\nu}$. For the x component, this gives

$$\begin{aligned}
S^x(\boldsymbol{q}) &= \hbar \sum_{n\nu} S^x_{n\nu} e^{-i\boldsymbol{q}\cdot(\boldsymbol{R}_n+\nu\boldsymbol{\tau})} = \hbar \sum_{n\nu} (\sin\vartheta \tilde{S}^x_{n\nu} + \nu \cos\vartheta \tilde{S}^z_{n\nu}) e^{-i\boldsymbol{q}\cdot(\boldsymbol{R}_n+\nu\boldsymbol{\tau})} \\
&= \hbar \sum_{n\nu} \left[\sin\vartheta \sqrt{\frac{S}{2}} (b^\dagger_{n\nu} + b_{n\nu}) + \nu \cos\vartheta (S - b^\dagger_{n\nu} b_{n\nu}) \right] e^{-i\boldsymbol{q}\cdot(\boldsymbol{R}_n+\nu\boldsymbol{\tau})} \\
&= \hbar \sin\vartheta \sqrt{\mathcal{N} S} \left(b^\dagger_{-\boldsymbol{q}} + b_{\boldsymbol{q}} \right) = \hbar \sin\vartheta \sqrt{\mathcal{N} S}\, (u_{\boldsymbol{q}} - v_{\boldsymbol{q}}) \left(\beta^\dagger_{-\boldsymbol{q}} + \beta_{\boldsymbol{q}} \right).
\end{aligned}$$

This component vanishes in zero field, because the contributions of the two sublattices cancel. Similarly

$$\begin{aligned}
S^y(\boldsymbol{q}) &= \hbar \sum_{n\nu} \tilde{S}^y_{n\nu} e^{-i\boldsymbol{q}\cdot(\boldsymbol{R}_n+\nu\boldsymbol{\tau})} = -i\hbar \sqrt{\mathcal{N} S} \left(b^\dagger_{-\boldsymbol{q}} - b_{\boldsymbol{q}} \right) \\
&= -i\hbar \sqrt{\mathcal{N} S}\, (u_{\boldsymbol{q}} + v_{\boldsymbol{q}}) \left(\beta^\dagger_{-\boldsymbol{q}} - \beta_{\boldsymbol{q}} \right) \\
S^z(\boldsymbol{q}) &= \hbar \sum_{n\nu} (-\nu \cos\vartheta \tilde{S}^x_{n\nu} + \sin\vartheta \tilde{S}^z_{n\nu}) e^{-i\boldsymbol{q}\cdot(\boldsymbol{R}_n+\nu\boldsymbol{\tau})} \\
&= \hbar \sum_{n\nu} \left[-\nu \cos\vartheta \sqrt{\frac{S}{2}} (b^\dagger_{n\nu} + b_{n\nu}) + \sin\vartheta (S - b^\dagger_{n\nu} b_{n\nu}) \right] e^{-i\boldsymbol{q}\cdot(\boldsymbol{R}_n+\nu\boldsymbol{\tau})} \\
&= \hbar \sin\vartheta \left(2\mathcal{N} S \delta_{\boldsymbol{q}0} - \sum_{\boldsymbol{k}} b^\dagger_{\boldsymbol{k}} b_{\boldsymbol{k}+\boldsymbol{q}} \right) = 2\hbar \mathcal{N} S \sin\vartheta \delta_{\boldsymbol{q}0}.
\end{aligned}$$

The last equality is valid at lowest order in the $\beta_{\boldsymbol{k}}$ operators. These expressions are useful for expressing the correlation functions of the spin operators in terms of the magnon dispersion, as we do in section 10.3.2.

Appendix B2

Bonus B2.1 Thermodynamics of free quantum particles

The equation of state of a many-particle system in equilibrium is $p/(k_\mathrm{B}T) = \frac{\partial}{\partial \mathscr{V}} \ln Z$, as we saw in section 2.1.2. We consider here free spin-S particles, for which the equation of state can be worked out in closed form by means of special functions called *polylogarithms*. This will also give us the opportunity to encounter other special functions: the Euler gamma and Riemann zeta functions. For independent particles (section 2.1.3), the partition function is $Z = \prod_\alpha (1 - \eta e^{-\beta \xi_\alpha})^{-\eta}$. The single-particle energies $\xi_\alpha = \varepsilon_\alpha - \mu$ with $\varepsilon_\alpha \equiv \hbar^2 k^2/(2m)$ are independent of the volume but the number of these levels is proportional to the volume, such that $\ln Z$ is an extensive function proportional to the volume. We introduce the *fugacity* $z = \eta e^{\beta\mu}$ and we convert the sum to an integral:

$$\frac{p}{k_B T} = \frac{\partial}{\partial \mathscr{V}} \mathscr{V} \underbrace{\frac{1}{\mathscr{V}} \sum_\alpha (-\eta) \ln(1 - z e^{-\beta \varepsilon_\alpha})}_{\text{independent of } \mathscr{V}}$$
$$= -\eta(2S+1) \int_0^\infty d\varepsilon \, \tilde{N}_0(\varepsilon) \ln(1 - z e^{-\beta\varepsilon}). \tag{2.92}$$

$2S + 1$ is the spin degeneracy and $\tilde{N}_0(\varepsilon)$ is the DOS per unit volume for each spin species. On writing the sum as an integral, we have implicitly assumed that the logarithm varies slowly over energies corresponding to the typical inter-level spacing. This assumption breaks down when $ze^{-\beta\varepsilon}$ approaches unity, as occurs in the Bose–Einstein condensate. We shall keep this in mind but ignore it for the time being. For fermions, $z < 0$ and there is no concern. The DOS can be readily evaluated in dimension d using equation (11):

$$\tilde{N}_0(\varepsilon) = \int \frac{d^d k}{(2\pi)^d} \underbrace{\delta\left(\varepsilon - \frac{\hbar^2 k^2}{2m}\right)}_{\frac{m}{\hbar^2 k}\delta\left(k - \sqrt{\frac{2m\varepsilon}{\hbar^2}}\right)} = \frac{m}{\hbar^2} \frac{s_d}{(2\pi)^d} \int_0^\infty dk \frac{k^{d-1}}{k} \delta\left(k - \sqrt{\frac{2m\varepsilon}{\hbar^2}}\right).$$

s_d is the surface of the d-dimensional unit sphere, given in terms of the Euler gamma function by $s_d = 2\pi^{d/2}/\Gamma(d/2)$. The free-particle DOS in dimension d is therefore

$$\tilde{N}_0(\varepsilon) = \tilde{N}_d \varepsilon^{d/2-1}, \qquad \tilde{N}_d = \frac{1}{\Gamma(d/2)} \left(\frac{m}{2\pi\hbar^2}\right)^{d/2}. \tag{2.93}$$

Performing the energy integration by parts, we get for the pressure:

$$\frac{p}{k_B T} = \eta(2S+1)\tilde{N}_d \frac{2\beta}{d} \int_0^\infty d\varepsilon \frac{\varepsilon^{d/2}}{e^{\beta\varepsilon}/z - 1}. \tag{2.94}$$

The chemical potential entering the fugacity must be eliminated in favor of the particle density n which, starting from equation (2.14), can be recast into

$$n = (2S+1) \int_0^\infty d\varepsilon \frac{\tilde{N}_0(\varepsilon)}{e^{\beta\varepsilon}/(z/\eta) - \eta} = \eta(2S+1)\tilde{N}_d \int_0^\infty d\varepsilon \frac{\varepsilon^{d/2-1}}{e^{\beta\varepsilon}/z - 1}. \tag{2.95}$$

Again, special care must be taken when the distribution function varies rapidly, as it does for bosons at low temperature. We see that both equations (2.94) and (2.95) involve analogous integrals. The internal energy density $U/\mathscr{V}$ is obtained similarly: it is the same expression as the density except for one additional factor of ε in the integrand, such that the resulting integral is the same as the one for the pressure. This allows us to readily establish the relation

$$\frac{U}{\mathscr{V}} = \frac{d}{2} p, \tag{2.96}$$

that will be useful below to compute the specific heat. It is a good time for a little mathematical digression, for getting in touch with the polylogarithms and establish the relation

$$\int_0^\infty d\varepsilon \frac{\varepsilon^{p-1}}{e^{\beta\varepsilon}/z - 1} = \frac{\Gamma(p)}{\beta^p}\mathrm{Li}_p(z).$$

$\mathrm{Li}_p(x)$ is the polylogarithm, which may be defined by its Taylor expansion as $\mathrm{Li}_p(x) = \sum_{q=1}^{\infty} x^q/q^p$ with the useful property $d\mathrm{Li}_p(x)/dx = \mathrm{Li}_{p-1}(x)/x$. For the proof, we simply use the expansion $1/(x^{-1} - 1) = \sum_{q=1}^{\infty} x^q$ in order to split the integral and let the expansion of the polylogarithm appear:

$$\int_0^\infty d\varepsilon \frac{\varepsilon^{p-1}}{e^{\beta\varepsilon}/z - 1} = \sum_{q=1}^{\infty} z^q \underbrace{\int_0^\infty d\varepsilon \frac{\varepsilon^{p-1}}{e^{q\beta\varepsilon}}}_{\Gamma(p)/(q\beta)^p} = \frac{\Gamma(p)}{\beta^p}\sum_{q=1}^{\infty}\frac{z^q}{q^p}.$$

We can thus obtain the relation between the fugacity and the density:

$$\mathrm{Li}_{d/2}(z) = \eta\frac{n}{2S+1}\left(\frac{2\pi\hbar^2}{mk_\mathrm{B}T}\right)^{d/2}. \tag{2.97}$$

Finally, introducing the reciprocal function of the polylogarithm to solve equation (2.97) for z and using the relation $\Gamma(d/2+1) = (d/2)\Gamma(d/2)$, we find the equation of state

$$\frac{p}{k_\mathrm{B}T} = (2S+1)\left(\frac{mk_\mathrm{B}T}{2\pi\hbar^2}\right)^{d/2}\eta\mathrm{Li}_{d/2+1}\left\{\mathrm{Li}_{d/2}^{-1}\left[\eta\frac{n}{2S+1}\left(\frac{2\pi\hbar^2}{mk_\mathrm{B}T}\right)^{d/2}\right]\right\}. \tag{2.98}$$

Equation (2.98) is not quite transparent in this form, but has the merit of being the solution of our problem, valid both for bosons (above the condensation temperature) and for fermions, and in any dimension d. Knowing the pressure, we can deduce all thermodynamic functions by differentiation of the free-energy density $\varpi = -p$, see equation (2.8). Since we know the pressure as a function of the variables T and z, $p = p(T, z) \propto T^{d/2+1}\mathrm{Li}_{d/2+1}(z)$, and $z = z(T, \mu)$, the entropy density can be readily shown to be (see equation (2.7))

$$\frac{S}{\mathcal{V}} = \left.\frac{dp}{dT}\right)_\mu = \frac{\partial p}{\partial T} + \frac{\partial p}{\partial z}\frac{\partial z}{\partial T} = \left(\frac{d}{2}+1\right)\frac{p}{T} - nk_\mathrm{B}\ln(\eta z), \tag{2.99}$$

where we have used the relation (2.97) between n and $\mathrm{Li}_{d/2}(z)$ to eliminate the latter. Alternatively, we could obtain the same result using equation (2.10):

$$n = \left.\frac{dp}{d\mu}\right)_T = \frac{\partial p}{\partial z}\frac{\partial z}{\partial \mu}, \qquad \frac{\partial z}{\partial \mu} = \frac{z}{k_\mathrm{B}T}, \qquad \frac{\partial z}{\partial T} = -\frac{z\mu}{k_\mathrm{B}T^2} = -\frac{z}{T}\ln(\eta z),$$

which yields $(\partial p/\partial z)(\partial z/\partial T) = -nk_\mathrm{B}\ln(\eta z)$. The specific heat can be calculated from equation (2.9), but it is easier and equivalent to deduce it from the internal energy (2.96):

$$\frac{C_{\mathscr{V}}}{\mathscr{V}} = \left.\frac{d(U/\mathscr{V})}{dT}\right)_{\mathscr{V}} = \frac{d}{2}\left[\frac{\partial p}{\partial T} + \frac{\partial p}{\partial z}\left(\frac{\partial z}{\partial T} + \frac{\partial z}{\partial \mu}\frac{d\mu}{dT}\right)\right] = \frac{d}{2}\left(\frac{S}{\mathscr{V}} + n\frac{d\mu}{dT}\right).$$

We can express $d\mu/dT$ in terms of the isothermal compressibility given by equation (2.11). From the relation $\mu = k_{\mathrm{B}}T\ln(\eta z)$ we deduce

$$\frac{d\mu}{dT} = k_{\mathrm{B}}\ln(\eta z) + \frac{k_{\mathrm{B}}T}{z}\frac{dz}{dT}.$$

We now differentiate equation (2.97) with respect to T and get:

$$\frac{\mathrm{Li}_{d/2-1}(z)}{z}\frac{dz}{dT} = \eta\frac{n}{2S+1}\left(\frac{2\pi\hbar^2}{mk_{\mathrm{B}}T}\right)^{d/2}\frac{-d/2}{T}.$$

We then eliminate $\mathrm{Li}_{d/2-1}(z)$ in favor of the compressibility. Using the expression of the density $n = n(T, z)$ given by equation (2.97), we have

$$\kappa_T = \left.\frac{1}{n^2}\frac{dn}{d\mu}\right)_T = \frac{1}{n^2}\frac{\partial n}{\partial z}\frac{\partial z}{\partial \mu} = \frac{2S+1}{n^2 k_{\mathrm{B}}T}\left(\frac{mk_{\mathrm{B}}T}{2\pi\hbar^2}\right)^{d/2}\eta\mathrm{Li}_{d/2-1}(z). \tag{2.100}$$

Injecting this in the previous relation, we see that $(k_{\mathrm{B}}T/z)dz/dT = -d/(2n\kappa_T T)$, such that the specific heat, like the entropy (2.99) and the compressibility (2.100), becomes an explicit function of T and z:

$$\frac{C_{\mathscr{V}}}{\mathscr{V}} = \frac{d}{2}\left(\frac{S}{\mathscr{V}} + nk_{\mathrm{B}}\ln(\eta z) - \frac{d}{2\kappa_T T}\right). \tag{2.101}$$

We are now in a good position to plot the thermodynamic quantities as a function of temperature for a given density n. We first solve equation (2.97) to get z, and we can then compute the pressure with equation (2.98), the isothermal compressibility with equation (2.100), the entropy with equation (2.99), and the isochoric specific heat with equation (2.101). We will see below that all these quantities can be plotted as universal curves independent of the density and spin if we use the proper normalization. We will also add the correction required in order to describe the Bose–Einstein condensation.

Before doing so, let us derive a few asymptotic results. To this end, we rewrite equation (2.98) in the form

$$\frac{p}{k_{\mathrm{B}}T} = \tau\mathscr{L}_d^{\eta}\left(\frac{n}{\tau}\right), \quad \tau = (2S+1)\left(\frac{mk_{\mathrm{B}}T}{2\pi\hbar^2}\right)^{d/2}, \quad \mathscr{L}_d^{\eta}(x) = \eta\mathrm{Li}_{d/2+1}\left(\mathrm{Li}_{d/2}^{-1}(\eta x)\right).$$

We expect the gas to be classical at high temperature, that is, $p\mathscr{V} = Nk_{\mathrm{B}}T$ should be recovered irrespective of the particle statistics and spin. Indeed, because $\mathrm{Li}_p(x) = x$ for $x \to 0$, we have $\mathscr{L}_d^{\eta}(x \to 0) = x$ and for large T we get $p/(k_{\mathrm{B}}T) = n$. Corrections to this high-T classical behavior can be obtained by including higher-order terms in the expansion of $\mathrm{Li}_p(x)$. The leading correction turns out to be $\mathscr{L}_d^{\eta}(x \to 0) = x - \eta x^2/2^{d/2+1}$, yielding the high-$T$ expansion

$$p\mathscr{V} = Nk_{\rm B}T\left[1 - \frac{\eta}{2}\frac{N/\mathscr{V}}{2S+1}\left(\frac{\pi\hbar^2}{mk_{\rm B}T}\right)^{d/2} + \cdots\right].$$

For bosons, the effect of quantum statistics is to decrease the pressure with respect to a classical gas, while for fermions the pressure is increased due to Pauli exclusion. The formula nicely illustrates the general principle that quantum particles behave more classically at high temperature, high dimensionality, high mass, and high spin. In the high-T classical regime, the behavior of the compressibility is simply $\kappa_T = 1/p$, as can be deduced from equation (2.11) using $\mathscr{V} = Nk_{\rm B}T/p$. For the entropy, we can replace p/T in equation (2.99) by $nk_{\rm B}$ and use the fact that $\mathrm{Li}_{d/2}(z) \to 0$ (see equation (2.97))—which implies $z \to 0$ and therefore $\mathrm{Li}_{d/2}(z) = z$—to deduce

$$\frac{S}{Nk_{\rm B}} = \frac{d}{2} + 1 + \ln\left[\frac{2S+1}{n}\left(\frac{mk_{\rm B}T}{2\pi\hbar^2}\right)^{d/2}\right]. \tag{2.102}$$

This generalizes the Sackur–Tetrode equation, which gives the entropy of an ideal classical gas, to any dimension and to the case where the particles have a spin degree of freedom. Using equation (2.9), we can check that the classical equipartition result $C_{\mathscr{V}} = (d/2)Nk_{\rm B}$ is reached at high T.

The low-T regime of a degenerate gas is much more interesting, and allows one to explore the richness of quantum assemblies captured by that of polylogarithms. At low temperature, we are interested in the function $\mathscr{L}_d^{\eta}(x)$ at large x. This requires us to study $\mathrm{Li}_{d/2}^{-1}(x)$ at large positive x for bosons and at large negative x for fermions. We have to treat bosons and fermions separately from here on: let us start with fermions.

We can use the asymptotic form $\mathrm{Li}_p(-e^x) = -x^p/\Gamma(p+1)$ valid for $x \to \infty$ in order to deduce $\mathrm{Li}_{d/2}^{-1}(-x) = -\exp\{[\Gamma(d/2+1)x]^{2/d}\}$ in the same limit. Making use of the asymptotic expansion once more, we readily find

$$\mathscr{L}_d^{-}(x \to \infty) = \frac{[\Gamma(d/2+1)]^{2/d}}{d/2+1}x^{2/d+1},$$

such that the equation of state becomes independent of temperature:

$$p\mathscr{V} = Nk_{\rm B}T_0, \qquad T_0 = \frac{[\Gamma(d/2+1)]^{2/d}}{d/2+1}\frac{2\pi\hbar^2}{mk_{\rm B}}\left(\frac{n}{2S+1}\right)^{2/d} = \frac{T_{\rm F}}{d/2+1}.$$

Here we have defined some degeneracy temperature T_0 and introduced the Fermi temperature $T_{\rm F} = \varepsilon_{\rm F}/k_{\rm B}$, with $\varepsilon_{\rm F} = \frac{2\pi\hbar^2}{m}[\Gamma(d/2+1)n/(2S+1)]^{2/d}$ the value of the chemical potential at $T = 0$. While the pressure of a classical gas vanishes at zero temperature, in a gas of fermions the Pauli exclusion maintains a finite pressure even at $T = 0$. This 'Pauli pressure' $\propto n^{2/d+1}$ disappears when the density approaches zero, as one understands, or for a large spin as the fermions can circumvent the exclusion principle thanks to the additional spin degrees of freedom. The Pauli pressure is responsible for stabilizing the neutron stars against gravitational collapse. The low-T

expansions are difficult to perform rigorously[11], because the behavior of $\mathrm{Li}_p(-e^x)$ is unknown beyond the asymptotic form used above. Approximations schemes such as the Sommerfeld expansion allow one to obtain asymptotic results that are correct up to a contribution which is exponentially small in $k_\mathrm{B}T/\varepsilon_\mathrm{F}$. For example, equation (2.56) obtained by neglecting the temperature dependence of the chemical potential is correct to linear order in T, because the variation of μ is second-order in T apart from this exponentially small term. Equation (2.56) can be recast as

$$\frac{C_{\mathcal{V}}}{Nk_\mathrm{B}} = \frac{d}{d+2}\frac{\pi^2}{3}\frac{T}{T_0}.$$

Since this expression also neglects the energy dependence of the DOS, it is accurate for $d = 2$ where the DOS is constant, but deviations from linearity at low-T are clearly visible in figure 2.5 for $d = 1$ and $d = 3$.

Let us now turn to bosons. The function $\mathrm{Li}_p(x)$ is monotonic and has the sign of its argument. It diverges at $x = 1$ for $p \leqslant 1$, such that $\mathrm{Li}_p^{-1}(x)$ is well-defined for any positive x if $p \leqslant 1$, that is, in dimensions one and two. If $p > 1$, however, $\mathrm{Li}_p(1) = \zeta(p)$, where $\zeta(x)$ is the Riemann zeta function, and $\mathrm{Li}_p(x > 1)$ is imaginary. Therefore, in three dimensions $\mathrm{Li}_{3/2}^{-1}(x)$ is not defined for $x > \zeta(3/2)$. This signals the Bose–Einstein condensation. For $n/\tau > \zeta(3/2)$, we can no longer ignore—as we did with the continuous approximation—that the occupation of the lowest-energy state becomes large, that is, of order N. Solving the equation $n/\tau = \zeta(3/2)$ for T yields the well-known expression for the condensation temperature:

$$T_c = \frac{1}{[\zeta(3/2)]^{2/3}}\frac{2\pi\hbar^2}{mk_\mathrm{B}}\left(\frac{n}{2S+1}\right)^{2/3} = 1.09T_0.$$

In dimensions one and two, this criterion gives a vanishing T_c for the Bose–Einstein condensation.

We now wish to plot the various thermodynamic quantities and compare with the asymptotic results. We must correct our expressions in order to take into account the macroscopic occupation of the ground state. The contribution of the state with $\varepsilon_\alpha = 0$ to the pressure is (see equation (2.92))

$$\frac{p_0}{k_\mathrm{B}T} = -\frac{2S+1}{\mathcal{V}}\eta\ln(1-z).$$

This contribution disappears in the thermodynamic limit, unless z becomes unity. Because $\mu \leqslant 0$ for bosons ($z \leqslant 1$), the factor $\ln(1 - ze^{-\beta\varepsilon_\alpha})$ can only diverge if $\varepsilon_\alpha = 0$, justifying to treat this level as discrete while using a continuous description for all other levels. The contribution p_0 was altogether discarded in dimension $d = 3$ on converting the sum over states to an integral, because the DOS $\tilde{N}_0(0) = 0$. We must therefore reintroduce it explicitly. For $d = 2$ and $d = 1$, the correction is in principle not needed, but adding it is harmless. Hence we add the term p_0 to the pressure in

[11] See Weinstock R 1969 *Am. J. Phys.* **37** 1273.

what follows. It is convenient to measure the temperatures in units of the degeneracy temperature, so we introduce

$$t = \frac{T}{T_0}, \qquad T_0 = A_d^{2/d}\frac{2\pi\hbar^2}{mk_B}\left(\frac{n}{2S+1}\right)^{2/d}, \qquad A_d = \frac{\Gamma(d/2+1)}{(d/2+1)^{d/2}}. \tag{2.103}$$

It is then easy to check that the pressure measured in units of nk_BT_0 is a function that depends on the density only implicitly through t:

$$\frac{p\mathscr{V}}{Nk_BT_0} = A_d t^{d/2+1}\eta\mathrm{Li}_{d/2+1}(z) - \frac{2S+1}{N}t\eta\ln(1-z). \tag{2.104}$$

The last term is p_0, which of course depends explicitly on the number of particles. Evaluating the density $n = (\partial p/\partial z)(\partial z/\partial\mu)$, we find the equation giving the fugacity as a function of the reduced temperature:

$$1 = A_d\, t^{d/2}\,\eta\mathrm{Li}_{d/2}(z) + \frac{2S+1}{N}\frac{\eta z}{1-z}. \tag{2.105}$$

The compressibility $\kappa_T = (1/n^2)(\partial n/\partial z)(\partial z/\partial\mu)$ is also a universal function of t and z if measured in units of $(nk_BT_0)^{-1}$:

$$\kappa_T nk_BT_0 = A_d t^{d/2-1}\eta\mathrm{Li}_{d/2-1}(z) + \frac{2S+1}{N}\frac{1}{t}\frac{\eta z}{(1-z)^2}. \tag{2.106}$$

The entropy (2.99), including the correction due to p_0 and measured in units of Nk_B, becomes

$$\frac{S}{Nk_B} = \frac{d/2+1}{t}\frac{p\mathscr{V}}{Nk_BT_0} - \ln(\eta z) + \frac{d}{2}\frac{2S+1}{N}\eta\ln(1-z). \tag{2.107}$$

The specific heat is finally obtained following the same logic:

$$\frac{C_{\mathscr{V}}}{Nk_B} = d/2\left[\frac{S}{Nk_B} + \ln(\eta z) - \frac{d}{2\kappa_T nk_BT_0}\frac{1}{t}\left(1 - \frac{2S+1}{N}\frac{\eta z}{1-z}\right)\right]. \tag{2.108}$$

The *Mathematica* program displayed in figure 2.4 implements these equations and outputs the data represented in figure 2.5 for the pressure, compressibility, entropy, and specific heat of fermions and bosons in dimensions one, two, and three. The known $T = 0$ results and high-T asymptotic behaviors are also shown for comparison.

Bonus B2.2 Empirical pseudopotentials for semiconductors

In semiconductors, the potential felt by valence and conduction electrons varies relatively slowly in space, in contrast to the potential felt by core electrons: the potential for valence and conduction electrons can be represented by a small number of components in the expansion (2.51). The idea of the empirical pseudopotential method is to determine these few components $V(\boldsymbol{G})$ of the potential in such a way

```
A[d_]:=Gamma[d/2+1]/(d/2+1)^(d/2)
z[eta_,d_,t_,N0_:10^6]     :=Re[x/.FindRoot[1-A[d]*t^(d/2)*eta*PolyLog[d/2,x]
                              -If[eta>0,1/N0*x/(1-x),0],{x,If[eta>0,1-1/N0,-2]}]]
p[eta_,d_,t_,N0_:10^6,z_]:=A[d]*t^(d/2+1)*eta*PolyLog[d/2+1,z]-If[eta>0,t/N0*Log[1-z],0]
p[eta_,d_,t_,N0_:10^6]     :=Module[{z=z[eta,d,t,N0]},p[eta,d,t,N0,z]]
k[eta_,d_,t_,N0_:10^6,z_]:=A[d]*t^(d/2-1)*eta*PolyLog[d/2-1,z]+If[eta>0,1/N0*1/t*z/(1-z)^2,0]
k[eta_,d_,t_,N0_:10^6]     :=Module[{z=z[eta,d,t,N0]},k[eta,d,t,N0,z]]
s[eta_,d_,t_,N0_:10^6,z_]:=(d/2+1)*p[eta,d,t,N0,z]/t-Log[eta*z]+If[eta>0,1/N0*d/2*Log[1-z],0]
s[eta_,d_,t_,N0_:10^6]     :=Module[{z=z[eta,d,t,N0]},s[eta,d,t,N0,z]]
c[eta_,d_,t_,N0_:10^6]     :=Module[{z=z[eta,d,t,N0]},d/2*(s[eta,d,t,N0,z]+Log[eta*z]
                              -d/(2*k[eta,d,t,N0,z]*t)*(1-If[eta>0,1/N0*z/(1-z),0]))]

Export[NotebookDirectory[]<>"fermions.dat",Union[N[{{0,1,1,1,1/3,1/2,3/5,0,0,0,0,0,0}}],
  Chop[Table[{t,p[-1,1,t],p[-1,2,t],p[-1,3,t],k[-1,1,t],k[-1,2,t],k[-1,3,t],
  s[-1,1,t],s[-1,2,t],s[-1,3,t],c[-1,1,t],c[-1,2,t],c[-1,3,t]},{t,0.02,4,0.02}]]]];
Export[NotebookDirectory[]<>"bosons.dat",Union[N[{{0,0,0,0,0,0,0,0,0,0,0,0,0}}],
  Chop[Table[{t,p[1,1,t],p[1,2,t],p[1,3,t],1/k[1,1,t],1/k[1,2,t],1/k[1,3,t],
  s[1,1,t],s[1,2,t],s[1,3,t],c[1,1,t],c[1,2,t],c[1,3,t],c[1,3,t,10^2]},{t,0.02,4,0.02}]]]];
```

Figure 2.4. Plotting the thermodynamic functions of free particles with *Mathematica*.

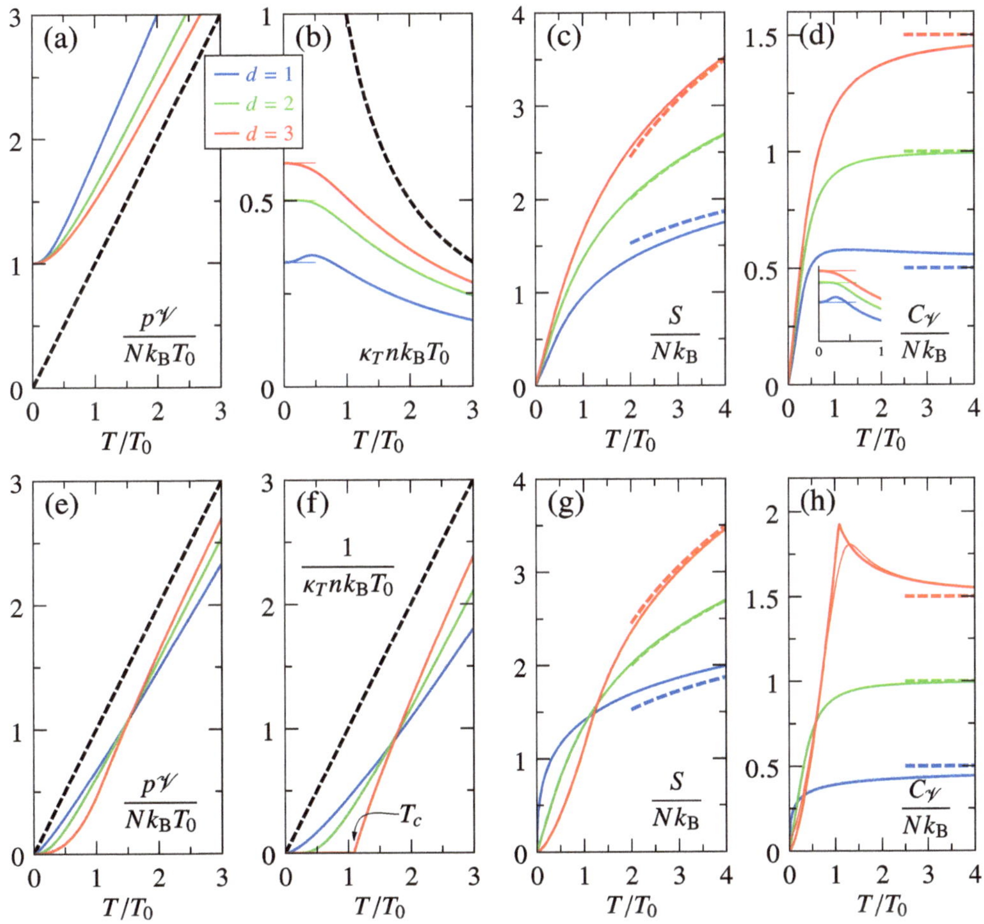

Figure 2.5. Thermodynamic functions of free fermions, (a)–(d), and free bosons, (e)–(h), in dimensions one (blue), two (green), and three (red). The thick dashed lines show the results for the ideal classical gas. In (b), the zero-temperature value of the compressibility is $d/(d+2)$ shown by the horizontal rules. The inset in (d) shows $C_{\mathscr{V}}/(Nk_B) \times T_0/T$ and the limiting values $d/(d+2) \times \pi^2/3$. Note that (f) shows the inverse of the compressibility. All results in (e)–(h) are calculated for $N/(2S+1) = 10^6$, except the thin red line in (h) for which $N/(2S+1) = 10^2$.

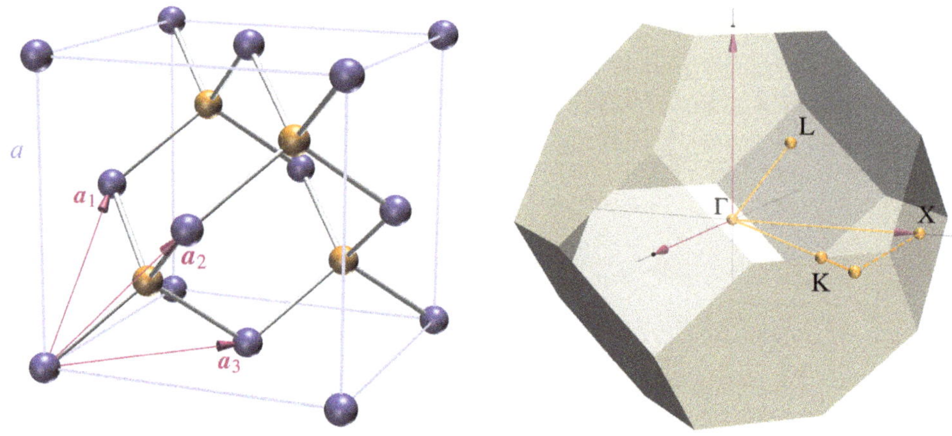

Figure 2.6. Left: zinc blende crystal structure. The two fcc sublattices are shown with different colors. The basis of the elementary cell and the fcc lattice parameter are also indicated. Right: Brillouin zone of the fcc lattice. The high-symmetry points are $\Gamma = (0,0,0)$, $X = (0, 1, 0)$, $L = (\frac{1}{2}, \frac{1}{2}, \frac{1}{2})$, and $K = (\frac{3}{4}, \frac{3}{4}, 0)$, in units of $2\pi/a$.

that the resulting bands fit some prominent experimental features like the band gap. This procedure has been applied by Cohen and Bergstresser[12] to fourteen semiconductors of the IV–IV (Si, Ge, Sn), III–V (AlSb, GaAs, GaP, GaSb, InAs, InP, InSb), and II–VI (CdTe, ZnS, ZnSe, ZnTe) families. The result is a very easy method to calculate the bands in these semiconductors.

These semiconductors crystallize in the zinc blende structure shown in figure 2.6. There are two inequivalent atomic sites in the unit cell, each forming an fcc sublattice. The two fcc sublattices are a vector $\boldsymbol{\tau} = a(\frac{1}{4}, \frac{1}{4}, \frac{1}{4})$ apart, with a the fcc lattice parameter. Denoting $V_1(\boldsymbol{G})$ and $V_2(\boldsymbol{G})$ the potentials originating from the two atomic sites and locating the origin between the two atoms, we have

$$\begin{aligned} V(\boldsymbol{r}) &= V_1(\boldsymbol{r} + \boldsymbol{\tau}/2) + V_2(\boldsymbol{r} - \boldsymbol{\tau}/2) \\ &= \sum_{\boldsymbol{G}} [V_1(\boldsymbol{G}) e^{i\boldsymbol{G}\cdot(\boldsymbol{r}+\boldsymbol{\tau}/2)} + V_2(\boldsymbol{G}) e^{i\boldsymbol{G}\cdot(\boldsymbol{r}-\boldsymbol{\tau}/2)}] \\ &= \sum_{\boldsymbol{G}} [V_1(\boldsymbol{G}) e^{i\boldsymbol{G}\cdot\boldsymbol{\tau}/2} + V_2(\boldsymbol{G}) e^{-i\boldsymbol{G}\cdot\boldsymbol{\tau}/2}] e^{i\boldsymbol{G}\cdot\boldsymbol{r}} \\ &= \sum_{\boldsymbol{G}} [V_s(\boldsymbol{G}) \cos(\boldsymbol{G}\cdot\boldsymbol{\tau}/2) + i V_a(\boldsymbol{G}) \sin(\boldsymbol{G}\cdot\boldsymbol{\tau}/2)] e^{i\boldsymbol{G}\cdot\boldsymbol{r}}. \end{aligned}$$

We have introduced the symmetric and antisymmetric parts of the potential, $V_s(\boldsymbol{G}) = [V_1(\boldsymbol{G}) + V_2(\boldsymbol{G})]/2$ and $V_a(\boldsymbol{G}) = [V_1(\boldsymbol{G}) - V_2(\boldsymbol{G})]/2$. Clearly $V_a(\boldsymbol{G}) = 0$ for the nonpolar IV–IV compounds. Following Cohen and Bergstresser, we assume spherical symmetry for the potential such that $V_s(\boldsymbol{G}) \equiv V_s(G^2)$ and $V_a(\boldsymbol{G}) \equiv V_a(G^2)$. The Fourier components of the potential therefore read:

[12] Cohen M L and Bergstresser T K 1966 *Phys. Rev.* **141** 789.

```
1  P[AlSb]:={4.003,-2.86,+0.27,+0.82,+0.82,+0.54,+0.27}
2  P[CdTe]:={3.661,-2.72, 0.00,+0.54,+2.04,+1.22,+0.54}
3  P[GaAs]:={4.729,-3.13,+0.14,+0.82,+0.95,+0.68,+0.14}
4  P[GaP] :={5.083,-2.99,+0.41,+0.95,+1.63,+0.95,+0.27}
5  P[GaSb]:={4.016,-2.99, 0.00,+0.68,+0.82,+0.68,+0.14}
6  P[Ge]  :={4.695,-3.13,+0.14,+0.82, 0.00, 0.00, 0.00}
7  P[InAs]:={4.123,-2.99, 0.00,+0.68,+1.09,+0.68,+0.41}
8  P[InP] :={4.380,-3.13,+0.14,+0.82,+0.95,+0.68,+0.14}
9  P[InSb]:={3.582,-2.72, 0.00,+0.54,+0.82,+0.68,+0.14}
10 P[Si]  :={5.101,-2.86,+0.54,+1.09, 0.00, 0.00, 0.00}
11 P[Sn]  :={3.571,-2.72, 0.00,+0.54, 0.00, 0.00, 0.00}
12 P[ZnS] :={5.139,-2.99,+0.41,+0.95,+3.27,+1.90,+0.54}
13 P[ZnSe]:={4.712,-3.13,+0.14,+0.82,+2.45,+1.63,+0.41}
14 P[ZnTe]:={4.082,-2.99, 0.00,+0.68,+1.77,+1.36,+0.14}
15 P[S_]  :={5.101, 0.00, 0.00, 0.00, 0.00, 0.00, 0.00}

16 V[S_,G_]:=Vs[S,G.G]Cos[G.{1,1,1}Pi/4]+I*Va[S,G.G]Sin[G.{1,1,1}Pi/4]
17 Vs[S_,3]:=P[S][[2]];Vs[S_,8]:=P[S][[3]];Vs[S_,11]:=P[S][[4]];Vs[S_,G_]:=0
18 Va[S_,3]:=P[S][[5]];Va[S_,4]:=P[S][[6]];Va[S_,11]:=P[S][[7]];Va[S_,G_]:=0

19 Levels[S_,k_]:=Module[{G,h},
20   G=Select[Flatten[Table[{-i+j+l,i-j+l,i+j-l},{i,-4,4},{j,-4,4},{l,-4,4}],2],(k+#).(k+#)<21&];
21   h=Outer[If[#1==#2,P[S][[1]](k+#1).(k+#1),V[S,#1-#2]]&,G,G,1,1];
22   Sort[Re[Eigenvalues[h]]][[Range[20]]]
23 ]

24 K=N[Flatten[{Table[{Sqrt[3]i/17,{1,1,1}(17-i)/34},{i,0,17}],
25             Table[{Sqrt[3]+2*i/17,{1,0,0}i/17},{i,1,17}],
26             Table[{Sqrt[3]+2+Sqrt[8]*i/27,{1,1,0}(1-i/27)},{i,1,27}]},1]];

27 S=ZnSe;EF=Levels[S,{0,0,0}][[4]];
28 Export[ToString[S]<>".dat",Join[{#[[1]]},Levels[S,#[[2]]]-EF]&/@K];
```

Figure 2.7. Implementation of the empirical pseudopotential method in *Mathematica*.

$$V(\boldsymbol{G}) = V_s(G^2)\cos(\boldsymbol{G}\cdot\boldsymbol{\tau}/2) + iV_a(G^2)\sin(\boldsymbol{G}\cdot\boldsymbol{\tau}/2).$$

As illustrated in figure 2.6, the elementary cell of the direct lattice is defined by the vectors $\boldsymbol{a}_1 = a(0, \frac{1}{2}, \frac{1}{2})$, $\boldsymbol{a}_2 = a(\frac{1}{2}, 0, \frac{1}{2})$, and $\boldsymbol{a}_3 = a(\frac{1}{2}, \frac{1}{2}, 0)$. The corresponding reciprocal lattice vectors are $\boldsymbol{b}_1 = \frac{2\pi}{a}(-1, 1, 1)$, $\boldsymbol{b}_2 = \frac{2\pi}{a}(1, -1, 1)$, and $\boldsymbol{b}_3 = \frac{2\pi}{a}(1, 1, -1)$. Thus the $\boldsymbol{G}$ vectors take the form

$$\boldsymbol{G} = \frac{2\pi}{a}(-i + j + l,\ i - j + l,\ i + j - l), \qquad \{i, j, l\} \in \mathbb{N}.$$

In units of $(2\pi/a)^2$, the squares of these vectors are $G^2 = 0, 3, 4, 8, 11, 12\ldots$ The Fourier component $V(G^2 = 0)$ sets a global shift in energy and can be put to zero without loss of generality. It turns out that, for all vectors such that $G^2 = 4$, we have $\cos(\boldsymbol{G}\cdot\boldsymbol{\tau}/2) = 0$ and for all vectors such that $G^2 = 8$, we have $\sin(\boldsymbol{G}\cdot\boldsymbol{\tau}/2) = 0$. Hence we retain only the following six Fourier components to represent the potential: $V_s(3)$, $V_s(8)$, $V_s(11)$, $V_a(3)$, $V_a(4)$, and $V_a(11)$.

The *Mathematica* program reproduced in figure 2.7 implements the band-structure calculation for the fourteen semiconductors. Lines 1–14 define the parameters for each semiconductor. The first parameter is the quantity $(2\pi/a)^2\hbar^2/(2m)$ expressed in electron-volts (eV). The subsequent six parameters are the symmetric and antisymmetric Fourier components of the potential as given in table II of Cohen and Bergstresser, which we have converted from Rydbergs to eV. Line 15 defines default values for the parameters: the same lattice parameter as Si and a vanishing potential; this is useful for drawing the bands that would correspond

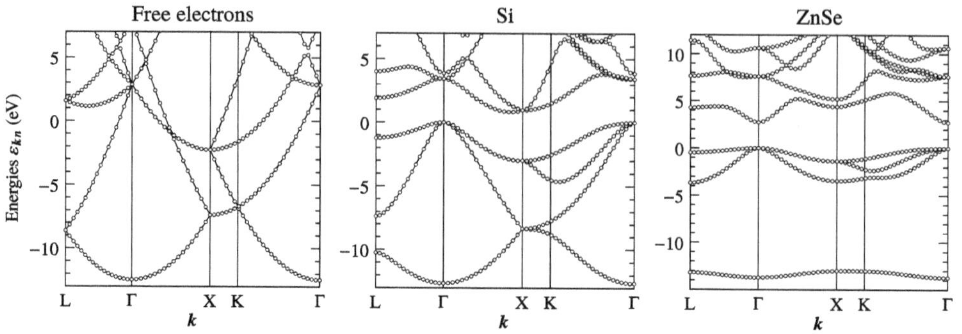

Figure 2.8. Band structures on the zinc blence lattice, calculated with the code of figure 2.7. Left: folded free-electron bands; the Fermi level is such that the electron density is that of Si, i.e. $n = 16/a_{\mathrm{Si}}^3$. Middle: Si, a typical nonpolar, small-indirect-band-gap compound. Right: ZnSe, a typical polar, wide-direct-band-gap compound.

to free electrons constrained to move on the Si lattice. Lines 16–18 define the potential $V(\boldsymbol{G})$: the generic formula is given at line 16, while lines 17 and 18 make the assignment of the symmetric and antisymmetric components, respectively. Unassigned components are set explicitly to zero by the instructions `Vs[S_,G_]:=0` and `Va[S_,G_]:=0`.

Lines 19–23 define the routine which computes the energy levels for one particular semiconductor (argument `S`) and one $\boldsymbol{k}$ point (argument `k`). In line 20, the set of $\boldsymbol{G}$ vectors is constructed: only vectors such that $|\boldsymbol{k}+\boldsymbol{G}|^2 < 21$—in units of $(2\pi/a)^2$—are retained in the basis. This is the same cutoff as in Cohen and Bergstresser; it leads to a basis of typically 100 $\boldsymbol{G}$ vectors. Line 21 builds the Hamiltonian given in the square bracket of equation (2.52). Since we work with vectors $\boldsymbol{k}$ and $\boldsymbol{G}$ expressed in units of $2\pi/a$, we must multiply the kinetic part $|\boldsymbol{k}+\boldsymbol{G}|^2$ by $(2\pi/a)^2\hbar^2/(2m)$. Line 22 returns the 20 lowest eigenvalues of the Hamiltonian. We take the real part in order to eliminate any small imaginary part that could appear due to numerical inaccuracy. Thus, for instance, the command `Levels[AlSb,{1,0,0}]` returns the AlSb bands in eV at $\boldsymbol{k} = \frac{2\pi}{a}(1, 0, 0)$.

Lines 24–26 are somewhat tedious: their role is to define the path in the Brillouin zone along which the bands are plotted. It is customary to plot the bands along high-symmetry lines of the fcc reciprocal lattice, as illustrated in figure 2.6. For our semiconductors, these lines are (i) from point L to point Γ, (ii) from Γ to X, and (iii) from the point $\frac{2\pi}{a}(1, 1, 0)$—which is outside the first Brillouin zone but has energy levels identical to those in X—to Γ, crossing the zone boundary at point K. The list `K` defined at line 24 is a series of elements $\{\ell, \{k_x, k_y, k_z\}\}$ where ℓ is a continuous abscissa to be used in the band diagram (see figure 2.8) and (k_x, k_y, k_z) is the corresponding $\boldsymbol{k}$ point.

Finally, lines 27 and 28 perform the actual calculation (for the semiconductor ZnSe in this case). The Fermi energy is set equal to the fourth level at the Γ point, i.e. the top of the valence band. Line 28 builds the list of levels associated with each abscissa ℓ_i, and saves it to file 'ZnSe.dat'. Executing this code on modern computers will take typically a few seconds.

In figure 2.8, we display the result for free electrons, Si, and ZnSe. The free-electron diagram shows the quadratic free-electron band folded to fit into the fcc Brillouin zone. Comparing with Si, one sees that much of the actual band structure of the small-band-gap semiconductors resembles the free-electron result. Si has an indirect band gap, the bottom of the conduction band being somewhere between Γ and X, while the top of the valence band is at Γ. In contrast, ZnSe has a direct band gap at Γ. One also sees that Si, being nonpolar, has additional degeneracies (in particular at the X point) as compared to the polar ZnSe.

Bonus B2.3 Spring models for phonons

Like for electronic bands, much of the apparently complicated phonon dispersion curves in real materials can be understood with the help of simple models that include the crystal symmetry. The simplest of these models envisions atoms connected by a network of classical springs. Each spring of constant K contributes a term $(1/2)Kx^2$ to the elastic energy, where x is the change in spring length. This model can offer a qualitatively correct approximation to the phonon spectrum based on just a small number of spring constants. We provide here a small code that solves any spring model in dimension d and we apply it to Si. According to equations (2.59) and (2.58), the dynamical matrix is:

$$D_{\nu i\mu j}(\boldsymbol{k}) = \frac{1}{\sqrt{M_\nu M_\mu}} \sum_m \left.\frac{\partial^2 U}{\partial u_{\nu i}(\boldsymbol{0})\partial u_{\mu j}(\boldsymbol{R}_m)}\right|_0 e^{i\boldsymbol{k}\cdot\boldsymbol{R}_m}.$$

The elastic energy is a sum of pairwise terms involving the displacements of two atoms connected by a spring. Only the terms including the displacement $\boldsymbol{u}_\nu(\boldsymbol{0})$ contribute to the dynamical matrix as written above. These terms are

$$\sum_{m'\mu'} \frac{1}{2} K_{|\boldsymbol{\tau}_\nu - \boldsymbol{R}_{m'} - \boldsymbol{\tau}_{\mu'}|}\big(|\boldsymbol{\tau}_\nu + \boldsymbol{u}_\nu(\boldsymbol{0}) - \boldsymbol{R}_{m'} - \boldsymbol{\tau}_{\mu'} - \boldsymbol{u}_{\mu'}(\boldsymbol{R}_{m'})| - |\boldsymbol{\tau}_\nu - \boldsymbol{R}_{m'} - \boldsymbol{\tau}_{\mu'}|\big)^2.$$

The sum runs over all atoms connected by a spring to the atom located at $\boldsymbol{\tau}_\nu$ in the central unit cell, $K_{|\dots|}$ is the corresponding spring constant which only depends on the rest length of the bond, and the quantity in parentheses is the change in bond length. When evaluating the derivatives, we must distinguish cases. If $\mu = \nu$ and $\boldsymbol{R}_m = \boldsymbol{0}$, we get contributions from all atoms connected to our central atom at $\boldsymbol{\tau}_\nu$. These contributions are

$$\sum_{m'\mu'} K_{|\boldsymbol{\tau}_\nu - \boldsymbol{R}_{m'} - \boldsymbol{\tau}_{\mu'}|} \frac{(\tau_{\nu i} - R_{m'i} - \tau_{\mu' i})(\tau_{\nu j} - R_{m'j} - \tau_{\mu' j})}{|\boldsymbol{\tau}_\nu - \boldsymbol{R}_{m'} - \boldsymbol{\tau}_{\mu'}|^2}.$$

If $\mu = \nu$ and $\boldsymbol{R}_m \neq \boldsymbol{0}$, the only bond that contributes is the one involving $\boldsymbol{u}_\nu(\boldsymbol{R}_m)$: it gives a term $-K_{|\boldsymbol{R}_m|} R_{mi} R_{mj}/|\boldsymbol{R}_m|^2$. The coefficients of the dynamical matrix corresponding to the atom ν are therefore

```
1  L2[x_]:=x.x
2  R[n_]:=Plus@@(n[[#]]a[#]&/@Range[d])
3  Dynamical[k_]:=ArrayFlatten[Table[1/Sqrt[M[mu]*M[nu]]*Plus@@(
4    K[L2[tau[nu]-#[[1]]-tau[#[[2]]]]]
5    *(KroneckerDelta[mu,nu]-KroneckerDelta[mu,#[[2]]]*Exp[I*k.#[[1]]])
6    *(tau[nu][[i]]-#[[1]][[i]]-tau[#[[2]]][[i]])
7    *(tau[nu][[j]]-#[[1]][[j]]-tau[#[[2]]][[j]])
8    /L2[tau[nu]-#[[1]]-tau[#[[2]]]]&/@
9    Select[{R[#[[1]]],#[[2]]}&/@
10   Tuples[{Tuples[Range[-Shells,Shells],d],Range[Nat]}]
11   ,tau[nu]-#[[1]]-tau[#[[2]]]!=0*Range[d]&])
12 ,{nu,1,Nat},{mu,1,Nat},{i,1,d},{j,1,d}]]

13 d=3; a[1]={0,1,1}/2; a[2]={1,0,1}/2; a[3]={1,1,0}/2;
14 Nat=2; M[1]=28; M[2]=28; tau[1]={0,0,0}; tau[2]={1,1,1}/4;
15 Shells=1; K[3/16]=1870; K[1/2]=214; K[L2_]:=0
16 DN[kx_,ky_,kz_]=N[Dynamical[{kx*2*Pi,ky*2*Pi,kz*2*Pi}]];
17 Phonons[kx_,ky_,kz_]:=Sort[Re[Sqrt[Eigenvalues[DN[kx,ky,kz]]]]]
18 BZ=N[Flatten[{Table[{Sqrt[3]i/17,{1,1,1}(17-i)/34},{i,0,17}],
19               Table[{Sqrt[3]+2*i/17,{1,0,0}i/17},{i,1,17}],
20               Table[{Sqrt[3]+2+Sqrt[8]*i/27,{1,1,0}(1-i/27)},{i,1,27}]},1]];
21 Export["Si.dat",Join[{#[[1]]},Phonons[#[[2,1]],#[[2,2]],#[[2,3]]]]&/@BZ];
```

Figure 2.9. Solving spring models with *Mathematica*. The lines 1–12 can provide the dynamical matrix analytically in any dimension d. The lines 13–21 calculate and save the phonon spectrum for Si.

$$D_{\nu i\nu j}(\boldsymbol{k}) = \frac{1}{M_\nu}\left[\sum_{m\mu} K_{|\boldsymbol{\tau}_\nu - \boldsymbol{R}_m - \boldsymbol{\tau}_\mu|}\frac{(\tau_{\nu i} - R_{mi} - \tau_{\mu i})(\tau_{\nu j} - R_{mj} - \tau_{\mu j})}{|\boldsymbol{\tau}_\nu - \boldsymbol{R}_m - \boldsymbol{\tau}_\mu|^2}\right.$$
$$\left. - \sum_m K_{|\boldsymbol{R}_m|}\frac{R_{mi}R_{mj}}{|\boldsymbol{R}_m|^2}e^{i\boldsymbol{k}\cdot\boldsymbol{R}_m}\right].$$

If $\mu \neq \nu$, we are again considering two different atoms such that only one bond contributes. The 'off-diagonal' dynamical matrix results as

$$D_{\nu i\mu j}(\boldsymbol{k}) = -\frac{1}{\sqrt{M_\nu M_\mu}}\sum_m K_{|\boldsymbol{\tau}_\nu - \boldsymbol{R}_m - \boldsymbol{\tau}_\mu|}\frac{(\tau_{\nu i} - R_{mi} - \tau_{\mu i})(\tau_{\nu j} - R_{mj} - \tau_{\mu j})}{|\boldsymbol{\tau}_\nu - \boldsymbol{R}_m - \boldsymbol{\tau}_\mu|^2}e^{i\boldsymbol{k}\cdot\boldsymbol{R}_m}.$$

These formula for $D(\boldsymbol{k})$ depend on the atomic masses, spring constants, and equilibrium atomic positions and they fully determine the model. We can gather both of them again into the single expression

$$D_{\nu i\mu j}(\boldsymbol{k}) = \frac{1}{\sqrt{M_\nu M_\mu}}\sum_{m\mu'} K_{|\boldsymbol{\tau}_\nu - \boldsymbol{R}_m - \boldsymbol{\tau}_{\mu'}|}\left(\delta_{\mu\nu} - \delta_{\mu\mu'}e^{i\boldsymbol{k}\cdot\boldsymbol{R}_m}\right)$$
$$\times\frac{(\tau_{\nu i} - R_{mi} - \tau_{\mu' i})(\tau_{\nu j} - R_{mj} - \tau_{\mu' j})}{|\boldsymbol{\tau}_\nu - \boldsymbol{R}_m - \boldsymbol{\tau}_{\mu'}|^2}. \tag{2.109}$$

The code printed in figure 2.9 implements this expression. The code does not maximize readability because it was meant to minimize length. The parameters are `d` for the dimensionality, `a[1]` to `a[d]` the d-dimensional basis vectors of the lattice, `Nat` the number of atoms in the unit cell, `M[1]` to `M[Nat]` the masses of the atoms, `tau[1]` to `tau[Nat]` the d-dimensional vectors giving the positions of the atoms in the unit cell, and `K[]` a function giving the spring constants. The function `L2[]` defined at line 1 takes a vector and returns the square of its length. The function `R[]`

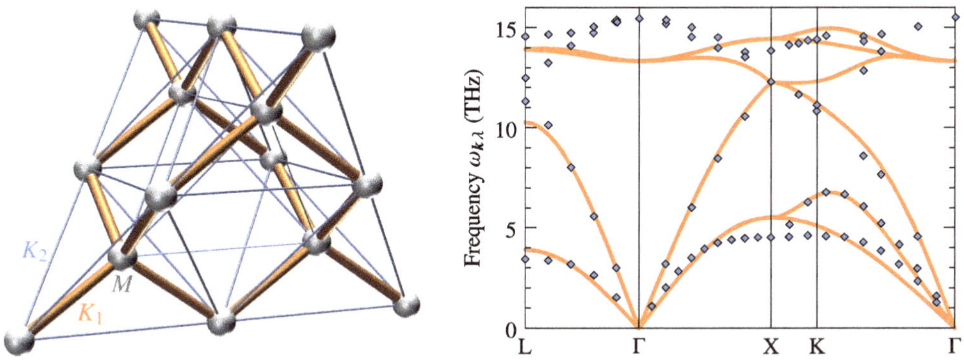

Figure 2.10. Left: spring model for the nonpolar zinc blende structure with atoms of mass M. The strong covalent bonds with spring constant K_1 and the weaker springs of strength K_2 are indicated. Right: phonon dispersion curves calculated with the code of figure 2.9 (orange) compared with experimental neutron-scattering data[13].

at line 2 takes a list of integers $\{n_1, \ldots, n_d\}$ and returns $\boldsymbol{R} = n_1\boldsymbol{a}_1 + \cdots + n_d\boldsymbol{a}_d$. The function `Table` at line 3, closed at line 12, builds the dynamical matrix. On lines 4–8, one recognizes the function that must be summed. This function is applied to a list of pairs $\{\boldsymbol{R}_m, \mu'\}$—each pair being addressed by #, such that `#[[1]]` stands for $\boldsymbol{R}_m$ and `#[[2]]` stands for μ'. The list is built on lines 9–11 using the functions `Tuples` and `Range`. This list must contain all atoms connected to the central unit cell. The integer `Shells` specifies how many layers of neighboring cells are included when building the list: `Shells=1` means that the various springs connect the central unit cell to its first neighbors at most. Finally, the `Select` on lines 9 and 11 removes the element $\{\mathbf{0}, \nu\}$ from the list, which would be suppressed anyway because $K_{|0|} = 0$, but produces a division by zero if left in place. All in all, the function `Dynamical[`$\{k_1, \ldots, k_d\}$`]` returns the dynamical matrix at the given wave vector.

The application to Si is done on lines 13–21. Note that the spring constant function `K[]` defined on line 15 takes as argument the square of the equilibrium length rather than the length; by default `K[]` returns zero, except for $l^2 = 3/16$ and $l^2 = 1/2$. The values K_1 and K_2 for these two cases are chosen in order to produce a good-looking result. Lines 18–20 define a path in the Brillouin zone like in figure 2.7. Figure 2.10 shows the resulting phonon dispersion curves, compared with experimental data measured by inelastic neutron scattering[13]. The agreement is of course not perfect; given the simplicity of a model with just two parameters, one can hardly expect more. The eigenvectors of $D(\boldsymbol{k})$ provide the displacements associated with each mode and allow one, in particular, to visualize the difference between acoustic and optical phonons.

Reference

Bruus H and Flensberg K 2004 *Condensed Matter Physics* (Oxford: Oxford University Press)

[13] Nilsson G and Nelin G 1972 *Phys. Rev.* B **6** 3777.

IOP Publishing

Spectroscopic Probes of Quantum Matter

Christophe Berthod

Chapter 3

Correlation functions: definitions and properties

The correlation functions encode the relationship between two observables at different times. This type of object is typically of the form $\langle A(t)B(t')\rangle$, where A and B are two operators and, for a system in thermal equilibrium, $\langle\cdots\rangle$ represents a thermal average. One is usually interested in steady states where these functions are invariant by translation in time, such that one can fix one of the times to zero without loss of generality. There are various kinds of correlation functions for each given pair of operators A and B. In this chapter, we review the diverse definitions that can be found in the condensed-matter literature. We emphasize the similarities and differences between them and we fix a notation in order to clear ambiguities. We present the Lehmann spectral representation and use it to derive a number of analytical properties and sum rules that the correlation functions must obey based on general principles.

3.1 A zoo of correlation functions

Depending upon the particular physical problem that we are considering, we may need to introduce different sorts of correlation functions. Although all of them share similarities and are related to one another, they differ in essential ways and should not be confused. We refer to the correlation function of the operators A and B in general with the symbol C_{AB} and we use superscripts to distinguish the various flavors. We conform to this notation whenever discussing general properties that are independent of A and B. Other notations will be introduced later for correlation functions of specific operators. The various types of correlation functions are:

Causal (*or 'time-ordered'*)
$$C_{AB}(t) = -\frac{i}{\hbar}\langle T_\eta\{A(t)B(0)\}\rangle \tag{3.1}$$

Retarded, first type
$$C^{\dagger}_{AB}(t) = -\frac{i}{\hbar}\theta(t)\langle A(t)B(0)\rangle \tag{3.2}$$

doi:10.1088/978-0-7503-1741-2ch3

Advanced, first type	$C^{-}_{AB}(t) = -\eta\frac{i}{\hbar}\theta(-t)\langle B(0)A(t)\rangle$	(3.3)
Retarded, second type (*or simply 'retarded'*)	$C^{R}_{AB}(t) = -\frac{i}{\hbar}\theta(t)\langle[A(t), B(0)]_{-\eta}\rangle$	(3.4)
Advanced, second type (*or simply 'advanced'*)	$C^{A}_{AB}(t) = \frac{i}{\hbar}\theta(-t)\langle[A(t), B(0)]_{-\eta}\rangle$	(3.5)
Greater	$C^{>}_{AB}(t) = -\frac{i}{\hbar}\langle A(t)B(0)\rangle$	(3.6)
Lesser	$C^{<}_{AB}(t) = -\eta\frac{i}{\hbar}\langle B(0)A(t)\rangle$	(3.7)
Keldysh	$C^{K}_{AB}(t) = -\frac{i}{\hbar}\langle[A(t), B(0)]_{+\eta}\rangle$	(3.8)
Imaginary-time (*or 'thermal' or 'Matsubara' or 'temperature'*)	$\mathscr{C}_{AB}(\tau) = -\langle T_\tau A(\tau)B(0)\rangle.$	(3.9)

The last zoo member—the somewhat eccentric imaginary-time correlation function—is listed here for completeness, but will be described in more detail in chapter 4. It is nothing but a convenient tool that helps to evaluate the real-time functions. The imaginary-time function is often easier to calculate than the real-time one and is especially well suited at finite temperature; furthermore, it allows one to recover the real-time functions by analytic continuation. In spite of its weird appearance, it will therefore become our favorite animal.

In these various definitions, the time evolution of the operators is meant in the Heisenberg picture like in equation (2.15), $\langle\cdots\rangle$ stands for the thermal average of equation (2.2), and T_η is the time ordering operator defined in equation (5), which sorts the operators on which it acts by order of decreasing times. Beware that the value of η in equations (3.1)–(3.9) is not determined by the particle statistics but by the type of operators A and B. For instance, even for fermions we must use $\eta = +1$ if we are considering 'bosonic' operators that involve products of two fermion operators, like spin or charge densities.

3.2 Lehmann spectral representation

The spectral representation is a way to express the time-Fourier transform of all correlation functions that we have just defined in term of only two functions of energy called the spectral-density functions, or simply the *spectral functions*. This representation is extremely helpful for studying the analytic properties of the correlation functions, finding out the mathematical relations between them, and performing various kinds of calculations. The spectral-density functions can be written down explicitly using a complete set of eigenstates.

All correlation functions that we have defined (except the imaginary-time function to be discussed later) can be expressed in terms of the two functions

$$F_1^{\pm}(t) = \mp\frac{i}{\hbar}\theta(\pm t)\langle A(t)B(0)\rangle \quad \text{and} \quad F_2^{\pm}(t) = \pm\eta\frac{i}{\hbar}\theta(\pm t)\langle B(0)A(t)\rangle.$$

For example, $C_{AB}(t) = F_1^+(t) + F_2^-(t)$, $C_{AB}^+(t) = F_1^+(t)$, etc. Consider first

$$F_1^{\pm}(t) = \mp\frac{i}{\hbar}\theta(\pm t)\frac{1}{Z}\mathrm{Tr}\ e^{-\beta K}e^{iKt/\hbar}Ae^{-iKt/\hbar}B. \tag{3.10}$$

We work as usual in the grand-canonical ensemble with $K = H - \mu N$. Introduce a complete set of states $|a\rangle$ such that $K|a\rangle = K_a|a\rangle$, insert twice the identity and use equation (8) to get (see Doc A3.1)

$$F_1^{\pm}(\omega) = \int_{-\infty}^{\infty} d\varepsilon \frac{\rho_{AB}^{>}(\varepsilon)}{\hbar\omega - \varepsilon \pm i0^+} \tag{3.11}$$

with

$$\boxed{\rho_{AB}^{>}(\varepsilon) = \frac{1}{Z}\sum_{ab} e^{-\beta K_a}\langle a|A|b\rangle\langle b|B|a\rangle\delta(\varepsilon + K_a - K_b).} \tag{3.12}$$

In very much the same way, we find (see Doc A3.1)

$$F_2^{\pm}(\omega) = \int_{-\infty}^{\infty} d\varepsilon \frac{\rho_{AB}^{<}(\varepsilon)}{\hbar\omega - \varepsilon \pm i0^+} \tag{3.13}$$

with

$$\boxed{\rho_{AB}^{<}(\varepsilon) = -\eta\frac{1}{Z}\sum_{ab} e^{-\beta K_b}\langle a|A|b\rangle\langle b|B|a\rangle\delta(\varepsilon + K_a - K_b).} \tag{3.14}$$

These results are summarized in the following table. It is seen that all correlators involved in the various real-time correlation functions can be deduced from the two spectral functions $\rho_{AB}^{>}(\varepsilon)$ and $\rho_{AB}^{<}(\varepsilon)$.

Function	Fourier transform
$-\frac{i}{\hbar}\theta(t)\langle A(t)B(0)\rangle$	$\int_{-\infty}^{\infty} d\varepsilon \frac{\rho_{AB}^{>}(\varepsilon)}{\hbar\omega - \varepsilon + i0^+}$
$\frac{i}{\hbar}\theta(-t)\langle A(t)B(0)\rangle$	$\int_{-\infty}^{\infty} d\varepsilon \frac{\rho_{AB}^{>}(\varepsilon)}{\hbar\omega - \varepsilon - i0^+}$
$\eta\frac{i}{\hbar}\theta(t)\langle B(0)A(t)\rangle$	$\int_{-\infty}^{\infty} d\varepsilon \frac{\rho_{AB}^{<}(\varepsilon)}{\hbar\omega - \varepsilon + i0^+}$
$-\eta\frac{i}{\hbar}\theta(-t)\langle B(0)A(t)\rangle$	$\int_{-\infty}^{\infty} d\varepsilon \frac{\rho_{AB}^{<}(\varepsilon)}{\hbar\omega - \varepsilon - i0^+}$

We can now write down, for definiteness and future reference, the spectral representation for each type of correlation function:

Causal $(F_1^+ + F_2^-)$

$$C_{AB}(t) = -\frac{i}{\hbar}\theta(t)\langle A(t)B(0)\rangle - \frac{i}{\hbar}\eta\theta(-t)\langle B(0)A(t)\rangle$$

$$C_{AB}(\omega) = \int_{-\infty}^{\infty} d\varepsilon \frac{\rho_{AB}^{>}(\varepsilon)}{\hbar\omega - \varepsilon + i0^+} + \int_{-\infty}^{\infty} d\varepsilon \frac{\rho_{AB}^{<}(\varepsilon)}{\hbar\omega - \varepsilon - i0^+} \tag{3.15}$$

Retarded, first type (F_1^+)

$$C_{AB}^{+}(t) = -\frac{i}{\hbar}\theta(t)\langle A(t)B(0)\rangle$$

$$C_{AB}^{+}(\omega) = \int_{-\infty}^{\infty} d\varepsilon \frac{\rho_{AB}^{>}(\varepsilon)}{\hbar\omega - \varepsilon + i0^+} \tag{3.16}$$

Advanced, first type (F_2^-)

$$C_{AB}^{-}(t) = -\eta\frac{i}{\hbar}\theta(-t)\langle B(0)A(t)\rangle$$

$$C_{AB}^{-}(\omega) = \int_{-\infty}^{\infty} d\varepsilon \frac{\rho_{AB}^{<}(\varepsilon)}{\hbar\omega - \varepsilon - i0^+} \tag{3.17}$$

Retarded, second type $(F_1^+ + F_2^+)$

$$C_{AB}^{R}(t) = -\frac{i}{\hbar}\theta(t)\langle A(t)B(0)\rangle + \eta\frac{i}{\hbar}\theta(t)\langle B(0)A(t)\rangle$$

$$C_{AB}^{R}(\omega) = \int_{-\infty}^{\infty} d\varepsilon \frac{\rho_{AB}^{>}(\varepsilon) + \rho_{AB}^{<}(\varepsilon)}{\hbar\omega - \varepsilon + i0^+} \tag{3.18}$$

Advanced, second type $(F_1^- + F_2^-)$

$$C_{AB}^{A}(t) = \frac{i}{\hbar}\theta(-t)\langle A(t)B(0)\rangle - \eta\frac{i}{\hbar}\theta(-t)\langle B(0)A(t)\rangle$$

$$C_{AB}^{A}(\omega) = \int_{-\infty}^{\infty} d\varepsilon \frac{\rho_{AB}^{>}(\varepsilon) + \rho_{AB}^{<}(\varepsilon)}{\hbar\omega - \varepsilon - i0^+} \tag{3.19}$$

Greater $(F_1^+ - F_1^-)$ (see *Doc A3.2*)

$$C_{AB}^{>}(t) = -\frac{i}{\hbar}\theta(t)\langle A(t)B(0)\rangle - \frac{i}{\hbar}\theta(-t)\langle A(t)B(0)\rangle$$

$$C_{AB}^{>}(\omega) = -2\pi i\rho_{AB}^{>}(\hbar\omega) \tag{3.20}$$

Lesser $(-F_2^+ + F_2^-)$ (see *Doc A3.2*)

$$C_{AB}^{<}(t) = -\eta\frac{i}{\hbar}\theta(t)\langle B(0)A(t)\rangle - \eta\frac{i}{\hbar}\theta(-t)\langle B(0)A(t)\rangle$$

$$C_{AB}^{<}(\omega) = 2\pi i\,\rho_{AB}^{<}(\hbar\omega) \tag{3.21}$$

Keldysh

$$C_{AB}^{K}(t) = C_{AB}^{>}(t) + C_{AB}^{<}(t)$$

$$C_{AB}^{K}(\omega) = -2\pi i[\rho_{AB}^{>}(\hbar\omega) - \rho_{AB}^{<}(\hbar\omega)]. \tag{3.22}$$

For convenience, we also define the total spectral function

$$\boxed{\rho_{AB}(\varepsilon) = \rho_{AB}^{>}(\varepsilon) + \rho_{AB}^{<}(\varepsilon).} \tag{3.23}$$

When using the definition of the spectral function, equations (3.12), (3.14), and (3.23), it is important to keep in mind that these formula are valid provided the states $|a\rangle$ and $|b\rangle$ are eigenstates of K with the eigenvalues K_a and K_b. If the operators A and B do not conserve the number of particles—for instance, single creation and annihilation operators do not—the states $|a\rangle$ and $|b\rangle$ have different particle numbers and the delta functions in equations (3.12) and (3.14) contain a term $-\mu(N_a - N_b)$.

3.3 Independent particles

In chapter 2, we saw how to describe many-body systems of independent particles within the second-quantization formalism. In this section, we will see how the spectral representation allows one to express the correlation functions of such independent-particle systems in terms of the single-particle energies and occupation numbers. We focus on two correlation functions of particular importance for electrons: the one-particle Green's function and the density–density correlation function. All correlation functions for independent particles can be obtained with the same method.

3.3.1 One-particle Green's function of independent electrons

The one-electron Green's function is the correlation function of the electron creation and annihilation operators. As we shall see in section 5.1.3.1, it has a simple physical interpretation and is the tool of choice to describe the one-electron spectroscopies such as photoemission and tunneling. For independent electrons, it is best to work with the one-electron basis formed by the eigenstates of the Hamiltonian, i.e. $K = \sum_\alpha \xi_\alpha c_\alpha^\dagger c_\alpha$. We obtain the Green's function in this representation from our previous general formulas by setting

$$A \equiv c_\alpha, \quad B \equiv c_\alpha^\dagger, \quad \text{and} \quad \eta = -1.$$

Since $B = A^\dagger$, the spectral functions are real. Exploiting the properties of creation and annihilation operators for independent fermions, we find (see Doc A3.3)

$$\rho^{>}_{c_\alpha c_\alpha^\dagger}(\varepsilon) = f(-\xi_\alpha)\delta(\varepsilon - \xi_\alpha) \quad \text{and} \quad \rho^{<}_{c_\alpha c_\alpha^\dagger}(\varepsilon) = f(\xi_\alpha)\delta(\varepsilon - \xi_\alpha), \tag{3.24}$$

such that the total spectral function—this is what is usually meant by 'spectral function' when dealing with the one-electron Green's function—is simply

$$\rho_{c_\alpha c_\alpha^\dagger}(\varepsilon) = \rho^{>}_{c_\alpha c_\alpha^\dagger}(\varepsilon) + \rho^{<}_{c_\alpha c_\alpha^\dagger}(\varepsilon) \equiv A_0(\alpha, \varepsilon) = \delta(\varepsilon - \xi_\alpha). \tag{3.25}$$

This is a very important first result expressing the fact that, in a system of independent particles, the one-particle excitations are eigenstates and therefore have an infinite life-time. We shall come back to this later. Using the spectral representation, we can write down the retarded Green's function for independent electrons in the basis that diagonalizes the Hamiltonian,

$$\boxed{G_0^R(\alpha, \varepsilon) \equiv C^R_{c_\alpha c_\alpha^\dagger}(\varepsilon/\hbar) = \int_{-\infty}^{\infty} d\varepsilon' \frac{A_0(\alpha, \varepsilon')}{\varepsilon - \varepsilon' + i0^+} = \frac{1}{\varepsilon - \xi_\alpha + i0^+}.} \tag{3.26}$$

Here, we regard the Green's function as a function of the energy ε measured from the chemical potential rather than a function of the frequency ω: this is often more convenient and saves plenty of $\hbar$ factors. It is customary to use the letter 'G' for the one-particle Green's function and the letter 'A' for its spectral function. The subscript '0' reminds us that we are dealing with independent particles. These expressions satisfy the properties

$$A(\alpha, \varepsilon) = -\frac{1}{\pi} \operatorname{Im} G^R(\alpha, \varepsilon), \quad \int_{-\infty}^{\infty} d\varepsilon\, A(\alpha, \varepsilon) = 1, \quad \text{and} \quad G^R(\alpha, |\varepsilon| \to \infty) = \frac{1}{\varepsilon},$$

which are also satisfied in the general case of interacting electrons, as we shall see shortly. The very simple form of equation (3.26) results because the operators $c_\alpha^\dagger$ create eigenstates of the Hamiltonian. This is also the reason why $C_{c_\alpha c_\beta^\dagger}(\omega) = 0$ for $\alpha \neq \beta$. Other one-particle Green's functions may be defined, for instance the momentum-space Green's function $G(\boldsymbol{k}, \varepsilon) = C_{c_k c_k^\dagger}(\varepsilon/\hbar)$, where $c_{\boldsymbol{k}}^\dagger$ is the creation operator for a plane wave, or the real-space Green's function $G(\boldsymbol{r}, \boldsymbol{r}', \varepsilon) = C_{c_r c_{r'}^\dagger}(\varepsilon/\hbar)$, where $c_{\boldsymbol{r}}^\dagger$ creates an electron at position $\boldsymbol{r}$. Because $c_{\boldsymbol{k}}^\dagger$ and $c_{\boldsymbol{r}}^\dagger$ are related to $c_\alpha^\dagger$ by a straightforward change of basis (see Doc A2.7), $G_0(\boldsymbol{k}, \varepsilon)$ and $G_0(\boldsymbol{r}, \boldsymbol{r}', \varepsilon)$ are also related to $G_0(\alpha, \varepsilon)$. For instance, because $c_{\boldsymbol{r}}^\dagger = \sum_\alpha \varphi_\alpha^*(\boldsymbol{r}) c_\alpha^\dagger$ where $\varphi_\alpha(\boldsymbol{r})$ is the wave function of the eigenstate created by $c_\alpha^\dagger$, we have

$$G_0^R(\boldsymbol{r}, \boldsymbol{r}', \varepsilon) = \sum_\alpha \frac{\varphi_\alpha(\boldsymbol{r})\varphi_\alpha^*(\boldsymbol{r}')}{\varepsilon - \xi_\alpha + i0^+} \tag{3.27}$$

for the real-space retarded Green's function of independent electrons. If the eigenstates are not known and an arbitrary basis has to be used, an inversion of the operator $\varepsilon\mathbb{1} - K$ will be necessary in order to obtain the Green's function, as we will see in section 5.2.2.1.

3.3.2 Density–density correlation function of free electrons

We have seen in section 1.1 that the density–density correlation function, i.e. the correlation of the operators $n(\boldsymbol{q})$ and $n(-\boldsymbol{q})$, can be measured experimentally by studying the scattering cross section for particles interacting with the material through short-range forces. For free electrons, the calculation of the density–density correlation is easily done by means of the spectral representation. We set

$$A \equiv n(\boldsymbol{q}) = \sum_{\boldsymbol{k}\sigma} c_{\boldsymbol{k}\sigma}^\dagger c_{\boldsymbol{k}+\boldsymbol{q}\sigma}, \qquad B \equiv n(-\boldsymbol{q}) = [n(\boldsymbol{q})]^\dagger, \qquad \eta = +1.$$

Equation (2.46) for the density is valid provided that $c_{\boldsymbol{k}\sigma}^\dagger$ is the creation operator for the plane-wave state (2.29) of wave vector $\boldsymbol{k}$. Since we are dealing with free electrons, these states also diagonalize the Hamiltonian. We find for the spectral functions (see Doc A3.3)

$$\rho^>_{n(\boldsymbol{q})n(-\boldsymbol{q})}(\varepsilon) = \sum_{\boldsymbol{k}\sigma} f(\xi_{\boldsymbol{k}}) f(-\xi_{\boldsymbol{k}+\boldsymbol{q}}) \delta(\varepsilon + \xi_{\boldsymbol{k}} - \xi_{\boldsymbol{k}+\boldsymbol{q}}) \tag{3.28a}$$

$$\rho^<_{n(\boldsymbol{q})n(-\boldsymbol{q})}(\varepsilon) = -\sum_{\boldsymbol{k}\sigma} f(-\xi_{\boldsymbol{k}}) f(\xi_{\boldsymbol{k}+\boldsymbol{q}}) \delta(\varepsilon + \xi_{\boldsymbol{k}} - \xi_{\boldsymbol{k}+\boldsymbol{q}}). \tag{3.28b}$$

The retarded density–density correlation function, also known as the *charge susceptibility*, is therefore simply (see Doc A3.3)

$$\boxed{\chi_{nn}^0(\boldsymbol{q}, \varepsilon) \equiv C^R_{n(\boldsymbol{q})n(-\boldsymbol{q})}(\varepsilon/\hbar) = \sum_{\boldsymbol{k}\sigma} \frac{f(\xi_{\boldsymbol{k}}) - f(\xi_{\boldsymbol{k}+\boldsymbol{q}})}{\varepsilon + \xi_{\boldsymbol{k}} - \xi_{\boldsymbol{k}+\boldsymbol{q}} + i0^+},} \tag{3.29}$$

where the superscript '0' again reminds us that this result applies to free electrons. This result is also very important because it is the basis for describing the collective charge behavior of interacting electrons (screening, plasmon, zero sound, etc) via the RPA approximation, as we shall see in section 5.1.4. This result can be painlessly generalized to the case of independent—but not free—electrons, e.g. Bloch electrons in periodic crystals (see Doc A3.4). In such situations, the correlation function also depends on the electronic wave functions, not only on the energies and occupation numbers like in equation (3.29).

3.4 Analytic properties and sum rules

Using the spectral representation, one can prove general analytic properties, symmetries, as well as several sum rules that the exact correlation functions must obey. These results are useful both for checking the consistency of approximations and for the analysis of experimental data.

3.4.1 General symmetry properties of the spectral functions

By exchanging the dummy indices a and b in the definitions of the spectral functions, equations (3.12) and (3.14), one easily finds that

$$\rho_{AB}^{>}(-\varepsilon) = -\eta\rho_{BA}^{<}(\varepsilon), \tag{3.30}$$

from which we deduce for the total spectral function:

$$\rho_{AB}(-\varepsilon) = -\eta\rho_{BA}(\varepsilon). \tag{3.31}$$

As a result, the spectral function is either odd or even when $A = B$, depending upon the value of η.

We will study below the perturbation theory which allows one to calculate the retarded and advanced correlation functions of the second type, i.e. to obtain the total spectral function ρ_{AB} (see equation (3.18)). On the other hand, we have seen that it is the greater function that is measured in scattering experiments. We therefore need a formula in order to compute the greater and lesser spectral functions from the total one. These relations can be deduced from equations (3.12), (3.14), and (3.23). They are (see Doc A3.5)

$$\rho_{AB}^{>}(\varepsilon) = -\eta d_{-\eta}(-\varepsilon)\rho_{AB}(\varepsilon), \qquad \rho_{AB}^{<}(\varepsilon) = -\eta d_{-\eta}(\varepsilon)\rho_{AB}(\varepsilon). \tag{3.32}$$

Hence the greater and lesser spectral functions are actually not independent, since they are related by

$$\rho_{AB}^{>}(\varepsilon) = -\eta e^{\beta\varepsilon}\rho_{AB}^{<}(\varepsilon), \qquad \rho_{AB}^{<}(\varepsilon) = -\eta e^{-\beta\varepsilon}\rho_{AB}^{>}(\varepsilon). \tag{3.33}$$

Lastly, by combining equations (3.30) and (3.33), we obtain relations known as the *detailed balance*:

$$\rho^{>}_{AB}(-\varepsilon) = e^{-\beta\varepsilon}\rho^{>}_{BA}(\varepsilon), \qquad \rho^{<}_{AB}(-\varepsilon) = e^{\beta\varepsilon}\rho^{<}_{BA}(\varepsilon). \tag{3.34}$$

3.4.2 Sum rule for the spectral function

Integrating the spectral functions over the energy, we obtain (see Doc A3.6)

$$\int_{-\infty}^{\infty} d\varepsilon\ \rho^{>}_{AB}(\varepsilon) = \langle AB\rangle \quad \text{and} \quad \int_{-\infty}^{\infty} d\varepsilon\ \rho^{<}_{AB}(\varepsilon) = -\eta\langle BA\rangle, \tag{3.35}$$

such that the spectral function satisfies what is sometimes called the f-sum rule:

$$\int_{-\infty}^{\infty} d\varepsilon\ \rho_{AB}(\varepsilon) = \langle [A,\ B]_{-\eta}\rangle. \tag{3.36}$$

In electronic systems, equation (3.35) can be for instance used to express the average *double occupancy* $\langle n_\uparrow n_\downarrow\rangle$ in terms of the density–density and spin–spin correlation functions (see Bonus B3.1). In the case of the one-particle Green's function ($A = c_\alpha$, $B = c_\alpha^\dagger$, $\eta = -1$), the operators anti-commute (equation (2.41)) and consequently the sum rule equals one:

$$\int_{-\infty}^{\infty} d\varepsilon\ A(\alpha,\ \varepsilon) = 1. \tag{3.37}$$

This sum rule expresses the conservation of spectral weight. When the particles interact one with the other, the spectral weight of the one-particle excitations is no longer concentrated at a single energy like in equation (3.25), but it is spread over a certain energy range. Nothing is lost, though, and the total weight remains one.

3.4.3 Sum rule for the occupation numbers

The average number of particles in the one-particle state φ_α is $\langle n_\alpha\rangle = \langle a_\alpha^\dagger a_\alpha\rangle$. Using the spectral representation, we can show that (see Doc A3.6)

$$\langle n_\alpha\rangle = \int_{-\infty}^{\infty} d\varepsilon\ d_{-\eta}(\varepsilon)\rho_{a_\alpha a_\alpha^\dagger}(\varepsilon). \tag{3.38}$$

The relation $\langle n_\alpha\rangle = d_{-\eta}(\xi_\alpha)$ valid for independent particles is recovered if one substitutes the independent-particle spectral function $\rho_{a_\alpha a_\alpha^\dagger}(\varepsilon) = \delta(\varepsilon - \xi_\alpha)$. For fermions, equation (3.38) can be used to relate the momentum distribution function $\langle n_{\boldsymbol{k}}\rangle$ to the spectral function:

$$\langle n_{\boldsymbol{k}}\rangle = \int_{-\infty}^{\infty} d\varepsilon\, f(\varepsilon)A(\boldsymbol{k},\ \varepsilon). \tag{3.39}$$

The latter sum rule is useful in the analysis of photoemission experiments.

3.4.4 Sum rule for the energy

For a general Hamiltonian of the form (2.45), it is possible to express the average energy $\langle K \rangle = \langle K_0 \rangle + \langle V \rangle$ in terms of the spectral function $\rho_{a_\alpha a_\alpha^\dagger}(\varepsilon)$:

$$\langle K \rangle = \frac{1}{2}\left[\langle K_0 \rangle + \sum_\alpha \int_{-\infty}^{\infty} d\varepsilon \; \varepsilon d_{-\eta}(\varepsilon) \rho_{a_\alpha a_\alpha^\dagger}(\varepsilon)\right]. \tag{3.40}$$

The derivation is performed in Doc A3.6. If the kinetic energy is diagonal in the basis of the $a_\alpha^\dagger$, its average value can also be expressed in terms of the spectral function using equation (3.38), $\langle K_0 \rangle = \sum_\alpha \xi_\alpha \int_{-\infty}^{\infty} d\varepsilon \; d_{-\eta}(\varepsilon) \rho_{a_\alpha a_\alpha^\dagger}(\varepsilon)$, such that equation (3.40) can be put in the form

$$\langle K \rangle = \sum_\alpha \int_{-\infty}^{\infty} d\varepsilon \; \frac{1}{2}(\xi_\alpha + \varepsilon) d_{-\eta}(\varepsilon) \rho_{a_\alpha a_\alpha^\dagger}(\varepsilon). \tag{3.41}$$

This is known as the Galitskii–Migdal formula. It is also interesting to check that in the case $V = 0$, one recovers from this expression the expected formula for non-interacting particles (see Doc A3.6):

$$\langle K_0 \rangle = \sum_\alpha \xi_\alpha d_{-\eta}(\xi_\alpha). \tag{3.42}$$

While the knowledge of the spectral function $\rho_{a_\alpha a_\alpha^\dagger}$ is sufficient to deduce the average energy, it is not sufficient to obtain the grand potential Ω and thus the thermodynamic properties. As shown by equation (1.9), in order to obtain Ω we have to know the spectral function for all values of the coupling constant (see Doc A5.19).

3.4.5 High-frequency behavior: moment expansion

At sufficiently high frequency, the correlation functions can be expanded in powers of $1/\omega$. This expansion is easily written down using the spectral representation. Consider for example the retarded function:

$$\begin{aligned} C_{AB}^R(\omega) &= \int_{-\infty}^{\infty} d\varepsilon \frac{\rho_{AB}(\varepsilon)}{\hbar\omega^+ - \varepsilon} = \frac{1}{\hbar\omega^+} \int_{-\infty}^{\infty} d\varepsilon \frac{\rho_{AB}(\varepsilon)}{1 - \varepsilon/\hbar\omega^+} \\ &= \sum_{n=0}^{\infty} \frac{1}{(\hbar\omega^+)^{n+1}} \int_{-\infty}^{\infty} d\varepsilon \; \varepsilon^n \rho_{AB}(\varepsilon), \end{aligned} \tag{3.43}$$

where we have set $\hbar\omega^+ = \hbar\omega + i0^+$ and used the expansion $1/(1-x) = 1 + x + x^2 + \cdots$ Hence the $(n+1)$th-order term in the $1/\omega$ expansion is given by the nth moment of the spectral function:

$$M_n = \int_{-\infty}^{\infty} d\varepsilon \; \varepsilon^n \rho_{AB}(\varepsilon). \tag{3.44}$$

At lowest order, we see from equation (3.36) that $M_0 = \langle [A, B]_{-\eta} \rangle$. Therefore, the asymptotic high-frequency behavior of the correlation function is

$$C_{AB}^R(\omega) = \frac{\langle [A, B]_{-\eta} \rangle}{\hbar\omega} \qquad \text{for} \qquad |\omega| \to \infty. \tag{3.45}$$

We could replace $\hbar\omega^+$ by $\hbar\omega$ because $1/\omega^+ = 1/\omega$ for $\omega \neq 0$. This result generalizes what we have obtained in section 3.3.1 for the Green's function of independent electrons. It also shows that if $[A, B]_{-\eta} = 0$, the correlation function vanishes at least as fast as $1/\omega^2$ at high frequency. Higher-order moments can also be expressed as average values of commutators. For instance (see Doc A3.7)

$$M_1 = \int_{-\infty}^{\infty} d\varepsilon \; \varepsilon \; \rho_{AB}(\varepsilon) = -\big\langle [[K, A], B]_{-\eta} \big\rangle. \tag{3.46}$$

The generalization to arbitrary order can easily be foreseen. Such commutators are often easier to calculate than the full correlation function and the high-frequency behavior can thus be obtained exactly.

3.4.6 Additional relations for adjoint operators

The spectral functions are, in general, complex: one readily checks from the definitions that $[\rho_{AB}^{>,<}(\varepsilon)]^* = \rho_{B^\dagger A^\dagger}^{>,<}(\varepsilon)$. In many cases of interest, though, we will deal with correlation functions of two adjoint operators, i.e. $B = A^\dagger$. Several additional relations can be proven in this particular case, owing to the fact that the spectral functions are real. Among these relations, the most useful is perhaps

$$\boxed{\rho_{AA^\dagger}(\varepsilon) = -\frac{1}{\pi}\mathrm{Im}\; C_{AA^\dagger}^R(\varepsilon/\hbar),} \tag{3.47}$$

which follows immediately from equations (3.18), (3.23), and (10). This relation is commonly called *fluctuation–dissipation theorem*. The spectral function $\rho_{AA^\dagger}(\varepsilon)$ indeed contains all information about the fluctuations of the observable A, while it can be shown that the quantity $(-1/\pi)\,\mathrm{Im}\; C_{AA^\dagger}^R(\omega)$ controls the energy dissipated when the observable A is excited by an external field of frequency ω (see Bonus B6.1). Equation (3.47) can also be regarded as a Kramers–Kronig relation expressing the fact that the retarded correlation functions are causal. Equation (3.47), together with the spectral representation of $C_{AA^\dagger}^R(\omega)$ in equation (3.18), indeed imply that the real and imaginary parts are related by the usual Kramers–Kronig relation:

$$\mathrm{Re}\; C_{AA^\dagger}^R(\omega) = -\frac{1}{\pi} \oint_{-\infty}^{\infty} d\omega' \frac{\mathrm{Im}\; C_{AA^\dagger}^R(\omega')}{\omega - \omega'}. \tag{3.48}$$

The document Bonus B3.2 discusses the relationships between analyticity of a function in the complex plane, causality, and the Kramers–Kronig relations.

Appendix A3

Doc A3.1 Spectral-density functions

We introduce twice the identity using equation (21) and write the trace as $\sum_c \langle c|\cdots|c\rangle$:

$$\begin{aligned}
F_1^{\pm}(t) &= \mp\frac{i}{\hbar}\theta(\pm t)\frac{1}{Z}\sum_{abc}\langle c|e^{-\beta K}e^{iKt/\hbar}|a\rangle\langle a|A|b\rangle\langle b|e^{-iKt/\hbar}B|c\rangle \\
&= \mp\frac{i}{\hbar}\theta(\pm t)\frac{1}{Z}\sum_{abc}e^{-\beta K_a}e^{i(K_a-K_b)t/\hbar}\underbrace{\langle c|a\rangle}_{\delta_{ca}}\langle a|A|b\rangle\langle b|B|c\rangle \\
&= \mp\frac{i}{\hbar}\theta(\pm t)\frac{1}{Z}\sum_{ab}e^{-\beta K_a}e^{i(K_a-K_b)t/\hbar}\langle a|A|b\rangle\langle b|B|a\rangle.
\end{aligned}$$

The time-Fourier transform is now readily performed using equation (8):

$$\begin{aligned}
F_1^{\pm}(\omega) = \int_{-\infty}^{\infty} dt\, e^{i\omega t}F_1^{\pm}(t) &= \frac{1}{Z}\sum_{ab}\frac{e^{-\beta K_a}}{\hbar\omega + K_a - K_b \pm i0^+}\langle a|A|b\rangle\langle b|B|a\rangle \\
&= \int_{-\infty}^{\infty} d\varepsilon \frac{\rho_{AB}^{>}(\varepsilon)}{\hbar\omega - \varepsilon \pm i0^+}
\end{aligned}$$

with $\rho_{AB}^{>}(\varepsilon)$ given by equation (3.12). In the same way, the function F_2 becomes:

$$\begin{aligned}
F_2^{\pm}(t) &= \pm\frac{i}{\hbar}\eta\theta(\pm t)\frac{1}{Z}\sum_{abc}\langle c|e^{-\beta K}Be^{iKt/\hbar}|a\rangle\langle a|A|b\rangle\langle b|e^{-iKt/\hbar}|c\rangle \\
&= \pm\frac{i}{\hbar}\eta\theta(\pm t)\frac{1}{Z}\sum_{ab}e^{-\beta K_b}e^{i(K_a-K_b)t/\hbar}\langle b|B|a\rangle\langle a|A|b\rangle,
\end{aligned}$$

such that the Fourier transform follows as

$$\begin{aligned}
F_2^{\pm}(\omega) &= -\eta\frac{1}{Z}\sum_{ab}\frac{e^{-\beta K_b}}{\hbar\omega + K_a - K_b \pm i0^+}\langle a|A|b\rangle\langle b|B|a\rangle \\
&= \int_{-\infty}^{\infty} d\varepsilon \frac{\rho_{AB}^{<}(\varepsilon)}{\hbar\omega - \varepsilon \pm i0^+}
\end{aligned}$$

with $\rho_{AB}^{<}(\varepsilon)$ given by equation (3.14).

Doc A3.2 Greater and lesser spectral functions

In Fourier space, the greater and lesser functions are simply proportional to the corresponding spectral functions. Indeed we have

$$\begin{aligned}
C_{AB}^{>}(\omega) &= \int_{-\infty}^{\infty} d\varepsilon \frac{\rho_{AB}^{>}(\varepsilon)}{\hbar\omega - \varepsilon + i0^+} - \int_{-\infty}^{\infty} d\varepsilon \frac{\rho_{AB}^{>}(\varepsilon)}{\hbar\omega - \varepsilon - i0^+} \\
&= \int_{-\infty}^{\infty} d\varepsilon \rho_{AB}^{>}(\varepsilon)\underbrace{\left(\frac{1}{\hbar\omega - \varepsilon + i0^+} - \frac{1}{\hbar\omega - \varepsilon - i0^+}\right)}_{-2\pi i\,\delta(\hbar\omega-\varepsilon)} = -2\pi i\,\rho_{AB}^{>}(\hbar\omega),
\end{aligned}$$

where we have used equation (9). Similarly,

$$C_{AB}^{<}(\omega) = \int_{-\infty}^{\infty} d\varepsilon\, \rho_{AB}^{<}(\varepsilon) \underbrace{\left(\frac{1}{\hbar\omega - \varepsilon - i0^+} - \frac{1}{\hbar\omega - \varepsilon + i0^+} \right)}_{2\pi i\, \delta(\hbar\omega - \varepsilon)}$$

$$= 2\pi i\, \rho_{AB}^{<}(\hbar\omega).$$

Doc A3.3 Spectral functions of independent fermions

For the one-electron Green's function, we must evaluate $\langle a|A|b\rangle\langle b|B|a\rangle$ for $A = c_\alpha$ and $B = c_\alpha^\dagger$: this reduces to $|\langle a|c_\alpha|b\rangle|^2$. The many-electron states $|a\rangle$ and $|b\rangle$ are Slater determinants for N_a and N_b electrons, respectively, constructed using the one-particle basis $\varphi_\alpha(1)$ that corresponds to the operators c_α and $c_\alpha^\dagger$. In the occupation-number representation, they take the form $|a\rangle = |n_1^{|a\rangle}, n_2^{|a\rangle}, \ldots, n_\alpha^{|a\rangle}, \ldots\rangle$ and $|b\rangle = |n_1^{|b\rangle}, n_2^{|b\rangle}, \ldots, n_\alpha^{|b\rangle}, \ldots\rangle$, with the occupation numbers satisfying the properties $\sum_i n_i^{|a\rangle} = N_a$, $\sum_i n_i^{|b\rangle} = N_b$, $\sum_i \varepsilon_i n_i^{|a\rangle} = E_a$, and $\sum_i \varepsilon_i n_i^{|b\rangle} = E_b$. Using equation (2.42), we see that

$$c_\alpha|b\rangle = (-1)^\gamma \sqrt{n_\alpha^{|b\rangle}}\, |\ldots, n_\alpha^{|b\rangle} - 1, \ldots\rangle$$

where $\gamma = \sum_{i<\alpha} n_i^{|b\rangle}$. As a result, we find

$$|\langle a|c_\alpha|b\rangle|^2 = n_\alpha^{|b\rangle} \delta_{n_1^{|a\rangle} n_1^{|b\rangle}} \delta_{n_2^{|a\rangle} n_2^{|b\rangle}} \cdots \delta_{n_\alpha^{|a\rangle} n_\alpha^{|b\rangle}-1} \cdots$$

$$= n_\alpha^{|b\rangle} \delta_{n_\alpha^{|a\rangle} n_\alpha^{|b\rangle}-1} \prod_{i\neq\alpha} \delta_{n_i^{|a\rangle} n_i^{|b\rangle}}.$$

In other words, all occupation numbers except n_α must be the same in the states $|a\rangle$ and $|b\rangle$ for the matrix element to be nonzero. Since, on the other hand, $n_\alpha^{|a\rangle} = n_\alpha^{|b\rangle} - 1$, we have $E_a = E_b - \varepsilon_\alpha$ and $N_a = N_b - 1$. In summary,

$$|\langle a|c_\alpha|b\rangle|^2 = \begin{cases} n_\alpha^{|b\rangle} & \text{if } n_i^{|a\rangle} = n_i^{|b\rangle} \text{ for } i \neq \alpha \text{ and } n_\alpha^{|a\rangle} = n_\alpha^{|b\rangle} - 1 \\ 0 & \text{otherwise.} \end{cases}$$

Thus, the spectral function equation (3.12) becomes

$$\rho_{c_\alpha c_\alpha^\dagger}^{>}(\varepsilon) = \frac{1}{Z} \sum_b e^{-\beta(E_b - \varepsilon_\alpha - \mu N_b + \mu)} n_\alpha^{|b\rangle} \delta(\varepsilon - \varepsilon_\alpha + \mu).$$

We note that $e^{-\beta(E_b - \mu N_b)} n_\alpha^{|b\rangle} = \langle b|e^{-\beta K} n_\alpha|b\rangle$ and introduce $\xi_\alpha \equiv \varepsilon_\alpha - \mu$ as usual. Therefore

$$\rho_{c_\alpha c_\alpha^\dagger}^{>}(\varepsilon) = e^{\beta\xi_\alpha} \delta(\varepsilon - \xi_\alpha) \underbrace{\frac{1}{Z} \sum_b \langle b|e^{-\beta K} n_\alpha|b\rangle}_{\mathrm{Tr}\, \rho n_\alpha = \langle n_\alpha\rangle = f(\xi_\alpha)}$$

$$= f(-\xi_\alpha)\delta(\varepsilon - \xi_\alpha),$$

as indicated in equation (3.24). The procedure for $\rho^<_{c_\alpha c_\alpha^\dagger}$ is similar:

$$\begin{aligned}\rho^<_{c_\alpha c_\alpha^\dagger}(\varepsilon) &= \frac{1}{Z}\sum_b e^{-\beta(E_b-\mu N_b)} n_\alpha^{|b\rangle}\delta(\varepsilon - \varepsilon_\alpha + \mu)\\ &= f(\xi_\alpha)\delta(\varepsilon - \xi_\alpha).\end{aligned}$$

The calculation is slightly more complicated for the density–density correlation function because, even if $B^\dagger = A$ in this case as well, the square $|\langle a|A|b\rangle|^2$ involves crossed terms that in the end turn out to vanish. We write

$$\begin{aligned}\langle a|A|b\rangle\langle b|B|a\rangle &= \langle a|A|b\rangle\langle a|B^\dagger|b\rangle^*\\ &= \sum_{\boldsymbol{k}\sigma}\sum_{\boldsymbol{k}'\sigma'}\langle a|c^\dagger_{\boldsymbol{k}\sigma}c_{\boldsymbol{k}+\boldsymbol{q}\sigma}|b\rangle\langle a|c^\dagger_{\boldsymbol{k}'\sigma'}c_{\boldsymbol{k}'+\boldsymbol{q}\sigma'}|b\rangle^*.\end{aligned}$$

The crossed terms are those with $\boldsymbol{k}\sigma \neq \boldsymbol{k}'\sigma'$. For convenience, we temporarily use the notations $(\boldsymbol{k}\sigma) \equiv \alpha$, $(\boldsymbol{k} + \boldsymbol{q}\sigma) \equiv \beta$, $(\boldsymbol{k}'\sigma') \equiv \alpha'$, and $(\boldsymbol{k}' + \boldsymbol{q}\sigma') \equiv \beta'$. Performing the same analysis as for the Green's function, we see that

$$\langle a|c^\dagger_\alpha c_\beta|b\rangle = \begin{cases}(-1)^\gamma\sqrt{1 - n_\alpha^{|b\rangle}}\sqrt{n_\beta^{|b\rangle}} & \text{if } n_i^{|a\rangle} = n_i^{|b\rangle} \text{ for } i \neq \alpha, \beta \text{ and}\\ & n_\alpha^{|a\rangle} = n_\alpha^{|b\rangle} + 1,\ n_\beta^{|a\rangle} = n_\beta^{|b\rangle} - 1\\ 0 & \text{otherwise,}\end{cases}$$

$$\langle a|c^\dagger_{\alpha'} c_{\beta'}|b\rangle^* = \begin{cases}(-1)^{\gamma'}\sqrt{1 - n_{\alpha'}^{|b\rangle}}\sqrt{n_{\beta'}^{|b\rangle}} & \text{if } n_i^{|a\rangle} = n_i^{|b\rangle} \text{ for } i \neq \alpha', \beta' \text{ and}\\ & n_{\alpha'}^{|a\rangle} = n_{\alpha'}^{|b\rangle} + 1,\ n_{\beta'}^{|a\rangle} = n_{\beta'}^{|b\rangle} - 1\\ 0 & \text{otherwise.}\end{cases}$$

For the product of the two matrix elements to be nonzero, we must clearly have either $\alpha' = \alpha$ and $\beta' = \beta$, or $\alpha' = \beta$ and $\beta' = \alpha$. If $\alpha' = \alpha$ and $\beta' = \beta$, the four additional conditions $n_\alpha^{|a\rangle} = n_\alpha^{|b\rangle} + 1$, $n_\beta^{|a\rangle} = n_\beta^{|b\rangle} - 1$, $n_{\alpha'}^{|a\rangle} = n_{\alpha'}^{|b\rangle} + 1$, and $n_{\beta'}^{|a\rangle} = n_{\beta'}^{|b\rangle} - 1$ can be satisfied together. On the contrary, they cannot be satisfied if $\alpha' = \beta$ and $\beta' = \alpha$: this second option must be discarded, and with it all the crossed terms. The constraints $\alpha' = \alpha$ and $\beta' = \beta$ imply $\gamma' = \gamma$ and consequently

$$\langle a|c^\dagger_\alpha c_\beta|b\rangle\langle a|c^\dagger_{\alpha'} c_{\beta'}|b\rangle^* = \begin{cases}(1 - n_\alpha^{|b\rangle})n_\beta^{|b\rangle} & \text{if } \alpha' = \alpha,\ \beta' = \beta,\ n_i^{|a\rangle} = n_i^{|b\rangle} \text{ for } i \neq \alpha, \beta\\ & \text{and } n_\alpha^{|a\rangle} = n_\alpha^{|b\rangle} + 1,\ n_\beta^{|a\rangle} = n_\beta^{|b\rangle} - 1\\ 0 & \text{otherwise.}\end{cases}$$

The relation between $|a\rangle$ and $|b\rangle$ implies $E_a = E_b - \varepsilon_{k+q} + \varepsilon_k$ and $N_a = N_b$. Using these conditions, we can finish the evaluation of the spectral function

$$\begin{aligned}
\rho^{>}_{n(q)n(-q)}(\varepsilon) &= \sum_{k\sigma} \frac{1}{Z} \sum_b e^{-\beta(E_b - \varepsilon_{k+q} + \varepsilon_k - \mu N_b)} (1 - n^{|b\rangle}_{k\sigma}) n^{|b\rangle}_{k+q\sigma} \delta(\varepsilon - \varepsilon_{k+q} + \varepsilon_k) \\
&= \sum_{k\sigma} e^{\beta(\varepsilon_{k+q} - \varepsilon_k)} \delta(\varepsilon - \varepsilon_{k+q} + \varepsilon_k) \underbrace{\frac{1}{Z} \sum_b \langle b | e^{-\beta K} (1 - n_{k\sigma}) n_{k+q\sigma} | b \rangle}_{\langle (1 - n_{k\sigma}) n_{k+q\sigma} \rangle = [1 - f(\xi_k)] f(\xi_{k+q})} \\
&= \sum_{k\sigma} \underbrace{e^{\beta \xi_{k+q}} f(\xi_{k+q})}_{f(-\xi_{k+q})} \underbrace{e^{-\beta \xi_k} f(-\xi_k)}_{f(\xi_k)} \delta(\varepsilon - \xi_{k+q} + \xi_k),
\end{aligned}$$

which is the first of equations (3.28). The second of equations (3.28) follows in exactly the same way (the minus sign comes from the $-\eta$ in equation (3.14), since we must set $\eta = +1$ here). Finally, the retarded correlation function takes the form

$$\begin{aligned}
\chi^0_{nn}(q, \varepsilon) &= \int_{-\infty}^{\infty} d\varepsilon' \frac{\rho^{>}_{n(q)n(-q)}(\varepsilon') + \rho^{<}_{n(q)n(-q)}(\varepsilon')}{\varepsilon - \varepsilon' + i0^+} \\
&= \sum_{k\sigma} \frac{f(\xi_k) f(-\xi_{k+q}) - f(-\xi_k) f(\xi_{k+q})}{\varepsilon + \xi_k - \xi_{k+q} + i0^+} \\
&= \sum_{k\sigma} \frac{f(\xi_k)[1 - f(\xi_{k+q})] - [1 - f(\xi_k)] f(\xi_{k+q})}{\varepsilon + \xi_k - \xi_{k+q} + i0^+} \\
&= \sum_{k\sigma} \frac{f(\xi_k) - f(\xi_{k+q})}{\varepsilon + \xi_k - \xi_{k+q} + i0^+}.
\end{aligned}$$

Doc A3.4 Density–density correlation function of independent electrons

Equation (3.29) is particularly simple because the wave functions of free electrons are simple. For electrons that are independent but not free, the wave functions carry information about the underlying potential and this has consequences for the density–density correlation function. In general, the Hamiltonian of independent electrons can be written as

$$K = \sum_{\alpha\sigma} \xi_\alpha \gamma^\dagger_{\alpha\sigma} \gamma_{\alpha\sigma},$$

where $\gamma^\dagger_{\alpha\sigma}$ is the operator that creates a spin-σ electron in the eigenstate α of the Hamiltonian, with excitation energy ξ_α and wave function $\varphi_\alpha(r)$. In order to express the density operator, we need the relation between $c^\dagger_{k\sigma}$ (the free-electron creation operator) and $\gamma^\dagger_{\alpha\sigma}$. Following equations (2.88) and (2.91), we get

$$\begin{aligned}
c_{k\sigma}^{\dagger} &= \frac{1}{\sqrt{\mathcal{V}}}\int d\boldsymbol{r}\; e^{i\boldsymbol{k}\cdot\boldsymbol{r}}\sum_{\alpha}\varphi_{\alpha}^{*}(\boldsymbol{r})\gamma_{\alpha\sigma}^{\dagger}\\
&= \frac{1}{\sqrt{\mathcal{V}}}\sum_{\alpha}\underbrace{\Big[\int d\boldsymbol{r}\;\varphi_{\alpha}(\boldsymbol{r})e^{-i\boldsymbol{k}\cdot\boldsymbol{r}}\Big]^{*}}_{\varphi_{\alpha}(\boldsymbol{k})}\gamma_{\alpha\sigma}^{\dagger}\\
&= \frac{1}{\sqrt{\mathcal{V}}}\sum_{\alpha}\varphi_{\alpha}^{*}(\boldsymbol{k})\gamma_{\alpha\sigma}^{\dagger}.
\end{aligned}$$

Inserting into equation (2.46) yields the expression of the density in terms of the γ's:

$$n(\boldsymbol{q}) = \sum_{\alpha\beta\sigma} S_{\alpha\beta}(\boldsymbol{q})\,\gamma_{\alpha\sigma}^{\dagger}\gamma_{\beta\sigma},\qquad S_{\alpha\beta}(\boldsymbol{q}) = \frac{1}{\mathcal{V}}\sum_{\boldsymbol{k}}\varphi_{\alpha}^{*}(\boldsymbol{k})\varphi_{\beta}(\boldsymbol{k}+\boldsymbol{q}).$$

Because the γ's have the same commutation rules than the c's, the calculation of the density–density correlation function proceeds exactly like for free electrons in Doc A3.3 with matrix elements like $\langle a|c_{\alpha}^{\dagger}c_{\beta}|b\rangle$ replaced by $\langle a|\gamma_{\alpha}^{\dagger}\gamma_{\beta}|b\rangle$. One thus finds

$$\rho_{n(\boldsymbol{q})n(-\boldsymbol{q})}^{>}(\varepsilon) = \sum_{\alpha\beta\sigma}\underbrace{S_{\alpha\beta}(\boldsymbol{q})S_{\beta\alpha}(-\boldsymbol{q})}_{|S_{\alpha\beta}(\boldsymbol{q})|^{2}}f(\xi_{\alpha})f(-\xi_{\beta})\delta(\varepsilon+\xi_{\alpha}-\xi_{\beta})$$

and a similar result for $\rho_{n(\boldsymbol{q})n(-\boldsymbol{q})}^{<}(\varepsilon)$, which finally gives

$$\chi_{nn}^{0}(\boldsymbol{q},\varepsilon) = \sum_{\alpha\beta\sigma}|S_{\alpha\beta}(\boldsymbol{q})|^{2}\frac{f(\xi_{\alpha})-f(\xi_{\beta})}{\varepsilon+\xi_{\alpha}-\xi_{\beta}+i0^{+}}.$$

Doc A3.5 Relations between spectral functions

Using equations (6), (3.12), (3.14), and (3.23), we can write

$$\begin{aligned}
-\eta d_{-\eta}(\varepsilon)\rho_{AB}(\varepsilon) &= \frac{-\eta}{e^{\beta\varepsilon}-\eta}\frac{1}{Z}\sum_{ab}e^{-\beta K_b}\underbrace{[e^{-\beta(K_a-K_b)}-\eta]}_{e^{\beta\varepsilon}-\eta}\\
&\quad\times\langle a|A|b\rangle\langle b|B|a\rangle\delta(\varepsilon+K_a-K_b)\\
&= \rho_{AB}^{<}(\varepsilon).
\end{aligned}$$

In the first line, we have introduced into the exponential the constraint imposed by the delta function. In a similar fashion:

$$\begin{aligned}
-\eta d_{-\eta}(-\varepsilon)\rho_{AB}(\varepsilon) &= \frac{1}{e^{-\beta\varepsilon}-\eta}\frac{1}{Z}\sum_{ab}e^{-\beta K_a}\underbrace{[-\eta+e^{\beta(K_a-K_b)}]}_{-\eta+e^{-\beta\varepsilon}}\\
&\quad\times\langle a|A|b\rangle\langle b|B|a\rangle\delta(\varepsilon+K_a-K_b)\\
&= \rho_{AB}^{>}(\varepsilon).
\end{aligned}$$

This establishes equation (3.32).

Doc A3.6 Elementary sum rules

Doc A3.6.1 Spectral function

The integration over ε simply removes the delta function in equation (3.12):

$$\int_{-\infty}^{\infty} d\varepsilon\, \rho_{AB}^{>}(\varepsilon) = \frac{1}{Z}\sum_{ab} e^{-\beta K_a}\langle a|A|b\rangle\langle b|B|a\rangle.$$

The b sum gives $\mathbb{1}$:

$$\begin{aligned}\int_{-\infty}^{\infty} d\varepsilon\, \rho_{AB}^{>}(\varepsilon) &= \frac{1}{Z}\sum_{a} e^{-\beta K_a}\langle a|AB|a\rangle = \frac{1}{Z}\sum_{a}\langle a|e^{-\beta K}AB|a\rangle \\ &= \mathrm{Tr}\,\rho AB = \langle AB\rangle.\end{aligned}$$

Similarly,

$$\begin{aligned}\int_{-\infty}^{\infty} d\varepsilon\, \rho_{AB}^{<}(\varepsilon) &= -\eta\frac{1}{Z}\sum_{ab} e^{-\beta K_b}\langle b|B|a\rangle\langle a|A|b\rangle \\ &= -\eta\frac{1}{Z}\sum_{b}\langle b|e^{-\beta K}BA|b\rangle \\ &= -\eta\mathrm{Tr}\,\rho BA = -\eta\langle BA\rangle.\end{aligned}$$

Doc A3.6.2 Occupation numbers

Using equations (3.7) and (3.21), we can write

$$\begin{aligned}\langle n_\alpha\rangle = \langle a_\alpha^\dagger a_\alpha\rangle &= \langle a_\alpha^\dagger(0)a_\alpha(0)\rangle = i\hbar\eta C_{a_\alpha a_\alpha^\dagger}^{<}(0) \\ &= i\hbar\eta\int_{-\infty}^{\infty}\frac{d\omega}{2\pi}\underbrace{C_{a_\alpha a_\alpha^\dagger}^{<}(\omega)}_{2\pi i\rho_{a_\alpha a_\alpha^\dagger}^{<}(\hbar\omega)} \\ &= -\eta\int_{-\infty}^{\infty} d\varepsilon\,\rho_{a_\alpha a_\alpha^\dagger}^{<}(\varepsilon).\end{aligned}$$

We can then use equation (3.32) and prove equation (3.38):

$$\langle n_\alpha\rangle = -\eta\int_{-\infty}^{\infty} d\varepsilon\,\rho_{a_\alpha a_\alpha^\dagger}^{<}(\varepsilon) = (-\eta)^2\int_{-\infty}^{\infty} d\varepsilon\, d_{-\eta}(\varepsilon)\rho_{a_\alpha a_\alpha^\dagger}(\varepsilon).$$

It is also interesting to derive equation (3.38) using the imaginary-time formalism. In this case, the trick is to permute the operators by taking advantage of the time ordering:

$$\langle n_\alpha\rangle = \langle T_\tau a_\alpha^\dagger(0)a_\alpha(0^-)\rangle = \eta\langle T_\tau a_\alpha(0^-)a_\alpha^\dagger(0)\rangle = -\eta\mathscr{C}_{a_\alpha a_\alpha^\dagger}(0^-).$$

To finish, we make use of equations (4.8), (4.11), and (16):

$$\langle n_\alpha \rangle = -\eta\frac{1}{\beta}\sum_{i\nu_n} e^{-i\nu_n 0^-}\underbrace{\mathscr{C}_{a_\alpha a_\alpha^\dagger}(i\nu_n)}_{\int_{-\infty}^{\infty} d\varepsilon \frac{\rho_{a_\alpha a_\alpha^\dagger}(\varepsilon)}{i\nu_n - \varepsilon}}$$

$$= -\eta\int_{-\infty}^{\infty} d\varepsilon\, \rho_{a_\alpha a_\alpha^\dagger}(\varepsilon)\underbrace{\frac{1}{\beta}\sum_{i\nu_n}\frac{e^{i\nu_n 0^+}}{i\nu_n - \varepsilon}}_{-\eta d_{-\eta}(\varepsilon)e^{i\varepsilon 0^+}}.$$

Doc A3.6.3 Energy

In order to prove equation (3.40), we start by noticing that

$$\sum_\alpha a_\alpha^\dagger[a_\alpha, K] = \sum_{\alpha\nu}\xi_{\alpha\nu}a_\alpha^\dagger a_\nu + \sum_{\alpha\nu\mu_1\mu_2} V_{\alpha\nu\mu_1\mu_2}a_\alpha^\dagger a_\nu^\dagger a_{\mu_2}a_{\mu_1}$$

$$= K_0 + 2V,$$

where we have used equation (25). On the other hand, we have

$$\langle a_\alpha^\dagger \underbrace{[a_\alpha, K]}_{i\hbar\partial_t a_\alpha|_{t=0}}\rangle = i\hbar\partial_t\langle a_\alpha^\dagger(0)a_\alpha(t)\rangle|_{t=0} = -\eta\hbar^2\partial_t C^<_{a_\alpha a_\alpha^\dagger}(t)|_{t=0}$$

$$= -\eta\hbar^2\partial_t\int_{-\infty}^{\infty}\frac{d\omega}{2\pi}\underbrace{C^<_{a_\alpha a_\alpha^\dagger}(\omega)}_{2\pi i\rho^<_{a_\alpha a_\alpha^\dagger}(\hbar\omega)} e^{-i\omega t}|_{t=0}$$

$$= -\eta\hbar^2\int_{-\infty}^{\infty} d\omega\, \omega\, \rho^<_{a_\alpha a_\alpha^\dagger}(\hbar\omega) = \int_{-\infty}^{\infty} d\varepsilon\, \varepsilon\, d_{-\eta}(\varepsilon)\rho_{a_\alpha a_\alpha^\dagger}(\varepsilon).$$

Combining these two results, we find

$$\langle K_0 + 2V\rangle = \sum_\alpha\int_{-\infty}^{\infty} d\varepsilon\, \varepsilon\, d_{-\eta}(\varepsilon)\rho_{a_\alpha a_\alpha^\dagger}(\varepsilon)$$

and equation (3.40) results by noticing that $\langle K\rangle = \frac{1}{2}\langle K_0 + 2V\rangle + \frac{1}{2}\langle K_0\rangle$. It is interesting to evaluate equation (3.40) for $V = 0$. In this latter case we get

$$\langle K_0\rangle = \sum_\alpha\int_{-\infty}^{\infty} d\varepsilon\, \varepsilon\, d_{-\eta}(\varepsilon)\rho^0_{a_\alpha a_\alpha^\dagger}(\varepsilon).$$

For independent fermions, we have already seen (equation (3.25)) that $\rho^0_{a_\alpha a_\alpha^\dagger}(\varepsilon) = \delta(\varepsilon - \xi_\alpha)$ and thus equation (3.42) follows at once. We now show that the same expression for the spectral function is also valid for independent bosons. Equations (3.12) and (3.14) give in our case

$$\rho^0_{b_\alpha b_\alpha^\dagger}(\varepsilon) = \frac{1}{Z}\sum_{ab}(e^{-\beta K_a} - e^{-\beta K_b})|\langle a|b_\alpha|b\rangle|^2\delta(\varepsilon + K_a - K_b).$$

Here b_α is the boson annihilation operator. Proceeding as in Doc A3.3, we see that

$$|\langle a|b_\alpha|b\rangle|^2 = \begin{cases} n_\alpha^{|b\rangle} & \text{if } n_i^{|a\rangle} = n_i^{|b\rangle} \text{ for } i \neq \alpha \text{ and } n_\alpha^{|a\rangle} = n_\alpha^{|b\rangle} - 1 \\ 0 & \text{otherwise,} \end{cases}$$

which requires that $E_a = E_b - \varepsilon_\alpha$ and $N_a = N_b - 1$:

$$\begin{aligned} \rho^0_{b_\alpha b_\alpha^\dagger}(\varepsilon) &= \frac{1}{Z}\sum_b [e^{-\beta(E_b-\varepsilon_\alpha-\mu N_b+\mu)} - e^{-\beta K_b}] n_\alpha^{|b\rangle}\delta(\varepsilon - \xi_\alpha) \\ &= \delta(\varepsilon - \xi_\alpha)(e^{\beta\xi_\alpha} - 1)\frac{1}{Z}\sum_b \langle b|e^{-\beta K} n_\alpha|b\rangle \\ &= \delta(\varepsilon - \xi_\alpha)\frac{1}{b(\xi_\alpha)}\mathrm{Tr}\,\rho\, n_\alpha = \delta(\varepsilon - \xi_\alpha), \end{aligned}$$

since $\mathrm{Tr}\,\rho\, n_\alpha = \langle n_\alpha\rangle = b(\xi_\alpha)$ for independent bosons, with $b(\varepsilon)$ the Bose–Einstein distribution function.

Doc A3.7 Moments of the spectral function

Here we show how the first moment of the spectral function—and the higher-order moments as well—can be evaluated using the spectral representation. We use equations (3.12), (3.14), and (3.23) to arrive at:

$$\begin{aligned} \int_{-\infty}^{\infty} d\varepsilon\, \varepsilon\, \rho_{AB}(\varepsilon) &= \frac{1}{Z}\sum_{ab}(e^{-\beta K_a} - \eta e^{-\beta K_b})\langle a|A|b\rangle\langle b|B|a\rangle(K_b - K_a) \\ &= \frac{1}{Z}\sum_{ab}(e^{-\beta K_a}K_b - e^{-\beta K_a}K_a - \eta e^{-\beta K_b}K_b + \eta e^{-\beta K_b}K_a) \\ &\quad \times \langle a|A|b\rangle\langle b|B|a\rangle. \end{aligned}$$

Each of the four terms can be rearranged in such a way that one of the sums gives the identity. For instance, the first term becomes

$$\begin{aligned} \frac{1}{Z}\sum_{ab}\langle a|e^{-\beta K_a}AK_b|b\rangle\langle b|B|a\rangle &= \frac{1}{Z}\sum_{ab}\langle a|e^{-\beta K}AK|b\rangle\langle b|B|a\rangle \\ &= \frac{1}{Z}\sum_{a}\langle a|e^{-\beta K}AKB|a\rangle = \mathrm{Tr}\,\rho AKB = \langle AKB\rangle. \end{aligned}$$

Considering all four terms we obtain equation (3.46):

$$\begin{aligned} \int_{-\infty}^{\infty} d\varepsilon\, \varepsilon\, \rho_{AB}(\varepsilon) &= \langle AKB\rangle - \langle KAB\rangle - \eta\langle BAK\rangle + \eta\langle BKA\rangle \\ &= -\left\langle [[K, A], B]_{-\eta}\right\rangle. \end{aligned}$$

Appendix B3

Bonus B3.1 Double occupancy

The double occupancy D measures the average occupation of the same point in space by two electrons of opposite spins:

$$D = \frac{1}{\mathcal{V}} \int d\boldsymbol{r}\, \langle n_\uparrow(\boldsymbol{r}) n_\downarrow(\boldsymbol{r}) \rangle = \frac{1}{\mathcal{V}} \int d\boldsymbol{r}\, \langle n_\downarrow(\boldsymbol{r}) n_\uparrow(\boldsymbol{r}) \rangle.$$

This quantity plays an important role in strongly correlated materials, where it is suppressed by the Coulomb repulsion. With the help of equation (3.35), we can relate D to the density–density and longitudinal spin–spin greater correlation functions. Equation (3.35) tells us that

$$\int_{-\infty}^{\infty} d\varepsilon\, \rho^{>}_{n(\boldsymbol{q})n(-\boldsymbol{q})}(\varepsilon) = \langle n(\boldsymbol{q}) n(-\boldsymbol{q}) \rangle$$
$$\int_{-\infty}^{\infty} d\varepsilon\, \rho^{>}_{S^z(\boldsymbol{q})S^z(-\boldsymbol{q})}(\varepsilon) = \langle S^z(\boldsymbol{q}) S^z(-\boldsymbol{q}) \rangle.$$

Summing over $\boldsymbol{q}$ and moving to real space, we get

$$\begin{aligned}\frac{1}{\mathcal{V}} \sum_{\boldsymbol{q}} \langle n(\boldsymbol{q}) n(-\boldsymbol{q}) \rangle &= \int d\boldsymbol{r}\, \langle n(\boldsymbol{r}) n(\boldsymbol{r}) \rangle \\ &= \int d\boldsymbol{r}\, \langle [n_\uparrow(\boldsymbol{r}) + n_\downarrow(\boldsymbol{r})][n_\uparrow(\boldsymbol{r}) + n_\downarrow(\boldsymbol{r})] \rangle,\end{aligned}$$

and similarly for $\langle S^z(\boldsymbol{q}) S^z(-\boldsymbol{q}) \rangle$,

$$\begin{aligned}\frac{1}{\mathcal{V}} \sum_{\boldsymbol{q}} \langle S^z(\boldsymbol{q}) S^z(-\boldsymbol{q}) \rangle &= \int d\boldsymbol{r}\, \langle S^z(\boldsymbol{r}) S^z(\boldsymbol{r}) \rangle \\ &= \left(\frac{\hbar}{2}\right)^2 \int d\boldsymbol{r}\, \langle [n_\uparrow(\boldsymbol{r}) - n_\downarrow(\boldsymbol{r})][n_\uparrow(\boldsymbol{r}) - n_\downarrow(\boldsymbol{r})] \rangle.\end{aligned}$$

Combined together, these relations lead to

$$D = \frac{1}{\mathcal{V}^2} \sum_{\boldsymbol{q}} \int_{-\infty}^{\infty} d\varepsilon\, \frac{1}{4}\left[\rho^{>}_{n(\boldsymbol{q})n(-\boldsymbol{q})}(\varepsilon) - \left(\frac{2}{\hbar}\right)^2 \rho^{>}_{S^z(\boldsymbol{q})S^z(-\boldsymbol{q})}(\varepsilon)\right].$$

Bonus B3.2 Analyticity, causality, and the Kramers–Kronig relations

Consider a complex function F of a complex variable z and denote $F^R(\omega)$ with $\omega \in \mathbb{R}$ the function just above the real axis: $F^R(\omega) \equiv F(z = \omega + i0^+)$. Assume now that the function is analytic in the upper half of the complex plane and vanishes in the upper half at least as $1/z$ for $|z| \to \infty$. These two conditions are sufficient for the function to obey Kramers–Kronig relations. Indeed, using equation (9) we have

$$\begin{aligned}\int_{-\infty}^{\infty} d\omega' \frac{F^R(\omega')}{\omega - \omega' + i0^+} &= \int_{-\infty}^{\infty} d\omega' [\mathrm{Re}\, F^R(\omega') + i\, \mathrm{Im}\, F^R(\omega')] \\ &\quad \times \left[\mathscr{P}\frac{1}{\omega - \omega'} - i\pi\delta(\omega - \omega') \right] \\ &= \fint_{-\infty}^{\infty} d\omega' \frac{\mathrm{Re}\, F^R(\omega')}{\omega - \omega'} + \pi\, \mathrm{Im}\, F^R(\omega) \\ &\quad + i\left[\fint_{-\infty}^{\infty} d\omega' \frac{\mathrm{Im}\, F^R(\omega')}{\omega - \omega'} - \pi\, \mathrm{Re}\, F^R(\omega) \right].\end{aligned}$$

The same integral can also be evaluated by closing the contour in the upper half:

$$\begin{aligned}\int_{-\infty}^{\infty} d\omega' \frac{F^R(\omega')}{\omega - \omega' + i0^+} &= \oint_C dz \frac{F(z)}{\omega - z + i0^+} = -\oint_C dz \frac{F(z)}{z - (\omega + i0^+)} \\ &= -2\pi i\, F(\omega + i0^+) = -2\pi i\, F^R(\omega) \\ &= 2\pi\, \mathrm{Im}\, F^R(\omega) - 2\pi i\, \mathrm{Re}\, F^R(\omega).\end{aligned}$$

We have used equation (7) and our assumptions of analyticity and behavior at infinity. Comparing the real and imaginary parts in the two results for the integral, we deduce the Kramers–Kronig relations:

$$\begin{aligned}\mathrm{Im}\, F^R(\omega) &= \frac{1}{\pi} \fint_{-\infty}^{\infty} d\omega' \frac{\mathrm{Re}\, F^R(\omega')}{\omega - \omega'} \\ \mathrm{Re}\, F^R(\omega) &= -\frac{1}{\pi} \fint_{-\infty}^{\infty} d\omega' \frac{\mathrm{Im}\, F^R(\omega')}{\omega - \omega'}.\end{aligned}$$

Kramers–Kronig consistency is tightly related to causality and sufficient to imply that the function $F^R(t)$ is proportional to $\theta(t)$. The two Kramers–Kronig relations can be recast into

$$\begin{aligned}F^R(\omega) &= \frac{i}{\pi} \fint_{-\infty}^{\infty} d\omega' \frac{\mathrm{Re}\, F^R(\omega') + i\, \mathrm{Im}\, F^R(\omega')}{\omega - \omega'} = \frac{i}{\pi} \fint_{-\infty}^{\infty} d\omega' \frac{F^R(\omega')}{\omega - \omega'} \\ &= \frac{i}{\pi} \int_{-\infty}^{\infty} d\omega'\, F^R(\omega') \left[\frac{1}{\omega - \omega' + i0^+} + i\pi\delta(\omega - \omega') \right] \\ &= \frac{i}{\pi} \int_{-\infty}^{\infty} d\omega' \frac{F^R(\omega')}{\omega - \omega' + i0^+} - F^R(\omega),\end{aligned}$$

from where it follows that

$$F^R(\omega) = \int_{-\infty}^{\infty} \frac{d\omega'}{2\pi}\, F^R(\omega') \frac{i}{\omega - \omega' + i0^+}.$$

Take the Fourier transform with the help of equation (8) (see Doc A1):

$$\begin{aligned}
F^R(t) &= \int_{-\infty}^{\infty} \frac{d\omega}{2\pi} F^R(\omega) e^{-i\omega t} \\
&= \int_{-\infty}^{\infty} \frac{d\omega'}{2\pi} F^R(\omega') \int_{-\infty}^{\infty} \frac{d\omega}{2\pi} e^{-i\omega t} \frac{i}{\omega - \omega' + i0^+} \\
&= \int_{-\infty}^{\infty} \frac{d\omega'}{2\pi} F^R(\omega') e^{-i\omega' t} \underbrace{\int_{-\infty}^{\infty} \frac{d\omega}{2\pi} e^{-i\omega t} \frac{i}{\omega + i0^+}}_{\theta(t)} \\
&= \theta(t) \int_{-\infty}^{\infty} \frac{d\omega}{2\pi} F^R(\omega) e^{-i\omega t}.
\end{aligned}$$

Here we have made implicit regularity assumptions about the function $F^R(\omega)$, such that the two integrals can be exchanged. Hence a Kramers–Kronig consistent function is causal. The converse is also true: given a causal function $F^R(t)$ and its Fourier transform $F^R(\omega)$, the continuation $F(z)$ of $F^R(\omega)$ in the upper half of the complex plane is analytic and decays at infinity, therefore Kramers–Kronig consistent. To see this, we note that, if $F^R(t) = \theta(t)\tilde{F}(t)$, the following property holds because $\theta^2(t) = \theta(t)$: $F^R(t) = F^R(t)\theta(t)$. As the Fourier transform of a product is a convolution, we have

$$\begin{aligned}
F^R(\omega) &= \int_{-\infty}^{\infty} \frac{d\omega'}{2\pi} F^R(\omega')\theta(\omega - \omega') \\
&= \int_{-\infty}^{\infty} \frac{d\omega'}{2\pi} F^R(\omega') \frac{i}{\omega - \omega' + i0^+},
\end{aligned}$$

were we have again used equation (8). The continuation in the complex plane is

$$F(z) = \frac{i}{2\pi} \int_{-\infty}^{\infty} d\omega' \frac{F^R(\omega')}{z - \omega' + i0^+}.$$

The poles of this function are at $z = \omega' - i0^+$, which is in the lower half of the complex plane. Hence the function is analytic in the upper half. It also clearly decays as $1/z$ at infinity.

Chapter 4

Imaginary-time formalism

The correlation functions that we are interested in are functions of time and temperature. The time dependence arises from the evolution operator, which typically implies exponentiating the Hamiltonian multiplied by $-it/\hbar$, while the temperature dependence stems from the statistical density matrix, which is an exponential of the Hamiltonian as well, but multiplied by $-\beta$. It turns out that the direct calculation of time-dependent correlation functions for interacting particles at finite temperature is hard. The very convenient tool of time ordering, which at zero temperature allows one to order the operators in the expansion of the time evolution, cannot order operators along both the $it/\hbar$ and β axes. The imaginary-time formalism introduced in this chapter gets around the difficulty by rotating the real-time axis by ninety degrees into the complex plane, where the evolution operator becomes identical to the statistical density matrix.

4.1 Motivation

In order to make progress beyond the independent-particle stage that we saw in section 3.3, we need a method for calculating correlation functions without completely neglecting the interactions. We must evaluate quantities of the kind

$$\langle A(t)B(0)\rangle = \mathrm{Tr}\,\rho A(t)B, \tag{4.1}$$

where (see equations (2.2), (2.15) and section 2.2.1)

$$\rho = \frac{e^{-\beta K}}{\mathrm{Tr}\,e^{-\beta K}}, \quad K = H - \mu N,$$
$$A(t) = U^{\dagger}(t)AU(t), \quad \text{and} \quad i\hbar\partial_t U(t) = KU(t).$$

One approach is perturbation theory, assuming that the Hamiltonian can be split in a part K_0 that can be solved and a part V that contains the difficult terms. We then seek an expansion in powers of V. Expanding $e^{-\beta K}$ and $e^{-iKt/\hbar}$ directly is not a good

doi:10.1088/978-0-7503-1741-2ch4

idea because this would mix together powers of K_0 and powers of V. There is in fact no reason to expand in powers of K_0. The solution is the interaction picture, in which the evolution operator is written as a product $U = U_0\hat{U}$ and the time evolution of the operators is governed by K_0 (see section 2.2.1). In the interaction picture, we have (equation (2.20))

$$\langle A(t)B(0)\rangle = \mathrm{Tr}\,\rho\ \hat{U}^{\dagger}(t)\hat{A}(t)\hat{U}(t)B \tag{4.2}$$

and we have seen that the evolution $\hat{U}(t)$ has a formal expansion in powers of V, equation (2.23), which can be made very elegant thanks to the time ordering operator.

The difficulty is that we do not have an equivalent expansion for ρ. The density matrix follows the equation of motion $-\partial_\beta\rho = (K - \langle K\rangle)\rho$—known as the Bloch equation—which is similar to the equation of motion of $U(t)$, however with the time variable t replaced by $-i\hbar\beta$. This leads to a simple expansion of ρ in powers of V, but this expansion involves an 'evolution' of the interaction along the imaginary variable $-i\hbar\beta$ rather than along the time axis. Mixing this expansion with the expansion of $\hat{U}(t)$ in real time leads to an intractable expression, mainly because there is no way to 'time order' mixed operators that evolve along two orthogonal axes.

The solution proposed by Matsubara[1] exploits the similarity between the operators $e^{-\beta K}$ and $e^{-iKt/\hbar}$ and considers an analytic continuation e^{-iKz} of both of them in the plane of complex times $\hbar z$. One then replaces the real-time evolution that $\hat{U}$ carries along t by an imaginary-time evolution along $-i\hbar\tau$, $\tau \in \mathbb{R}$: $e^{-iKt/\hbar} \to e^{-iK(-i\tau)} = e^{-K\tau}$. This is done by introducing new operators that evolve in imaginary rather than real time. Thanks to this trick, a tractable expansion of the correlation functions of these new operators evolving in imaginary time can be written down. Due to the analytic nature of e^{-iKz}, it turns out that the real-time functions in which we are interested can be recovered by the analytic continuation of the imaginary-time ones.

In order to achieve this program, we introduce the new operators

$$\boxed{A(\tau) \equiv e^{\tau K}Ae^{-\tau K},} \tag{4.3}$$

where τ is a real number in the range $-\beta < \tau < \beta$.[2] Note that the unit of τ is inverse energy—time divided by $\hbar$— rather than time. Now we ought to study the imaginary-time correlation functions of the kind

$$\langle A(\tau)B(0)\rangle = \mathrm{Tr}\,\rho A(\tau)B \tag{4.4}$$

instead of their real-time counterparts $\langle A(t)B(0)\rangle$. This additional complication is the price to pay for working at finite temperature, which implies using the statistical density matrix. At zero temperature, the traces involving ρ are replaced by ground-state expectation values and the problem does not occur. The imaginary time is

[1] Matsubara T 1955 *Prog. Theor. Phys.* **14** 351.

[2] The imaginary-time evolution is periodic with period 2β.

avoided in many textbooks that focus on the zero-temperature perturbation theory. While this is sufficient for many purposes, it lacks the generality of a finite-temperature formalism. After all, experiments are never performed at zero temperature.

Before moving on, we point out two reasons not to spend too much time studying the zero-temperature formalism. First, this formalism can give the wrong answer: in some instances, the $T = 0$ results do not coincide with the $T \to 0$ limit of the finite-temperature ones. This is due to the discontinuity of the distribution functions at $T = 0$, a discontinuity that is absent as long as T remains finite[3]. Furthermore, the perturbation theory at $T = 0$ delivers the real-time ordered or causal correlation functions, equation (3.1), which are not very useful per se. In contrast, as we shall see, the imaginary-time ordered correlations functions provided by the Matsubara formalism give access to the retarded correlation functions, which are directly relevant for making contact with experiments.

4.2 Correlation functions in imaginary time

We define the imaginary-time correlation function of the operators A and B as

$$\boxed{\mathscr{C}_{AB}(\tau) = -\langle T_\tau A(\tau)B(0)\rangle,} \tag{4.5}$$

where T_τ is the τ-ordering operator, defined in complete analogy with equation (5):

$$T_\tau A(\tau)B(0) = \theta(\tau)A(\tau)B(0) + \eta\theta(-\tau)B(0)A(\tau). \tag{4.6}$$

The index τ in T_τ is there to avoid confusion with the temperature T; it does not mean that T_τ is a function of τ. Like for the real-time functions, the value of η is set by the statistics of the operators, not that of the particles (see section 3.1). We use script instead of roman symbols for imaginary-time objects and Greek letters instead or roman letters for the time argument. The first important property to notice is that the imaginary-time correlation functions are 'η-periodic' as a function of τ (see Doc A4.1):

$$\mathscr{C}_{AB}(\tau - \beta) = \eta\mathscr{C}_{AB}(\tau) \qquad 0 < \tau < \beta. \tag{4.7}$$

Hence $\mathscr{C}_{AB}(\tau)$ at negative imaginary time is not independent from $\mathscr{C}_{AB}(\tau)$ at positive time. An illustration is given in figure 4.1. Being periodic functions of τ, the correlation functions have a representation as a discrete Fourier series:

$$\boxed{\mathscr{C}_{AB}(\tau) = \frac{1}{\beta}\sum_{i\nu_n}\mathscr{C}_{AB}(i\nu_n)e^{-i\nu_n\tau}.} \tag{4.8}$$

[3] Kohn W and Luttinger J M 1960 *Phys. Rev.* **118** 41.

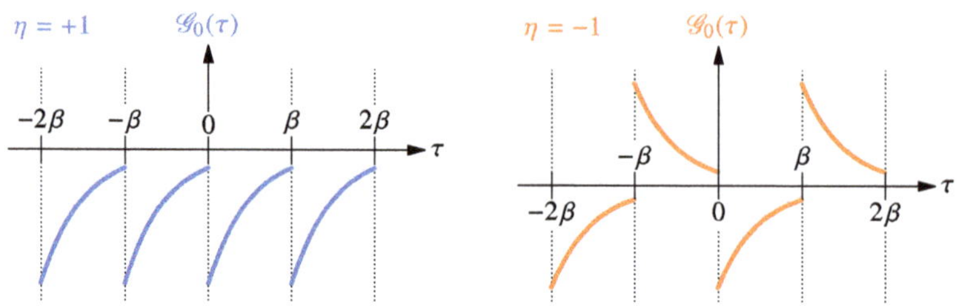

Figure 4.1. The imaginary-time correlation functions are real if the Hamiltonian is real, periodic over 2β, and periodic (anti-periodic) over β for bosonic (fermionic) operators. The example shown is the non-interacting single-particle Green's function given by equation (17) for $\varepsilon = 2k_\mathrm{B}T$; see also section 5.2.2.1. Due to the discontinuity at $\tau = 0$, it is meaningless to refer to correlation functions at $\tau = 0$ and we shall always refer to $\tau = 0^+$ or $\tau = 0^-$.

The η-periodicity requires that $e^{-i\nu_n(\tau-\beta)} = \eta e^{-i\nu_n\tau}$, in other words $e^{i\nu_n\beta} = \eta$, which means that

$$\boxed{\nu_n = \begin{cases} 2n\pi/\beta \equiv \Omega_n & \eta = +1 \\ (2n+1)\pi/\beta \equiv \omega_n & \eta = -1. \end{cases}} \tag{4.9}$$

The discrete quantities Ω_n and ω_n are called the bosonic and fermionic Matsubara frequencies, respectively. We use the notation ν_n when referring indistinctively to bosonic and fermionic frequencies. Note that the common practice is to call them frequencies although they are actually energies. Figure 4.2 shows that the analytic continuation $d_{-\eta}(z)$ of the distribution functions, equation (6), has poles at the Matsubara frequencies. The inverse Fourier transform reads[4]

$$\boxed{\mathcal{C}_{AB}(i\nu_n) = \int_0^\beta d\tau\, \mathcal{C}_{AB}(\tau)e^{i\nu_n\tau}.} \tag{4.10}$$

4.3 Analytic continuation

The most remarkable property of the imaginary-time ordered correlation functions is that they have the *same* spectral representation as the real-time retarded and advanced functions. To see this, we expand $\mathcal{C}_{AB}(\tau)$ as in section 3.2 using the complete set of eigenstates of H and we obtain (see Doc A4.2):

$$\mathcal{C}_{AB}(i\nu_n) = \int_{-\infty}^{\infty} d\varepsilon \frac{\rho_{AB}(\varepsilon)}{i\nu_n - \varepsilon}, \tag{4.11}$$

where $\rho_{AB}(\varepsilon)$ is given by equation (3.23). By comparing with the spectral representation of the retarded correlation function, equation (3.18), we see that

$$\boxed{C^R_{AB}(\omega) = \mathcal{C}_{AB}(i\nu_n \to \hbar\omega + i0^+).} \tag{4.12}$$

[4] The definition of the discrete Fourier transform implies integration over the whole domain as $\frac{1}{2}\int_{-\beta}^{\beta} d\tau$, which can be recast in the form of equation (4.10) with the help of equation (4.7).

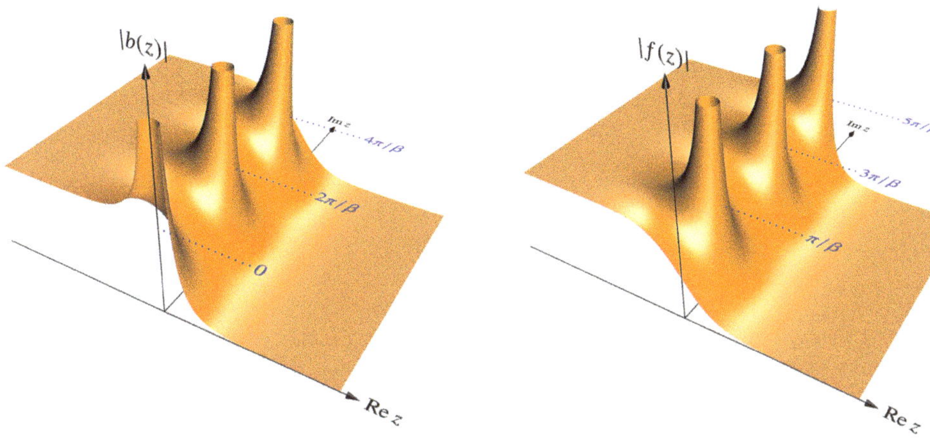

Figure 4.2. The analytic continuation of the Bose–Einstein (left) and Fermi–Dirac (right) distribution to the complex plane has poles along the imaginary axis at the Matsubara frequencies $i\Omega_n$ and $i\omega_n$, respectively.

In words, *the real-frequency retarded correlation function is obtained by replacing in the analytical expression of* $\mathscr{C}_{AB}(i\nu_n)$ *the complex frequency (energy)* $i\nu_n$ *by* $\hbar\omega+i0^+$.

This result immediately suggests to extend the function $\mathscr{C}_{AB}(i\nu_n)$ to the whole plane of complex energies by a straightforward analytic continuation:

$$\boxed{\mathscr{C}_{AB}(z) = \int_{-\infty}^{\infty} d\varepsilon \frac{\rho_{AB}(\varepsilon)}{z-\varepsilon}.} \tag{4.13}$$

The retarded, advanced, and Matsubara correlation functions then appear as particular instances of $\mathscr{C}_{AB}(z)$. The notation $\mathscr{C}_{AB}(i\nu_n)$ that we chose rather than $\mathscr{C}_{AB}(\nu_n)$ emphasizes the fact that these coefficients are values of $\mathscr{C}_{AB}(z)$ taken along the imaginary axis. From the definition, we see that $\mathscr{C}_{AB}(z)$ is analytic everywhere in the complex plane except on the real axis $\text{Im}\, z = 0$, where it has a discontinuity proportional to the spectral function: equation (4.13) together with equation (9) indeed imply that $\mathscr{C}_{AB}(\varepsilon + i0^+) - \mathscr{C}_{AB}(\varepsilon - i0^+) = -2\pi i\rho_{AB}(\varepsilon)$. An illustration is given in figure 4.3. Because, on the other hand, $\mathscr{C}_{AB}(\varepsilon + i0^+) = C^R_{AB}(\varepsilon/\hbar)$ and $\mathscr{C}_{AB}(\varepsilon - i0^+) = C^A_{AB}(\varepsilon/\hbar)$, we can express the spectral function in terms of the retarded and advanced correlation functions and deduce a variant of equation (4.13) that is sometimes useful:

$$\mathscr{C}_{AB}(z) = \frac{i}{2\pi}\int_{-\infty}^{\infty} d\varepsilon \frac{C^R_{AB}(\varepsilon/\hbar) - C^A_{AB}(\varepsilon/\hbar)}{z-\varepsilon}. \tag{4.14}$$

One further step allows one to express $\mathscr{C}_{AB}(z)$ only in terms of retarded functions. From the definitions (3.12) and (3.14), we see that $\rho_{AB}(\varepsilon) = \rho^*_{B^\dagger A^\dagger}(\varepsilon)$, from where it follows that $C^A_{AB}(\omega) = [C^R_{B^\dagger A^\dagger}(\omega)]^*$ and finally:

$$\mathscr{C}_{AB}(z) = \frac{i}{2\pi}\int_{-\infty}^{\infty} d\varepsilon \frac{C^R_{AB}(\varepsilon/\hbar) - [C^R_{B^\dagger A^\dagger}(\varepsilon/\hbar)]^*}{z-\varepsilon}. \tag{4.15}$$

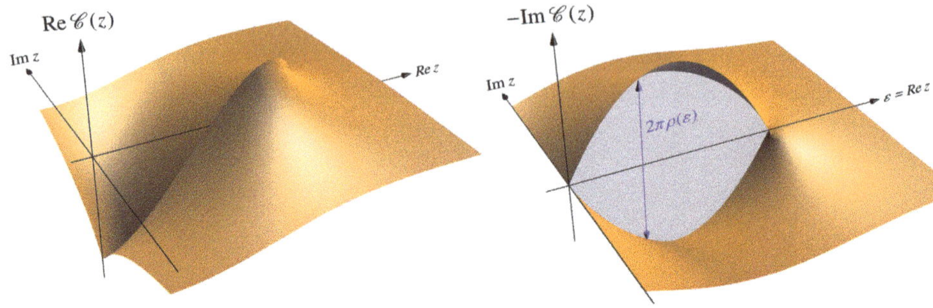

Figure 4.3. The imaginary-time correlation functions $\mathscr{C}(z)$ are analytic in the complex plane except on the real axis, where they have a discontinuity proportional to the spectral function. The example shown is the density–density correlation function of free electrons in three dimensions at $q = 2k_{\mathrm{F}}$. The spectral function is real, such that the discontinuity is only in the imaginary part of $\mathscr{C}(z)$. The spectral function gives the number of particle–hole excitations, which increases linearly at small ε and vanishes for $\varepsilon > 8\varepsilon_{\mathrm{F}}$; see section 5.1.4.5.

Appendix A4

Doc A4.1 (Anti)-periodicity of imaginary-time functions

It is sufficient to prove equation (4.7) in the case $\tau > 0$. Since $\tau < \beta$, this means that $\tau - \beta < 0$ and therefore

$$\begin{aligned}\mathscr{C}_{AB}(\tau-\beta) &= -\langle T_\tau A(\tau-\beta)B(0)\rangle = -\eta\langle B(0)A(\tau-\beta)\rangle\\ &= -\eta\frac{1}{Z}\mathrm{Tr}\, e^{-\beta K}Be^{(\tau-\beta)K}Ae^{-(\tau-\beta)K}\\ &= -\eta\frac{1}{Z}\mathrm{Tr}\, e^{-\beta K}e^{\tau K}Ae^{-\tau K}e^{\beta K}e^{-\beta K}B\\ &= -\eta\frac{1}{Z}\mathrm{Tr}\, e^{-\beta K}\underbrace{A(\tau)B(0)}_{=T_\tau A(\tau)B(0)} = -\eta\langle T_\tau A(\tau)B(0)\rangle = \eta\mathscr{C}_{AB}(\tau).\end{aligned}$$

At the third line, we have used the property $\mathrm{Tr}\,ABC = \mathrm{Tr}\,BCA$ called the cyclic property of the trace and at the fourth line, we have used the fact that $\tau > 0$ and reintroduced the imaginary-time ordering.

Doc A4.2 Spectral representation of imaginary-time functions

According to equation (4.10), it is sufficient to know $\mathscr{C}_{AB}(\tau)$ at positive times in order to calculate $\mathscr{C}_{AB}(i\nu_n)$. Using equations (2.2) and (4.3) we have

$$\begin{aligned}\mathscr{C}_{AB}(\tau > 0) &= -\frac{1}{Z}\sum_a \langle a|e^{-\beta K}e^{\tau K}Ae^{-\tau K}B|a\rangle\\ &= -\frac{1}{Z}\sum_{ab}\langle a|e^{-\beta K}e^{\tau K}A|b\rangle\langle b|e^{-\tau K}B|a\rangle\\ &= -\frac{1}{Z}\sum_{ab} e^{-\beta K_a}e^{\tau(K_a-K_b)}\langle a|A|b\rangle\langle b|B|a\rangle.\end{aligned}$$

We now perform the time integration in equation (4.10):

$$\int_0^\beta d\tau\, e^{(i\nu_n+K_a-K_b)\tau} = \frac{e^{(i\nu_n+K_a-K_b)\tau}\,\big|_0^\beta}{i\nu_n + K_a - K_b} = \frac{e^{(i\nu_n+K_a-K_b)\beta} - 1}{i\nu_n + K_a - K_b}$$
$$= \frac{\eta e^{\beta(K_a-K_b)} - 1}{i\nu_n + K_a - K_b},$$

where we have taken into account that $e^{i\nu_n\beta} = \eta$. Collecting the terms, we find

$$\mathscr{C}_{AB}(i\nu_n) = -\frac{1}{Z}\sum_{ab} e^{-\beta K_a}\frac{\eta e^{\beta(K_a-K_b)} - 1}{i\nu_n + K_a - K_b}\langle a|A|b\rangle\langle b|B|a\rangle$$
$$= \frac{1}{Z}\sum_{ab}\frac{e^{-\beta K_a} - \eta e^{-\beta K_b}}{i\nu_n + K_a - K_b}\langle a|A|b\rangle\langle b|B|a\rangle$$
$$= \int_{-\infty}^{\infty} d\varepsilon \frac{\rho_{AB}(\varepsilon)}{i\nu_n - \varepsilon}$$

with $\rho_{AB}(\varepsilon)$ defined by equations (3.12), (3.14), and (3.23).

IOP Publishing

Spectroscopic Probes of Quantum Matter

Christophe Berthod

Chapter 5

Calculating correlation functions

This chapter presents two methods that allow one to obtain approximate analytical and numerical results for interacting many-particle systems. These methods are the perturbation theory and associated Feynman-diagram techniques and the equation of motion with the usual mean-field decoupling schemes. The perturbation theory is reviewed in detail and applied to several condensed-matter problems involving one- and two-particle correlation functions. The equation of motion is derived in its most general form and decoupled in order to produce the Hartree–Fock–Gor'kov approximation. From there, the main results for a spin-singlet superconductor with or without translation invariance are recalled.

5.1 Perturbation theory and Feynman diagrams

5.1.1 Expansion of the correlation functions

Let us come back to the problem of expanding a correlation function as a power series in the interaction V, as presented in section 4.1. Working in the interaction picture and following the procedure used for real time in section 2.2.2, we write the imaginary-time evolution operator $U(\tau) = e^{-\tau K}$ as $U(\tau) = U_0(\tau)\hat{U}(\tau)$. The resulting equation of motion for $\hat{U}$ is $-\partial_\tau \hat{U}(\tau) = \hat{V}(\tau)\hat{U}(\tau)$, in complete analogy with the real-time derivation. Hence a similar expansion follows, with the difference that the time domain starts at $\tau = 0$ instead of $t = -\infty$:

$$\hat{U}(\tau) = \sum_{n=0}^{\infty} \frac{(-1)^n}{n!} \int_0^\tau d\tau_1 \cdots d\tau_n \, T_\tau \hat{V}(\tau_1) \cdots \hat{V}(\tau_n) \tag{5.1}$$

or in symbolic form,

$$\hat{U}(\tau) = T_\tau \exp\left\{-\int_0^\tau d\tau' \, \hat{V}(\tau')\right\}. \tag{5.2}$$

doi:10.1088/978-0-7503-1741-2ch5

Inserting this expansion in the expression of the imaginary-time correlation function, we obtain (see Doc A5.1)

$$\mathscr{C}_{AB}(\tau) = -\frac{\sum_{n=0}^{\infty}\frac{(-1)^n}{n!}\int_0^\beta d\tau_1 \cdots d\tau_n \, \langle T_\tau \hat{V}(\tau_1) \cdots \hat{V}(\tau_n)\hat{A}(\tau)B(0)\rangle_0}{\sum_{n=0}^{\infty}\frac{(-1)^n}{n!}\int_0^\beta d\tau_1 \cdots d\tau_n \, \langle T_\tau \hat{V}(\tau_1) \cdots \hat{V}(\tau_n)\rangle_0}. \tag{5.3}$$

The index 0 in the average values $\langle \cdots \rangle_0$ means that they must be evaluated using the density matrix in the absence of perturbation:

$$\langle A\rangle_0 = \frac{\mathrm{Tr} e^{-\beta K_0} A}{\mathrm{Tr} e^{-\beta K_0}} = \frac{\sum_a e^{-\beta K_a}\langle a|A|a\rangle}{\sum_a e^{-\beta K_a}}, \tag{5.4}$$

where K_a and $|a\rangle$ are the eigenvalues and many-body eigenvectors of K_0. Since K_a and $|a\rangle$ are known by assumption, these traces can be calculated—at least in principle. The expression (5.3) formally solves our problem: it gives a systematic recipe for evaluating the correlation function by computing an infinite series of terms. Each term involves operators in the interaction picture. Again, those can be computed because they evolve in time with K_0: $\hat{A}(\tau) = e^{\tau K_0} A e^{-\tau K_0} = \sum_{ab} e^{\tau(K_a - K_b)}|a\rangle\langle a|A|b\rangle\langle b|$. Incidentally, since the average values are evaluated in the absence of interaction, the actual value of $\hat{A}(\tau)$ is the same as the value A would take in the Heisenberg picture, but in the absence of interaction. We can therefore remove the hats in equation (5.3) if we remember that all time evolutions—not only the thermodynamic averages—are controlled by K_0.

It is important to realize that three essential ingredients have been necessary in order to obtain a workable perturbation expansion of the correlation functions in powers of V: (i) the imaginary-time formalism allowed us to use the same expansion for the density matrix ρ and the evolution U; (ii) the time ordering allowed us to rewrite this expansion in a simple way; and (iii) the interaction picture allowed us to rewrite traces over $K_0 + V$ as traces over K_0 only. Ingredient (i) is crucial if we want to work at finite temperature; ingredient (ii) explains why the perturbation expansion is possible only for the time-ordered correlation functions (and not for the retarded or advanced ones); finally, ingredient (iii) makes the practical calculations possible.

Although equation (5.3) gives a recipe, it does not mean that this recipe is easy to realize in practice: there are infinitely many terms, each of them looking exceedingly complicated with many operators, time ordering, and multiple time integrations. Three additional ingredients will be needed to move forward: (i) Wick's theorem, once generalized to finite-temperature thermodynamic averages, will allow us to rewrite an average of a product of many operators as a sum of products of averages of only two operators; (ii) the 'linked cluster theorem' will allow us to eliminate many of the terms and keep only the so-called connected ones; finally (iii) the

Feynman diagrams will allow us to interpret and manipulate the remaining terms of the expansion by replacing the complicated formulas with equivalent but much more insightful drawings.

5.1.2 Wick's theorem and cancellation of vacuum diagrams

Wick's theorem states that, *for a quadratic Hamiltonian* (recall that the perturbation expansion of $\mathscr{C}_{AB}$ involves average values with respect to the quadratic Hamiltonian K_0), high-order products of creation and annihilation operators can be factorized into lower-order products, eventually into all possible pairwise products of only two operators. We refer the reader to Bruus & Flensberg (2004, p 198) for a proof.

Following Wick's theorem, the term of order n in the numerator of equation (5.3) generates several kinds of contributions, some of which have the form

$$\langle V(\tau_1) \cdots V(\tau_j)A(\tau)B(0)\rangle\langle V(\tau_{j+1}) \cdots V(\tau_n)\rangle.$$

We omit the T_τ operators and the $\langle \cdots \rangle_0$ index in this qualitative discussion for brevity; we also remove the hats keeping in mind that the operators evolve in time with K_0. Such terms are called *disconnected*, because they are the product of a contribution already present at some order lower than n (order j in this example) and a contribution appearing in the denominator of equation (5.3). The term of order $j + 1$ in the numerator will also give rise among others to the disconnected term

$$\langle V(\tau_1) \cdots V(\tau_j)A(\tau)B(0)\rangle\langle V(\tau_{j+1})\rangle,$$

and so on. It turns out that if we collect all the factors appearing in front of the connected term $\langle V(\tau_1) \cdots V(\tau_j)A(\tau)B(0)\rangle$ in the expansion of the numerator, we obtain exactly the same terms as those forming the denominator—they are called vacuum diagrams—with the same prefactors (see Bruus & Flensberg (2004, p 239)). The numerator of equation (5.3) can therefore be rewritten as the sum of all connected terms multiplied by a factor that cancels the denominator exactly. Moreover, each connected term of order n is generated by Wick's theorem in $n!$ equivalent versions that differ only by a permutation of the internal time arguments. Retaining only one of these *topologically equivalent* terms, we can remove the $n!$ in the expansion and get

$$\boxed{\mathscr{C}_{AB}(\tau) = -\sum_{n=0}^{\infty}(-1)^n \int_0^\beta d\tau_1 \cdots d\tau_n \, \langle T_\tau V(\tau_1) \cdots V(\tau_n)A(\tau)B(0)\rangle_0^{\text{con-diff}}.} \tag{5.5}$$

Equation (5.5) is the starting point for all further developments presented in this section and we shall repeatedly return to it. As we have skipped the details that conduct from equation (5.3) to equation (5.5), the exact meaning of 'con-diff', which stands for 'connected and topologically different', remains somewhat obscure at this stage. The practical applications of equation (5.5) will hopefully clarify these notions. In order to proceed, we have to specify the operators A and B as well as the interaction V.

5.1.3 One-particle Green's function

The one-particle Green's function is the simplest of all correlation functions: it gives the correlation of the operators a_α and $a_\beta^\dagger$ and is commonly denoted by the letter G in honor of George Green (1793–1841):

$$\boxed{\mathscr{G}_{\alpha\beta}(\tau) \equiv \mathscr{C}_{a_\alpha a_\beta^\dagger}(\tau) = -\langle T_\tau a_\alpha(\tau) a_\beta^\dagger(0)\rangle.} \tag{5.6}$$

It goes without saying that the index β in this expression is not the inverse temperature but a one-particle state index.

5.1.3.1 Physical interpretation

The one-particle Green's function contains the complete information about the one-particle excitations in a quantum system. To see this, we consider for simplicity the case $\beta = \alpha$ and $\tau > 0$. $\mathscr{G}_{\alpha\alpha}(\tau > 0)$ is a thermodynamic average of terms like

$$\langle a|a_\alpha(\tau)a_\alpha^\dagger(0)|a\rangle = \underbrace{\langle a|e^{\tau K}a_\alpha}_{\langle\Psi_2(\tau)|}\underbrace{e^{-\tau K}a_\alpha^\dagger|a\rangle}_{|\Psi_1(\tau)\rangle}.$$

$|\Psi_1(\tau)\rangle$ is the many-particle wave-function obtained by starting from the many-body eigenstate $|a\rangle$, adding a particle in state α at time 0, and letting this evolve until time τ. $|\Psi_2(\tau)\rangle$ also starts from $|a\rangle$ but lets $|a\rangle$ evolve from time 0 to time τ before creating a particle in state α at time τ. A large overlap between $|\Psi_1\rangle$ and $|\Psi_2\rangle$ means that creating the particle α at time 0 or at another time does not make much difference: once this particle has been created, it remains stable. On the contrary, a small overlap means that the particle α created at time 0 has almost disappeared at time τ.

Assume that the excitation described by $|\Psi_1\rangle$ is stable. This happens if $a_\alpha^\dagger|a\rangle$ is an eigenstate of K. In other words $|\Psi_1(\tau)\rangle = e^{-(K_a+\xi_\alpha)\tau}a_\alpha^\dagger|a\rangle$. Likewise, this implies that $\langle\Psi_2(\tau)| = e^{K_a\tau}\langle a|a_\alpha$. As a result, the overlap is $\langle\Psi_2(\tau)|\Psi_1(\tau)\rangle \sim e^{-\xi_\alpha\tau}$ or, in the frequency domain, $\mathscr{G}_{\alpha\alpha}(i\nu_n) \sim (i\nu_n - \xi_\alpha)^{-1}$. Thus, if the one-particle excitations produced by $a_\alpha^\dagger$ are stable, the Green's function exhibits poles at energies corresponding to the one-particle excitations energies ξ_α. In this case, the spectral function is just a delta function as we already saw for independent particles in section 3.3.1: $A(\alpha, \varepsilon) \equiv \rho_{a_\alpha a_\alpha^\dagger}(\varepsilon) = (-1/\pi)\text{Im}\,\mathscr{G}_{\alpha\alpha}(i\nu_n \to \varepsilon + i0^+) = \delta(\varepsilon - \xi_\alpha)$.

Assume now that the excitation described by $|\Psi_1\rangle$ is damped, for example due to scattering on impurities, collective excitations, other particles, etc. Then $|\Psi_1(\tau)\rangle \sim e^{-(K_a+\xi_\alpha-i\Gamma_\alpha)\tau}a_\alpha^\dagger|a\rangle$ with $\Gamma_\alpha > 0$[1]. As a result, we would have in the frequency domain $\mathscr{G}_{\alpha\alpha}(i\nu_n) \sim (i\nu_n - \xi_\alpha + i\Gamma_\alpha)^{-1}$. The spectral function is no longer a delta function but gets broadened into a Lorentzian function of width Γ_α (see figure 5.1): $A(\alpha, \varepsilon) = (\Gamma_\alpha/\pi)/[(\varepsilon - \xi_\alpha)^2 + \Gamma_\alpha^2]$. $\hbar/\Gamma_\alpha$ clearly represents the life-time of the excitation. This life-time in general depends on the energy ε. Furthermore, for the Green's function to remain causal as it should, the quantity Γ_α must have both real

[1] This corresponds, in real time $\tau \to it/\hbar$, to an overlap decreasing exponentially like $e^{-\Gamma_\alpha t/\hbar}$.

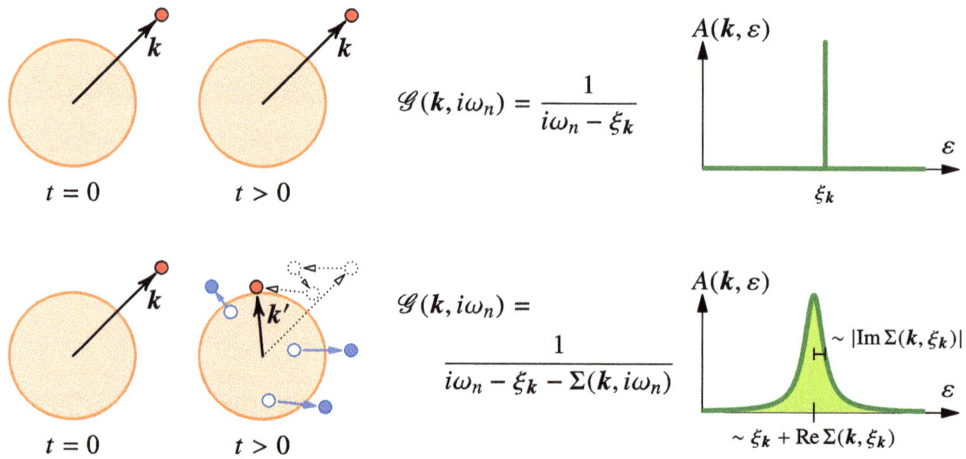

Figure 5.1. Physical meaning of the one-particle Green's function. If a particle injected at time $t = 0$ (in this example, an electron with wave vector $\boldsymbol{k}$) keeps its identity at time $t > 0$, then the Green's function has a pole at the corresponding excitation energy and the spectral function is a delta function. If the injected electron loses momentum and energy by exciting electron–hole pairs across the Fermi surface, then the Green's function acquires a self-energy correction and the broadened spectral function has a Lorentzian-like shape.

and imaginary parts. The generalized inverse life-time, or scattering rate, is denoted $\Sigma_\alpha(\varepsilon)$ and is called the self-energy. Our task is to calculate the self-energy.

5.1.3.2 Density of states and local density of states

Beside the self-energy, which gives access to the dynamical properties of the one-particle excitations, a very useful quantity encoded in the one-particle Green's function is the *density of states* (DOS) and its local counterpart in systems without translational invariance, the *local density of states* (LDOS). In a system of independent particles, the DOS counts the number of one-particle eigenstates whose energy is ε (per unit volume), and is simply given by $N(\varepsilon) = \frac{1}{\mathscr{V}}\sum_\alpha \delta(\varepsilon - \xi_\alpha)$. In an interacting system, there are no one-particle eigenstates anymore but we can still calculate the total spectral weight contributed by the one-particle excitations at energy ε as

$$\begin{aligned} N(\varepsilon) &= \frac{1}{\mathscr{V}}\sum_\alpha \rho_{a_\alpha a_\alpha^\dagger}(\varepsilon) = \frac{1}{\mathscr{V}}\sum_\alpha\left(-\frac{1}{\pi}\right)\mathrm{Im}\; G^R_{\alpha\alpha}(\varepsilon) \\ &= \frac{1}{\mathscr{V}}\sum_\alpha\left(-\frac{1}{\pi}\right)\mathrm{Im}\;\mathscr{G}_{\alpha\alpha}(i\nu_n \to \varepsilon + i0^+), \end{aligned} \tag{5.7}$$

where we have made use of equations (3.47) and (4.12). The DOS will be discussed more extensively in section 9.4. The documents Bonus B5.2 and Bonus B5.3 show how one calculates the DOS of superconductors using the BCS and strong-coupling Eliashberg theories, respectively.

In a system where the translation invariance is broken, it is necessary to consider how the spectral weight of the one-particle excitations is distributed *locally*. For this purpose we have to choose the real-space representation and work with the operator

$a_{r\sigma}^{\dagger}$, which creates a particle of spin σ (or any other property needed to characterize the particle) at point $\boldsymbol{r}$. The corresponding Green's function is $\mathscr{G}_{\sigma\sigma'}(\boldsymbol{r}, \boldsymbol{r}', \tau) = -\langle T_\tau a_{r\sigma}(\tau) a_{r'\sigma'}^{\dagger}(0)\rangle$ and it allows us to define the LDOS as

$$N(\boldsymbol{r}, \varepsilon) = \sum_{\sigma\sigma'} \left(-\frac{1}{\pi}\right) \text{Im}\, \mathscr{G}_{\sigma\sigma'}(\boldsymbol{r}, \boldsymbol{r}, i\nu_n \to \varepsilon + i0^+). \tag{5.8}$$

We will see in section 9.5 that the LDOS can be measured by scanning tunneling microscopy. Bonus B5.1 and Bonus B9.1 provide practical implementations of equation (5.8).

5.1.3.3 Diagrams for a one-body operator

The easiest way to get in touch with diagrams is to consider as our interaction V a one-body operator, like for instance the static potential of an impurity, which has the general form

$$V = \sum_{\alpha\beta} V_{\alpha\beta} a_\alpha^{\dagger} a_\beta \equiv \tag{5.9}$$

The in-going and out-going arrows in the picture represent the annihilation and creation operator, respectively, and the star represents the matrix element. We are interested in calculating the Green's function (5.6). As we just discussed, the Green's function describes the propagation of a particle from its creation at time 0 in a state β to its annihilation at time τ in a state α. We will represent this by a double line with an arrow going from β to α:

$$\mathscr{G}_{\alpha\beta}(\tau) = \tag{5.10}$$

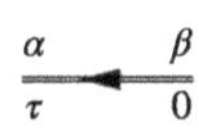

We adopt the convention to orient the arrow from right to left such that the order of the indices is the same as in the formula for the Green's function; the rightmost operator acts first (here: $a_\beta^{\dagger}$). According to equation (5.5), the expansion of $\mathscr{G}$ reads

$$\begin{aligned}\mathscr{G}_{\alpha\beta}(\tau) = &-\sum_{n=0}^{\infty}(-1)^n \int_0^\beta d\tau_1 \cdots d\tau_n \sum_{\alpha_1\beta_1} V_{\alpha_1\beta_1} \cdots \sum_{\alpha_n\beta_n} V_{\alpha_n\beta_n} \\ &\times \langle T_\tau a_{\alpha_1}^{\dagger}(\tau_1) a_{\beta_1}(\tau_1) \cdots a_{\alpha_n}^{\dagger}(\tau_n) a_{\beta_n}(\tau_n) a_\alpha(\tau) a_\beta^{\dagger}(0)\rangle_0^{\text{con-diff}}.\end{aligned} \tag{5.11}$$

Let us start with the zeroth-order term:

$$n = 0: \quad -\langle T_\tau a_\alpha(\tau) a_\beta^{\dagger}(0)\rangle_0 = \mathscr{G}_{\alpha\beta}^0(\tau) = \tag{5.12}$$

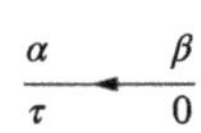

This is no surprise. We represent the Green's function in the absence of interaction (often called the *free propagator*) by a single line. There are more terms to handle at order one; let us use a slightly simplified notation by omitting T_τ and the index $\langle \cdots \rangle_0$. The term in equation (5.11) with $n = 1$ is

$$n = 1: \quad +\int_0^\beta d\tau_1 \sum_{\alpha_1\beta_1} V_{\alpha_1\beta_1} \langle a_{\alpha_1}^{\dagger}(\tau_1) a_{\beta_1}(\tau_1) a_\alpha(\tau) a_\beta^{\dagger}(0)\rangle^{\text{con-diff}}. \tag{5.13}$$

The Wick theorem applied to the average generates $\langle a^\dagger_{\alpha_1}(\tau_1)a_{\beta_1}(\tau_1)\rangle\langle a_\alpha(\tau)a^\dagger_\beta(0)\rangle$, $\eta\langle a^\dagger_{\alpha_1}(\tau_1)a_\alpha(\tau)\rangle\langle a_{\beta_1}(\tau_1)a^\dagger_\beta(0)\rangle$, and $\langle a^\dagger_{\alpha_1}(\tau_1)a^\dagger_\beta(0)\rangle\langle a_{\beta_1}(\tau_1)a_\alpha(\tau)\rangle$. The first of these terms is a disconnected one and the last term is anomalous, i.e. it involves an operator that does not conserve the number of particles and therefore is zero on average[2]. One single term remains, which can be rewritten as (see Doc A5.2)

$$\int_0^\beta d\tau_1 \sum_{\alpha_1\beta_1} \mathcal{G}^0_{\alpha\alpha_1}(\tau-\tau_1)V_{\alpha_1\beta_1}\mathcal{G}^0_{\beta_1\beta}(\tau_1) = \tag{5.14}$$

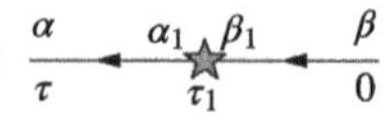

The meaning of this term is obvious: it accounts for processes in which, during the trip from state β at time 0 to state α at time τ, the particle first goes to state β_1 from which it is scattered by the potential at time τ_1 into state α_1, before continuing towards state α. These *internal variables*—β_1, τ_1, and α_1—are summed in order to take into account all possible processes of that kind. At order $n = 2$, we have the average

$$\left\langle a^\dagger_{\alpha_1}(\tau_1)a_{\beta_1}(\tau_1)a^\dagger_{\alpha_2}(\tau_2)a_{\beta_2}(\tau_2)a_\alpha(\tau)a^\dagger_\beta(0)\right\rangle,$$

which generates 15 terms. Among them, nine vanish because they are anomalous with average values of the kind $\langle a^\dagger a^\dagger\rangle$. Four of the remaining six terms are disconnected, namely

$$\begin{gathered}\left\langle a^\dagger_{\alpha_1}(\tau_1)a_{\beta_1}(\tau_1)\right\rangle\left\langle a^\dagger_{\alpha_2}(\tau_2)a_{\beta_2}(\tau_2)\right\rangle\left\langle a_\alpha(\tau)a^\dagger_\beta(0)\right\rangle\\ \eta\left\langle a^\dagger_{\alpha_1}(\tau_1)a_{\beta_1}(\tau_1)\right\rangle\left\langle a^\dagger_{\alpha_2}(\tau_2)a_\alpha(\tau)\right\rangle\left\langle a_{\beta_2}(\tau_2)a^\dagger_\beta(0)\right\rangle\\ \left\langle a^\dagger_{\alpha_1}(\tau_1)a_{\beta_2}(\tau_2)\right\rangle\left\langle a_{\beta_1}(\tau_1)a^\dagger_{\alpha_2}(\tau_2)\right\rangle\left\langle a_\alpha(\tau)a^\dagger_\beta(0)\right\rangle\\ \eta\left\langle a^\dagger_{\alpha_1}(\tau_1)a_\alpha(\tau)\right\rangle\left\langle a_{\beta_1}(\tau_1)a^\dagger_\beta(0)\right\rangle\left\langle a^\dagger_{\alpha_2}(\tau_2)a_{\beta_2}(\tau_2)\right\rangle,\end{gathered}$$

and the two remaining terms are *topologically equivalent*, i.e. one becomes the other upon exchange of the internal variables $(\alpha_1, \beta_1, \tau_1)$ with $(\alpha_2, \beta_2, \tau_2)$:

$$\begin{gathered}\left\langle a^\dagger_{\alpha_1}(\tau_1)a_{\beta_2}(\tau_2)\right\rangle\left\langle a_{\beta_1}(\tau_1)a^\dagger_\beta(0)\right\rangle\left\langle a^\dagger_{\alpha_2}(\tau_2)a_\alpha(\tau)\right\rangle\\ \eta\left\langle a^\dagger_{\alpha_1}(\tau_1)a_\alpha(\tau)\right\rangle\left\langle a_{\beta_1}(\tau_1)a^\dagger_{\alpha_2}(\tau_2)\right\rangle\left\langle a_{\beta_2}(\tau_2)a^\dagger_\beta(0)\right\rangle.\end{gathered} \tag{5.15}$$

Thus we obtain at second order (see Doc A5.2):

$$+\int_0^\beta d\tau_1 d\tau_2 \sum_{\alpha_1\beta_1}\sum_{\alpha_2\beta_2} \mathcal{G}^0_{\alpha\alpha_1}(\tau-\tau_1)V_{\alpha_1\beta_1}\mathcal{G}^0_{\beta_1\alpha_2}(\tau_1-\tau_2)V_{\alpha_2\beta_2}\mathcal{G}^0_{\beta_2\beta}(\tau_2) = \tag{5.16}$$

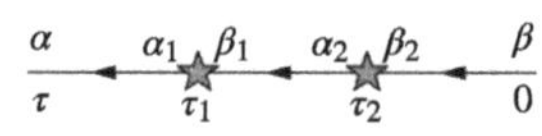

[2] This is not true in general and would be wrong in the BCS model of a superconductor, for example. More generally, when K_0 does not conserve the number of particles ($[K_0, N] \neq 0$), one must reconsider the present derivation and introduce so-called *anomalous propagators*. See also section 5.2.2.3 and Bonus B5.3.

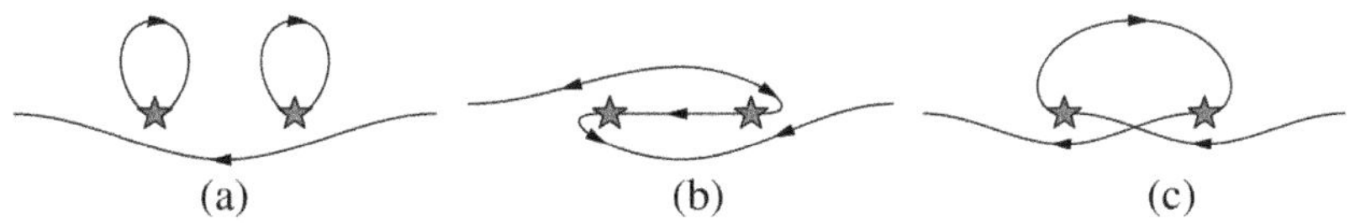

Figure 5.2. Examples of second-order diagrams that are (a) disconnected, (b) anomalous, and (c) topologically equivalent to the diagram in equation (5.16).

The interpretation of this term goes again without problem. Inspecting the diagrams in equations (5.14) and (5.16), one sees that the notions of disconnected, anomalous, and topologically equivalent terms of the Wick expansion acquire an eloquent graphical meaning. The diagram representing a disconnected term is composed of disjointed pieces: for instance, by connecting β with α, α_1 with β_1, and α_2 with β_2 one obtains the first of the disconnected terms listed above and displayed in figure 5.2(a). Anomalous diagrams have conflicting connections in the form of lines carrying opposite arrows, as shown in figure 5.2(b). Finally, topologically equivalent diagrams can be continuously deformed to become identical, like the ones in equation (5.16) and figure 5.2(c).

The generalization to higher-order is now obvious: the nth order term involves n scattering events represented by matrix elements $V_{\alpha_i\beta_i}$ and $n+1$ free propagator lines connecting the scattering events together, with the *external lines* starting at $(\beta,0)$ and ending at (α, τ). There is only one diagram at each order because all the different ways of connecting together n scattering events are topologically equivalent.

One easily performs the time integrations in each diagram by moving from the imaginary time to imaginary Matsubara frequencies. Since the expressions at all orders are just convolutions in the time arguments, they become simple products in frequency space (see Doc A5.3). The following expression results for $\mathcal{G}$:

$$\mathcal{G}_{\alpha\beta}(i\nu_n) = \mathcal{G}^0_{\alpha\beta}(i\nu_n) + \sum_{\alpha_1\beta_1} \mathcal{G}^0_{\alpha\alpha_1}(i\nu_n) V_{\alpha_1\beta_1} \mathcal{G}^0_{\beta_1\beta}(i\nu_n) + \cdots \tag{5.17}$$

If we define $\mathcal{G}$ (without indices) as the matrix made of the elements $\mathcal{G}_{\alpha\beta}$, and likewise for the matrix V with matrix elements $V_{\alpha\beta}$, then, clearly, equation (5.17) can be written and solved in a compact form using matrix products:

$$\begin{aligned}\mathcal{G}(i\nu_n) &= \mathcal{G}_0(i\nu_n) + \mathcal{G}_0(i\nu_n)V\mathcal{G}_0(i\nu_n) + \mathcal{G}_0(i\nu_n)V\mathcal{G}_0(i\nu_n)V\mathcal{G}_0(i\nu_n) + \cdots \\ &= \mathcal{G}_0(i\nu_n) + \mathcal{G}_0(i\nu_n)V\mathcal{G}(i\nu_n) \\ &= [\mathcal{G}_0^{-1}(i\nu_n) - V]^{-1}.\end{aligned} \tag{5.18a}$$

The last line was obtained by multiplying the second from the left by $\mathcal{G}_0^{-1}$ and from the right by $\mathcal{G}^{-1}$. We could have done the same using diagrams:

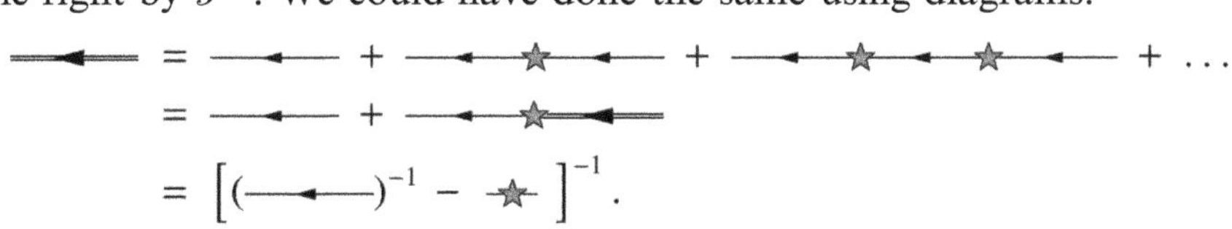

(5.18*b*)

We see that the problem of calculating the Green's function in the presence of a local potential (or, more generally, any perturbation represented by a one-body operator) boils down to the inversion of a matrix (equation (5.18)). This is the fate of all quadratic problems: the particles are independent such that a closed solution exists and coincides with the complete sum of the perturbation series. We revisit in section 5.2.2.1 the relation between the Green's function and the Hamiltonian inverse. To make further progress toward explicit solutions, we would have to specify the unperturbed system and the potential. We shall return to this problem later when studying the resistivity of metals (section 8.3). The document Bonus B5.1 proposes a practical implementation of equation (5.18) for the problem of one-dimensional electrons in a potential.

5.1.3.4 Diagrams for a two-body operator

We turn now to the more interesting and difficult case of a perturbation that is a two-body operator like the Coulomb interaction. The general form is

$$V = \frac{1}{2}\sum_{\alpha\beta\gamma\delta} V_{\alpha\beta\gamma\delta}a^\dagger_\alpha a^\dagger_\beta a_\delta a_\gamma \equiv \quad . \tag{5.19}$$

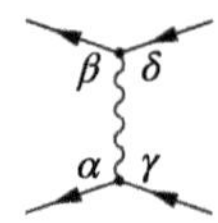

Note that the labelling of the picture is not arbitrary and respects equation (2.44), which implies that the interaction scatters the particle initially in state γ to state α and the one initially in state δ to state β. Using again equation (5.5), we can write the expansion of $\mathscr{G}$, which is just slightly more complicated than equation (5.11):

$$\begin{aligned}\mathscr{G}_{\alpha\beta}(\tau) = &-\sum_{n=0}^{\infty}\left(-\frac{1}{2}\right)^n \int_0^\beta d\tau_1 \cdots d\tau_n \sum_{\alpha_1\beta_1\gamma_1\delta_1} V_{\alpha_1\beta_1\gamma_1\delta_1} \cdots \sum_{\alpha_n\beta_n\gamma_n\delta_n} V_{\alpha_n\beta_n\gamma_n\delta_n}\\ &\times \langle T_\tau a^\dagger_{\alpha_1}(\tau_1)a^\dagger_{\beta_1}(\tau_1)a_{\delta_1}(\tau_1)a_{\gamma_1}(\tau_1) \cdots a^\dagger_{\alpha_n}(\tau_n)a^\dagger_{\beta_n}(\tau_n)a_{\delta_n}(\tau_n)a_{\gamma_n}(\tau_n)\\ &\times a_\alpha(\tau)a^\dagger_\beta(0)\rangle_0^{\text{con-diff}}.\end{aligned} \tag{5.20}$$

At order $n = 0$ we get $\mathscr{G}_0$, as for the one-body operator. But the term of order $n = 1$ is already more rich, since

$$\left\langle a^\dagger_{\alpha_1}(\tau_1)a^\dagger_{\beta_1}(\tau_1)a_{\delta_1}(\tau_1)a_{\gamma_1}(\tau_1)a_\alpha(\tau)a^\dagger_\beta(0)\right\rangle$$

generates 15 terms. Nine of those vanish due to anomalous averages and two are disconnected:

$$\eta\left\langle a^\dagger_{\alpha_1}(\tau_1)a_{\delta_1}(\tau_1)\right\rangle\left\langle a^\dagger_{\beta_1}(\tau_1)a_{\gamma_1}(\tau_1)\right\rangle\left\langle a_\alpha(\tau)a^\dagger_\beta(0)\right\rangle$$

$$\left\langle a^\dagger_{\alpha_1}(\tau_1)a_{\gamma_1}(\tau_1)\right\rangle\left\langle a^\dagger_{\beta_1}(\tau_1)a_{\delta_1}(\tau_1)\right\rangle\left\langle a_\alpha(\tau)a^\dagger_\beta(0)\right\rangle.$$

The remaining four terms come in two pairs. Each pair has two terms that are not topologically equivalent (they differ by more than a permutation of the time arguments) but give the same contribution due to the symmetry of the interaction

—$V(\boldsymbol{r}, \boldsymbol{r}') = V(\boldsymbol{r}', \boldsymbol{r})$ implying $V_{\alpha\beta\gamma\delta} = V_{\beta\alpha\delta\gamma}$ according to equation (2.44). The first such pair is

$$\begin{aligned}&\eta\left\langle a^{\dagger}_{\alpha_1}(\tau_1)a_{\gamma_1}(\tau_1)\right\rangle\left\langle a^{\dagger}_{\beta_1}(\tau_1)a_{\alpha}(\tau)\right\rangle\left\langle a_{\delta_1}(\tau_1)a^{\dagger}_{\beta}(0)\right\rangle\\&\eta\left\langle a^{\dagger}_{\alpha_1}(\tau_1)a_{\alpha}(\tau)\right\rangle\left\langle a^{\dagger}_{\beta_1}(\tau_1)a_{\delta_1}(\tau_1)\right\rangle\left\langle a_{\gamma_1}(\tau_1)a^{\dagger}_{\beta}(0)\right\rangle\end{aligned} \tag{5.21}$$

and it gives rise to the following contribution to the Green's function (see Doc A5.4):

$$-\eta\int_0^{\beta} d\tau_1 \sum_{\alpha_1\beta_1\gamma_1\delta_1} \mathscr{G}^0_{\alpha\alpha_1}(\tau-\tau_1)\mathscr{G}^0_{\delta_1\beta_1}(\tau_1-\tau_1^{+})V_{\alpha_1\beta_1\gamma_1\delta_1}\mathscr{G}^0_{\gamma_1\beta}(\tau_1) =$$

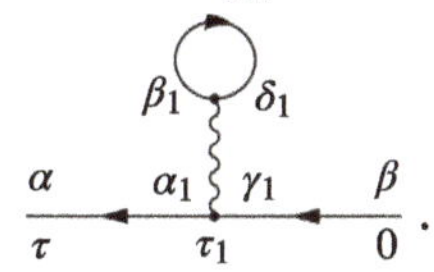

(5.22)

The factor 1/2 is cancelled owing to the two equivalent diagrams. This term represents a process in which the particle goes from state β to state γ_1 at time τ_1, where it interacts with a particle–hole pair and is scattered to state α_1 before evolving to reach state α at time τ (see figure 5.3). Note that the Green's function of the particle–hole pair must be evaluated at time 0^- (see Doc A5.4 and figure 4.1). In the context of the Coulomb interaction, this term is called the Hartree term because it accounts for the direct (classical) Hartree interaction between an electron and the average electron density. The second pair of terms is

$$\begin{aligned}&\left\langle a^{\dagger}_{\alpha_1}(\tau_1)a_{\delta_1}(\tau_1)\right\rangle\left\langle a^{\dagger}_{\beta_1}(\tau_1)a_{\alpha}(\tau)\right\rangle\left\langle a_{\gamma_1}(\tau_1)a^{\dagger}_{\beta}(0)\right\rangle\\&\left\langle a^{\dagger}_{\alpha_1}(\tau_1)a_{\alpha}(\tau)\right\rangle\left\langle a^{\dagger}_{\beta_1}(\tau_1)a_{\gamma_1}(\tau_1)\right\rangle\left\langle a_{\delta_1}(\tau_1)a^{\dagger}_{\beta}(0)\right\rangle\end{aligned} \tag{5.23}$$

which in turn gives (see Doc A5.4)

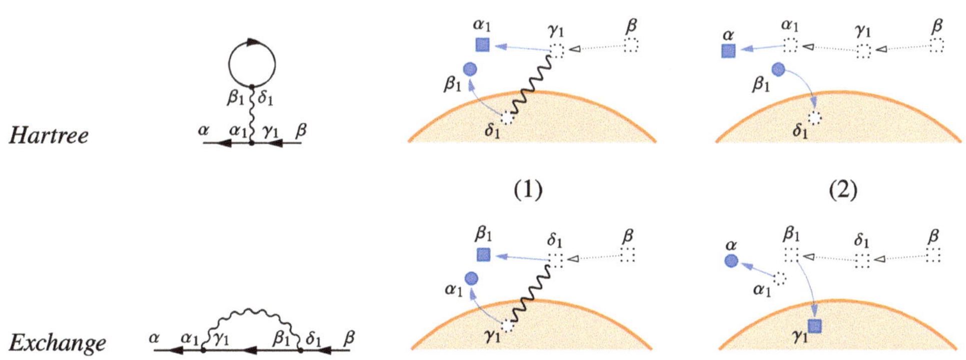

Figure 5.3. Interpretation of the Hartree and exchange diagrams for fermions. (1) The incident electron represented by a dotted black square interacts with another electron represented by a dotted circle in the Fermi sea. The latter can only be excited to a state outside the Fermi sea due to Pauli exclusion. (2) The excited electron can either recombine with the hole while the incident electron continues its exciting life (Hartree term), or continue to live outside the Fermi sea while the incident electron takes a rest by recombining with the hole (exchange term).

$$-\int_0^\beta d\tau_1 \sum_{\alpha_1\beta_1\gamma_1\delta_1} \mathscr{G}^0_{\alpha\alpha_1}(\tau-\tau_1)\mathscr{G}^0_{\gamma_1\beta_1}(\tau_1-\tau_1^+)V_{\alpha_1\beta_1\gamma_1\delta_1}\mathscr{G}^0_{\delta_1\beta}(\tau_1) =$$

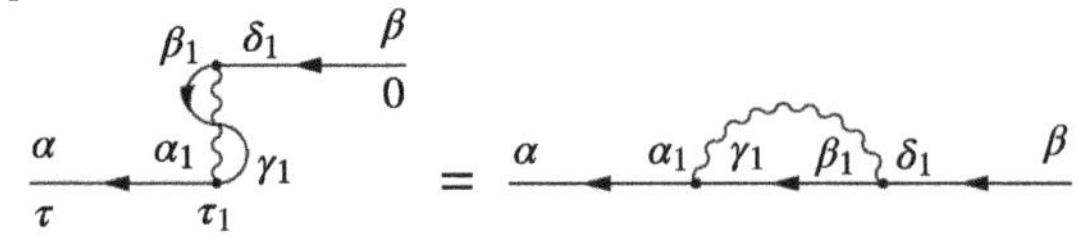

(5.24)

In this process, the particle evolves from state β to state δ_1 at time τ_1 where it is scattered into state β_1 within a particle–hole pair and gets annihilated in state γ_1; at the same time, the particle of the particle–hole pair is scattered from state γ_1 to state α_1 and then evolves onto state α at time τ (see figure 5.3). Thus, the incoming particle has been exchanged with another particle and this term is therefore called the exchange term. The second diagrammatic representation is somewhat easier to remember but abandons a sense of chronological order that is preserved in the first representation.

Before moving on to the second order, we point out that the two first-order diagrams become matrix products in the frequency domain, exactly like in the case of a one-body operator (see Doc A5.5):

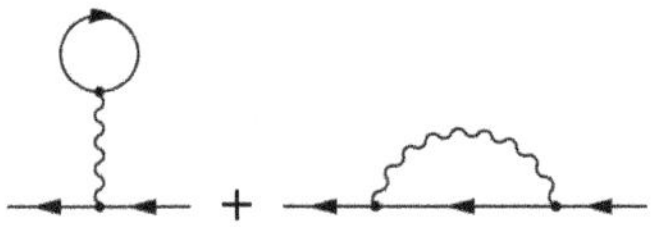

$$= \mathscr{G}^0(i\nu_n)\Sigma^{(1)}(i\nu_n)\mathscr{G}^0(i\nu_n)$$

$$\Sigma^{(1)}_{\alpha\beta}(i\nu_n) = \frac{1}{\beta}\sum_{i\bar{\nu}_1}\sum_{\gamma_1\delta_1} \mathscr{G}^0_{\delta_1\gamma_1}(i\bar{\nu}_1)e^{i\bar{\nu}_1 0^+}(-\eta V_{\alpha\gamma_1\beta\delta_1} - V_{\alpha\gamma_1\delta_1\beta}). \quad (5.25)$$

The same is true for all diagrams of all orders. For the notation $i\bar{\nu}_1$, see remark at the end of Doc A5.3. We will see that $\Sigma^{(1)}_{\alpha\beta}(i\nu_n)$ is an example of the self-energy that was introduced in section 5.1.3.1.

The complete analysis of the second-order term in equation (5.20) becomes obviously more tedious and we leave it to the adventurous reader. Up to here we have drawn diagrams in order to illustrate formula; this is the point where we start using diagrams to effectively replace cumbersome calculations. Here are some figures: $9 \times 7 \times 5 \times 3 = 945$ terms of second order are generated by Wick's theorem, 825 are anomalous and 40 are disconnected. The remaining 80 are grouped in ten families of $2! \times 2^2 = 8$ equivalent members ($n!$ for topological equivalence and 2^n for symmetry of interaction). The ten relevant diagrams of second order can be constructed either by combining first-order diagrams:

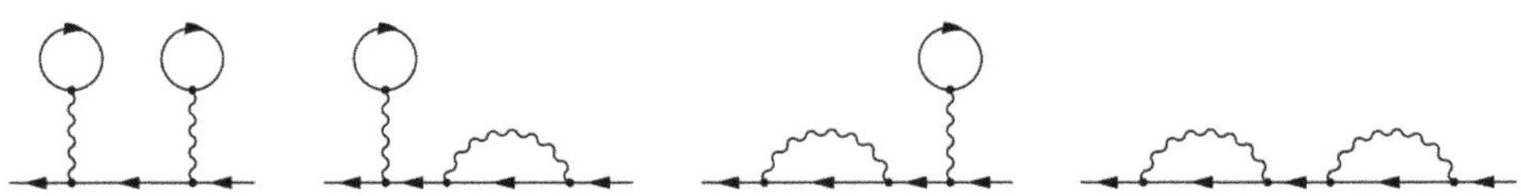

(2.a) (2.b) (2.c) (2.d)

by 'decorating' internal particle lines in first-order diagrams with first-order Hartree or exchange corrections:

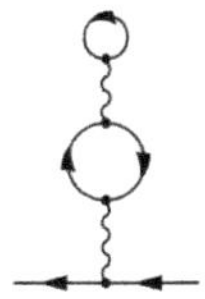

(2.e)

(2.f)

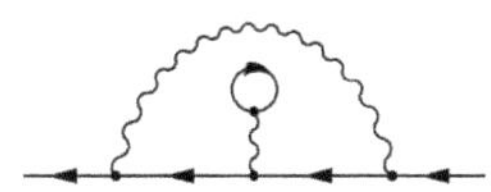

(2.g)

(2.h)

by decorating interaction lines in first-order diagrams with particle loops like in (2.i)—note that diagram (2.e) can also be obtained in this way—or by decorating vertices in first-order diagrams like in (2.j)—note that diagrams (2.f) and (2.g) can also be seen as vertex corrections. Finally, another recipe to generate some of the diagrams of order $n + 1$ from diagrams of order n is to connect all pairs of particle lines with an interaction line. Thus (2.g) and (2.i) can be obtained from the first-order Hartree diagram while (2.h) and (2.j) follow from the first-order exchange diagram.

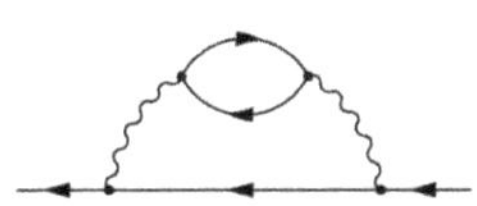

(2.i)

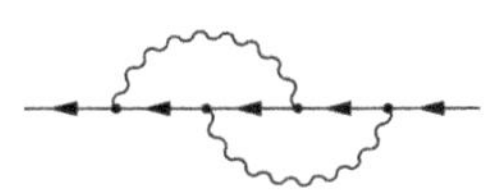

(2.j)

Two important remarks must be made at this point. The first concerns the sign of the diagrams. Each term generated by Wick's theorem has a sign that reflects the number of individual permutations needed to generate this term starting from the un-decoupled average—this sign is always +1 for bosons. How can we know this sign for fermions by just looking at the diagram? The trick is to count the number of fermion loops, i.e. any continuous closed path along fermion lines. We will not try to prove this here but rather trust the generations of physicists who checked that this gives the correct sign (Fetter & Walecka, 1971, p 98). Another sign comes from the prefactor $(-1)^n$ with n the order of the diagram. Hence the sign of a diagram is $(-1)^n\eta^L$ with L the number of loops.

The second remark concerns the translation of diagrams from the time domain to the frequency domain. In practice, we shall mostly work with Matsubara frequencies such that each particle line in a diagram represents a free propagator $\mathscr{G}^0_{\alpha\beta}(i\bar{\nu}_j)$ carrying a different frequency. We must, however, satisfy the constraint that the total frequency entering and leaving each interaction line is the same. This can be understood by considering an interaction at some time τ:

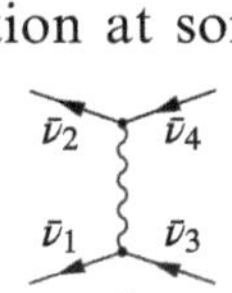

$$\rightarrow \quad \int_0^\beta d\tau\, e^{-i\bar{\nu}_1\tau} e^{-i\bar{\nu}_2\tau} e^{-i\bar{\nu}_3(-\tau)} e^{-i\bar{\nu}_4(-\tau)} = \beta\, \delta_{\bar{\nu}_1+\bar{\nu}_2, \bar{\nu}_3+\bar{\nu}_4}.$$

The Fourier transforms of the propagators on the four lines yield four phase factors and there is no other dependence on τ; after integrating on τ, which is an internal variable, there results a constraint on the frequencies.

We can now pursue the expansion. At third order the number of diagrams increases drastically. Here are two examples:

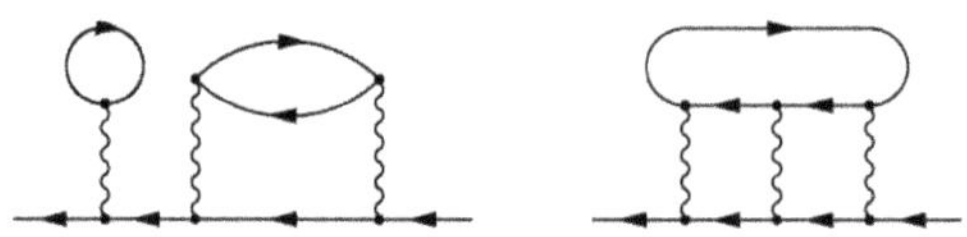

(3.a) (3.b)

(3.a) is a combination of the first-order Hartree diagram with (2.i) while (3.b) is obtained by connecting two particle lines in (2.i). How many are they? 74: Wick's theorem generates $13 \times 11 \times 9 \times 7 \times 5 \times 3 = 135135$ terms, 130095 of which vanish, 1488 are disconnected and 3552 remain in 74 families of $3! \times 2^3 = 48$ equivalent pieces. We are slowly getting buried under diagrams and we need a new idea.

5.1.3.5 Dyson equation and self-energy

This new idea comes from the observation that there are two kinds of second- and higher-order diagrams. Some of them like the series (2.a–2.d) and (3.a) are (matrix) products of lower-order diagrams. Such diagrams are called *reducible*. Graphically, reducible diagrams can be split in two parts by cutting just one particle line. On the contrary, the diagrams of the series (2.e–2.j) and (3.b) are *irreducible*. This simple observation allows us to make tremendous progress because infinite series of diagrams can be resumed at once. Consider for instance the following approximation for $\mathscr{G}_{\alpha\beta}$, which we could call the 'infinite-order Hartree' approximation:

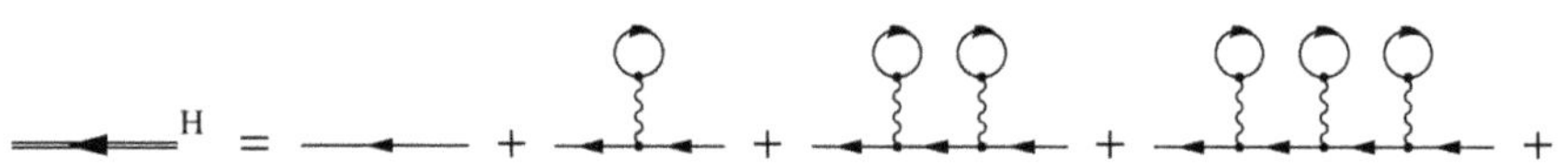 (5.26)

It is obvious that the complete series can be rewritten as

 (5.27)

as can be verified by direct substitution. To solve the latter equation, one multiplies from the left by the matrix 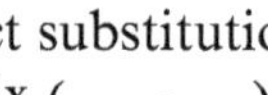 and from the right by 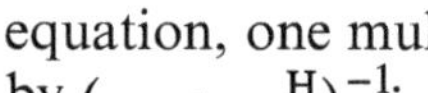:

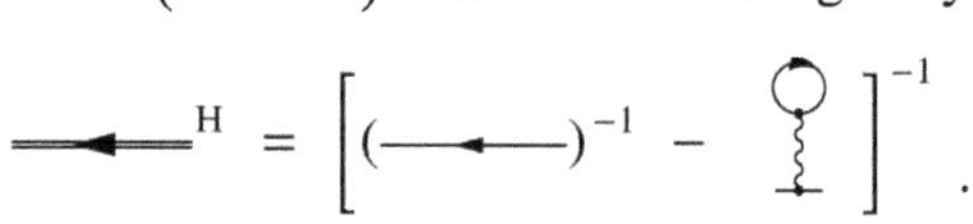 (5.28)

In this way we have obtained an approximation for $\mathscr{G}$ that contains terms up to infinite order in the interaction, but is not harder to evaluate than the first-order approximation 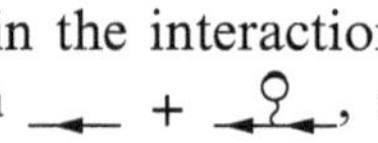, since in both cases only one non-trivial diagram is needed. The trick can easily be generalized. For instance, the 'infinite-order Hartree plus exchange' approximation would read

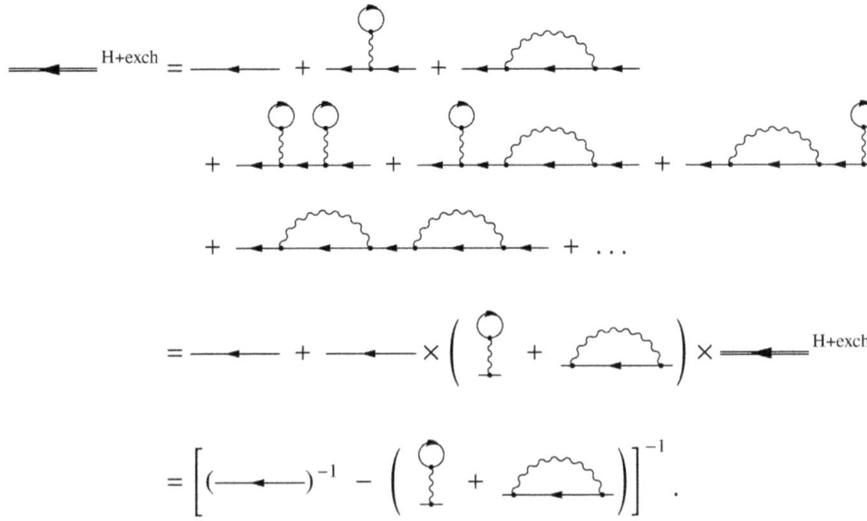

(5.29)

Note that the diagrams containing the interaction in equations (5.28) and (5.29) have no external lines. Proceeding with the same logic, we see that by including all irreducible parts in the parenthesis of equation (5.29), the whole series for $\mathscr{G}_{\alpha\beta}$ is indeed generated:

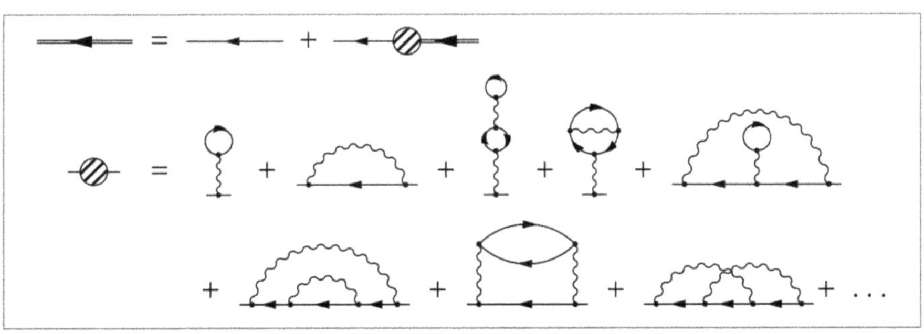

(5.30)

In mathematical terms and matrix form, this expression becomes

$$\mathscr{G}(i\nu_n) = \mathscr{G}_0(i\nu_n) + \mathscr{G}_0(i\nu_n)\Sigma(i\nu_n)\mathscr{G}(i\nu_n) = [\mathscr{G}_0^{-1}(i\nu_n) - \Sigma(i\nu_n)]^{-1}. \tag{5.31}$$

This is known as the Dyson equation. The quantity Σ is the self-energy that we have introduced in the qualitative discussion of section 5.1.3.1. We have already encountered another example of self-energy in the section about one-body operators: equation (5.18). The self-energy was in that case simply given by the perturbation V itself. For a two-body operator, equation (5.30) shows that the self-energy is obtained by taking all irreducible diagrams of the Green's function and removing the two external lines.

There are two irreducible self-energy diagrams at first order, six at second order, 42 at third order, etc. In most textbooks, many of these diagrams will not even be displayed for the following reasons. As we will see in section 5.1.3.7 devoted to the Coulomb interaction, some diagrams are real and furthermore yield a constant result (i.e. they depend neither on α or β, nor on the energy $i\nu_n$). This happens for instance to the first, third, and fourth diagrams in the right-hand side of equation (5.30). A real and constant self-energy is equivalent to a shift of the chemical potential. Therefore, the contribution of these real constant diagrams can be absorbed in a redefinition of the chemical potential, which must anyhow be determined self-consistently in order to fix the density. We also see that the fifth diagram in the right-hand side of equation (5.30) has an *internal* line that is modified by a real and constant self-energy. This diagram is also automatically taken into account by the adjustment of the chemical potential: the latter is carried by $\mathscr{G}_0$ and will therefore be reflected on all internal lines. Moreover, there are diagrams such as the second in the right-hand side of equation (5.30) that turn out to be real and independent of frequency. These diagrams can be absorbed in a redefinition of the one-particle energies ξ_α. Likewise, the sixth diagram is accounted for by redefining the one-particle energies on the internal line. Finally, the only really interesting diagrams in equation (5.30) are the last two.

In summary, thanks to Dyson's equation, the calculation of the one-particle Green's function by perturbation theory is reduced to the evaluation of all irreducible self-energy diagrams. The crucial point is that any low-order approximation to the self-energy automatically leads to a much better infinite-order approximation to the Green's function. Second-order perturbation theory often leads to divergences and the only way to cure these divergences is to include high-order terms The self-energy and Dyson's equation are very convenient tools for achieving this goal.

5.1.3.6 Self-energy diagrams for impurity scattering

The scattering of electrons on defects is the process that dominates the resistivity of metals at low temperature. The defects break translation invariance and locally perturb the electronic structure. We model this perturbation by a local potential acting on the conduction electrons. This potential alone cannot explain resistivity: a local potential induces elastic (energy-conserving) scattering and no dissipation can result. In other words, the life-time of the conduction electrons remains infinite consistently with the fact that the self-energy is real: see equation (5.18*a*) and section 8.2. The scattering is elastic provided that the whole pattern of quantum interferences generated at each defect remains coherent over the entire volume. In real materials, other scattering mechanisms blur this long-range coherence and defect-induced resistivity becomes possible over length scales exceeding the typical coherence volume. A convenient way of modeling the loss of coherence is to wash out any information about the relative distances between the defects by averaging over their positions. The so-called *impurity average*[3] must be performed after having

[3] Kohn W and Luttinger J M 1957 *Phys. Rev.* **108** 590.

calculated the Green's function for each configuration of the impurities. The strategy will be to write down the Green's function for one particular configuration of the impurities, their positions being chosen at random, and then to average over all configurations in order to recover a translation-invariant Green's function. For simplicity, we will consider only one type of impurity characterized by a potential $v(\boldsymbol{r})$ such that the total potential is

$$V(\boldsymbol{r}) = \sum_{\ell=1}^{N_i} v(\boldsymbol{r} - \boldsymbol{R}_\ell) \tag{5.32}$$

with $\boldsymbol{R}_\ell$ the positions of the N_i impurities. We start in the real-space representation where the potential is diagonal. Following our results for the one-body operators, the self-energy is also diagonal:

$$\Sigma_{\alpha\beta} = V_{\alpha\beta} \to V_{\boldsymbol{r}\boldsymbol{r}'} = \delta(\boldsymbol{r} - \boldsymbol{r}')V(\boldsymbol{r}). \tag{5.33}$$

As a result, if we assume that the unperturbed system described by $\mathscr{G}_0$ is invariant by translation, the Dyson equation takes the form

$$\begin{aligned}\mathscr{G}(\boldsymbol{r}, \boldsymbol{r}', i\nu_n) = {}& \mathscr{G}_0(\boldsymbol{r} - \boldsymbol{r}', i\nu_n) \\ &+ \int d\boldsymbol{r}_1\, \mathscr{G}_0(\boldsymbol{r} - \boldsymbol{r}_1, i\nu_n)V(\boldsymbol{r}_1)\mathscr{G}_0(\boldsymbol{r}_1 - \boldsymbol{r}', i\nu_n) \\ &+ \int d\boldsymbol{r}_1 d\boldsymbol{r}_2\, \mathscr{G}_0(\boldsymbol{r} - \boldsymbol{r}_1, i\nu_n)V(\boldsymbol{r}_1) \\ &\times \mathscr{G}_0(\boldsymbol{r}_1 - \boldsymbol{r}_2, i\nu_n)V(\boldsymbol{r}_2)\mathscr{G}_0(\boldsymbol{r}_2 - \boldsymbol{r}', i\nu_n) + \cdots \end{aligned} \tag{5.34}$$

We can now perform the impurity average of each term in equation (5.34) by means of the following prescription:

$$\langle V(\boldsymbol{r}_1) \cdots V(\boldsymbol{r}_n)\rangle_{\text{imp}} = \frac{1}{\mathscr{V}^{N_i}} \int d\boldsymbol{R}_1 \cdots d\boldsymbol{R}_{N_i}\, V(\boldsymbol{r}_1) \cdots V(\boldsymbol{r}_n). \tag{5.35}$$

Starting with $n = 1$, we have simply (see Doc A5.6)

$$\langle V(\boldsymbol{r}_1)\rangle_{\text{imp}} = n_i v(\boldsymbol{q} = 0), \tag{5.36}$$

with n_i the impurity concentration and $v(\boldsymbol{q})$ the Fourier transform of the impurity potential. This is a constant that gives the first-order correction to the chemical potential in the presence of the impurities: the chemical potential must adjust in order to compensate the overall shift of energy levels due to the superposition of impurity potentials. Since $\langle V(\boldsymbol{r}_1)\rangle_{\text{imp}}$ does not depend on $\boldsymbol{r}_1$ any longer, the second term in the right-hand side of equation (5.34) becomes a function of $\boldsymbol{r} - \boldsymbol{r}'$, hence translation invariance is restored as expected. Fourier transforming this term, we obtain (see Doc A5.6)

$$n_i v(\boldsymbol{q} = 0)[\mathscr{G}_0(\boldsymbol{k}, i\nu_n)]^2 \; = \tag{5.37}$$

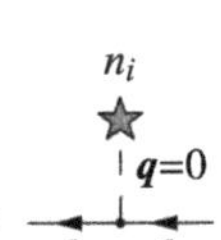

If not for the replacement of the potential by its spatial average, this result is the same as equation (5.14). For $n = 2$ we find (see Doc A5.6)

$$\langle V(\boldsymbol{r}_1)V(\boldsymbol{r}_2)\rangle_{\text{imp}} = n_i(n_i - \mathcal{V}^{-1})[v(\boldsymbol{q} = 0)]^2 + n_i \int d\boldsymbol{r}\; v(\boldsymbol{r})v(\boldsymbol{r} + \boldsymbol{r}_1 - \boldsymbol{r}_2), \quad (5.38)$$

which is a function of $\boldsymbol{r}_1 - \boldsymbol{r}_2$. The first term is the second-order correction to the chemical potential while the second term describes processes in which the particle scatters two times on the same impurity. Assuming $n_i(n_i - \mathcal{V}^{-1}) \approx n_i^2$—in other words, $n_i \gg 1/\mathcal{V}$, there are many impurities in the volume $\mathcal{V}$—and Fourier-transforming, the third term in the right-hand side of equation (5.34) yields

$$n_i^2 v^2(\boldsymbol{0})[\mathcal{G}_0(\boldsymbol{k}, i\nu_n)]^3 + n_i \frac{1}{\mathcal{V}} \sum_{\boldsymbol{q}} \mathcal{G}_0(\boldsymbol{k}, i\nu_n)v(\boldsymbol{q})\mathcal{G}_0(\boldsymbol{k} - \boldsymbol{q}, i\nu_n)v(-\boldsymbol{q})\mathcal{G}_0(\boldsymbol{k}, i\nu_n) =$$

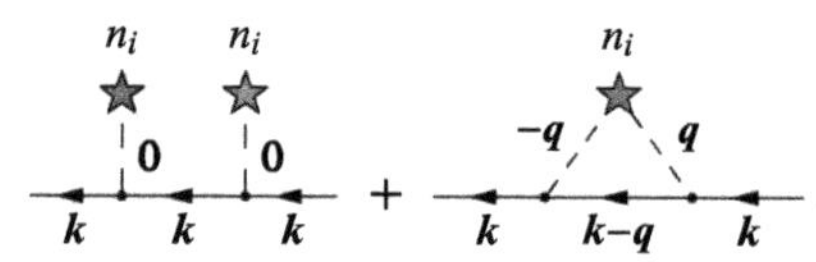

(5.39)

Here the impurity average produces something new compared with equation (5.16): the second term describes an electron scattering twice on the same impurity with exactly opposite exchanges of momenta, an outcome of the restored translation invariance. The momentum exchanges would be independent from each other without impurity average.

The analysis can be continued at $n = 3$ and gives five terms:

Like for the two-body operator, we get reducible and irreducible diagrams. It is then straightforward to perform the infinite sums of reducible diagrams like in section 5.1.3.5 and deduce a Dyson equation in which the self-energy is made of all irreducible terms:

$\Sigma(\boldsymbol{k}, i\nu_n) =$

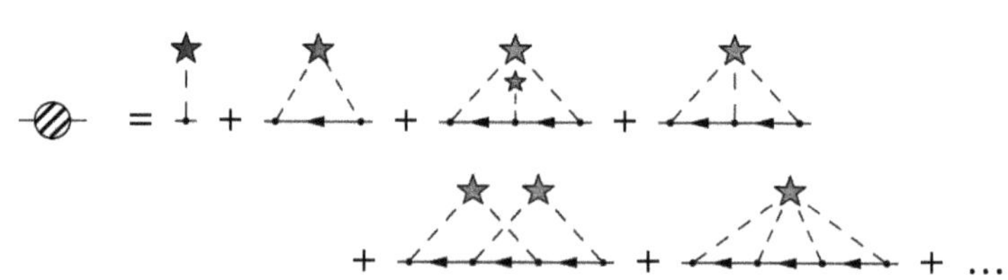

(5.40)

At lowest order, the self-energy is simply the constant $n_i v(\boldsymbol{0})$. The net effect is therefore a shift of the chemical potential that compensates the average potential induced by the distribution of impurities. For all practical purposes, we can forget this term and consider that it is included in the definition of the chemical potential. The second term in the self-energy yields the *first Born approximation* (1BA):

$$\Sigma^{1\mathrm{BA}}(\boldsymbol{k}, i\nu_n) =$$

$$= n_i \frac{1}{\mathcal{V}} \sum_{\boldsymbol{q}} \frac{|v(\boldsymbol{q})|^2}{i\nu_n - \xi_{\boldsymbol{k}-\boldsymbol{q}}}. \tag{5.41}$$

We shall come back to this expression and evaluate it explicitly in section 8.3 when we study the resistivity. An improved version consists of taking into account all diagrams of lowest order in the impurity concentration n_i, which are expected to dominate since generally n_i is much smaller than the particle density. These diagrams describe repeated scattering on one impurity (see Doc A5.7):

$$\Sigma^{\mathrm{FBA}}(\boldsymbol{k}, i\nu_n) =$$

$$+$$

$$+$$

$$+ \dots$$

$$= n_i \frac{1}{\mathcal{V}} \sum_{\boldsymbol{q}} \frac{v(-\boldsymbol{q}) w(\boldsymbol{q}; \boldsymbol{k}, i\nu_n)}{i\nu_n - \xi_{\boldsymbol{k}-\boldsymbol{q}}} =$$

$$\tag{5.42a}$$

$$w(\boldsymbol{q};\, \boldsymbol{k},\, i\nu_n) = v(\boldsymbol{q}) + \frac{1}{\mathcal{V}} \sum_{\boldsymbol{q}'} \frac{v(\boldsymbol{q} - \boldsymbol{q}') w(\boldsymbol{q}';\, \boldsymbol{k},\, i\nu_n)}{i\nu_n - \xi_{\boldsymbol{k}-\boldsymbol{q}}}. \tag{5.42b}$$

This is the *full Born approximation*. Finally, the *self-consistent Born approximation* is obtained by replacing the free propagator $(i\nu_n - \xi_{\boldsymbol{k}})^{-1}$ in equation (5.42) by the full propagator, $\mathcal{G}(\boldsymbol{k},\, i\nu_n) = [i\nu_n - \xi_{\boldsymbol{k}} - \Sigma^{\mathrm{SCBA}}(\boldsymbol{k},\, i\nu_n)]^{-1}$. In practice, the first Born approximation is sufficient for many purposes: the correction to $v(\boldsymbol{q})$ in $w(\boldsymbol{q};\, \boldsymbol{k},\, i\nu_n)$ turns out to be small (Bruus & Flensberg, 2004, p 222).

5.1.3.7 Self-energy diagrams for the Coulomb interaction

In this section, we use the methods of diagrammatic perturbation theory to calculate the self-energy corrections due to the Coulomb interaction in fermionic systems. It may appear dubious at first sight to use perturbation theory for the Coulomb interaction, as this interaction is not at all weak compared with the typical energy scale of a fermion gas, which is the Fermi energy. Among the effects associated with the Coulomb interaction, some are one-particle effects that can be captured by mean-field, independent-electron approximations. Once those are subtracted out, the remaining interaction is ready for perturbation theory, at least for systems that are not strongly correlated. In the latter, the Coulomb interaction leads to radical change in behaviors such as a phase transition toward a new ground state and the perturbation theory is not appropriate.

In order to describe the Coulomb interaction, we work in the plane-wave basis, equation (2.29), such that the matrix elements of the interaction in equation (5.19) become (see Doc A5.8)

$$V_{\alpha\beta\gamma\delta} \longrightarrow \langle \boldsymbol{k}_1\sigma_1\boldsymbol{k}_2\sigma_2|V|\boldsymbol{k}_3\sigma_3\boldsymbol{k}_4\sigma_4\rangle$$

$$= \delta_{\sigma_1\sigma_3}\delta_{\sigma_2\sigma_4}\delta_{\boldsymbol{k}_1+\boldsymbol{k}_2,\boldsymbol{k}_3+\boldsymbol{k}_4}\frac{1}{\mathcal{V}}V(\boldsymbol{k}_1-\boldsymbol{k}_3) = \tag{5.43}$$

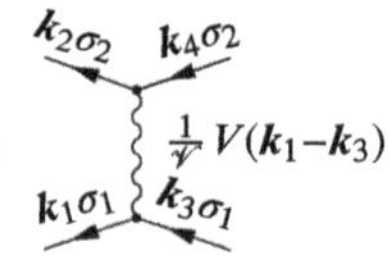

where $V(\boldsymbol{q}) = e^2/(\epsilon_0 q^2)$ is the Fourier transform of the Coulomb potential. Note that we did not take into account the constraint on momenta when labeling the diagram. In view of the discussion in the preceding paragraph, one is tempted to use for $V(\boldsymbol{q})$ the *screened* Coulomb interaction. This procedure is not quite rigorous, though, and induces a risk of counting some diagrams twice. There is nevertheless a well-controlled way of replacing the bare Coulomb interaction by the screened one within perturbation theory, as we shall see in section 5.1.4.6.

We assume that the system under consideration is translation-invariant and non-magnetic, such that the matrix Green's function in equation (5.20) and the matrix self-energy in equation (5.31) are diagonal in the momentum basis and independent of the spin σ:

$$\mathscr{G}_{\alpha\beta}(i\nu_n) \longrightarrow \delta_{\sigma\sigma'}\,\delta_{\boldsymbol{k}\boldsymbol{k}'}\mathscr{G}(\boldsymbol{k}, i\omega_n), \qquad \Sigma_{\alpha\beta}(i\nu_n) \longrightarrow \delta_{\sigma\sigma'}\,\delta_{\boldsymbol{k}\boldsymbol{k}'}\Sigma(\boldsymbol{k}, i\omega_n). \tag{5.44}$$

The Dyson equation (5.31) reduces to an algebraic equation for the diagonal elements $\mathscr{G}(\boldsymbol{k}, i\omega_n)$,

$$\mathscr{G}(\boldsymbol{k}, i\omega_n) = \frac{1}{i\omega_n - \xi_{\boldsymbol{k}} - \Sigma(\boldsymbol{k}, i\omega_n)}, \tag{5.45}$$

where we have used the fact that $\mathscr{G}_0^{-1}(\boldsymbol{k}, i\omega_n) = i\omega_n - \xi_{\boldsymbol{k}}$.

Let us start with the Hartree contribution to the self-energy $\Sigma(\boldsymbol{k}, i\omega_n)$. Following the rules for calculating diagrams, which are summarized in Doc A5.9, we find (see Doc A5.9)

$$= nV(\boldsymbol{q} = 0), \tag{5.46}$$

with $n = N/\mathcal{V}$ the electron density. As anticipated, this is a real constant value (independent of $\boldsymbol{k}$ and $i\omega_n$). One should not be troubled by the fact that this constant is infinite! It is actually expected, since we have been considering a many-electrons system without positive ions to ensure neutrality. The infinite Hartree self-energy is the shift of chemical potential needed to compensate the electrostatic shift of the energy levels, which is infinite in the thermodynamic limit. In a more complete model—the so-called *jellium model*—there would be a uniform density n of positive ions, whose mutual repulsion also produces an electrostatic shift $+nV(\boldsymbol{q} = 0)$, but whose attractive interaction with the electrons induces a negative shift $-2nV(\boldsymbol{q} = 0)$, such that these purely electrostatic contributions exactly cancel in a neutral system.

The exchange contribution to the self-energy is (see Doc A5.9)

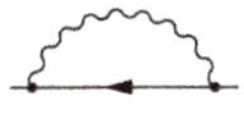

$$= -\frac{1}{\mathscr{V}}\sum_{k_1} V(\boldsymbol{k}_1 - \boldsymbol{k}) f(\xi_{\boldsymbol{k}_1}) \equiv \Sigma^{\text{exch}}(\boldsymbol{k}). \tag{5.47}$$

This term is also real, shows a dependence on $\boldsymbol{k}$ but is independent of the frequency $i\omega_n$. The net effect of the exchange term is to 'renormalize' the one-particle energies according to $\xi_{\boldsymbol{k}} \rightarrow \xi_{\boldsymbol{k}}^* = \xi_{\boldsymbol{k}} + \Sigma^{\text{exch}}(\boldsymbol{k})$ as can be seen from equation (5.45). It is noteworthy that this renormalization is negative, which reflects the energy gain due to exchange. The renormalized excitation with $\xi_{\boldsymbol{k}}^* < \xi_{\boldsymbol{k}}$ therefore has a larger effective mass $m^* > m$.

We see that no life-time effects are found at first order: the exchange term changes the one-particle energies but does not lead to a damping of the excitations because the corresponding self-energy has no imaginary part. In order to find life-time effects, we must go to second order in the self-energy. We obtain for the two second-order diagrams (see Doc A5.9):

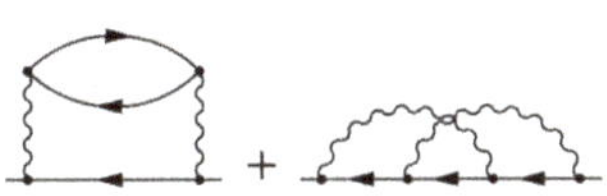

$$= \frac{1}{\mathscr{V}^2}\sum_{\boldsymbol{k}_1\boldsymbol{k}_2} \left[2|V(\boldsymbol{k}_2 - \boldsymbol{k})|^2 - V(\boldsymbol{k}_2 - \boldsymbol{k})V(\boldsymbol{k}_1 - \boldsymbol{k}_2)\right] \times \frac{f(\xi_{\boldsymbol{k}_1}) f(-\xi_{\boldsymbol{k}_2}) f(-\xi_{\boldsymbol{k}+\boldsymbol{k}_1-\boldsymbol{k}_2}) + f(-\xi_{\boldsymbol{k}_1}) f(\xi_{\boldsymbol{k}_2}) f(\xi_{\boldsymbol{k}+\boldsymbol{k}_1-\boldsymbol{k}_2})}{i\omega_n + \xi_{\boldsymbol{k}_1} - (\xi_{\boldsymbol{k}_2} + \xi_{\boldsymbol{k}+\boldsymbol{k}_1-\boldsymbol{k}_2})}. \tag{5.48}$$

These two terms depend on both $\boldsymbol{k}$ and $i\omega_n$ and do have an imaginary part. Figure 5.4 proposes a step by step interpretation of these scattering processes. We will come back to the evaluation of these terms in section 8.4 when discussing the resistivity of metals. The crucial property is that the imaginary part of these diagrams behaves as

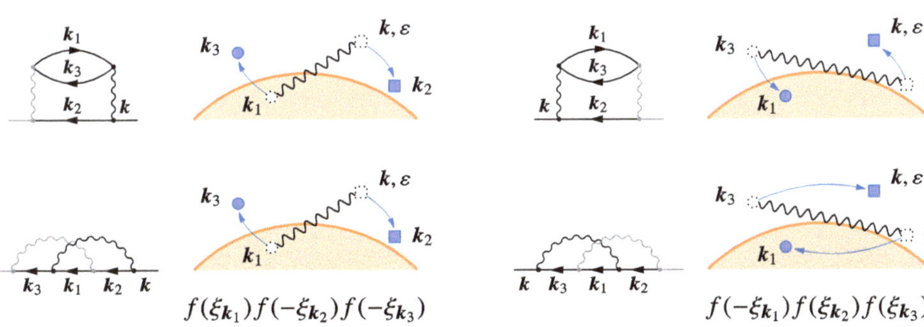

Figure 5.4. Interpretation of the second-order Coulomb self-energy diagrams. The rightmost interaction vertex in the diagrams describes a process where the incoming electron (momentum $\boldsymbol{k}$ and energy ε) interacts with an electron in the Fermi sea (momentum $\boldsymbol{k}_1$ and energy $\xi_{\boldsymbol{k}_1}$). The incoming electron is scattered to momentum $\boldsymbol{k}_2$ and the Fermi-sea electron to momentum $\boldsymbol{k}_3 = \boldsymbol{k} + \boldsymbol{k}_1 - \boldsymbol{k}_2$. The energy cost is $\varepsilon + \xi_{\boldsymbol{k}_1} - (\xi_{\boldsymbol{k}_2} + \xi_{\boldsymbol{k}_3})$, which explains the energy denominator in equation (5.48); for the process to be possible, $\boldsymbol{k}_1$ must be an occupied state and $\boldsymbol{k}_2$, $\boldsymbol{k}_3$ must be empty states, hence the indicated combination of Fermi factors. The leftmost vertex in the diagrams describes the process where the two virtual particles interact again and are scattered back to $\boldsymbol{k}$ and $\boldsymbol{k}_1$ without (top) or with (bottom) an exchange of the electrons.

$\varepsilon^2 + (\pi k_B T)^2$ at low energy, which, as we shall see, is the reason for the $\sim T^2$ scaling of the electron–electron scattering contribution to the resistivity of metals.

5.1.3.8 Self-energy diagrams for the electron–phonon interaction

The interaction of electrons with phonons in solids leads to inelastic scattering processes in which an electron emits or absorbs a phonon, thus changing its momentum and energy. As a result, the energy and momentum of an electron excited above the Fermi surface is progressively redistributed to phonons and other electrons; after a while, everything is dissolved into an incoherent soup of low-energy excitations. The resulting finite life-time of excited electrons is encoded in the self-energy. In this section, we use perturbation theory to calculate the self-energy due to the electron–phonon coupling. Unlike in the previous sections, we have to deal here with a system containing two kinds of particles: electrons and phonons. When the two families of particles are decoupled, there is no worry: the wave functions are simple products of wave functions describing particles living in orthogonal Hilbert spaces. When the particles interact, this nice property breaks down. Fortunately, in perturbation theory, all time evolutions and thermal averages must be evaluated with the coupling set to zero.

The electron–phonon coupling Hamiltonian is given in equation (2.73). This coupling involves pairs of phonon operators in the form $b_{q\lambda} + b^\dagger_{-q\lambda}$. It is therefore convenient to introduce the creation operator

$$B^\dagger_{q\lambda} = b^\dagger_{q\lambda} + b_{-q\lambda}, \tag{5.49}$$

which adds momentum $\boldsymbol{q}$ to the phonon system, either by creating a phonon of polarization λ with momentum $\boldsymbol{q}$ or by removing a phonon of polarization λ and momentum $-\boldsymbol{q}$. A quantity of importance in the following discussion is the correlation function of the operators $B_{q\lambda}$ in the absence of coupling. We find (see Doc A5.10)

$$\mathscr{C}^0_{B_{q\lambda}B^\dagger_{q'\lambda'}}(\tau) = \delta_{qq'}\delta_{\lambda\lambda'}\mathscr{D}^0_\lambda(\boldsymbol{q}, \tau), \tag{5.50}$$

where $\mathscr{D}^0_\lambda(\boldsymbol{q}, \tau) = -\langle T_\tau B_{q\lambda}(\tau) B^\dagger_{q\lambda}(0)\rangle$ is the *free-phonon propagator*. In the frequency domain, this propagator has the form (see Doc A5.10)

$$\boxed{\mathscr{D}^0_\lambda(\boldsymbol{q}, i\Omega_n) = \frac{1}{i\Omega_n - \hbar\omega_{q\lambda}} - \frac{1}{i\Omega_n + \hbar\omega_{q\lambda}} = \frac{2\hbar\omega_{q\lambda}}{(i\Omega_n)^2 - (\hbar\omega_{q\lambda})^2}.} \tag{5.51}$$

Following equation (5.5), the expansion of the Green's function for electrons coupled to phonons reads:

$$\mathscr{G}(\boldsymbol{k}, \tau) = -\sum_{n=0}^{\infty}(-1)^n \int_0^\beta d\tau_1 \cdots d\tau_n \left\langle T_\tau H_{\text{el-ph}}(\tau_1) \cdots H_{\text{el-ph}}(\tau_n) c_{k\sigma}(\tau) c^\dagger_{k\sigma}(0)\right\rangle_0^{\text{con-diff}}. \tag{5.52}$$

As shown in Doc A5.11, this can be transformed into the form

$$\mathcal{G}(\boldsymbol{k}, \tau) = -\sum_{n=0}^{\infty}(-1)^n \int_0^\beta d\tau_1 \cdots d\tau_n \langle T_\tau V_{\text{el-el}}^{\text{ph}}(\tau_1) \cdots V_{\text{el-el}}^{\text{ph}}(\tau_n) c_{\boldsymbol{k}\sigma}(\tau) c_{\boldsymbol{k}\sigma}^\dagger(0)\rangle_0^{\text{con-diff}}, \quad (5.53)$$

where $V_{\text{el-el}}^{\text{ph}}$ is an effective electron–electron interaction given by

$$V_{\text{el-el}}^{\text{ph}}(\tau) = \frac{1}{2}\sum_{\boldsymbol{k}_1\sigma_1}\sum_{\boldsymbol{k}_2\sigma_2}\sum_{\boldsymbol{q}\lambda}|g_{\boldsymbol{q}\lambda}|^2 \int_0^\beta d\tau' \, \mathcal{D}_\lambda^0(\boldsymbol{q}, \tau - \tau') \times c_{\boldsymbol{k}_1+\boldsymbol{q}\sigma_1}^\dagger(\tau) c_{\boldsymbol{k}_2-\boldsymbol{q}\sigma_2}^\dagger(\tau') c_{\boldsymbol{k}_2\sigma_2}(\tau') c_{\boldsymbol{k}_1\sigma_1}(\tau). \quad (5.54)$$

This interaction is very similar in its form to the Coulomb interaction which, from equations (5.43) and (2.44) follows as

$$V_{\text{Cb}}(\tau) = \frac{1}{2}\sum_{\boldsymbol{k}_1\sigma_1}\sum_{\boldsymbol{k}_2\sigma_2}\sum_{\boldsymbol{q}} \frac{V(\boldsymbol{q})}{\mathcal{V}} c_{\boldsymbol{k}_1+\boldsymbol{q}\sigma_1}^\dagger(\tau) c_{\boldsymbol{k}_2-\boldsymbol{q}\sigma_2}^\dagger(\tau) c_{\boldsymbol{k}_2\sigma_2}(\tau) c_{\boldsymbol{k}_1\sigma_1}(\tau). \quad (5.55)$$

The main differences between $V_{\text{el-el}}^{\text{ph}}$ and V_{Cb} are that $V_{\text{el-el}}^{\text{ph}}$ has an explicit time dependence encoded in the τ' integral and is of second order in the electron–phonon coupling $g_{\boldsymbol{q}\lambda}$. The physical interpretation is that the effective interaction between electrons is *mediated by the exchange of phonons*: a phonon is emitted (absorbed) at some time by the electronic system and absorbed (emitted) at some later time, leading to a *retarded interaction*. If the phonon propagation were instantaneous, i.e. $\mathcal{D}_\lambda^0(\boldsymbol{q}, \tau - \tau') \propto \delta(\tau - \tau')$, then $V_{\text{el-el}}^{\text{ph}}$ would be instantaneous as well, like the Coulomb interaction[4]. But the phonons have their own dynamics, as shown in equation (5.51), which leads to a non-trivial frequency dependence of the phonon-mediated interaction. This brings us to another difference between $V_{\text{el-el}}^{\text{ph}}$ and V_{Cb}, one with spectacular consequences: while V_{Cb} is positive (repulsive), $V_{\text{el-el}}^{\text{ph}}$ is attractive at low energy because $D_\lambda^{0R}(\boldsymbol{q}, \varepsilon) = \mathcal{D}_\lambda^0(\boldsymbol{q}, i\Omega_n \to \varepsilon + i0^+)$ is negative for $|\varepsilon| < \hbar\omega_{\boldsymbol{q}\lambda}$ (see figure 5.5). Superconductivity is known to result from this attractive interaction.

The formal similarity between $V_{\text{el-el}}^{\text{ph}}$ and V_{Cb} allows us to use the results of section 5.1.3.7 in order to calculate the self-energy due to the electron–phonon interaction, with only a few adjustments. We represent the interaction vertex like this:

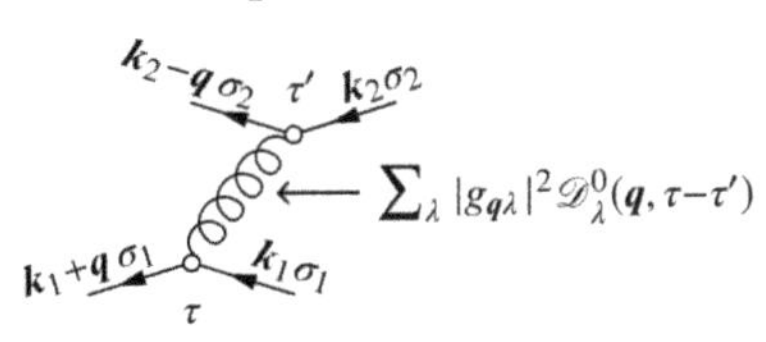

(5.56)

Like for the Coulomb interaction, the total momentum is conserved and a momentum $\boldsymbol{q}$ is exchanged between the electrons[5]. However, unlike for the

[4] Of course, in reality the Coulomb interaction is retarded because the photons propagate at the speed of light. We do not consider relativistic effects in this book, however, but instead assume that V_{Cb} is instantaneous.

[5] In a periodic lattice, the momentum conservation is to be enforced modulo a vector $\boldsymbol{G}$ of the reciprocal lattice (see equation (2.51)). Processes in which the total momentum of incoming and outgoing electrons differ by some vector $\boldsymbol{G}$ are called *Umklapp* processes and they dominate at large momentum transfers.

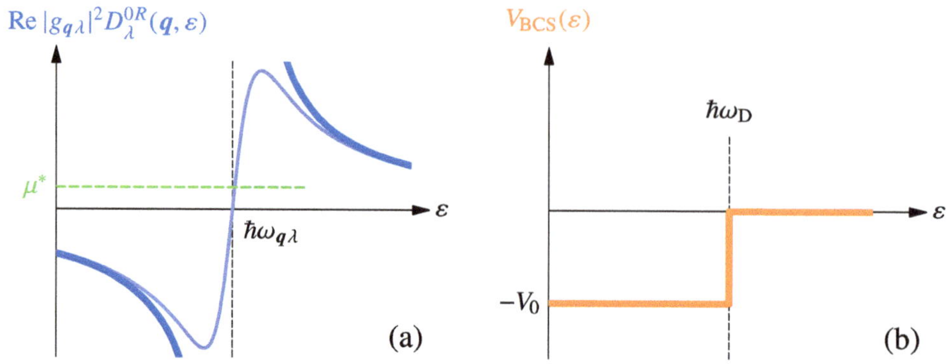

Figure 5.5. (a) Energy dependence of the real part of the phonon Green's function (thick line). This quantity controls the energy dependence of $V_{\text{el–el}}^{\text{ph}}$ and is negative for $\varepsilon < \hbar\omega_{q\lambda}$. For real, i.e. non-free phonons, the singularity at $\hbar\omega_{q\lambda}$ is suppressed (thin line). μ^* indicates the average value of the screened Coulomb repulsion on the Fermi surface: if μ^* is sufficiently small, the net interaction between electrons is attractive on the Fermi surface. (b) Simplified Bardeen–Cooper–Schrieffer interaction: the interaction is attractive up to a typical phonon frequency, taken as the Debye frequency ω_{D}.

Coulomb interaction, an energy $i\Omega_n$ is also exchanged between the electrons, reflecting the retarded nature of the interaction. A simplification with respect to V_{Cb} is that diagrams involving contributions of the kind

$$\propto |g_{q=0,\lambda}|^2 \tag{5.57}$$

where the shaded circle represents any connected sub-diagram, vanish because the exchanged momentum is $\boldsymbol{q} = 0$ and the electron–phonon coupling vanishes at $\boldsymbol{q} = 0$ (see equation (2.74)). At first order in $V_{\text{el–el}}^{\text{ph}}$, we only have to consider the exchange diagram in the self-energy equation (5.30), which reads (see Doc A5.12):

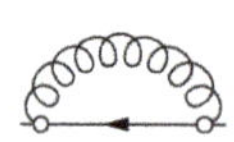

$$= \sum_{q\lambda} |g_{q\lambda}|^2 \left[\frac{f(-\xi_{k-q}) + b(\hbar\omega_{q\lambda})}{i\omega_n - (\hbar\omega_{q\lambda} + \xi_{k-q})} + \frac{f(\xi_{k-q}) + b(\hbar\omega_{q\lambda})}{i\omega_n + \hbar\omega_{q\lambda} - \xi_{k-q}} \right]. \tag{5.58}$$

The interpretation of this result in formulated in figure 5.6. The existence of two terms, representing two different processes, is a consequence of the retarded nature of the interaction: we can have $\tau > \tau'$ or $\tau' > \tau$ at the vertices (see equation (5.56)). The first term in the square brackets corresponds to the process in which the incoming electron of momentum $\boldsymbol{k}$ and energy ε first emits a phonon of momentum $\boldsymbol{q}$ and frequency $\omega_{q\lambda}$ and then re-absorbs this same phonon at a later time. After emitting the phonon, the electron has momentum $\boldsymbol{k} - \boldsymbol{q}$: for this to be possible, the state $\boldsymbol{k} - \boldsymbol{q}$ must be empty, which explains the factor $f(-\xi_{k-q})$ in equation (5.58). The energy difference, on the other hand, is $\varepsilon - (\hbar\omega_{q\lambda} + \xi_{k-q})$ explaining the denominator. The correction $b(\hbar\omega_{q\lambda})$ represents the physical phenomenon of

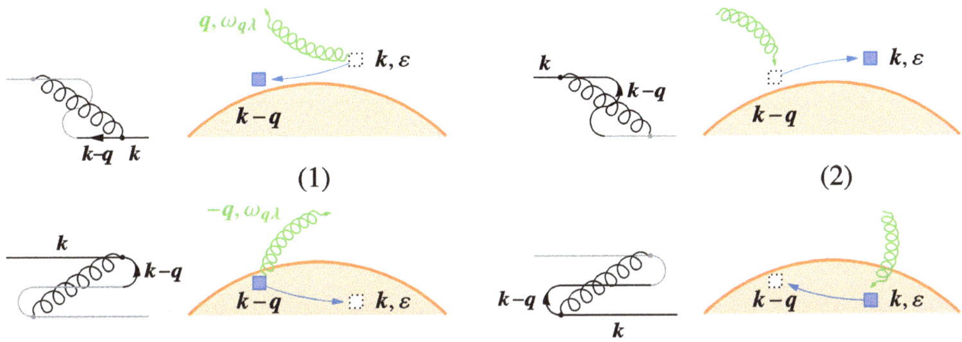

Figure 5.6. Interpretation of equation (5.58). The first term corresponds to emission and re-absorption of a phonon by an electron (top). The second term is the equivalent process for a hole (bottom).

stimulated emission: the emission of phonons is enhanced if phonons are already present. The second term in the square brackets contributes mostly for negative energies ε and is therefore best visualized as a process for a hole at momentum $\boldsymbol{k}$ and energy ε. This hole is first 'filled' by an electron of higher energy, hence a factor $f(\xi_{k-q})$, leaving a hole in state $\boldsymbol{k}-\boldsymbol{q}$; the phonon is then absorbed by the electron in state $\boldsymbol{k}$, which returns to $\boldsymbol{k}-\boldsymbol{q}$ and leaves the initial hole in state $\boldsymbol{k}$. The fact that the intermediate state is a hole (the $\boldsymbol{k}-\boldsymbol{q}$ hole) is seen in the diagram in the fact that the corresponding propagator goes 'back in time'.

These various mechanisms contribute to 'dress' the electrons with a cloud of phonons, making the electrons heavier with an effective mass $m^* > m$ and short-lived with a finite life-time. Mathematically, the effective mass and the life-time are seen in the facts that the self-energy depends on energy and possesses an imaginary part. These important aspects are discussed further in section 7.3. In Bonus B7.5, equation (5.58) is evaluated analytically in the simpler case of a phonon spectrum with a single dispersionless optical phonon. For acoustic phonons, the imaginary part of equation (5.58) goes like $|\varepsilon|^3$ and explains the dominant contribution $\sim T^3$ of the electron–phonon interaction to the resistivity of metals. Higher-order terms can be evaluated in the same way and describe the so-called multi-phonon processes. In the case of acoustic phonons, it turns out that the second-order terms are smaller than the first-order term, equation (5.58), by a factor $(m/M)^{1/2} \sim 10^{-3}$, where m and M are the electron and nucleus masses, respectively[6]. Therefore, the higher-order processes can generally be ignored.

5.1.4 Two-particle correlation functions

The Green's function studied in the previous section, equation (5.6), is a one-particle correlation function in the sense that it describes the time evolution of a single particle created at time zero. Two-particle correlation functions describe the evolution of a pair of particles or, as will be discussed in this section, the evolution of a particle–hole pair. These correlation functions are of the kind:

[6] Migdal A B 1958 *Soviet Phys. JETP* **7** 996. See also Schrieffer (1964 p 156).

$$\mathscr{C}_{\alpha\beta\gamma\delta}(\tau) = -\langle T_\tau a^\dagger_\alpha(\tau) a_\beta(\tau) a^\dagger_\gamma(0) a_\delta(0)\rangle = - \tag{5.59}$$

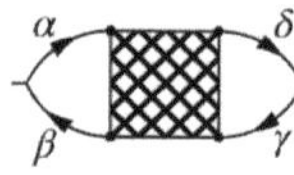

As in equation (5.10), the right vertex corresponds to imaginary time zero and the left vertex to imaginary time τ. The shaded box represents all interactions that can take place like the double line in equation (5.10) and the minus sign in front of the diagram will be explained below.

5.1.4.1 *Physical interpretation*

The object defined in equation (5.59) describes the propagation of a particle–hole pair created at time 0—the particle being created in state γ and the hole being created by annihilating a particle in state δ—and annihilated in another state at time τ. Like the Green's function for single-particle excitations, this propagator will therefore have poles at energies corresponding to stable particle–hole excitations.

Usually, one considers correlators constructed by summing over the indices α, β, γ, and δ. One thus describes the propagation of a superposition of many particle–hole pairs: such superpositions are collective excitations. For example, the density–density correlation function $\mathscr{C}_{n(\boldsymbol{q})n(-\boldsymbol{q})}(\tau)$ with $n(\boldsymbol{q}) = \sum_{\boldsymbol{k}\sigma} a^\dagger_{\boldsymbol{k}\sigma} a_{\boldsymbol{k}+\boldsymbol{q}\sigma}$ describes the propagation of density fluctuations, i.e. collective excitations made of particle–hole pairs with the same relative wave vector $\boldsymbol{q}$. It is sometimes called the *density-fluctuation propagator*. We will see that $\mathscr{C}_{n(\boldsymbol{q})n(-\boldsymbol{q})}(i\Omega_n)$ has poles at energies and wave vectors corresponding to the resonant density oscillations, like for instance the plasmon.

As a first step, we look at the lowest-order term in the perturbation series for $\mathscr{C}_{\alpha\beta\gamma\delta}$. This term is know as the *particle–hole bubble*.

5.1.4.2 *Particle–hole bubble*

The zeroth-order term in the perturbation series equation (5.5) for $\mathscr{C}_{\alpha\beta\gamma\delta}(\tau)$ is simply (see Doc A5.13)

$$\mathscr{C}^0_{\alpha\beta\gamma\delta}(\tau) = -\eta\, \mathscr{G}^0_{\beta\gamma}(\tau)\mathscr{G}^0_{\delta\alpha}(-\tau) = - \tag{5.60}$$

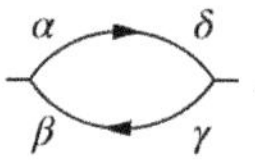

The minus sign in front of the diagram comes directly from the definition (5.59) while the η factor is accounted for by the diagrammatic rules because there is one particle loop. It is customary to work in the representation in which the Green's function is diagonal, i.e. $\mathscr{G}_{\alpha\beta} \propto \delta_{\alpha\beta}$. In that case, the particle–hole bubble takes a familiar form in the frequency domain (see Doc A5.13):

$$\mathscr{C}^0_{\alpha\beta\gamma\delta}(i\Omega_n) = \delta_{\alpha\delta}\delta_{\beta\gamma} \frac{d_{-\eta}(\xi_\alpha) - d_{-\eta}(\xi_\beta)}{i\Omega_n + \xi_\alpha - \xi_\beta}. \tag{5.61}$$

One easily sees how the free-electron density–density correlation function (3.29) can be deduced from equation (5.61). $\mathscr{C}^0_{\alpha\beta\gamma\delta}$ plays, in the diagrammatic perturbation theory for two-particle correlation functions, a role similar to $\mathscr{G}^0_{\alpha\beta}$ in the diagrammatic expansion of the one-particle Green's function. Unlike $\mathscr{G}^0_{\alpha\beta}$, $\mathscr{C}^0_{\alpha\beta\gamma\delta}$ has an

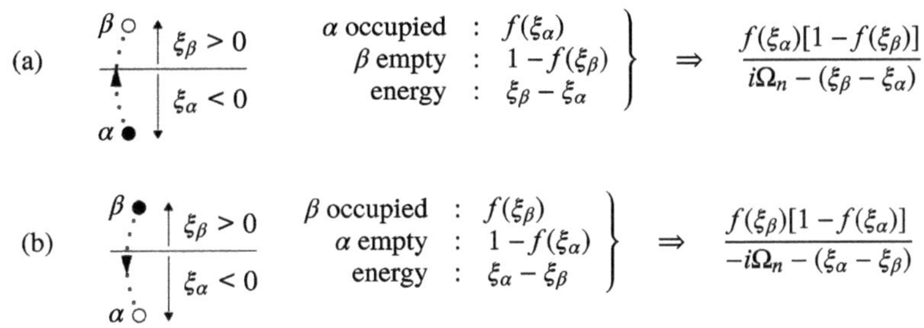

Figure 5.7. Interpretation of the particle–hole bubble for fermions. In the particle–hole emission process (a), one expects a pole when the injected energy $i\Omega_n$ equals $\xi_\beta - \xi_\alpha$, provided that state α is occupied and state β is empty. In the recombination process (b), the energy $i\Omega_n$ is released (minus sign) by the transition of energy $\xi_\alpha - \xi_\beta$ if state β is occupied and state α is empty. Summing the two terms, we obtain equation (5.61) since $f(\xi_\alpha)[1 - f(\xi_\beta)] - f(\xi_\beta)[1 - f(\xi_\alpha)] = f(\xi_\alpha) - f(\xi_\beta)$.

explicit temperature dependence due to the distribution functions in the numerator. The reason is that particle–hole excitations depend on the occupation of single-particle states, as illustrated in figure 5.7. In contrast, the injection described by $\mathscr{G}^0_{\alpha\beta}$ of a new particle in a system is unrelated to the occupation numbers in that system.

5.1.4.3 Diagrams for a one-body operator

Like for the Green's function, we start with the case of a one-body operator since this is the simplest case and one can solve it exactly. The general formula (5.5) gives:

$$\mathscr{C}_{\alpha\beta\gamma\delta}(\tau) = -\sum_{n=0}^{\infty}(-1)^n \int_0^\beta d\tau_1 \cdots d\tau_n \sum_{\alpha_1\beta_1} V_{\alpha_1\beta_1} \cdots \sum_{\alpha_n\beta_n} V_{\alpha_n\beta_n}$$
$$\langle T_\tau a^\dagger_{\alpha_1}(\tau_1) a_{\beta_1}(\tau_1) \cdots a^\dagger_{\alpha_n}(\tau_n) a_{\beta_n}(\tau_n) a^\dagger_\alpha(\tau) a_\beta(\tau) a^\dagger_\gamma(0) a_\delta(0)\rangle_0^{\text{con-diff}}. \tag{5.62}$$

At order $n = 0$, we find the particle–hole bubble $\mathscr{C}^0_{\alpha\beta\gamma\delta}(\tau)$ that we have just encountered. At order $n = 1$, the Wick theorem generates two terms that are nonzero, connected, and topologically different, namely (omitting summations for simplicity):

$$V_{\alpha_1\beta_1}\, \eta \underbrace{\langle T_\tau a^\dagger_{\alpha_1}(\tau_1) a_\beta(\tau)\rangle_0}_{-\eta\mathscr{G}^0_{\beta\alpha_1}(\tau-\tau_1)} \underbrace{\langle T_\tau a_{\beta_1}(\tau_1) a^\dagger_\gamma(0)\rangle_0}_{-\mathscr{G}^0_{\beta_1\gamma}(\tau_1)} \underbrace{\langle T_\tau a^\dagger_\alpha(\tau) a_\delta(0)\rangle_0}_{-\eta\mathscr{G}^0_{\delta\alpha}(-\tau)} = -$$
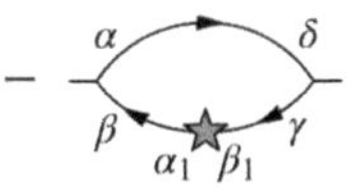

$$V_{\alpha_1\beta_1} \underbrace{\langle T_\tau a^\dagger_{\alpha_1}(\tau_1) a_\delta(0)\rangle_0}_{-\eta\mathscr{G}^0_{\delta\alpha_1}(-\tau_1)} \underbrace{\langle T_\tau a_{\beta_1}(\tau_1) a^\dagger_\alpha(\tau)\rangle_0}_{-\mathscr{G}^0_{\beta_1\alpha}(\tau_1-\tau)} \underbrace{\langle T_\tau a_\beta(\tau) a^\dagger_\gamma(0)\rangle_0}_{-\mathscr{G}^0_{\beta\gamma}(\tau)} = -$$
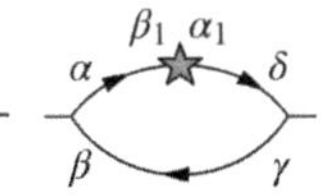

.

It is not difficult to foresee what will happen at orders $n > 1$: terms with increasing numbers of scattering events on the particle and hole lines will be generated, such that by summing all these terms we simply get a 'renormalized' particle–hole bubble in which the two free Green's functions in equation (5.60) are replaced by the exact Green's function given in equation (5.18*b*):

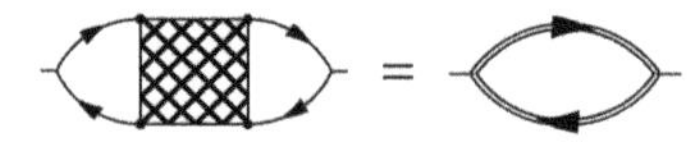

(5.63)

In this case, the particle and the hole in the particle–hole pair propagate and are scattered by the potential independently. This is, of course, due to the absence of interaction between the particle and the hole. Quite generally, in a non-interacting system, the many-particle correlation functions will be simple products of the individual one-particle propagators: this is just another way of stating Wick's theorem. In this particularly simple situation, we can use the spectral representation of the exact Green's functions to express the correlator in the frequency domain as (see Doc A5.13):

$$\mathscr{C}_{\alpha\beta\gamma\delta}(i\Omega_n) = \int_{-\infty}^{\infty} d\varepsilon_1 d\varepsilon_2\, \rho_{a_\delta a_\alpha^\dagger}(\varepsilon_1)\rho_{a_\beta a_\gamma^\dagger}(\varepsilon_2)\frac{d_{-\eta}(\varepsilon_1) - d_{-\eta}(\varepsilon_2)}{i\Omega_n + \varepsilon_1 - \varepsilon_2}. \tag{5.64}$$

Equation (5.64) is the expected generalization of equation (5.61), to which it reduces if the spectral functions are replaced by delta functions.

For non-interacting particles subject to a one-body potential, equations (5.63) and (5.64) provide the exact result. This is of course not true for particles interacting via two-body forces. In the latter case, two new types of particle–hole diagrams emerge as will be seen in the next section, which are collectively called *vertex corrections*. Nevertheless, in many problems and actual calculations it is customary to neglect the vertex corrections and use equation (5.63) as an approximation to the exact particle–hole propagator, with the full line taken as the appropriate Green's function for the system under consideration.

5.1.4.4 Diagrams for a two-body operator

For a two-body operator, we find as for a one-body operator diagrams corresponding to a 'decoration' of the particle or hole, such as:

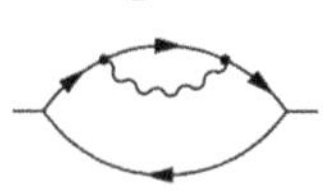

or

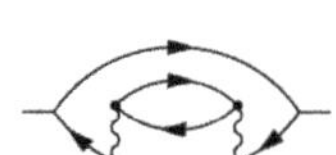

(5.65)

Diagrams of this kind describe processes in which the particle and/or the hole encounter scattering events in their propagation but without influencing one another. The sum of all such diagrams is generated by replacing the free particle and hole lines in the particle–hole bubble by the renormalized lines, like in equation (5.63). However, there are also two other types of diagrams which do not appear in the case of a one-body operator. The first such type involves interactions between the particle and the hole, for instance:

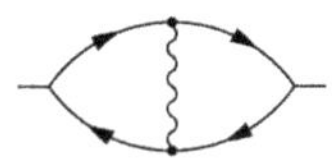
(5.66)

The second type involves diagrams connecting together different bubbles with interaction lines, for instance:

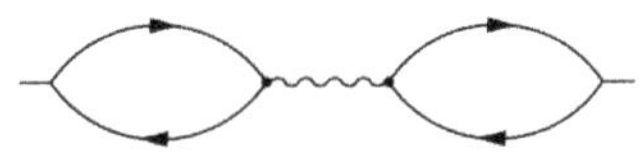
(5.67)

These latter diagrams describe processes in which the particle–hole pair interacts with another particle–hole pair and recombines. This corresponds to the (classical) interaction between density fluctuations; these diagrams are therefore crucial to describe electrostatic effects such as screening.

Here, very much like for the single-particle Green's function, it is possible to perform the summation of infinite series of diagrams and define the analogous of an irreducible self-energy. Indeed, one notices that the diagram in equation (5.67), unlike those in equations (5.65) and (5.66), can be separated in two disconnected pieces by cutting just one interaction line. We will call such particle–hole diagrams reducible, and the others irreducible. Then, if

$\Pi_{\alpha\beta\gamma\delta} =$ 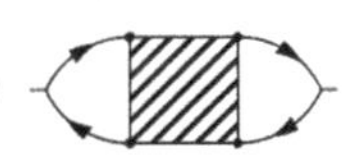
(5.68)

represents the sum of all irreducible particle–hole diagrams, we can write

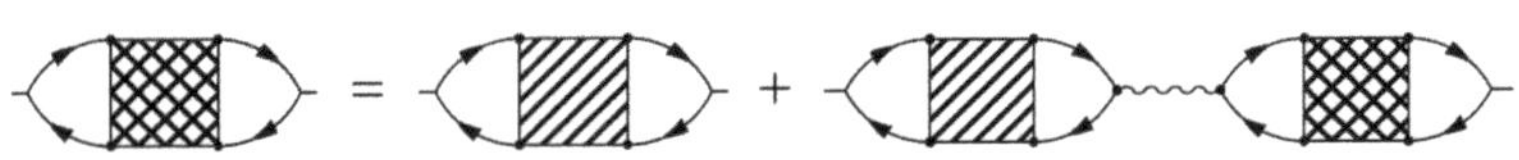
(5.69)

as can be checked by direct inspection. This is the counterpart for two-particle correlators of Dyson's equation (5.31). The quantity Π is called the *polarization*. We will see in section 5.1.4.6 that it indeed describes the polarization of the system, which has the effect of screening the bare interaction. Please note that there is no minus sign in front of the diagram in the definition of the polarization (5.68), unlike for the other two-particle correlation functions (5.59).

The sum of all irreducible diagrams entering the polarization $\Pi_{\alpha\beta\gamma\delta}$ can also be written in a suggestive way as:

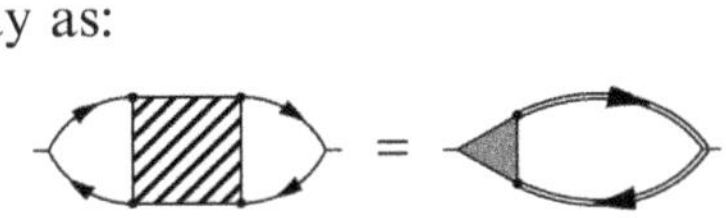
(5.70)

In this representation, one can see the appearance of a *renormalized vertex* given by the sum of irreducible vertex corrections:

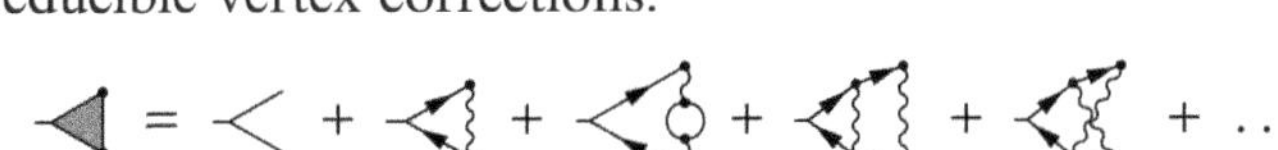
(5.71)

It is important to notice that only one of the two vertices must be renormalized in equation (5.70). Otherwise most diagrams would be counted twice. On the contrary, both propagator lines are renormalized because they are not topologically equivalent, one representing a particle and the other a hole.

A very common scheme, known as the *random-phase approximation* (RPA), consists in using the particle–hole bubble as an approximation for the full polarization:

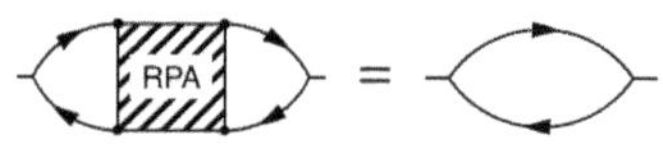

(5.72)

Obviously, this approximation neglects both the propagator renormalization and the vertex corrections. However, when used in equation (5.69) it allows one to take into account the complete series of diagrams of the kind (5.67). An improved version, the self-consistent RPA, makes use of the renormalized particle–hole bubble:

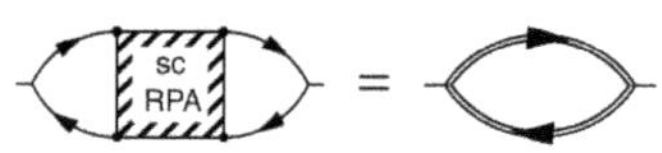

(5.73)

5.1.4.5 Density–density correlation function

Generalities We have already encountered the density–density correlation function in sections 1.1 and 3.3.2. In imaginary time and for a translation-invariant system, this function is defined as

$$\chi_{nn}(\boldsymbol{q},\,\tau) = -\langle T_\tau n(\boldsymbol{q},\,\tau) n(-\boldsymbol{q},\,0)\rangle \tag{5.74}$$

with $n(\boldsymbol{q})$ the density operator given in equation (2.46). The density–density correlation function is a superposition of particle–hole diagrams like (5.59) with a specific relation between the particle and the hole, namely a difference $\boldsymbol{q}$ in their wave vectors:

$\chi_{nn}(\boldsymbol{q},\tau) = -$

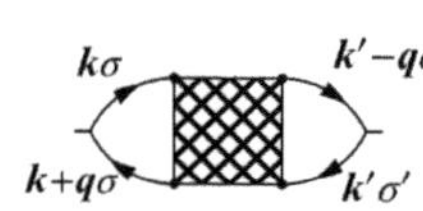

(5.75)

In this diagram and in the subsequent ones of the same kind, unlike in equation (5.59), the variables ($\boldsymbol{k}\sigma$) and ($\boldsymbol{k}'\sigma'$) are summed over as in the definition of the density. The variables that remain, $\boldsymbol{q}$ and τ, belong to the external 'connectors'. One can indeed see that there is a flow of momentum $\hbar\boldsymbol{q}$ from right to left in the diagram. A density fluctuation of wave vector $\boldsymbol{q}$ is injected from the right (at time zero) by kicking a hole from $\boldsymbol{k}' - \boldsymbol{q}$ to $\boldsymbol{k}'$ and released from the left (at time τ) with the recombination of an electron from $\boldsymbol{k} + \boldsymbol{q}$ to $\boldsymbol{k}$. In a system of free particles, the expression of χ^0_{nn} is easily worked out using equation (5.61) and gives

$$\boxed{\chi^0_{nn}(\boldsymbol{q},\,i\Omega_n) = \sum_{\boldsymbol{k}\sigma} \frac{d_{-\eta}(\xi_{\boldsymbol{k}}) - d_{-\eta}(\xi_{\boldsymbol{k}+\boldsymbol{q}})}{i\Omega_n + \xi_{\boldsymbol{k}} - \xi_{\boldsymbol{k}+\boldsymbol{q}}},} \tag{5.76}$$

as we have already found using the spectral representation (see equation (3.29)). In the particular case of free fermions, we can perform the momentum sum exactly at zero temperature. The resulting expression for the retarded function is somewhat complicated (see Doc A5.14) but contains much useful information. We only quote here the behavior of $\chi^0_{nn}(\boldsymbol{q}, \varepsilon)$ in a few important limits—in the following, we use the notation χ^0_{nn} rather than χ^{0R}_{nn} for the retarded function:

$$\left.\begin{array}{ll} \boldsymbol{q}=\varepsilon=0 & \chi^0_{nn}(0,0)=-N_0^{\rm el}(0)\quad [\text{first } \varepsilon=0, \text{ then } \boldsymbol{q}\to 0] \\ \varepsilon=0 & \chi^0_{nn}(\boldsymbol{q},0)=-N_0^{\rm el}(0)\left[\dfrac{1}{2}+\dfrac{1}{4}\left(\dfrac{1}{x}-x\right)\ln\left|\dfrac{1+x}{1-x}\right|\right] \\ \boldsymbol{q}\to 0 & \chi^0_{nn}(\boldsymbol{q}\to 0, \varepsilon\propto q)=-N_0^{\rm el}(0)\left(1-\dfrac{\zeta}{2}\ln\left|\dfrac{1+\zeta}{1-\zeta}\right|\right) \\ \varepsilon\to\infty & \chi^0_{nn}(\boldsymbol{q},\varepsilon)=\dfrac{\frac{4}{3}N_0^{\rm el}(0)\varepsilon_{\rm F}\varepsilon_{\boldsymbol{q}}}{\varepsilon^2} \end{array}\right\} \quad (5.77)$$

Free fermions at $T = 0$

$$x=\frac{q}{2k_{\rm F}},\quad \zeta=\frac{m\varepsilon}{\hbar^2 k_{\rm F}q}\quad \varepsilon_{\rm F}=\frac{\hbar^2k_{\rm F}^2}{2m},\quad \varepsilon_{\boldsymbol{q}}=\frac{\hbar^2q^2}{2m}.$$

In these expressions, $N_0^{\rm el}(0)$ stands for the density of states (DOS) at the Fermi energy, namely $N_0^{\rm el}(0) = mk_{\rm F}\mathscr{V}/(\pi^2\hbar^2)$ for free electrons in 3D. The result $\chi^0_{nn}(0, 0) = -N_0^{\rm el}(0)$ has a wider validity, though, and holds for any system of independent fermions (see Doc A5.14). The high-energy behavior $\propto \varepsilon^{-2}$ is also a general property valid for all two-particle auto-correlation functions, due to the fact that their spectral function is an odd function of ε. As a result, all odd powers of $1/\varepsilon$ disappear in the moment expansion (3.43). That the term in ε^{-1} disappears can also be directly seen from equation (3.45), since $[n(\boldsymbol{q}), n(-\boldsymbol{q})]_- = 0$. Therefore, while the high-energy behavior of the single-particle Green's function is always exactly $1/\varepsilon$, the high-energy behavior of the retarded density–density correlation function is controlled by the first moment of its spectral function:

$$\chi^R_{nn}(\boldsymbol{q}, \varepsilon\to\infty) = \frac{1}{\varepsilon^2}\int_{-\infty}^{\infty} d\varepsilon\ \varepsilon\rho_{n(\boldsymbol{q})n(-\boldsymbol{q})}(\varepsilon). \quad (5.78)$$

These various limits of $\chi^0_{nn}(\boldsymbol{q}, \varepsilon)$ will be used below to discuss the general features of the charge-excitation spectrum.

Let us consider now the case of interacting particles. Following the reasoning of the previous section, we can rewrite $\chi_{nn}(\boldsymbol{q}, i\Omega_n)$ in terms of the irreducible polarization as

$$\boxed{\chi_{nn}(\boldsymbol{q},\, i\Omega_n) = \frac{-\Pi(\boldsymbol{q},\, i\Omega_n)}{1 + \frac{1}{\mathcal{V}} V(\boldsymbol{q})\Pi(\boldsymbol{q},\, i\Omega_n)}.} \tag{5.79}$$

In translating equation (5.69), we must be careful about the signs:  $= -\chi$, 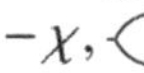 $= \Pi$, and $= -V/\mathcal{V}$, the latter sign being the one associated with each power of the interaction in equation (5.5). Equation (5.79) is the basis for studying the collective charge excitations in a system of interacting particles. Stable collective modes correspond to poles of $\chi_{nn}(\boldsymbol{q},\, i\Omega_n)$ and are therefore determined by finding the values of $\boldsymbol{q}$ and ε that satisfy the equation

$$1 + \frac{1}{\mathcal{V}} V(\boldsymbol{q})\Pi(\boldsymbol{q},\, i\Omega_n \to \varepsilon + i0^+) = 0. \tag{5.80}$$

Clearly, this equation can only be satisfied in regions of the $(\boldsymbol{q},\, \varepsilon)$ space where $\mathrm{Im}\,\Pi(\boldsymbol{q},\, \varepsilon + i0^+) = 0$. One can distinguish four kinds of solutions. Solutions such as $\varepsilon = \hbar v q$ describe sound-like waves propagating in the system. An example is the Landau zero sound with $v \geqslant v_\mathrm{F}$, which shows up when the interaction V is short-ranged like in ^{3}He. There are also solutions of the form $\varepsilon = \varepsilon_0 + Aq^2$. Those correspond to the plasmon which appears when the interaction is long-ranged and which, like an optical phonon, has a finite energy at $\boldsymbol{q} = 0$. There might also be solutions such that $\varepsilon = 0$ and $\boldsymbol{q} \neq 0$. The latter correspond to static ($\varepsilon = 0$) and periodic ($\boldsymbol{q} \neq 0$) charge modulations known as charge-density waves (CDW). If a charge-density wave solution exists, this generally means that the system is unstable toward the formation of an ordered ground state in which this density wave is realized. An example is the Wigner crystal. Finally, a solution at $\varepsilon = 0$ and $\boldsymbol{q} = 0$ also signals an instability of the system toward an ordered ground state in which the density is reorganized, like e.g. in a ferroelectric material.

Density excitation spectrum in RPA, plasmon and zero sound Since in the RPA approximation one takes (see equation (5.72))

$$\Pi^{\mathrm{RPA}}(\boldsymbol{q},\, i\Omega_n) = -\chi^0_{nn}(\boldsymbol{q},\, i\Omega_n), \tag{5.81}$$

the density–density correlation function in the RPA approximation reads

$$\chi^{\mathrm{RPA}}_{nn}(\boldsymbol{q},\, i\Omega_n) = \frac{\chi^0_{nn}(\boldsymbol{q},\, i\Omega_n)}{1 - \frac{1}{\mathcal{V}} V(\boldsymbol{q})\chi^0_{nn}(\boldsymbol{q},\, i\Omega_n)}. \tag{5.82}$$

Based on this expression, we can draw a graph representing the density excitations of the interacting homogeneous fermion gas (figure 5.8). It is convenient to measure the wave vectors in units of $2k_\mathrm{F}$ and the energies in units of $4\varepsilon_\mathrm{F}$, i.e. to work with the variables $x = q/(2k_\mathrm{F})$ and $y = \varepsilon/(4\varepsilon_\mathrm{F})$. The region of the $(x,\, y)$ plane where particle–hole excitations do exist corresponds to the region where $\mathrm{Im}\,\chi_{nn}(x,\, y) \neq 0$. In the RPA approximation, this region coincides with the region where $\mathrm{Im}\,\chi^0_{nn}(x,\, y) \neq 0$. We see in Doc A5.14 that the imaginary part of χ^0_{nn} is proportional to y if

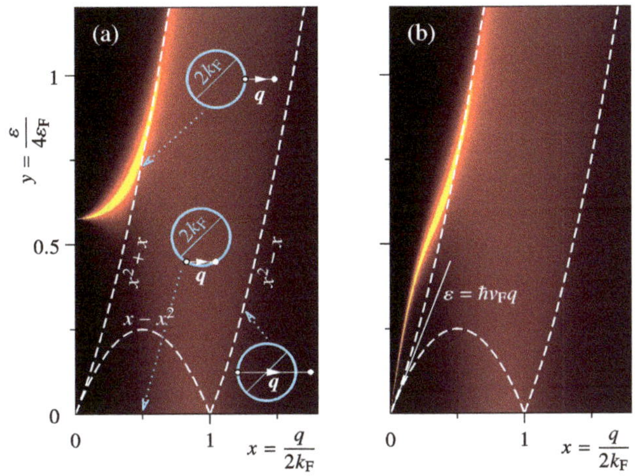

Figure 5.8. Density excitation spectrum in the 3D homogeneous fermion gas for (a) long-range and (b) short-range interaction. The color scale represents the value of $|\chi_{nn}^{\rm RPA}(\boldsymbol{q}, \varepsilon)|$, and shows the plasmon in (a) and the zero sound in (b) as bright lines. Typical particle–hole excitations are sketched in (a).

$y < x - x^2$. This defines a first region in the graph below the line $x - x^2$, where electron–hole excitations prevent the formation of collective modes[7]. More precisely, collective modes that would exist in this region are strongly damped by exchanging energy with incoherent electron–hole excitations and therefore would have a short life-time. This attenuation of collective modes is known as *Landau damping*. Particle–hole excitations are also present in the region between the lines $x^2 + x$ and $x^2 - x$, where $\operatorname{Im}\chi_{nn}^0(x, y)$ decreases as y increases. $x^2 + x$ is the highest energy that can be reached for a particle–hole excitation of momentum x, i.e. when the wave vector is normal to the Fermi surface, while $x^2 - x$ is the minimal energy attainable with a wave vector larger than $2k_{\rm F}$ (see figure 5.8).

The type of collective excitation that can exist in the region where $\operatorname{Im}\chi_{nn}(x, y) = 0$ depends on the range of the interaction V. In the limit $q \to 0$ ($\zeta^{-1} = x/y \to 0$), we can see from equation (5.77) that

$$\chi_{nn}^0(\boldsymbol{q} \to 0, \varepsilon) = \frac{1}{3}N_0^{\rm el}(0)\left(\frac{x}{y}\right)^2 + \mathcal{O}\left[\left(\frac{x}{y}\right)^4\right] \tag{5.83}$$

and therefore χ_{nn}^0 increases like $(q/\varepsilon)^2$. If the interaction is long-ranged like the Coulomb interaction, i.e. $V(\boldsymbol{q} \to 0) \propto q^{-2}$, then $V(0)\chi_{nn}^0(0, \varepsilon > 0)$ is finite and the equation $1 - \frac{1}{\mathcal{V}}V(0)\chi_{nn}^0(0, \varepsilon) = 0$ admits a solution for a finite $\varepsilon \equiv \hbar\omega_p$. In the homogeneous electron gas, one thus obtains the plasmon excitation and its dispersion as (see Doc A5.15)

[7] In one dimension, $\chi_{nn}^0(x, y) = 0$ for $y < x - x^2$ because in this case the Fermi surface reduces to two points and therefore only excitations with $q \approx 0$ and $q \approx 2k_{\rm F}$ are possible close to $\varepsilon = 0$.

$$\boxed{\omega_{\text{plasmon}}(\boldsymbol{q}) = \omega_p + \frac{3v_{\text{F}}^2}{10\omega_p}q^2 + \mathcal{O}(q^4), \qquad \omega_p^2 = \frac{e^2n}{\epsilon_0 m}.} \tag{5.84}$$

The plasmon and its dispersion are visible in figure 5.8(a) as a bright line. We also see in the figure how Landau damping destroys the plasmon, which dies out upon entering the continuum of particle–hole excitations at high energies. The plasmon and its quadratic dispersion have been observed in many simple metals and found to be in good agreement with the RPA result (5.84) in Na,[8] Mg,[9] Al,[10] and even Si,[11] while some deviations were found in K, Rb, and Cs.[8]

If the interaction is short-ranged, $V(\boldsymbol{q} = 0)$ is finite. We then see that the equation $1 - \frac{1}{\mathcal{V}}V(\boldsymbol{q})\chi_{nn}^0(\boldsymbol{q}, \varepsilon) = 0$ can still be satisfied in the limit $\boldsymbol{q} \to 0$, provided that the excitation is a kind of sound with $\varepsilon = \hbar v_0 q$ or $y = \zeta x$, such that x/y becomes a constant. In this case, the parameter ζ measures the wave velocity relative to the Fermi velocity, $\zeta = v_0/v_{\text{F}}$, with $v_{\text{F}} = \hbar k_{\text{F}}/m$. We can no longer rely on the expansion (5.83) because x/y is not a small number. Instead, we must solve the full $1 - \frac{1}{\mathcal{V}}V(0)\chi_{nn}^0(\boldsymbol{q} \to 0, \varepsilon) = 0$ which, using equation (5.77), can be recast as

$$-1 + \frac{\zeta}{2}\ln\left|\frac{1+\zeta}{1-\zeta}\right| = \frac{1}{\frac{1}{\mathcal{V}}V(0)N_0^{\text{el}}(0)}. \tag{5.85}$$

The graphical solution of this equation is illustrated in figure 5.9. We see that there are always two solutions, one with $\zeta < 1$ and one with $\zeta > 1$. The first solution implies $v_0 < v_{\text{F}}$ and corresponds to a damped sound: the line $y = \zeta x$ in figure 5.8(b) lies within the region where $\text{Im}\,\chi_{nn} \neq 0$ if $\zeta < 1$. On the contrary, the solution with $v_0 > v_{\text{F}}$ corresponds to a propagating sound which appears clearly in figure 5.8(b). The velocity of this sound approaches v_{F} in the limit of vanishingly small interactions. This sound is quite different from the ordinary sound in materials and has been called the *zero sound.* The driving force for ordinary sound is the pressure gradient generated by oscillations of the density away from thermodynamic equilibrium. Therefore, local equilibrium must be restored in the time interval between two density oscillations for a normal sound to be able to propagate. This is only possible if the typical collision time τ_c is short compared with the sound frequency, i.e. $\omega\tau_c \ll 1$. As τ_c typically increases like T^{-2} with decreasing temperature (see section 8.4), the condition $\omega\tau_c \ll 1$ is no longer satisfied as $T \to 0$ and the ordinary sound disappears at low temperature. On the contrary, the driving force for the zero sound is the direct interaction between particles and, for this sound not to be damped by collisions, we must be in the opposite limit, $\omega\tau_c \gg 1$. One thus sees that nothing prevents the propagation of the zero sound down to the lowest temperatures, hence its name. The zero sound was discovered by Landau and has been

[8] vom Felde A, Sprösser-Prou J and Fink J 1989 *Phys. Rev.* B **40** 10181.

[9] Chen C H 1976 *J. Phys. C: Solid State Phys.* **9** L321.

[10] Sprösser-Prou J, vom Felde A and Fink J 1989 *Phys. Rev.* B **40** 5799.

[11] Chen C H, Meixner A E and Kincaid B M 1980 *Phys. Rev. Lett.* **44** 951.

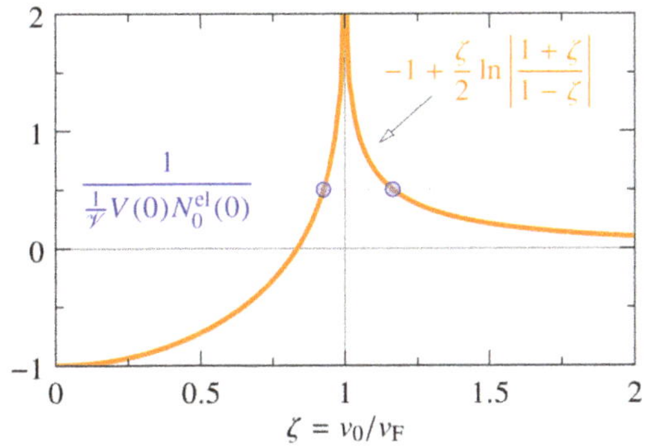

Figure 5.9. Graphical solution of the equation giving the velocity of the zero sound in the RPA approximation for a gas of fermions interacting with short-range interactions. Only the solution with $v_0 > v_{\rm F}$ corresponds to a propagating sound, the other excitation being damped by particle–hole excitations.

observed in ^{3}He, which behaves like a liquid of neutral fermions interacting via short-range forces[12].

Charge-density wave instabilities and nesting It may happen that the density–density correlation function exhibits poles at $\varepsilon = 0$ for a finite wave vector $\boldsymbol{Q}$, even in the absence of interaction. As is clear from equation (5.76), if $\xi_{\boldsymbol{k}} = \xi_{\boldsymbol{k}+\boldsymbol{Q}}$ for some vector $\boldsymbol{Q}$ and sufficiently many vectors $\boldsymbol{k}$ in the Brillouin zone, $\chi^0_{nn}(\boldsymbol{Q}, 0)$ can diverge. The property $\xi_{\boldsymbol{k}} = \xi_{\boldsymbol{k}+\boldsymbol{Q}}$ is a special case of *nesting* and $\boldsymbol{Q}$ is called a *nesting vector*. As an illustration of this behavior, figure 5.10 shows $\chi^0_{nn}(\boldsymbol{Q}, 0)$ for the two-dimensional square lattice with a tight-binding nearest-neighbor dispersion $\xi_{\boldsymbol{k}} = 2t(\cos k_x a + \cos k_y a) - \mu$, where a is the lattice parameter. If $\mu = 0$, the band is half-filled and the density amounts to one electron per lattice site. In these conditions, the Fermi surface ($\xi_{\boldsymbol{k}} = 0$) is a square as indicated in figure 5.10(a). This square is perfectly nested with a nesting vector $\boldsymbol{Q} = (\pm\pi/a, \pm\pi/a)$. The corresponding free density–density bubble (5.76) at $\varepsilon = 0$ and zero temperature is displayed in figure 5.10(b). One can see the divergence at $\boldsymbol{Q} = (\pi/a, \pi/a)$, which results from the nesting of the Fermi surface. Since poles in the density–density correlation function correspond to stable excitations (in other words, eigenstates) of the system, a pole at $\varepsilon = 0$ indicates that there exists a competing ground state characterized by a periodic modulation of the density. The uniform ground state—the one that was initially assumed to be the true ground state when calculating χ^0_{nn}—is said to be *unstable* towards the formation of an ordered ground state that breaks translational symmetry. Such a symmetry-breaking state is called charge-density wave (CDW). The mechanism is analogous to the Peierls instability of one-dimensional systems.

This kind of instability will generally be suppressed by electron–electron interactions since, according to equation (5.82), $\chi_{nn}(\boldsymbol{Q}, 0)$ approaches the finite

[12] Abel W R, Anderson A C and Wheatley J C 1966 *Phys. Rev. Lett.* **17** 74. Sköld K, Pelizzari C A, Kleb R and Ostrowski G E 1976 *Phys. Rev. Lett.* **37** 842. Hilton P A, Cowley R A, Scherm R and Stirling W G 1980 *J. Phys. C: Solid State Phys.* **13** L295.

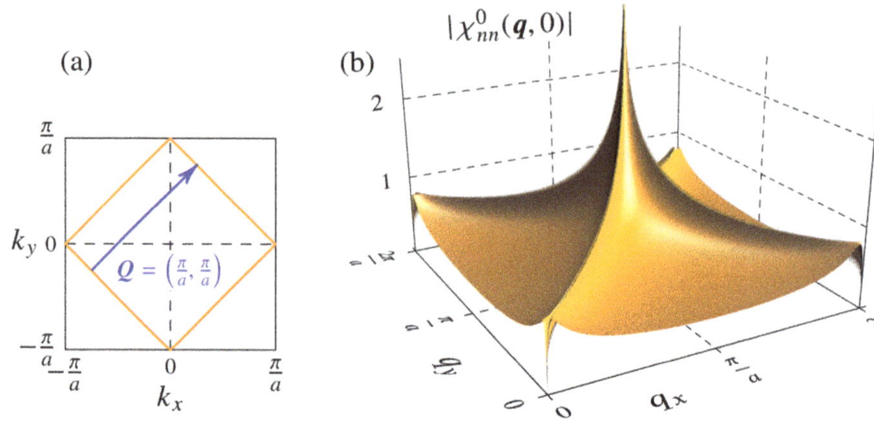

Figure 5.10. Charge-density wave instability in the 2D square lattice at half filling. The 45°-rotated square in (a) shows the perfectly nested Fermi surface. The nesting vector $\boldsymbol{Q} = (\frac{\pi}{a}, \frac{\pi}{a})$ is also indicated. The corresponding zero-energy density–density bubble is plotted in (b) and shows a logarithmic singularity at $\boldsymbol{Q}$.

value $-\mathscr{V}/V(\boldsymbol{Q})$ if $\chi^0_{nn}(\boldsymbol{Q}, 0)$ diverges. But the interaction can also trigger other CDW instabilities, which correspond (in the RPA approximation) to solutions of the equation $\chi^0_{nn}(\boldsymbol{Q}, 0) = \mathscr{V}/V(\boldsymbol{Q})$. Since $\chi^0_{nn}(\boldsymbol{Q}, 0)$ varies with temperature, the solution, if any, occurs at one particular temperature to be interpreted as the critical temperature for the CDW phase transition. Let us consider a simple case of nesting where this can happen, namely $\xi_{\boldsymbol{k}+\boldsymbol{Q}} = -\xi_{\boldsymbol{k}}$ for all $\boldsymbol{k}$. This condition is obeyed at half-filling ($\mu = 0$) by the two-dimensional tight-binding dispersion discussed above. It is easy to see that the nesting condition implies

$$\chi^0_{nn}(\boldsymbol{Q}, 0) = -\mathscr{V}\int_{-\infty}^{\infty} d\varepsilon\, N_0^{\text{el}}(\varepsilon)\frac{1 - 2f(\varepsilon)}{2\varepsilon} < 0, \qquad (\xi_{\boldsymbol{k}+\boldsymbol{Q}} = -\xi_{\boldsymbol{k}}). \tag{5.86}$$

Because $\chi^0_{nn}(\boldsymbol{Q}, 0) < 0$, there is no instability for a repulsive interaction. If the interaction is attractive, like the phonon-mediated interaction of equation (5.54), there will be a CDW instability but also, of course, a superconducting instability. The system is said to have *competing ground states*: only the instability occurring at the highest temperature has a chance to be realized in practice. In order to estimate the CDW transition temperature T_c^{CDW}, we may assume the simple form $N_0^{\text{el}}(\varepsilon) = N_0^{\text{el}}(0)$ over the bandwidth extending from $-W/2$ to $W/2$, which leads to the following equation for the critical temperature:

$$\begin{aligned}\frac{\mathscr{V}}{V(\boldsymbol{Q})} = \chi^0_{nn}(\boldsymbol{Q}, 0) &= -\mathscr{V}N_0^{\text{el}}(0)\int_{-W/2}^{W/2}\frac{d\varepsilon}{2\varepsilon}\tanh\left(\frac{\varepsilon}{2k_{\text{B}}T_c^{\text{CDW}}}\right)\\ &\approx -\mathscr{V}N_0^{\text{el}}(0)\ln\left(0.567\frac{W}{k_{\text{B}}T_c^{\text{CDW}}}\right),\end{aligned} \tag{5.87}$$

where we have used equation (20), assuming $k_B T_c^{\mathrm{CDW}} \ll W$. The solution is

$$k_B T_c^{\mathrm{CDW}} \approx 0.567\, W e^{1/[V(\boldsymbol{Q}) N_0^{\mathrm{el}}(0)]}. \tag{5.88}$$

This expression is similar to the BCS superconducting temperature (5.144): the half-bandwidth $W/2$ replaces the Debye energy $\hbar\omega_D$ and the total DOS replaces the DOS per spin (the BCS pairing interaction only acts between electrons of opposite spins). If $T_c^{\mathrm{CDW}} > T_c$, the occurrence of the CDW might prevent the formation of the superconducting state at lower temperatures.

CDW are most often observed in one-dimensional and quasi-one-dimensional materials like $NbSe_3$, $TaSe_3$, or the organic conductors TMTSF or TTF-TCNQ. In such materials, the CDW state is usually accompanied by a lattice distortion which stabilizes the CDW[13]. CDW also often occur on surfaces, which are realizations of two-dimensional systems analogous to the square lattice of figure 5.10. The CDWs on surfaces are most commonly called *surface reconstructions*.

5.1.4.6 Polarization and dielectric screening

In addition to being useful for determining the collective density excitations and the charge instabilities of a system, the density-polarization $\Pi(\boldsymbol{q}, i\Omega_n)$ appearing in equation (5.79) is closely related to dielectric screening. To see this, we note that it is possible to define a Dyson-like equation for the screened interaction, in complete analogy with equations (5.30) and (5.69):

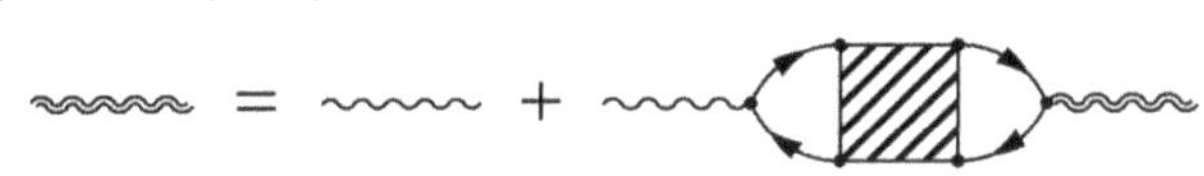

(5.89)

The right-hand side of this equation generates all possible decorations of the interaction line ∼∼∼ obtained by polarization insertions. Like the phonon-mediated interaction (5.54), the screened interaction $W(\boldsymbol{q}, i\Omega_n)$ is energy-dependent and can be deduced by solving equation (5.89). The latter reads, schematically, $-W/\mathcal{V} = -V/\mathcal{V} + (-V/\mathcal{V})\Pi(-W/\mathcal{V})$, which gives (we assume translation invariance in this section):

$$\boxed{W(\boldsymbol{q}, i\Omega_n) = \frac{V(\boldsymbol{q})}{1 + \dfrac{1}{\mathcal{V}} V(\boldsymbol{q})\Pi(\boldsymbol{q}, i\Omega_n)}.} \tag{5.90}$$

If we replace ∼∼∼ by ≈≈≈ in our diagrams, we are performing yet another partial resummation of diagrams up to infinite order. In doing so, we must be careful not to count some diagrams twice. For instance if we use the *screened exchange* self-energy diagram

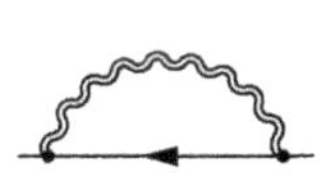

[13] See e.g. Grüner G 1988 *Rev. Mod. Phys.* **60** 1129.

we should not also retain the first of the second-order contributions in equation (5.48), since the latter is already accounted for in the screened exchange[14].

By definition, the ratio between the bare and screened interactions is the longitudinal dielectric function $\epsilon_{\|}(\boldsymbol{q}, \omega)$ (see Doc A5.16). From equation (5.90), we read:

$$\boxed{\epsilon_{\|}(\boldsymbol{q}, \omega) = 1 + \frac{1}{\mathcal{V}} V(\boldsymbol{q}) \Pi^{R}(\boldsymbol{q}, \hbar\omega).} \tag{5.91}$$

This relation is sometimes written as

$$\epsilon_{\|}^{-1}(\boldsymbol{q}, \omega) = 1 + \frac{1}{\mathcal{V}} V(\boldsymbol{q}) \chi_{nn}^{R}(\boldsymbol{q}, \hbar\omega). \tag{5.92}$$

Using equation (5.79), one easily checks that these two definitions are indeed equivalent. Working within the RPA approximation, we can now recover a few well-known expressions for the dielectric function. In the RPA approximation, we have equation (5.81) and it follows that:

$$\epsilon_{\|}^{\mathrm{RPA}}(\boldsymbol{q}, \omega) = 1 - \frac{1}{\mathcal{V}} V(\boldsymbol{q}) \chi_{nn}^{0}(\boldsymbol{q}, \hbar\omega). \tag{5.93}$$

In the homogeneous interacting electron gas, we therefore get in the static limit and at zero temperature, using equation (5.77):

$$\epsilon_{\|}^{\mathrm{RPA}}(\boldsymbol{q}, 0) = 1 + \frac{k_{\mathrm{TF}}^2}{q^2}\left[\frac{1}{2} + \frac{1}{4}\left(\frac{2k_{\mathrm{F}}}{q} - \frac{q}{2k_{\mathrm{F}}}\right)\ln\left|\frac{1 + \frac{q}{2k_{\mathrm{F}}}}{1 - \frac{q}{2k_{\mathrm{F}}}}\right|\right], \tag{5.94}$$

where the Thomas–Fermi wave vector is given by $k_{\mathrm{TF}}^2 = \frac{e^2 N_0^{\mathrm{el}}(0)}{\epsilon_0 \mathcal{V}}$. In the long-wavelength limit, we recover the famous result which can also be derived from the Thomas–Fermi approximation:

$$\epsilon_{\|}^{\mathrm{RPA}}(\boldsymbol{q} \to 0, 0) = 1 + \frac{k_{\mathrm{TF}}^2}{q^2}. \tag{5.95}$$

In the same limit, the screened interaction is therefore

---

[14] A popular and successful approximation to the self-energy of interacting fermionic systems is the so-called 'GW' approximation. In the GW approximation, both the interaction screening and the propagator renormalization are taken into account by using the diagram

$\Sigma^{\mathrm{GW}} =$ 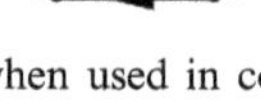

This approximation solves the band-gap problem when used in conjunction with density-functional theory (DFT) and can produce accurate quasi-particle energies. See Hedin L 1999 *J. Phys.: Condens. Matter* **11** R489.

$$W^{\mathrm{RPA}}(\boldsymbol{q} \to 0, 0) = \frac{e^2}{\epsilon_0} \frac{1}{q^2 + k_{\mathrm{TF}}^2}. \tag{5.96}$$

Fourier transforming with the help of equation (14), we find that the screened interaction is short-ranged:

$$W^{\mathrm{RPA}}(\boldsymbol{r} \to \infty, 0) = \frac{e^2}{4\pi\epsilon_0} \frac{e^{-k_{\mathrm{TF}} r}}{r}. \tag{5.97}$$

A more accurate calculation that takes into account the complete Lindhard function (5.94) leads to a screened interaction displaying Friedel oscillations at large distances:

$$W^{\mathrm{RPA}}(\boldsymbol{r} \to \infty, 0) = \frac{e^2}{4\pi\epsilon_0}\left[\frac{e^{-k_{\mathrm{TF}} r}}{r} + \frac{2k_{\mathrm{F}}^3 k_{\mathrm{TF}}^2}{(8k_{\mathrm{F}}^2 + k_{\mathrm{TF}}^2)^2} \frac{\cos 2k_{\mathrm{F}} r}{(k_{\mathrm{F}} r)^3} + \mathcal{O}(r^{-4})\right]. \tag{5.98}$$

These oscillations result from the discontinuity of the occupation numbers at the Fermi-surface, which in turn leads to the logarithmic singularity at $q = 2k_{\mathrm{F}}$ in equation (5.94). The oscillations are therefore damped at finite temperature. Finally, we quote the standard expression in the high-frequency limit, which can be derived using equation (5.77):

$$\epsilon_{\parallel}^{\mathrm{RPA}}(\boldsymbol{q}, \omega \to \infty) = 1 - \frac{\omega_p^2}{\omega^2}. \tag{5.99}$$

These few simple results show how perturbation theory, once properly resumed in order to include an infinite series of diagrams, allows one to recover approximate expressions which can also be obtained by other methods. The advantages of perturbation theory is that it provides a fully first-principle approach to the various problems, and indicates directions to develop systematic improvements over existing approximations.

5.1.4.7 Spin–spin correlation function

Analogous to the density–density correlation function defined in equation (5.74), the spin–spin correlation function is

$$\chi_{ss}(\boldsymbol{q}, \tau) = -\langle T_\tau \boldsymbol{S}(\boldsymbol{q}, \tau) \cdot \boldsymbol{S}(-\boldsymbol{q}, 0)\rangle, \tag{5.100}$$

with $\boldsymbol{S}(\boldsymbol{q})$ the Fourier transform of the spin-density operator given in the plane-wave basis by equation (2.47). The Pauli matrices τ^x, τ^y, and τ^z are given in equation (22). It is customary to introduce longitudinal and transverse components for this correlator, where 'longitudinal' in this context means parallel to the spin quantization axis taken as the z axis. As shown in Doc A5.17, for electrons interacting via spin-conserving forces like the Coulomb force, we have

$$\chi_{ss}(\boldsymbol{q}, i\Omega_n) = \left(\frac{\hbar}{2}\right)^2 \left[\chi_{\parallel}(\boldsymbol{q}, i\Omega_n) + \chi_{\perp}(\boldsymbol{q}, i\Omega_n)\right] \tag{5.101a}$$

$$\chi_{\|}(\boldsymbol{q}, i\Omega_n) = -\frac{\Pi(\boldsymbol{q}, i\Omega_n) + 4\dfrac{V(\boldsymbol{q})}{\mathcal{V}}\Pi_{\uparrow\uparrow}(\boldsymbol{q}, i\Omega_n)\Pi_{\downarrow\downarrow}(\boldsymbol{q}, i\Omega_n)}{1 + \dfrac{V(\boldsymbol{q})}{\mathcal{V}}\Pi(\boldsymbol{q}, i\Omega_n)} \tag{5.101b}$$

$$\chi_{\perp}(\boldsymbol{q}, i\Omega_n) = -2[\Pi_{\uparrow\downarrow}(\boldsymbol{q}, i\Omega_n) + \Pi_{\downarrow\uparrow}(\boldsymbol{q}, i\Omega_n)], \tag{5.101c}$$

where the spin-dependent polarizations are

$$\Pi_{\sigma\sigma'}(\boldsymbol{q}, i\Omega_n) = \tag{5.102}$$

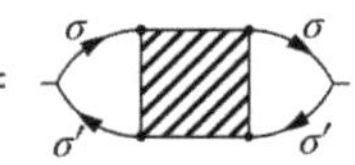

and $\Pi \equiv \Pi_{\uparrow\uparrow} + \Pi_{\downarrow\downarrow}$ is the total polarization entering equation (5.79).

For non-interacting and non-spin polarized electrons, we have $V(\boldsymbol{q}) = 0$ and $\Pi_{\uparrow\uparrow} = \Pi_{\downarrow\downarrow} = \Pi_{\uparrow\downarrow} = \Pi_{\downarrow\uparrow} = \frac{1}{2}\Pi^0 = -\frac{1}{2}\chi^0_{nn}$. We therefore obtain

$$\left(\frac{2}{\hbar}\right)^2 \chi^0_{ss}(\boldsymbol{q}, i\Omega_n) = 3\chi^0_{nn}(\boldsymbol{q}, i\Omega_n). \tag{5.103}$$

Up to numbers, the free density–density and spin–spin correlation functions are identical. The factor 3 reflects the three components of the spin. This result is expected because the density and spin excitations are both carried by the same particles and, in the absence of interactions, the propagation of these excitations is governed by the propagation of that free particle. The situation is going to be different in the presence of interactions because a density excitation produces a charge disturbance and is therefore sensitive to the Coulomb interaction, while a spin excitation can be charge-neutral.

We can see this by considering the case of an interacting paramagnetic system for which $\Pi_{\uparrow\uparrow} = \Pi_{\downarrow\downarrow} = \Pi_{\uparrow\downarrow} = \Pi_{\downarrow\uparrow} = \frac{1}{2}\Pi$, such that equation (5.101) gives a result similar to that of independent particles:

$$\left(\frac{2}{\hbar}\right)^2 \chi_{ss}(\boldsymbol{q}, i\Omega_n) = -3\Pi(\boldsymbol{q}, i\Omega_n) \qquad \text{(paramagnetic system).} \tag{5.104}$$

This is quite different from the density–density result (5.79) and it shows that the interaction does not lead to such things as spin plasmons or spin-waves in a paramagnetic electronic system. The existence of static spin-density waves (SDW) is still possible, though. At the RPA level, for instance, we have simply $\chi^{\mathrm{RPA}}_{ss} \propto \chi^0_{nn}$ such that the system of figure 5.10 has a SDW instability at $\boldsymbol{Q} = (\frac{\pi}{a}, \frac{\pi}{a})$, which indicates the existence of a competing anti-ferromagnetic ground state.

In a fully polarized system, one can to first approximation assume that only one spin state is available ($\Pi_{\uparrow\uparrow} = \Pi$ and $\Pi_{\downarrow\downarrow} = \Pi_{\uparrow\downarrow} = \Pi_{\downarrow\uparrow} = 0$). Since in such a system the density $n = n_{\uparrow}$, the existence of the density plasmon also implies the existence of a spin plasmon, an oscillation of the local magnetization. Hence one expects to have a pole in the spin susceptibility. We indeed find from equation (5.101) that, in this case,

$$\left(\frac{2}{\hbar}\right)^2 \chi_{ss}(\boldsymbol{q}, i\Omega_n) = -\frac{\Pi(\boldsymbol{q}, i\Omega_n)}{1 + \frac{V(\boldsymbol{q})}{\mathcal{V}}\Pi(\boldsymbol{q}, i\Omega_n)} = \chi_{nn}(\boldsymbol{q}, i\Omega_n) \tag{5.105}$$

(fully polarized system),

such that the spin and density excitations have identical structures as expected.

The formulas given in this section were derived for an interaction that does not depend on the spin, like the Coulomb and phonon-mediated interactions. One of the most popular model used to investigate the properties of strongly-correlated electrons, the Hubbard model (2.79), represents the interaction by a contact potential $V(\boldsymbol{r}, \boldsymbol{r}') = U\delta(\boldsymbol{r} - \boldsymbol{r}')$, or $V(\boldsymbol{q}) = U$. As a result, the interaction becomes spin-dependent because of the Pauli exclusion principle: electrons with parallel spins cannot occupy the same position in space and therefore cannot feel the interaction. For the Hubbard model, the interaction vertex of equation (5.43) carries one additional constraint, namely $\delta_{\sigma_1, -\sigma_2}$. The analysis of the spin susceptibility must be revisited accordingly: in the diagrammatic equation of Doc A5.17, the second term on the right-hand side disappears. Repeating the analysis, we then find

$$\chi_{\parallel}^{\text{Hubbard}} = \frac{-\Pi - 2\frac{U}{\mathcal{V}}\Pi_{\uparrow\uparrow}\Pi_{\downarrow\downarrow}}{1 - \left(\frac{U}{\mathcal{V}}\right)^2 \Pi_{\uparrow\uparrow}\Pi_{\downarrow\downarrow}}, \tag{5.106}$$

while the expression of $\chi_{\perp}$ remains the same as above. In a paramagnetic state, the spin–spin correlation function becomes

$$\left(\frac{2}{\hbar}\right)^2 \chi_{ss}^{\text{Hubbard}}(\boldsymbol{q}, i\Omega_n) = \frac{-\Pi(\boldsymbol{q}, i\Omega_n)\left[3 - \frac{U}{\mathcal{V}}\Pi(\boldsymbol{q}, i\Omega_n)\right]}{1 - \frac{1}{2}\frac{U}{\mathcal{V}}\Pi(\boldsymbol{q}, i\Omega_n)} \tag{5.107}$$

(paramagnetic state).

Collective excitations are now possible, unlike in equation (5.104), if the denominator $1 - \frac{1}{2}\frac{U}{\mathcal{V}}\text{Re}\,\Pi(\boldsymbol{q}, \varepsilon + i0^+)$ vanishes. Note the sign difference with respect to equation (5.80). These opposite signs for the charge and spin susceptibilities imply, in particular, that in the case of the nesting condition considered in equation (5.86), where we have seen that no CDW instability occurs for a repulsive interaction, a SDW instability *does* occur, corresponding to an ordering of the spins with wave vector $\boldsymbol{Q}$.

This very brief overview should make it clear that the spin–spin correlation function can exhibit a rich variety of behaviors. We will return to it in chapter 10.

5.2 Equation-of-motion method

The perturbation theory is useful when there exists a small parameter and expanding in powers of this parameter makes sense. In other situations, the equation-of-motion method is an alternate analytical approach to evaluate correlation functions. This method provides a direct and elegant solution to the problem of independent particles. Furthermore, it allows one to easily derive the conventional mean-field theories of Hartree–Fock and BCS–Gor'kov and to give them a transparent interpretation in terms of neglected high-order correlations.

5.2.1 The equation of motion

The equation of motion of the imaginary-time correlation function of the operators A and B, $\mathscr{C}_{AB}(\tau) = -\langle T_\tau A(\tau)B(0)\rangle$, is readily found to be (see Doc A5.18)

$$\boxed{-\partial_\tau \mathscr{C}_{AB}(\tau) = \delta(\tau)\langle [A, B]_{-\eta}\rangle + \mathscr{C}_{[A,\, K]B}(\tau).} \tag{5.108}$$

The first term in the right-hand side arises due to the time derivative of the theta function implied by the imaginary-time ordering, while the second term is due to the time derivative of $A(\tau)$. As we see, solving this equation for $\mathscr{C}_{AB}(\tau)$ requires another correlation function $\mathscr{C}_{[A,\, K]B}(\tau)$. This is just the beginning of an endless series of coupled equations, because the equation of motion of $\mathscr{C}_{[A,\, K]B}(\tau)$ in turn depends on $\mathscr{C}_{[[A,\, K],\, K]B}(\tau)$, etc. The obvious strategy within the equation-of-motion framework is therefore to cut this infinite series by approximating the high-order correlation functions by products of lower-order ones—as if Wick's theorem would apply—and thus obtain a closed set of equations. For independent particles, this truncation is exact due to the absence of correlations. Therefore, although the equation-of-motion method does not provide a practical recipe for calculating correlation functions exactly, it gives a strategy for building a better and better approximation by decoupling the correlation functions at higher and higher order. From now on, we focus on the one-particle Green's function. The treatment of two-particle functions follows the same procedure.

5.2.2 One-particle Green's function

Let us recall that the one-particle Green's function is the correlation function of the creation and annihilation operators and thus corresponds to taking $A = a_\alpha$ and $B = a_\beta^\dagger$, in which case $[A, B]_{-\eta}$ becomes $[a_\alpha, a_\beta^\dagger]_{-\eta} = \delta_{\alpha\beta}$ according to equation (2.41). The Green's function is denoted $\mathscr{G}_{\alpha\beta}(\tau)$ as in equation (5.6):

$$\mathscr{G}_{\alpha\beta}(\tau) \equiv \mathscr{C}_{a_\alpha a_\beta^\dagger}(\tau) = -\left\langle T_\tau a_\alpha(\tau) a_\beta^\dagger(0)\right\rangle. \tag{5.109}$$

5.2.2.1 Independent particles

For independent particles, the Hamiltonian generically has the form of the first term in equation (2.45), $K_0 = \sum_{\alpha\beta}\xi_{\alpha\beta}a_\alpha^\dagger a_\beta$, and the commutator of a_α and K_0 that we need in the equation of motion (5.108) is given in equation (25): $[a_\alpha, K_0] = \sum_\gamma \xi_{\alpha\gamma}a_\gamma$. The equation of motion therefore reads

$$-\partial_\tau \mathscr{G}^0_{\alpha\beta}(\tau) = \delta(\tau)\delta_{\alpha\beta} + \sum_\gamma \xi_{\alpha\gamma} \underbrace{\mathscr{C}^0_{a_\gamma a^\dagger_\beta}(\tau)}_{\mathscr{G}^0_{\gamma\beta}(\tau)}. \tag{5.110}$$

A superscript 0 is attached to $\mathscr{G}$ in order to stress that these and the following relations are only valid for independent particles. We see that the equation of motion involves only $\mathscr{G}^0$ in this case and can therefore be solved in closed form. Moving to the frequency domain where $-\partial_\tau \to i\nu_n$ and $\delta(\tau) \to 1$, we have $i\nu_n \mathscr{G}^0_{\alpha\beta}(i\nu_n) = \delta_{\alpha\beta} + \sum_\gamma \xi_{\alpha\gamma} \mathscr{G}^0_{\gamma\beta}(i\nu_n)$ or, in other words,

$$\sum_\gamma \underbrace{\left[i\nu_n\delta_{\alpha\gamma} - \xi_{\alpha\gamma}\right]}_{[i\nu_n\mathbb{1} - K_0]_{\alpha\gamma}} \underbrace{\mathscr{G}^0_{\gamma\beta}(i\nu_n)}_{[\mathscr{G}_0(i\nu_n)]_{\gamma\beta}} = \underbrace{\delta_{\alpha\beta}}_{[\mathbb{1}]_{\alpha\beta}}. \tag{5.111}$$

It is natural to introduce a 'matrix Green's function' as we did in section 5.1.3.3. We then see that equation (5.111) reduces to the matrix equation $(i\nu_n\mathbb{1} - K_0)\mathscr{G}_0(i\nu_n) = \mathbb{1}$, that is

$$\boxed{\mathscr{G}_0(i\nu_n) = (i\nu_n\mathbb{1} - K_0)^{-1}.} \tag{5.112}$$

This equation is very convenient, as it provides a recipe for evaluating the Green's function of independent particles through a simple matrix inversion. It is the same result that we have obtained in section 5.1.3.3 by summing all diagrams in perturbation theory for a one-body operator. If the one-particle states φ_α corresponding to $a^\dagger_\alpha$ are chosen as the eigenstates of K_0, then $K_0 = \sum_\alpha \xi_\alpha a^\dagger_\alpha a_\alpha$ is diagonal and the matrix Green's function is also diagonal[15]:

$$\mathscr{G}^0_{\alpha\beta}(i\nu_n) = \frac{\delta_{\alpha\beta}}{i\nu_n - \xi_\alpha}. \tag{5.113}$$

---

[15] Formally, we could write using Dirac's notation,

$$i\nu_n\mathbb{1} - K_0 = \sum_{\alpha\beta} |\alpha\rangle \underbrace{\langle\alpha|i\nu_n\mathbb{1} - K_0|\beta\rangle}_{\delta_{\alpha\beta}(i\nu_n - \xi_\alpha)} \langle\beta| = \sum_\alpha (i\nu_n - \xi_\alpha)|\alpha\rangle\langle\alpha|.$$

This expresses the fact that $i\nu_n\mathbb{1} - K_0$ is represented numerically by a diagonal matrix. The matrix representing the inverse is therefore simply the diagonal matrix with inverse matrix elements, namely

$$\mathscr{G}_0(i\nu_n) = (i\nu_n\mathbb{1} - K_0)^{-1} = \sum_\alpha \frac{|\alpha\rangle\langle\alpha|}{i\nu_n - \xi_\alpha},$$

and the matrix elements of $\mathscr{G}_0$ follow:

$$\mathscr{G}^0_{\alpha\beta}(i\nu_n) = \langle\alpha|\mathscr{G}_0(i\nu_n)|\beta\rangle = \sum_\gamma \frac{\langle\alpha|\gamma\rangle\langle\gamma|\beta\rangle}{i\nu_n - \xi_\gamma} = \frac{\delta_{\alpha\beta}}{i\nu_n - \xi_\alpha}.$$

The analytic continuation to real energy gives the retarded function that we have already encountered in equation (3.26).

5.2.2.2 Interacting particles

If the Hamiltonian involves an interaction term like the second term in equation (2.45), the equation of motion requires the commutator $[a_\alpha, V]= \sum_{\gamma\mu_1\mu_2} V_{\alpha\gamma\mu_1\mu_2} a_\gamma^\dagger a_{\mu_2} a_{\mu_1}$ (see equation (25)) in addition to the contribution from $[a_\alpha, K_0]$. It follows that

$$-\partial_\tau \mathcal{G}_{\alpha\beta}(\tau) = \delta(\tau)\delta_{\alpha\beta} + \sum_\gamma \xi_{\alpha\gamma}\mathcal{G}_{\gamma\beta}(\tau) + \sum_{\gamma\mu_1\mu_2} V_{\alpha\gamma\mu_1\mu_2}\mathcal{C}_{a_\gamma^\dagger a_{\mu_2} a_{\mu_1},\, a_\beta^\dagger}(\tau). \tag{5.114}$$

This is the general equation of motion of the one-particle Green's function. Its physical meaning is that the relaxation of single-particle excitations described by $\mathcal{G}_{\alpha\beta}(\tau)$ occurs via the two-body interaction through processes involving the creation of particle–hole pairs. The equation of motion of $\mathcal{C}_{a_\gamma^\dagger a_{\mu_2} a_{\mu_1},\, a_\beta^\dagger}(\tau)$ describes the dynamics of these particle–hole pairs, which is controlled by higher-order processes and so on and so forth along the infinite chain of coupled equations mentioned in section 5.2.1. The equation of motion (5.114) cannot be solved exactly for a general interaction V, but it can be used to prove exact relations involving the Green's function. An example is equation (1.9) relating the grand potential Ω to $\mathcal{G}_{\alpha\beta}(\tau)$, the proof of which is given in Doc A5.19.

If Wick's theorem did apply, $\mathcal{C}_{a_\gamma^\dagger a_{\mu_2} a_{\mu_1},\, a_\beta^\dagger}(\tau)$ in equation (5.114) could be factorized into products of correlation functions of only two field operators, i.e. Green's functions. In this way, the infinite chain of equations of motion could be solved in closed form. The Wick theorem does not hold if a two-body interaction is present, however, such that $\mathcal{C}_{a_\gamma^\dagger a_{\mu_2} a_{\mu_1},\, a_\beta^\dagger}(\tau)$ cannot be factorized. The essence of the mean-field theories discussed in the next paragraph is precisely to neglect the correlations present in $\mathcal{C}_{a_\gamma^\dagger a_{\mu_2} a_{\mu_1},\, a_\beta^\dagger}(\tau)$ and to treat this quantity as if particles were independent.

5.2.2.3 Hartree–Fock–Gor'kov decoupling

5.2.2.3.1 Gor'kov equations

In this approximation, one assumes that the correlation function in the last term of the equation of motion of the Green's function can be 'decoupled' using Wick's theorem, as if the particles were independent. This approximation thus neglects non-trivial two-particle correlations and is, therefore, an independent-particle or mean-field approximation. It is in fact the most general static mean-field approximation one can think of. In the absence of pairing it is equivalent to the Hartree–Fock approximation, while in the presence of pairing it is just the BCS theory. We derive it here in a somewhat abstract way in order to emphasize the generality of the approach. Specific applications will be considered towards the end of this section. The approximation reads:

$$\begin{aligned}\mathscr{C}_{a_\gamma^\dagger a_{\mu_2} a_{\mu_1},\, a_\beta^\dagger}(\tau) \approx &- \left\langle T_\tau a_\gamma^\dagger(\tau) a_{\mu_2}(\tau)\right\rangle \left\langle T_\tau a_{\mu_1}(\tau) a_\beta^\dagger(0)\right\rangle \\ &- \eta \left\langle T_\tau a_\gamma^\dagger(\tau) a_{\mu_1}(\tau)\right\rangle \left\langle T_\tau a_{\mu_2}(\tau) a_\beta^\dagger(0)\right\rangle \\ &- \left\langle T_\tau a_\gamma^\dagger(\tau) a_\beta^\dagger(0)\right\rangle \left\langle T_\tau a_{\mu_2}(\tau) a_{\mu_1}(\tau)\right\rangle \\ = &\left\langle a_\gamma^\dagger a_{\mu_2}\right\rangle \mathscr{G}_{\mu_1\beta}(\tau) + \eta \left\langle a_\gamma^\dagger a_{\mu_1}\right\rangle \mathscr{G}_{\mu_2\beta}(\tau) + \left\langle a_{\mu_2} a_{\mu_1}\right\rangle \mathscr{F}^\dagger_{\gamma\beta}(\tau). \qquad (5.115)\end{aligned}$$

The new correlation function introduced here, namely

$$\boxed{\mathscr{F}^\dagger_{\alpha\beta}(\tau) \equiv \mathscr{C}_{a_\alpha^\dagger a_\beta^\dagger}(\tau) = -\left\langle T_\tau a_\alpha^\dagger(\tau) a_\beta^\dagger(0)\right\rangle,} \qquad (5.116)$$

is called the *anomalous Green's function*. The notation $\mathscr{F}^\dagger$ does not have the usual meaning of hermitic conjugation but stands as a reminder that this is the correlation function $\langle a^\dagger a^\dagger \rangle$. The anomalous Green's function vanishes in normal metals in which the number of particles is conserved ($[K, N] = 0$), but can be nonzero in BCS models of superconductors when a finite density of Cooper pairs exists. In the mean-field approximation, the anomalous propagator is directly related to the superconducting order parameter. The equation of motion of $\mathscr{F}^\dagger_{\alpha\beta}$ is readily written down using the general equation of motion (5.108), the commutator (26), and the same type of decoupling as equation (5.115):

$$\begin{aligned}-\partial_\tau \mathscr{F}^\dagger_{\alpha\beta}(\tau) &= -\sum_\gamma \xi^*_{\alpha\gamma} \mathscr{F}^\dagger_{\gamma\beta}(\tau) - \sum_{\gamma\mu_1\mu_2} V^*_{\alpha\gamma\mu_1\mu_2} \mathscr{C}_{a_{\mu_1}^\dagger a_{\mu_2}^\dagger a_\gamma,\, a_\beta^\dagger}(\tau) \\ &\approx -\sum_\gamma \xi^*_{\alpha\gamma} \mathscr{F}^\dagger_{\gamma\beta}(\tau) - \sum_{\gamma\mu_1\mu_2} V^*_{\alpha\gamma\mu_1\mu_2} \Big[\langle a_{\mu_1}^\dagger a_{\mu_2}^\dagger\rangle \mathscr{G}_{\gamma\beta}(\tau) \\ &\quad + \eta \langle a_{\mu_1}^\dagger a_\gamma\rangle \mathscr{F}^\dagger_{\mu_2\beta}(\tau) + \langle a_{\mu_2}^\dagger a_\gamma\rangle \mathscr{F}^\dagger_{\mu_1\beta}(\tau)\Big]. \qquad (5.117)\end{aligned}$$

Note the absence of a term proportional to $\delta(\tau)$, owing to the fact that $[a_\alpha^\dagger, a_\beta^\dagger]_{-\eta} = 0$. Collecting the equations of motion of $\mathscr{G}$ and $\mathscr{F}^\dagger$, using a matrix notation and moving to the frequency domain, we obtain the two coupled equations known as the Gor'kov equations (see Doc A5.20):

$$\boxed{\left(i\nu_n \mathbb{1} - \tilde{K}_0\right)\mathscr{G}(i\nu_n) - \Delta \mathscr{F}^\dagger(i\nu_n) = \mathbb{1}} \qquad (5.118a)$$

$$\boxed{\left(i\nu_n \mathbb{1} + \tilde{K}_0^*\right)\mathscr{F}^\dagger(i\nu_n) + \Delta^* \mathscr{G}(i\nu_n) = 0.} \qquad (5.118b)$$

We have introduced the modified self-consistent one-particle Hamiltonian $\tilde{K}_0 = K_0 + V_\mathrm{H} + V_\mathrm{x}$ as well as the mean fields

$$[V_\mathrm{H}]_{\alpha\beta} = \sum_{\mu_1\mu_2} V_{\alpha\mu_1\beta\mu_2} \langle a_{\mu_1}^\dagger a_{\mu_2}\rangle = -\eta \sum_{\mu_1\mu_2} V_{\alpha\mu_1\beta\mu_2} \mathscr{G}_{\mu_2\mu_1}(\tau = 0^-) \qquad (5.119a)$$

$$[V_{\rm x}]_{\alpha\beta} = \eta \sum_{\mu_1\mu_2} V_{\alpha\mu_1\mu_2\beta}\langle a^\dagger_{\mu_1} a_{\mu_2}\rangle = -\sum_{\mu_1\mu_2} V_{\alpha\mu_1\mu_2\beta}\mathscr{G}_{\mu_2\mu_1}(\tau = 0^-) \tag{5.119b}$$

$$\Delta_{\alpha\beta} = \sum_{\mu_1\mu_2} V_{\alpha\beta\mu_1\mu_2}\langle a_{\mu_2} a_{\mu_1}\rangle = -\eta \sum_{\mu_1\mu_2} V_{\alpha\beta\mu_1\mu_2}\left[\mathscr{F}^\dagger_{\mu_2\mu_1}(\tau = 0^-)\right]^*. \tag{5.119c}$$

As shown in Doc A5.20, $V_{\rm H}$ and $V_{\rm x}$ are the usual Hartree and exchange potentials if V is the Coulomb interaction and they play similar roles for other types of model or effective interactions. In particular, the exchange potential is negative (attractive) for fermions if the interaction V is repulsive. The Gor'kov equations form a set of self-consistent one-particle equations which can be solved using standard linear-algebra techniques. In the absence of superconducting pairing ($\mathscr{F}^\dagger = \Delta = 0$), they reduce to the well-known Hartree–Fock approximation. In the traditional formulation, one derives the latter by approximating the ground state with a single Slater determinant like equation (2.32) and optimizing the one-particle wave functions φ_α that enter this determinant such as to minimize the ground-state energy. The resulting variational equations determining the φ_α turn out to be the same as the equation $(i\nu_n\mathbb{1} - \tilde{K}_0)\mathscr{G}(i\nu_n) = \mathbb{1}$. The present formulation widens the meaning of that approximation from a variational theory of the ground state to a consistent theory of the full single-particle excitation spectrum that neglects two-particle correlations beyond the correlation entailed by the exclusion principle—that *are* accounted for in the Hartree–Fock approximation.

The Gor'kov equations are formulated in a variety of different but equivalent ways in the literature, as we briefly discuss now.

5.2.2.3.2 Dyson-like formulation

In the Dyson equation (5.31), the Green's function is expressed in terms of a bare part $\mathscr{G}_0$ and a self-energy Σ. The same thing may be achieved here by eliminating $\mathscr{F}^\dagger$ from the two Gor'kov equations and solving for $\mathscr{G}$. We thus obtain

$$\mathscr{G}(i\nu_n) = [\mathscr{G}_0^{-1}(i\nu_n) - \Sigma(i\nu_n)]^{-1} \tag{5.120a}$$

where

$$\mathscr{G}_0^{-1}(i\nu_n) = i\nu_n\mathbb{1} - \tilde{K}_0 \qquad \text{and} \qquad \Sigma(i\nu_n) = \Delta\mathscr{G}_0^T(-i\nu_n)\Delta^*. \tag{5.120b}$$

Here 'T' means transposition and we used $\tilde{K}_0^* = \tilde{K}_0^T$. This formulation is useful when the mean-field (or *order parameter*) Δ is known and the single-particle excitation spectrum or the local density of states (LDOS) is to be calculated. A particularly simple example is the case of a homogeneous spin-singlet superconductor with an order parameter $\Delta_{\boldsymbol{k}}$: in this case the matrix $\Sigma(i\omega_n)$ is replaced by the function $\Sigma(\boldsymbol{k}, i\omega_n) = |\Delta_{\boldsymbol{k}}|^2/(i\omega_n + \xi_{-\boldsymbol{k}})$; see section 5.2.2.3.5. Another example is a vortex in a superconductor, where the variation of Δ in real space is approximately known such that equations (5.120) provide a straightforward way to evaluate the LDOS (see Bonus B9.1).

5.2.2.3.3 Nambu formalism

In the Nambu formalism, the two Gor'kov equations are recast into a single matrix equation. This can also be achieved by introducing two-component field operators known as the *Nambu spinors*. It is readily verified that the two Gor'kov equations and their complex conjugate can be collected together in the matrix form

$$\begin{pmatrix} i\nu_n - \tilde{K}_0 & -\Delta \\ \Delta^* & i\nu_n + \tilde{K}_0^* \end{pmatrix}\begin{pmatrix} \mathscr{G} & -\mathscr{F} \\ \mathscr{F}^\dagger & -\mathscr{G}^* \end{pmatrix} = \begin{pmatrix} 1 & 0 \\ 0 & 1 \end{pmatrix}, \tag{5.121}$$

where $\mathscr{F}$ is the complex conjugate of $\mathscr{F}^\dagger$. It is then natural to introduce the *Nambu matrix Green's function*

$$\hat{\mathscr{G}} = \begin{pmatrix} \mathscr{G} & -\mathscr{F} \\ \mathscr{F}^\dagger & -\mathscr{G}^* \end{pmatrix} \tag{5.122}$$

as well as the Nambu Hamiltonian

$$\hat{K} = \begin{pmatrix} \tilde{K}_0 & \Delta \\ -\Delta^* & -\tilde{K}_0^* \end{pmatrix} \tag{5.123}$$

such that equation (5.121) simplifies to

$$(i\nu_n \mathbb{1} - \hat{K})\hat{\mathscr{G}}(i\nu_n) = \mathbb{1}. \tag{5.124}$$

This is just the same equation as equation (5.112) for independent particles. The independent-particle nature of the mean-field approximation thus appears very explicitly in the Nambu notation: the two Gor'kov equations are rewritten as an independent-particle problem at the expense of promoting the Green's function to a matrix with one additional 'Nambu index'. Physically, this means that the elementary excitations that diagonalize the Hamiltonian at the mean-field level and behave as independent objects in a superconductor are not ordinary particles or holes like in conventional metals, but a specific particle–hole admixture depending on the properties of the superconducting order parameter, as we shall see shortly.

The Nambu formalism is especially useful in conjunction with perturbation theory. The latter must be reconsidered in superconductors because, in addition to the usual diagrams, there are now diagrams involving the anomalous propagators $\mathscr{F}^\dagger$ and $\mathscr{F}$, i.e. lines with either two particles created or annihilated at both ends like in figure 5.2(b). This complication can be elegantly sorted out within the Nambu formalism by promoting the propagator lines to the level of 2×2 matrices: the lines in the usual diagrams thus correspond to the matrix $\hat{\mathscr{G}}$ just defined rather than to the *function* $\mathscr{G}$. The vertices must also be replaced by 2×2 Pauli matrices. In this way, all anomalous diagrams are automatically included. For details, see Schrieffer 1964 and Bonus B5.3, where the Eliashberg equations and an example of solution are presented.

5.2.2.3.4 Bogoliubov–de Gennes equations

The famous Bogoliubov–de Gennes equations for superconductors were originally introduced in order to deal with situations in which the superconducting order parameter or the potential vary in space. Rather than focussing on the Green's function, they focus on the wave functions of the elementary excitations in the superconductor. Let us therefore define the eigenvalues and eigenvectors of the 'Nambu Hamiltonian' introduced above: $\hat{K}|\gamma\rangle = E_\gamma|\gamma\rangle$. These are the Bogoliubov–de Gennes equations. They are usually written in 2×2 matrix form as[16]

$$\begin{pmatrix} \tilde{K}_0 & \Delta \\ -\Delta^* & -\tilde{K}_0^* \end{pmatrix}\begin{pmatrix} |u_\gamma\rangle \\ |v_\gamma\rangle \end{pmatrix} = E_\gamma\begin{pmatrix} |u_\gamma\rangle \\ |v_\gamma\rangle \end{pmatrix}. \tag{5.125}$$

In this formulation, the problem becomes an eigenvalue problem to be solved for $|u_\gamma\rangle$, $|v_\gamma\rangle$, and E_γ. $u_\gamma(\boldsymbol{r})$ and $v_\gamma(\boldsymbol{r})$ are the particle and hole parts of the elementary excitations that diagonalize $\hat{K}$. The spectrum of eigenvalues E_γ is even with respect to $E_\gamma = 0$: it is indeed easy to check that, if (u_γ, v_γ) is a solution with energy E_γ, (v_γ^*, u_γ^*) is also a solution with energy $-E_\gamma$. For a complete self-consistent solution of the Bogoliubov–de Gennes system, we need to relate the mean fields $V_{\rm H}$, $V_{\rm x}$, and Δ to the amplitudes u_γ, v_γ, and the energies E_γ. This can be done simply by noting that the Green's function is diagonal in the representation of the $|\gamma\rangle$:

$$\hat{\mathscr{G}}(i\nu_n) = (i\nu_n\mathbb{1} - \hat{K})^{-1} = \sum_\gamma \frac{|\gamma\rangle\langle\gamma|}{i\nu_n - E_\gamma} = \sum_\gamma \frac{1}{i\nu_n - E_\gamma}\begin{pmatrix} |u_\gamma\rangle\langle u_\gamma| & |u_\gamma\rangle\langle v_\gamma| \\ |v_\gamma\rangle\langle u_\gamma| & |v_\gamma\rangle\langle v_\gamma| \end{pmatrix}. \tag{5.126}$$

Projecting onto the original representation of the one-particle states $|\alpha\rangle$, we get

$$\mathscr{G}_{\alpha\beta}(i\nu_n) = \langle\alpha|\hat{\mathscr{G}}_{11}(i\nu_n)|\beta\rangle = \sum_\gamma \frac{\langle\alpha|u_\gamma\rangle\langle u_\gamma|\beta\rangle}{i\nu_n - E_\gamma} \tag{5.127}$$

$$\mathscr{F}^\dagger_{\alpha\beta}(i\nu_n) = \langle\alpha|\hat{\mathscr{G}}_{21}(i\nu_n)|\beta\rangle = \sum_\gamma \frac{\langle\alpha|v_\gamma\rangle\langle u_\gamma|\beta\rangle}{i\nu_n - E_\gamma}. \tag{5.128}$$

[16] Please note the sign difference between equation (5.125) and the Bogoliubov–de Gennes equations as they are commonly written, with $+\Delta^*$ instead of $-\Delta^*$. This is not a misprint. The 'plus' sign arises when these equations are specialized to the case of spin-singlet pairing of electrons, i.e. when only electrons of opposite spins pair. In this case $\Delta \to \Delta_{\uparrow\downarrow}$ and $-\Delta^* \to \Delta^*_{\uparrow\downarrow}$ (see section 5.2.2.3.5 and Doc A5.21). Equation (5.125) applies to spin-singlet pairing of electrons as well as to bosons and spin-triplet pairing.

These expressions lead to the mean fields in equation (5.119) with the help of equation (17):

$$\mathscr{G}_{\alpha\beta}(\tau = 0^-) = -\eta \sum_\gamma \langle\alpha|u_\gamma\rangle\langle u_\gamma|\beta\rangle d_{-\eta}(E_\gamma) \tag{5.129}$$

$$\mathscr{F}^\dagger_{\alpha\beta}(\tau = 0^-) = -\eta \sum_\gamma \langle\alpha|v_\gamma\rangle\langle u_\gamma|\beta\rangle d_{-\eta}(E_\gamma). \tag{5.130}$$

5.2.2.3.5 Solution for spin-singlet pairing of electrons

The derivations until now are general and somewhat formal: here we detail the application to the most common situation, which is spin-singlet pairing of electrons in a non-magnetic system. We thus recover the basic BCS equations and their extension to inhomogeneous superconductivity. We start in the real-space representation and assume that a pairing interaction $V(\boldsymbol{r}, \boldsymbol{r}')$ acts between electrons. In this situation, we find that the pairing field $\Delta_{\alpha\beta}$ becomes (see Doc A5.21) $\Delta_{\alpha\beta} \to \Delta_{\boldsymbol{r}\sigma\boldsymbol{r}'\sigma'} = -V(\boldsymbol{r}, \boldsymbol{r}')\langle c_{\boldsymbol{r}\sigma}c_{\boldsymbol{r}'\sigma'}\rangle$. If the electron pairing takes place in the singlet channel only, then $\Delta_{\uparrow\uparrow} = \Delta_{\downarrow\downarrow} = 0$ and we define

$$\Delta(\boldsymbol{r}, \boldsymbol{r}') \equiv \Delta_{\boldsymbol{r}\uparrow\boldsymbol{r}'\downarrow} = -V(\boldsymbol{r}, \boldsymbol{r}')\langle c_{\boldsymbol{r}\uparrow}c_{\boldsymbol{r}'\downarrow}\rangle. \tag{5.131}$$

We similarly have $\mathscr{F}^\dagger_{\uparrow\uparrow} = \mathscr{F}^\dagger_{\downarrow\downarrow} = 0$ and define

$$\mathscr{F}^\dagger(\boldsymbol{r}, \boldsymbol{r}', \tau) \equiv \mathscr{F}^\dagger_{\downarrow\uparrow}(\boldsymbol{r}, \boldsymbol{r}', \tau). \tag{5.132}$$

The assumption of a non-magnetic system implies that the Hamiltonian has no spin-flip term and that the Green's function is diagonal in the spin indices: $[\tilde{K}_0]_{\boldsymbol{r}\sigma\boldsymbol{r}'\sigma'} = \delta_{\sigma\sigma'}\tilde{K}_0(\boldsymbol{r}, \boldsymbol{r}')$ and $\mathscr{G}_{\boldsymbol{r}\sigma\boldsymbol{r}'\sigma'}(\tau) \equiv \delta_{\sigma\sigma'}\mathscr{G}(\boldsymbol{r}, \boldsymbol{r}', \tau)$. As a result we find that the Gor'kov equations take the form (see Doc A5.21)

$$\begin{aligned}\int d\boldsymbol{r}_1[i\omega_n\delta(\boldsymbol{r} - \boldsymbol{r}_1) - \tilde{K}_0(\boldsymbol{r}, \boldsymbol{r}_1)]\mathscr{G}(\boldsymbol{r}_1, \boldsymbol{r}', i\omega_n)& \\ - \int d\boldsymbol{r}_1\, \Delta(\boldsymbol{r}, \boldsymbol{r}_1)\mathscr{F}^\dagger(\boldsymbol{r}_1, \boldsymbol{r}', i\omega_n) &= \delta(\boldsymbol{r} - \boldsymbol{r}')\end{aligned} \tag{5.133a}$$

$$\begin{aligned}\int d\boldsymbol{r}_1\Big[i\omega_n\delta(\boldsymbol{r} - \boldsymbol{r}_1) + \tilde{K}_0^*(\boldsymbol{r}, \boldsymbol{r}_1)\Big]\mathscr{F}^\dagger(\boldsymbol{r}_1, \boldsymbol{r}', i\omega_n)& \\ - \int d\boldsymbol{r}_1\, \Delta^*(\boldsymbol{r}_1, \boldsymbol{r})\mathscr{G}(\boldsymbol{r}_1, \boldsymbol{r}', i\omega_n) &= 0,\end{aligned} \tag{5.133b}$$

and they are supplemented by the gap equation

$$\Delta(\boldsymbol{r}, \boldsymbol{r}') = V(\boldsymbol{r}, \boldsymbol{r}')[\mathscr{F}^\dagger(\boldsymbol{r}', \boldsymbol{r}, \tau = 0^+)]^*. \tag{5.134}$$

These equations do not assume translational invariance and can therefore be used to investigate non-homogeneous systems like vortices, vortex lattices, surfaces, interfaces, defects, grain boundaries etc. The corresponding equations in the Dyson, Nambu, and Bogoliubov–de Gennes formulations are readily obtained too (see Doc A5.21).

In case of translation invariance, the Gor'kov equations become algebraic because $\tilde{K}_0$, $\mathcal{G}$, Δ, and $\mathcal{F}^\dagger$ are all functions of the difference in coordinates and Fourier transformation immediately yields

$$(i\omega_n - \xi_k)\mathcal{G}(\boldsymbol{k}, i\omega_n) - \Delta_k \mathcal{F}^\dagger(\boldsymbol{k}, i\omega_n) = 1 \tag{5.135a}$$

$$(i\omega_n + \xi_{-k})\mathcal{F}^\dagger(\boldsymbol{k}, i\omega_n) - \Delta_k^* \mathcal{G}(\boldsymbol{k}, i\omega_n) = 0, \tag{5.135b}$$

where ξ_k is the Fourier transform of $\tilde{K}_0(\boldsymbol{r} - \boldsymbol{r}')$, etc. We can solve by eliminating $\mathcal{F}^\dagger$ to find

$$\boxed{\mathcal{G}_{\text{BCS}}(\boldsymbol{k}, i\omega_n) = \frac{1}{i\omega_n - \xi_k - \dfrac{|\Delta_k|^2}{i\omega_n + \xi_{-k}}}.} \tag{5.136}$$

This is the Green's function for a homogeneous spin-singlet BCS superconductor. It may be used to compute the DOS using equation (5.7). The document Bonus B5.2 shows how this calculation can be performed for arbitrary gap symmetries in dimension two. In the particular case of an *s*-wave symmetry, i.e. a momentum-independent gap $\Delta_k \equiv \Delta$, the calculation can be completed analytically and yields, for a flat normal-state DOS $N_0^{\text{el}}(0)$, the well-known expression (see Bonus B5.2)

$$N_{\text{BCS}}^{(s-\text{wave})}(\varepsilon) = N_0^{\text{el}}(0)\text{Re}\left(\frac{|\varepsilon|}{\sqrt{\varepsilon^2 - \Delta^2}}\right). \tag{5.137}$$

This function is plotted in figure 5.11(c). A slightly more complicated formula results if $N_0^{\text{el}}(\varepsilon)$ is not flat, as reported in Bonus B5.2. The BCS Green's function (5.136) can also be recast in a form which shows explicitly the electron–hole and long-lived nature of the elementary excitations, the so-called *Bogoliubov excitations*:

$$\mathcal{G}_{\text{BCS}}(\boldsymbol{k}, i\omega_n) = \frac{u_k^2}{i\omega_n - E_k} + \frac{v_k^2}{i\omega_n + E_k} \tag{5.138}$$

with

$$E_k = \sqrt{\xi_k^2 + |\Delta_k|^2}, \qquad u_k^2 = \frac{1}{2}\left(1 + \frac{\xi_k}{E_k}\right), \qquad v_k^2 = \frac{1}{2}\left(1 - \frac{\xi_k}{E_k}\right). \tag{5.139}$$

For simplicity, we assumed inversion symmetry—sometimes referred to as *time-reversal symmetry*—i.e. $\tilde{K}_0(\boldsymbol{r}) = \tilde{K}_0(-\boldsymbol{r})$, which implies $\xi_k = \xi_{-k}$. In the form of equation (5.138), it appears clearly that $\mathcal{G}_{\text{BCS}}$ is the superposition of two poles at energies $+E_k$ and $-E_k$ with weights u_k^2 and v_k^2, respectively. Since $E_k > 0$, the pole at $+E_k$ represents an electron-like excitation: if $\xi_k \gg |\Delta_k|$, we indeed have $E_k \approx \xi_k$, $u_k^2 \approx 1$, and $v_k^2 \approx 0$, such that only this part subsists. Inversely, if $\xi_k \ll -|\Delta_k|$ we have $u_k^2 \approx 0$, $v_k^2 \approx 1$, and only the second term in equation (5.138) remains, which represents a hole-like excitation at negative energy $-E_k$. At low energies $\xi_k \approx \Delta_k$, the excitations are a mixing of electron-like and hole-like parts. The conservation of the

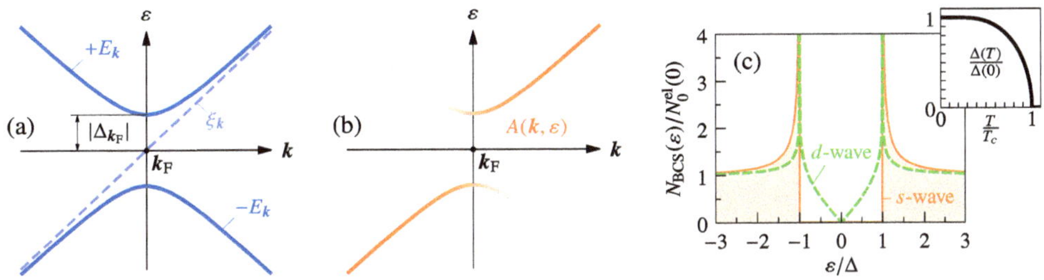

Figure 5.11. Gapped excitation spectrum of a spin-singlet BCS superconductor. (a) Excitation energies $\pm E_{\boldsymbol{k}}$ compared with the bare dispersion $\xi_{\boldsymbol{k}}$. At $\boldsymbol{k} = \boldsymbol{k}_F$, the minimal excitation energy is $\pm|\Delta_{\boldsymbol{k}_F}|$. (b) Representation of the spectral function: at each $\boldsymbol{k}$, the spectral function is the superposition of two delta peaks of weights $u_{\boldsymbol{k}}^2$ (upper branch) and $v_{\boldsymbol{k}}^2$ (lower branch). The weight on each branch is indicated by the color scale. At $\boldsymbol{k} = \boldsymbol{k}_F$, both branches have the same weight of 1/2. (c) BCS density of states for *s*-wave and *d*-wave gap symmetries (see Bonus B5.2). Inset: Temperature dependence of the BCS gap from equation (5.142).

total spectral weight is guaranteed by the property $u_{\boldsymbol{k}}^2 + v_{\boldsymbol{k}}^2 = 1$. At the location of the Fermi surface ($\xi_{\boldsymbol{k}} = 0$), we have $E_{\boldsymbol{k}} = |\Delta_{\boldsymbol{k}}|$: the Bogoliubov excitations are gapped and the minimal excitation energy at a given Fermi-surface wave vector is $|\Delta_{\boldsymbol{k}}|$, which illustrates why the order parameter Δ is commonly called *the gap*. The shape of the excitation spectrum is illustrated in figure 5.11, which also displays the spectral function

$$A_{BCS}(\boldsymbol{k}, \varepsilon) = u_{\boldsymbol{k}}^2\delta(\varepsilon - E_{\boldsymbol{k}}) + v_{\boldsymbol{k}}^2\delta(\varepsilon + E_{\boldsymbol{k}}). \tag{5.140}$$

We conclude this section with a discussion of the gap equation, which in the present context reads (see Doc A5.21)

$$\Delta_{\boldsymbol{k}} = -\frac{1}{\mathcal{V}}\sum_{\boldsymbol{k}'} V(\boldsymbol{k} - \boldsymbol{k}')\frac{\Delta_{\boldsymbol{k}'}}{2E_{\boldsymbol{k}'}}\tanh\left(\frac{\beta}{2}E_{\boldsymbol{k}'}\right). \tag{5.141}$$

This is a non-linear equation because $E_{\boldsymbol{k}'}$ in the integrand depends on $\Delta_{\boldsymbol{k}'}$ (equation (5.139)). Simplified forms can be written down at $T = 0$ [$\tanh(\cdots) \equiv 1$] and $T = T_c$ ($\Delta_{\boldsymbol{k}} \to 0$). Let us continue with an interaction of the BCS weak-coupling type as depicted in figure 5.5(b): $V(\boldsymbol{k} - \boldsymbol{k}') = -V_0$ if $\boldsymbol{k}$ and $\boldsymbol{k}'$ are wave vectors close to the Fermi surface—specifically, if $|\xi_{\boldsymbol{k}}|$ and $|\xi_{\boldsymbol{k}'}|$ are smaller than $\hbar\omega_D$—and $V(\boldsymbol{k} - \boldsymbol{k}') = 0$ otherwise. This is the simplest way of representing an attraction that is only effective inside a shell of thickness $2\hbar\omega_D$ around the Fermi surface, as e.g. the phonon-mediated interaction of equation (5.54). The BCS interaction, together with equation (5.141), imply that $\Delta_{\boldsymbol{k}} = 0$ if $|\xi_{\boldsymbol{k}}| > \hbar\omega_D$. We now look for an *s*-wave solution of the kind $\Delta_{\boldsymbol{k}} = \Delta$ for $|\xi_{\boldsymbol{k}}| < \hbar\omega_D$. The gap equation simplifies to

$$\frac{1}{V_0} = \frac{1}{\mathcal{V}}\sum_{|\xi_{\boldsymbol{k}'}|<\hbar\omega_D}\frac{\tanh\left(\frac{\beta}{2}E_{\boldsymbol{k}'}\right)}{2E_{\boldsymbol{k}'}} = \int_{-\hbar\omega_D}^{\hbar\omega_D} d\xi\, \tilde{N}_0^{el}(\xi)\frac{\tanh\left(\frac{\beta}{2}\sqrt{\xi^2 + \Delta^2}\right)}{2\sqrt{\xi^2 + \Delta^2}}. \tag{5.142}$$

$\tilde{N}_0^{el}$ is the non-interacting density of states *per spin direction*. If the latter can be considered constant over the energy range $\pm\hbar\omega_D$, the equation reduces at $T = 0$ to $1/[V_0\tilde{N}_0^{el}(0)] = \int_0^{\hbar\omega_D} d\xi/\sqrt{\xi^2 + \Delta^2} = \sinh^{-1}(\hbar\omega_D/\Delta)$, or

$$\Delta_{\rm BCS}(T=0) = \hbar\omega_{\rm D}/\sinh\left(\frac{1}{V_0\tilde{N}_0^{\rm el}(0)}\right) \approx 2\,\hbar\omega_{\rm D}e^{-1/[V_0\tilde{N}_0^{\rm el}(0)]}. \tag{5.143}$$

The last approximation is justified because the interaction is usually weak, such that $V_0\tilde{N}_0^{\rm el}(0) \ll 1$. At $T = T_c$, where Δ approaches zero, we have instead $1/[V_0\tilde{N}_0^{\rm el}(0)] = \int_0^{\hbar\omega_{\rm D}} d\xi\, \tanh(\frac{\beta_c}{2}\xi)/\xi$. Assuming $k_{\rm B}T_c \ll \hbar\omega_{\rm D}$ and using equation (20), this becomes $1/[V_0\tilde{N}_0^{\rm el}(0)] \approx \ln[(2e^\gamma/\pi)\,\beta_c\hbar\omega_{\rm D}]$. The solution is

$$k_{\rm B}T_c \approx \frac{2e^\gamma}{\pi}\,\hbar\omega_{\rm D}e^{-1/[V_0\tilde{N}_0^{\rm el}(0)]}. \tag{5.144}$$

Hence $2\Delta_{\rm BCS}(T=0)/(k_{\rm B}T_c) \approx 2\pi/e^\gamma = 3.53$, a ratio known as the *BCS coupling ratio*. The temperature dependence of the BCS gap $\Delta(T)$ obtained by solving numerically equation (5.142) is shown in figure 5.11(c). The curve $\Delta(T)/\Delta(0)$ versus T/T_c is said to be *universal*, because in the regime where it is valid, it does not depend on the model parameters that appear in equation (5.142): V_0, $\tilde{N}_0^{\rm el}(0)$, and $\hbar\omega_{\rm D}$.

This brief overview illustrates the power of the Gor'kov formalism: we have been able to derive all the main results of the BCS theory of superconductivity without even mentioning the celebrated BCS ground-state wave-function and the associated lengthy operator algebra. Moreover, these results all follow rather directly by simple deductions from the general equation of motion (5.114) and its decoupling (5.115), which is the only non-trivial and intellectually challenging step of the whole derivation.

Appendix A5

Doc A5.1 Expansion of correlation functions in imaginary time

The imaginary-time-ordered correlation function, equation (4.5), reads

$$\mathscr{C}_{AB}(\tau) = -\langle T_\tau A(\tau)B(0)\rangle = -\frac{1}{Z}{\rm Tr}e^{-\beta K}T_\tau e^{\tau K}Ae^{-\tau K}B(0).$$

In the interaction picture, the imaginary-time evolution operator $e^{-\tau K}$ is factorized as $e^{-\tau K_0}\hat{U}(\tau)$ in complete analogy with the real-time evolution (see section 2.2.1), such that $e^{-\beta K} = e^{-\beta K_0}\hat{U}(\beta)$. Using these expressions we find

$$\begin{aligned}\mathscr{C}_{AB}(\tau) &= -\frac{1}{Z}{\rm Tr}e^{-\beta K_0}\hat{U}(\beta)T_\tau\hat{U}^{-1}(\tau)\underbrace{e^{\tau K_0}Ae^{-\tau K_0}}_{\hat{A}(\tau)}\hat{U}(\tau)B(0)\\ &= -\frac{1}{Z}{\rm Tr}e^{-\beta K_0}\hat{U}(\beta)\Big[\theta(\tau)\hat{U}^{-1}(\tau)\hat{A}(\tau)\hat{U}(\tau)B(0) + \eta\theta(-\tau)B(0)\hat{U}^{-1}(\tau)\hat{A}(\tau)\hat{U}(\tau)\Big].\end{aligned}$$

The first term in the square brackets involves the product $\hat{U}^{-1}(\tau)\hat{A}(\tau)\hat{U}(\tau)B(0)$ for a positive time τ. This term is in fact equal to $\hat{A}(\tau)B(0)$, as can be seen by taking advantage of the time ordering:

$$\begin{aligned}\hat{U}^{-1}(\tau)\hat{A}(\tau)\hat{U}(\tau)B(0) &= T_\tau\hat{U}^{-1}(\tau^+)\hat{A}(\tau)\hat{U}(\tau^-)B(0)\\ &= T_\tau\hat{A}(\tau)\hat{U}^{-1}(\tau^+)\hat{U}(\tau^-)B(0) = T_\tau\hat{A}(\tau)B(0) = A(\tau)B(0).\end{aligned}$$

We have shifted the time arguments by infinitesimal amounts, $\tau^+ > \tau > \tau^- > 0$, such that the time ordering retrieves the operators in the correct order. Then, we have used our right to permute operators under the time ordering; the sign implied by our permutation is +1 because the evolution $\hat{U}(\tau)$ involves an even number of creation or annihilation operators. Finally, in order to remove the time ordering, we have used the facts that $\tau > 0$ and $\hat{U}^{-1}(\tau^+)\hat{U}(\tau^-) = \mathbb{1}$ because $\hat{U}(\tau)$ is continuous. A completely analogous manipulation shows that for the second term in the square brackets, we also have $B(0)\hat{U}^{-1}(\tau)\hat{A}(\tau)\hat{U}(\tau) = B(0)\hat{A}(\tau)$. It is worthwhile to realize how crucial the time ordering is in this step of the derivation. We have thus obtained

$$\begin{aligned}\mathscr{C}_{AB}(\tau) &= -\frac{1}{Z}\mathrm{Tr}e^{-\beta K_0}\hat{U}(\beta)T_\tau\hat{A}(\tau)B(0) = -\frac{1}{Z}\mathrm{Tr}e^{-\beta K_0}T_\tau\hat{U}(\beta)\hat{A}(\tau)B(0)\\ &= -\frac{\mathrm{Tr}e^{-\beta K_0}}{Z}\frac{\mathrm{Tr}e^{-\beta K_0}T_\tau\hat{U}(\beta)\hat{A}(\tau)B(0)}{\mathrm{Tr}e^{-\beta K_0}} = -\frac{\mathrm{Tr}e^{-\beta K_0}}{Z}\langle T_\tau\hat{U}(\beta)\hat{A}(\tau)B(0)\rangle_0,\end{aligned}$$

where the second equality holds because $\beta > \tau$ and where we have introduced the notation $\langle X\rangle_0$ for the thermodynamic average of an operator X in the ensemble of states defined by K_0: $\langle X\rangle_0 = \mathrm{Tr}\,(e^{-\beta K_0}X)/\mathrm{Tr}e^{-\beta K_0}$. We finally rewrite the partition function

$$Z = \mathrm{Tr}e^{-\beta K} = \mathrm{Tr}e^{-\beta K_0}\hat{U}(\beta) = \mathrm{Tr}e^{-\beta K_0}\frac{\mathrm{Tr}e^{-\beta K_0}\hat{U}(\beta)}{\mathrm{Tr}e^{-\beta K_0}} = \mathrm{Tr}e^{-\beta K_0}\langle\hat{U}(\beta)\rangle_0$$

and obtain $\mathscr{C}_{AB}(\tau) = -\langle T_\tau\hat{U}(\beta)\hat{A}(\tau)B(0)\rangle_0/\langle\hat{U}(\beta)\rangle_0$. Together with equation (5.1), this yields equation (5.3).

Doc A5.2 One-body potential: terms of the expansion

The non-vanishing contribution of order $n = 1$ in equation (5.13) is

$$\begin{aligned}&\eta\int_0^\beta d\tau_1\sum_{\alpha_1\beta_1}V_{\alpha_1\beta_1}\underbrace{\langle a_{\alpha_1}^\dagger(\tau_1)a_\alpha(\tau)\rangle}_{-\eta\mathscr{G}^0_{\alpha\alpha_1}(\tau-\tau_1)}\underbrace{\langle a_{\beta_1}(\tau_1)a_\beta^\dagger(0)\rangle}_{-\mathscr{G}^0_{\beta_1\beta}(\tau_1)}\\ &= \int_0^\beta d\tau_1\sum_{\alpha_1\beta_1}\mathscr{G}^0_{\alpha\alpha_1}(\tau-\tau_1)V_{\alpha_1\beta_1}\mathscr{G}^0_{\beta_1\beta}(\tau_1),\end{aligned}$$

consistently with equation (5.14). At order $n = 2$, the two terms in equation (5.15) yield

$$-\eta \int_0^\beta d\tau_1 d\tau_2 \sum_{\alpha_1\beta_1} \sum_{\alpha_2\beta_2} V_{\alpha_1\beta_1} V_{\alpha_2\beta_2} \times \underbrace{\langle a_{\alpha_1}^\dagger(\tau_1) a_\alpha(\tau)\rangle}_{-\eta\mathcal{G}^0_{\alpha\alpha_1}(\tau-\tau_1)} \underbrace{\langle a_{\beta_1}(\tau_1) a_{\alpha_2}^\dagger(\tau_2)\rangle}_{-\mathcal{G}^0_{\beta_1\alpha_2}(\tau_1-\tau_2)} \underbrace{\langle a_{\beta_2}(\tau_2) a_\beta^\dagger(0)\rangle}_{-\mathcal{G}^0_{\beta_2\beta}(\tau_2)}$$

$$= \int_0^\beta d\tau_1 d\tau_2 \sum_{\alpha_1\beta_1} \sum_{\alpha_2\beta_2} \mathcal{G}^0_{\alpha\alpha_1}(\tau-\tau_1) V_{\alpha_1\beta_1} \mathcal{G}^0_{\beta_1\alpha_2}(\tau_1-\tau_2) V_{\alpha_2\beta_2} \mathcal{G}^0_{\beta_2\beta}(\tau_2),$$

which is just equation (5.16).

Doc A5.3 One-body diagrams in the frequency domain

The contribution of order n to $\mathcal{G}_{\alpha\beta}(\tau)$ in equation (5.10) is

$$\mathcal{G}^{(n)}_{\alpha\beta}(\tau) = \int_0^\beta d\tau_1 \cdots d\tau_n \sum_{\alpha_1\beta_1} \cdots \sum_{\alpha_n\beta_n} \mathcal{G}^0_{\alpha\alpha_1}(\tau-\tau_1) V_{\alpha_1\beta_1} \mathcal{G}^0_{\beta_1\alpha_2}(\tau_1-\tau_2) V_{\alpha_2\beta_2} \cdots V_{\alpha_n\beta_n} \mathcal{G}^0_{\beta_n\beta}(\tau_n).$$

Performing the Fourier transform by means of equations (4.10) and (4.8), we get

$$\begin{aligned}
\mathcal{G}^{(n)}_{\alpha\beta}(i\nu_n) &= \int_0^\beta d\tau \mathcal{G}^{(n)}_{\alpha\beta}(\tau) e^{i\nu_n\tau} \\
&= \sum_{\alpha_1\beta_1} \cdots \sum_{\alpha_n\beta_n} \frac{1}{\beta} \sum_{i\bar{\nu}_0} \cdots \frac{1}{\beta} \sum_{i\bar{\nu}_n} \mathcal{G}^0_{\alpha\alpha_1}(i\bar{\nu}_0) V_{\alpha_1\beta_1} \mathcal{G}^0_{\beta_1\alpha_2}(i\bar{\nu}_1) V_{\alpha_2\beta_2} \cdots V_{\alpha_n\beta_n} \mathcal{G}^0_{\beta_n\beta}(i\bar{\nu}_n) \\
&\quad \times \underbrace{\int_0^\beta d\tau d\tau_1 \cdots d\tau_n \, e^{i\nu_n\tau} e^{-i\bar{\nu}_0(\tau-\tau_1)} e^{-i\bar{\nu}_1(\tau_1-\tau_2)} \cdots e^{-i\bar{\nu}_n\tau_n}}_{\beta^{n+1}\delta_{\nu_n\bar{\nu}_0}\delta_{\bar{\nu}_0\bar{\nu}_1}\cdots\delta_{\bar{\nu}_{n-1}\bar{\nu}_n}} \\
&= \sum_{\alpha_1\beta_1} \cdots \sum_{\alpha_n\beta_n} \mathcal{G}^0_{\alpha\alpha_1}(i\nu_n) V_{\alpha_1\beta_1} \mathcal{G}^0_{\beta_1\alpha_2}(i\nu_n) V_{\alpha_2\beta_2} \cdots V_{\alpha_n\beta_n} \mathcal{G}^0_{\beta_n\beta}(i\nu_n),
\end{aligned}$$

which proves equation (5.17). The notation $i\bar{\nu}_j$ used here is a short-hand for $i\nu_n^{(j)}$. $i\bar{\nu}_j$ represents a full set of Matsubara frequencies, like $i\nu_n$, and should not be confused with the jth frequency $i\nu_j$.

Doc A5.4 Two-body potential: terms of the expansion

The two Hartree terms in equation (5.21) contribute the following to $\mathcal{G}_{\alpha\beta}(\tau)$ in equation (5.20):

$$-\left(-\frac{1}{2}\right)2\eta\int_0^\beta d\tau_1\sum_{\alpha_1\beta_1\gamma_1\delta_1}V_{\alpha_1\beta_1\gamma_1\delta_1}\underbrace{\langle a^\dagger_{\alpha_1}(\tau_1)a_\alpha(\tau)\rangle}_{-\eta\mathcal{G}^0_{\alpha\alpha_1}(\tau-\tau_1)}\underbrace{\langle a^\dagger_{\beta_1}(\tau_1)a_{\delta_1}(\tau_1)\rangle}_{-\eta\mathcal{G}^0_{\delta_1\beta_1}(\tau_1-\tau_1^+)}\underbrace{\langle a_{\gamma_1}(\tau_1)a^\dagger_\beta(0)\rangle}_{-\mathcal{G}^0_{\gamma_1\beta}(\tau_1)}$$
$$=-\eta\int_0^\beta d\tau_1\sum_{\alpha_1\beta_1\gamma_1\delta_1}\mathcal{G}^0_{\alpha\alpha_1}(\tau-\tau_1)\mathcal{G}^0_{\delta_1\beta_1}(\tau_1-\tau_1^+)V_{\alpha_1\beta_1\gamma_1\delta_1}\mathcal{G}^0_{\gamma_1\beta}(\tau_1).$$

In the average $\langle a^\dagger_{\beta_1}(\tau_1)a_{\delta_1}(\tau_1)\rangle$, the two operators act at the same time. We must therefore shift one time by an infinitesimal amount to ensure that the time ordering in $\mathcal{G}^0_{\delta_1\beta_1}(\tau_1-\tau_1^+)$ returns the operators in the correct order. Hence we get equation (5.22). The only difference between the Hartree terms in equation (5.21) and the exchange terms in equation (5.23) is the interchange of the indices γ_1 and δ_1 and the resulting factor η:

$$-\left(-\frac{1}{2}\right)2\int_0^\beta d\tau_1\sum_{\alpha_1\beta_1\gamma_1\delta_1}V_{\alpha_1\beta_1\gamma_1\delta_1}\underbrace{\langle a^\dagger_{\alpha_1}(\tau_1)a_\alpha(\tau)\rangle}_{-\eta\mathcal{G}^0_{\alpha\alpha_1}(\tau-\tau_1)}\underbrace{\langle a^\dagger_{\beta_1}(\tau_1)a_{\gamma_1}(\tau_1)\rangle}_{-\eta\mathcal{G}^0_{\gamma_1\beta_1}(\tau_1-\tau_1^+)}\underbrace{\langle a_{\delta_1}(\tau_1)a^\dagger_\beta(0)\rangle}_{-\mathcal{G}^0_{\delta_1\beta}(\tau_1)}$$
$$=-\int_0^\beta d\tau_1\sum_{\alpha_1\beta_1\gamma_1\delta_1}\mathcal{G}^0_{\alpha\alpha_1}(\tau-\tau_1)\mathcal{G}^0_{\gamma_1\beta_1}(\tau_1-\tau_1^+)V_{\alpha_1\beta_1\gamma_1\delta_1}\mathcal{G}^0_{\delta_1\beta}(\tau_1).$$

Doc A5.5 Two-body diagrams in the frequency domain

Summing equations (5.22) and (5.24), swapping the dummy indices γ_1 and δ_1 in the exchange term, and performing the Fourier transform, we get

$$\sum_{\alpha_1\beta_1\gamma_1\delta_1}\frac{1}{\beta^3}\sum_{i\bar{\nu}_0 i\bar{\nu}_1 i\bar{\nu}_2}\mathcal{G}^0_{\alpha\alpha_1}(i\bar{\nu}_0)\mathcal{G}^0_{\delta_1\beta_1}(i\bar{\nu}_1)\mathcal{G}^0_{\gamma_1\beta}(i\bar{\nu}_2)(-\eta V_{\alpha_1\beta_1\gamma_1\delta_1}-V_{\alpha_1\beta_1\delta_1\gamma_1})$$
$$\times\underbrace{\int_0^\beta d\tau d\tau_1\, e^{i\nu_n\tau}e^{-i\bar{\nu}_0(\tau-\tau_1)}e^{-i\bar{\nu}_1(\tau_1-\tau_1^+)}e^{-i\bar{\nu}_2\tau_1}}_{\beta^2\delta_{\bar{\nu}_0\nu_n}\delta_{\bar{\nu}_0\bar{\nu}_2}e^{i\bar{\nu}_1 0^+}}$$
$$=\sum_{\alpha_1\beta_1\gamma_1\delta_1}\frac{1}{\beta}\sum_{i\bar{\nu}_1}\mathcal{G}^0_{\alpha\alpha_1}(i\nu_n)\mathcal{G}^0_{\delta_1\beta_1}(i\bar{\nu}_1)e^{i\bar{\nu}_1 0^+}\mathcal{G}^0_{\gamma_1\beta}(i\nu_n)(-\eta V_{\alpha_1\beta_1\gamma_1\delta_1}-V_{\alpha_1\beta_1\delta_1\gamma_1}).$$

This expression can be recast in the form

$$\sum_{\alpha_1\gamma_1}\mathcal{G}^0_{\alpha\alpha_1}(i\nu_n)\left[\frac{1}{\beta}\sum_{i\bar{\nu}_1}\sum_{\beta_1\delta_1}\mathcal{G}^0_{\delta_1\beta_1}(i\bar{\nu}_1)e^{i\bar{\nu}_1 0^+}(-\eta V_{\alpha_1\beta_1\gamma_1\delta_1}-V_{\alpha_1\beta_1\delta_1\gamma_1})\right]\mathcal{G}^0_{\gamma_1\beta}(i\nu_n)$$
$$=\sum_{\alpha_1\gamma_1}\mathcal{G}^0_{\alpha\alpha_1}(i\nu_n)\Sigma^{(1)}_{\alpha_1\gamma_1}(i\nu_n)\mathcal{G}^0_{\gamma_1\beta}(i\nu_n)$$

with $\Sigma^{(1)}_{\alpha_1\gamma_1}$ as in equation (5.25).

Doc A5.6 Impurity average

At order $n = 1$, the impurity average defined by equation (5.35) gives:

$$\langle V(\boldsymbol{r}_1)\rangle_{\text{imp}} = \frac{1}{\mathcal{V}^{N_i}}\int d\boldsymbol{R}_1\cdots d\boldsymbol{R}_{N_i}\sum_{\ell=1}^{N_i} v(\boldsymbol{r}_1-\boldsymbol{R}_\ell) = \sum_{\ell=1}^{N_i}\frac{1}{\mathcal{V}}\int d\boldsymbol{R}\; v(\boldsymbol{r}_1-\boldsymbol{R})$$
$$= \frac{N_i}{\mathcal{V}}\int d\boldsymbol{r}\; v(\boldsymbol{r}) = n_i v(\boldsymbol{q}=0),$$

with $n_i = N_i/\mathcal{V}$ the impurity concentration and $v(\boldsymbol{q})$ the Fourier transform of the impurity potential $v(\boldsymbol{r})$. Replacing $V(\boldsymbol{r}_1)$ by $\langle V(\boldsymbol{r}_1)\rangle_{\text{imp}}$ in the first-order term on the right-hand side of equation (5.34) and taking the Fourier transform, we find

$$\int d\boldsymbol{r} e^{-i\boldsymbol{k}\cdot(\boldsymbol{r}-\boldsymbol{r}')}\int d\boldsymbol{r}_1\, \mathscr{G}_0(\boldsymbol{r}-\boldsymbol{r}_1, i\nu_n)\, n_i v(\boldsymbol{0})\, \mathscr{G}_0(\boldsymbol{r}_1-\boldsymbol{r}', i\nu_n)$$
$$= n_i v(\boldsymbol{0})\frac{1}{\mathcal{V}^2}\sum_{\boldsymbol{k}_1\boldsymbol{k}_2}\mathscr{G}_0(\boldsymbol{k}_1, i\nu_n)\mathscr{G}_0(\boldsymbol{k}_2, i\nu_n)\underbrace{\int d\boldsymbol{r} d\boldsymbol{r}_1 e^{-i\boldsymbol{k}\cdot(\boldsymbol{r}-\boldsymbol{r}')}e^{i\boldsymbol{k}_1\cdot(\boldsymbol{r}-\boldsymbol{r}_1)}e^{i\boldsymbol{k}_2\cdot(\boldsymbol{r}_1-\boldsymbol{r}')}}_{\mathcal{V}^2\delta_{\boldsymbol{k}\boldsymbol{k}_1}\delta_{\boldsymbol{k}_1\boldsymbol{k}_2}}$$
$$= n_i v(\boldsymbol{0})[\mathscr{G}_0(\boldsymbol{k}, i\nu_n)]^2.$$

At second order, the impurity average yields two terms:

$$\langle V(\boldsymbol{r}_1)V(\boldsymbol{r}_2)\rangle_{\text{imp}} = \frac{1}{\mathcal{V}^{N_i}}\int d\boldsymbol{R}_1\cdots d\boldsymbol{R}_{N_i}\sum_{\ell=1}^{N_i} v(\boldsymbol{r}_1-\boldsymbol{R}_\ell)\sum_{\ell'=1}^{N_i} v(\boldsymbol{r}_2-\boldsymbol{R}_{\ell'})$$
$$= \sum_{\ell,\ell'=1}^{N_i}\frac{1}{\mathcal{V}^2}\int d\boldsymbol{R}_\ell d\boldsymbol{R}_{\ell'}\; v(\boldsymbol{r}_1-\boldsymbol{R}_\ell)v(\boldsymbol{r}_2-\boldsymbol{R}_{\ell'})$$
$$= \sum_{\ell\neq\ell'}^{N_i}\frac{1}{\mathcal{V}^2}\int d\boldsymbol{R} d\boldsymbol{R}'\; v(\boldsymbol{r}_1-\boldsymbol{R})v(\boldsymbol{r}_2-\boldsymbol{R}')$$
$$+ \sum_{\ell=\ell'}^{N_i}\frac{1}{\mathcal{V}}\int d\boldsymbol{R}\; v(\boldsymbol{r}_1-\boldsymbol{R})v(\boldsymbol{r}_2-\boldsymbol{R})$$
$$= \frac{N_i(N_i-1)}{\mathcal{V}^2}[v(\boldsymbol{q}=0)]^2 + n_i\int d\boldsymbol{r}\; v(\boldsymbol{r})v(\boldsymbol{r}+\boldsymbol{r}_1-\boldsymbol{r}_2).$$

Hence two terms result by replacing $V(\boldsymbol{r}_1)V(\boldsymbol{r}_2)$ by $\langle V(\boldsymbol{r}_1)V(\boldsymbol{r}_2)\rangle_{\text{imp}}$ in the second-order term on the right-hand side of equation (5.34). In Fourier space, the first of those terms is $n_i(n_i-\mathcal{V}^{-1})v^2(\boldsymbol{0})[\mathscr{G}_0(\boldsymbol{k}, i\nu_n)]^3$ and the second is $n_i\frac{1}{\mathcal{V}}\sum_{\boldsymbol{q}}\mathscr{G}_0(\boldsymbol{k}, i\nu_n)v(\boldsymbol{q})$ $\mathscr{G}_0(\boldsymbol{k}-\boldsymbol{q}, i\nu_n)v(-\boldsymbol{q})\mathscr{G}_0(\boldsymbol{k}, i\nu_n)$, as can be readily verified by performing the Fourier transform as we did at order $n = 1$.

Doc A5.7 Full Born approximation

The full Born approximation takes into account all diagrams of order n_i and can be rewritten in a compact way by introducing an effective impurity potential, $w(\boldsymbol{q};\, \boldsymbol{k},\, i\nu_n)$, in the following way:

$\Sigma^{\mathrm{FBA}}(\boldsymbol{k}, i\nu_n) =$ 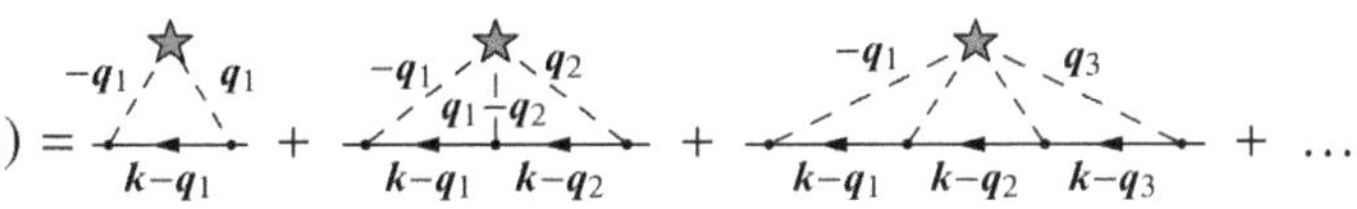

$$
\begin{aligned}
&= n_i \frac{1}{\mathcal{V}} \sum_{\boldsymbol{q}_1} v(-\boldsymbol{q}_1)\mathcal{G}_0(\boldsymbol{k}-\boldsymbol{q}_1, i\nu_n)\Big\{ v(\boldsymbol{q}_1) + \\
&\quad + \frac{1}{\mathcal{V}} \sum_{\boldsymbol{q}_2} v(\boldsymbol{q}_1-\boldsymbol{q}_2)\mathcal{G}_0(\boldsymbol{k}-\boldsymbol{q}_2, i\nu_n)\Big[v(\boldsymbol{q}_2) + \\
&\quad + \frac{1}{\mathcal{V}} \sum_{\boldsymbol{q}_3} v(\boldsymbol{q}_2-\boldsymbol{q}_3)\mathcal{G}_0(\boldsymbol{k}-\boldsymbol{q}_3, i\nu_n)\Big(v(\boldsymbol{q}_3) + \ldots \Big)\Big]\Big\} \\
&= n_i \frac{1}{\mathcal{V}} \sum_{\boldsymbol{q}_1} v(-\boldsymbol{q}_1)\mathcal{G}_0(\boldsymbol{k}-\boldsymbol{q}_1, i\nu_n) w(\boldsymbol{q}_1; \boldsymbol{k}, i\nu_n)
\end{aligned}
$$

$$
w(\boldsymbol{q}_1; \boldsymbol{k}, i\nu_n) = v(\boldsymbol{q}_1) + \frac{1}{\mathcal{V}} \sum_{\boldsymbol{q}_2} v(\boldsymbol{q}_1-\boldsymbol{q}_2)\mathcal{G}_0(\boldsymbol{k}-\boldsymbol{q}_2, i\nu_n) w(\boldsymbol{q}_2; \boldsymbol{k}, i\nu_n).
$$

We see that the final expression for the full Born approximation is very similar to the first Born approximation (5.41), the only difference being that the potential $v(\boldsymbol{q})$ is replaced by an effective momentum and energy-dependent potential $w(\boldsymbol{q};\, \boldsymbol{k},\, i\nu_n)$.

Doc A5.8 Coulomb interaction in the plane-wave basis

The matrix elements of the Coulomb interaction are evaluated by inserting the basis functions of equation (2.29) into equation (2.44):

$$
\begin{aligned}
&\langle \boldsymbol{k}_1\sigma_1\boldsymbol{k}_2\sigma_2 | V | \boldsymbol{k}_3\sigma_3\boldsymbol{k}_4\sigma_4 \rangle \\
&= \int d\boldsymbol{r}d\boldsymbol{r}' \sum_{\sigma\sigma'} \frac{\delta_{\sigma\sigma_1}}{\sqrt{\mathcal{V}}} e^{-i\boldsymbol{k}_1\cdot\boldsymbol{r}} \frac{\delta_{\sigma'\sigma_2}}{\sqrt{\mathcal{V}}} e^{-i\boldsymbol{k}_2\cdot\boldsymbol{r}'} \underbrace{V(\boldsymbol{r}-\boldsymbol{r}')}_{\frac{1}{\mathcal{V}}\sum_{\boldsymbol{q}} V(\boldsymbol{q}) e^{i\boldsymbol{q}\cdot(\boldsymbol{r}-\boldsymbol{r}')}} \frac{\delta_{\sigma\sigma_3}}{\sqrt{\mathcal{V}}} e^{i\boldsymbol{k}_3\cdot\boldsymbol{r}} \frac{\delta_{\sigma'\sigma_4}}{\sqrt{\mathcal{V}}} e^{i\boldsymbol{k}_4\cdot\boldsymbol{r}'} \\
&= \delta_{\sigma_1\sigma_3}\delta_{\sigma_2\sigma_4} \frac{1}{\mathcal{V}} \sum_{\boldsymbol{q}} V(\boldsymbol{q}) \delta_{\boldsymbol{k}_1,\, \boldsymbol{k}_3+\boldsymbol{q}} \delta_{\boldsymbol{k}_2,\, \boldsymbol{k}_4-\boldsymbol{q}} = \delta_{\sigma_1\sigma_3}\delta_{\sigma_2\sigma_4} \delta_{\boldsymbol{k}_1+\boldsymbol{k}_2,\, \boldsymbol{k}_3+\boldsymbol{k}_4} \frac{1}{\mathcal{V}} V(\boldsymbol{k}_1-\boldsymbol{k}_3),
\end{aligned}
$$

consistently with equation (5.43).

Doc A5.9 Diagrammatic rules for the Coulomb interaction

The recipe for translating a Coulomb-interaction self-energy diagram into the corresponding mathematical formula is the following.

1. Label the diagram:

- Attach to the external legs the variables $(\boldsymbol{k}, \sigma, \omega_n)$.
- Attach to each internal fermion line a set of variables $(\boldsymbol{k}_j, \sigma_j, \bar{\omega}_j)$.
- Impose spin conservation at each vertex.
- Impose momentum and energy conservation for each interaction line.

2. Translate the diagram:
   - A factor $(-1)^n(-1)^L$, where n is the order of the diagram (number of interaction lines), and L the number of closed fermion loops.
   - A factor $\frac{1}{\mathscr{V}}V(\boldsymbol{k}_1 - \boldsymbol{k}_3)$ for each interaction line.
   - A factor $\mathscr{G}_0(\boldsymbol{k}_j, i\bar{\omega}_j) = 1/(i\bar{\omega}_j - \xi_{\boldsymbol{k}_j})$ for each internal fermion line.
   - A sum over all $\boldsymbol{k}_j$, σ_j, $\bar{\omega}_j$, and a factor $\frac{1}{\beta}$ for each Matsubara sum.

Applying these rules, we obtain for the Hartree diagram:

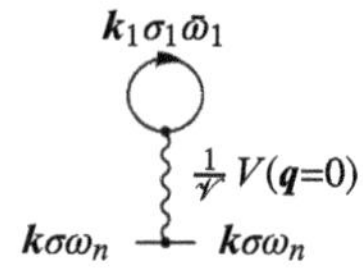

$$= (-1)^1(-1)^1 \sum_{\boldsymbol{k}_1\sigma_1} \frac{1}{\beta} \sum_{i\bar{\omega}_1} \frac{1}{\mathscr{V}} V(\boldsymbol{q} = 0)\mathscr{G}_0(\boldsymbol{k}_1, i\bar{\omega}_1)$$

$$= V(\boldsymbol{q} = 0)\frac{1}{\mathscr{V}} \sum_{\boldsymbol{k}_1\sigma_1} \underbrace{\frac{1}{\beta} \sum_{i\bar{\omega}_1} \frac{1}{i\bar{\omega}_1 - \xi_{\boldsymbol{k}_1}}}_{f(\xi_{\boldsymbol{k}_1})}$$

$$= V(\boldsymbol{q} = 0)\frac{1}{\mathscr{V}} \underbrace{\sum_{\boldsymbol{k}_1\sigma_1} f(\xi_{\boldsymbol{k}_1})}_{N} = nV(\boldsymbol{q} = 0).$$

At the second line, we have used equation (16) and at the last line n is the electron density $N/\mathscr{V}$. Note that we have omitted to write a factor $e^{i\bar{\omega}_1 0^+}$, which has its origin in the infinitesimal shift of time implied by the bubble (see section 5.1.3.4) and is required to ensure the convergence of the Matsubara sum—and to grant us the right to use equation (16). The calculation of the exchange diagram is similar. Note the absence of spin sum, because the spin is fixed on the internal fermion line:

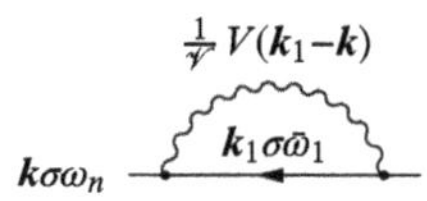

$$= (-1)^1(-1)^0 \sum_{\boldsymbol{k}_1} \frac{1}{\beta} \sum_{i\bar{\omega}_1} \frac{1}{\mathscr{V}} V(\boldsymbol{k}_1 - \boldsymbol{k})\mathscr{G}_0(\boldsymbol{k}_1, i\bar{\omega}_1)$$

$$= -\frac{1}{\mathscr{V}} \sum_{\boldsymbol{k}_1} V(\boldsymbol{k}_1 - \boldsymbol{k}) f(\xi_{\boldsymbol{k}_1}).$$

At second order, we have the diagram

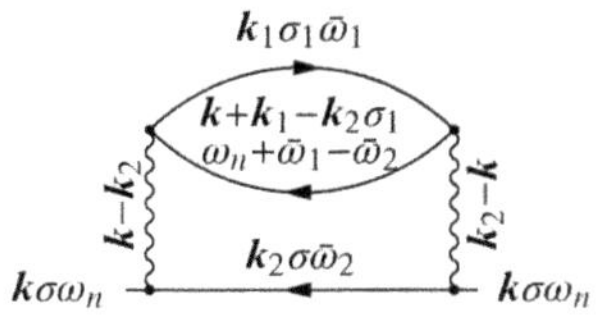

$$= (-1)^2(-1)^1 \sum_{\boldsymbol{k}_1\boldsymbol{k}_2\sigma_1} \frac{1}{\beta^2} \sum_{i\varpi_1 i\varpi_2} \frac{V(\boldsymbol{k}_2-\boldsymbol{k})}{\mathcal{V}} \frac{V(\boldsymbol{k}-\boldsymbol{k}_2)}{\mathcal{V}}$$
$$\times \mathcal{G}_0(\boldsymbol{k}_1, i\varpi_1)\mathcal{G}_0(\boldsymbol{k}_2, i\varpi_2)$$
$$\times \mathcal{G}_0(\boldsymbol{k}+\boldsymbol{k}_1-\boldsymbol{k}_2, i\omega_n + i\varpi_1 - i\varpi_2)$$

$$= -\frac{2}{\mathcal{V}^2} \sum_{\boldsymbol{k}_1\boldsymbol{k}_2} |V(\boldsymbol{k}_2-\boldsymbol{k})|^2 \frac{1}{\beta^2} \sum_{i\varpi_1 i\varpi_2}$$
$$\times \frac{1}{i\varpi_1 - \xi_{\boldsymbol{k}_1}} \frac{1}{i\varpi_2 - \xi_{\boldsymbol{k}_2}} \frac{1}{i\omega_n + i\varpi_1 - i\varpi_2 - \xi_{\boldsymbol{k}+\boldsymbol{k}_1-\boldsymbol{k}_2}}.$$

We first perform the Matsubara sum on $i\varpi_1$ using equation (16) and we introduce temporarily the notation $\boldsymbol{k}_3 = \boldsymbol{k} + \boldsymbol{k}_1 - \boldsymbol{k}_2$:

$$\frac{1}{\beta}\sum_{i\varpi_1} \frac{1}{i\varpi_1 - \xi_{\boldsymbol{k}_1}} \frac{1}{i\varpi_1 + i\omega_n - i\varpi_2 - \xi_{\boldsymbol{k}_3}} =$$
$$\frac{f(\xi_{\boldsymbol{k}_1})}{\xi_{\boldsymbol{k}_1} + i\omega_n - i\varpi_2 - \xi_{\boldsymbol{k}_3}} + \frac{f(\xi_{\boldsymbol{k}_3} + i\varpi_2 - i\omega_n)}{\xi_{\boldsymbol{k}_3} + i\varpi_2 - i\omega_n - \xi_{\boldsymbol{k}_1}} = \frac{f(\xi_{\boldsymbol{k}_3}) - f(\xi_{\boldsymbol{k}_1})}{i\varpi_2 - i\omega_n - \xi_{\boldsymbol{k}_1} + \xi_{\boldsymbol{k}_3}}.$$

We could replace $f(\xi_{\boldsymbol{k}_3} + i\varpi_2 - i\omega_n)$ by $f(\xi_{\boldsymbol{k}_3})$ because $i\varpi_2 - i\omega_n$ is an *even* Matsubara frequency, such that in the denominator of the Fermi function we have $e^{\beta(i\varpi_2 - i\omega_n)} = 1$. We next perform the $i\varpi_2$ sum:

$$\frac{1}{\beta}\sum_{i\varpi_2} \frac{1}{i\varpi_2 - \xi_{\boldsymbol{k}_2}} \frac{1}{i\varpi_2 - i\omega_n - \xi_{\boldsymbol{k}_1} + \xi_{\boldsymbol{k}_3}} = \frac{f(\xi_{\boldsymbol{k}_2})}{\xi_{\boldsymbol{k}_2} - i\omega_n - \xi_{\boldsymbol{k}_1} + \xi_{\boldsymbol{k}_3}}$$
$$+ \frac{f(i\omega_n + \xi_{\boldsymbol{k}_1} - \xi_{\boldsymbol{k}_3})}{i\omega_n + \xi_{\boldsymbol{k}_1} - \xi_{\boldsymbol{k}_3} - \xi_{\boldsymbol{k}_2}} = \frac{-b(\xi_{\boldsymbol{k}_1} - \xi_{\boldsymbol{k}_3}) - f(\xi_{\boldsymbol{k}_2})}{i\omega_n + \xi_{\boldsymbol{k}_1} - \xi_{\boldsymbol{k}_2} - \xi_{\boldsymbol{k}_3}}.$$

Here we have replaced $f(i\omega_n + \xi)$ by $-b(\xi)$ with b the Bose–Einstein distribution, since $e^{\beta i\omega_n} = -1$. Finally we use the mathematical identity

$$b(\xi_1 + \xi_2) = -\frac{f(\xi_1)f(\xi_2)}{f(\xi_1) - f(-\xi_2)}$$

and rearrange terms to obtain the value of the diagram as

$$\frac{2}{\mathcal{V}^2}\sum_{\boldsymbol{k}_1\boldsymbol{k}_2} |V(\boldsymbol{k}_2 - \boldsymbol{k})|^2 \frac{f(\xi_{\boldsymbol{k}_1})f(-\xi_{\boldsymbol{k}_2})f(-\xi_{\boldsymbol{k}_3}) + f(-\xi_{\boldsymbol{k}_1})f(\xi_{\boldsymbol{k}_2})f(\xi_{\boldsymbol{k}_3})}{i\omega_n + \xi_{\boldsymbol{k}_1} - \xi_{\boldsymbol{k}_2} - \xi_{\boldsymbol{k}_3}},$$

in agreement with equation (5.48). The other diagram at second order is obtained similarly:

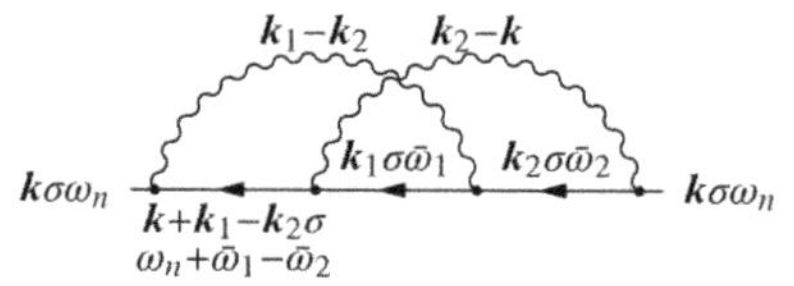

$$= \sum_{k_1 k_2} \frac{1}{\beta^2} \sum_{i\bar{\omega}_1 i\bar{\omega}_2} \frac{V(k_2 - k)}{\mathscr{V}} \frac{V(k_1 - k_2)}{\mathscr{V}} \times \mathscr{G}_0(k_1, i\bar{\omega}_1)\mathscr{G}_0(k_2, i\bar{\omega}_2) \times \mathscr{G}_0(k + k_1 - k_2, i\omega_n + i\bar{\omega}_1 - i\bar{\omega}_2).$$

Since the three Green's functions are the same as in the first diagram, the Matsubara sum yields the same expression and the diagram is the second term of equation (5.48).

Doc A5.10 Free phonon propagator

Using the definition of the phonon operator, $B^\dagger_{q\lambda} = b^\dagger_{q\lambda} + b_{-q\lambda}$, the fact that correlation functions of the kinds $\langle bb \rangle_0$ and $\langle b^\dagger b^\dagger \rangle_0$ vanish[17], and working in the representation that diagonalizes the phonon Hamiltonian, we find:

$$\begin{aligned}
\mathscr{C}^0_{B_{q_1\lambda_1} B^\dagger_{q_2\lambda_2}}(\tau) &= - \langle T_\tau B_{q_1\lambda_1}(\tau) B^\dagger_{q_2\lambda_2}(0) \rangle_0 \\
&= - \langle T_\tau b_{q_1\lambda_1}(\tau) b^\dagger_{q_2\lambda_2}(0) \rangle_0 - \langle T_\tau b^\dagger_{-q_1\lambda_1}(\tau) b_{-q_2\lambda_2}(0) \rangle_0 \\
&= - \delta_{q_1 q_2} \delta_{\lambda_1\lambda_2} \left[\langle T_\tau b_{q_1\lambda_1}(\tau) b^\dagger_{q_1\lambda_1}(0) \rangle_0 + \langle T_\tau b^\dagger_{-q_1\lambda_1}(\tau) b_{-q_1\lambda_1}(0) \rangle_0 \right] \\
&= - \delta_{q_1 q_2} \delta_{\lambda_1\lambda_2} \langle T_\tau B_{q_1\lambda_1}(\tau) B^\dagger_{q_1\lambda_1}(0) \rangle_0 \\
&\equiv \delta_{q_1 q_2} \delta_{\lambda_1\lambda_2} \mathscr{D}^0_{\lambda_1}(q_1, \tau),
\end{aligned}$$

where we have defined $\mathscr{D}_\lambda(q, \tau) = -\langle T_\tau B_{q\lambda}(\tau) B^\dagger_{q\lambda}(0) \rangle$. In order to calculate the spectral function of $\mathscr{D}^0_\lambda(q, \tau)$, we note that for independent bosons the spectral function is $\rho_{b_\alpha b^\dagger_\alpha}(\varepsilon) = \delta(\varepsilon - \xi_\alpha)$ with ξ_α the boson energies (see Doc A3.6.3). Hence

$$\rho_{B_{q\lambda} B^\dagger_{q\lambda}}(\varepsilon) = \rho_{b_{q\lambda} b^\dagger_{q\lambda}}(\varepsilon) + \underbrace{\rho_{b^\dagger_{-q\lambda} b_{-q\lambda}}(\varepsilon)}_{= -\rho_{b_{-q\lambda} b^\dagger_{-q\lambda}}(-\varepsilon)} = \delta(\varepsilon - \hbar\omega_{q\lambda}) - \delta(-\varepsilon - \hbar\omega_{-q\lambda}).$$

We have used equation (3.31) and introduced the phonon energies like in section 2.5.2. Since $\omega_{-q\lambda} = \omega_{q\lambda}$, we thus have $\rho_{B_{q\lambda} B^\dagger_{q\lambda}}(\varepsilon) = \delta(\varepsilon - \hbar\omega_{q\lambda}) - \delta(\varepsilon + \hbar\omega_{q\lambda})$. Finally, using equation (4.11), the expression of the free phonon propagator is found to be

$$\mathscr{D}^0_\lambda(q, i\Omega_n) = \frac{1}{i\Omega_n - \hbar\omega_{q\lambda}} - \frac{1}{i\Omega_n + \hbar\omega_{q\lambda}}.$$

---

[17] Without coupling to the electrons, the number of phonons is conserved, see equation (2.65).

Doc A5.11 Effective electron–electron interaction mediated by phonons

At order n, the thermodynamic average in equation (5.52) is of the form (schematically)

$$\langle H_{\text{el-ph}} \cdots H_{\text{el-ph}} cc^\dagger \rangle_0 = \langle c_1^\dagger c_1 B_1 \cdots c_n^\dagger c_n B_n cc^\dagger \rangle_0 = \langle B_1 \cdots B_n \rangle_0 \langle c_1^\dagger c_1 \cdots c_n^\dagger c_n cc^\dagger \rangle_0.$$

We can average on the phonons and electrons separately because the average is to be evaluated in the absence of coupling, i.e. in a system where the electrons and phonons live in two different Hilbert spaces. The phonon averages $\langle B_1 \cdots B_n \rangle_0$ vanish if the number of phonon operators is odd: $\langle B_1 \rangle_0 = \langle B_1 B_2 B_3 \rangle_0 = \ldots = 0$. As a result, all terms of odd order in equation (5.52) disappear. Writing $n = 2m$ and introducing a modified time-dependent electron–phonon Hamiltonian,

$$H'_{\text{el-ph}}(\tau) = \int_0^\beta d\tau' \sum_{\boldsymbol{k}_1\sigma_1} \sum_{\boldsymbol{q}_1\lambda_1} \sum_{\boldsymbol{k}_2\sigma_2} \sum_{\boldsymbol{q}_2\lambda_2} g_{\boldsymbol{q}_1\lambda_1} g_{\boldsymbol{q}_2\lambda_2} B_{\boldsymbol{q}_1\lambda_1}(\tau) B_{\boldsymbol{q}_2\lambda_2}(\tau')$$
$$\times\, c^\dagger_{\boldsymbol{k}_1+\boldsymbol{q}_1\sigma_1}(\tau) c^\dagger_{\boldsymbol{k}_2+\boldsymbol{q}_2\sigma_2}(\tau') c_{\boldsymbol{k}_2\sigma_2}(\tau') c_{\boldsymbol{k}_1\sigma_1}(\tau),$$

we can rewrite equation (5.52) as

$$\mathscr{G}(\boldsymbol{k}, \tau) = -\sum_{m=0}^{\infty} \int_0^\beta d\tau_1 \cdots d\tau_m \Big\langle T_\tau H'_{\text{el-ph}}(\tau_1) \cdots H'_{\text{el-ph}}(\tau_m) c_{\boldsymbol{k}\sigma}(\tau) c^\dagger_{\boldsymbol{k}\sigma}(0) \Big\rangle_0^{\text{con-diff}}.$$

In the definition of $H'_{\text{el-ph}}$, we have shifted the operator $c_{\boldsymbol{k}_1\sigma_1}$ from its original position—just after $c^\dagger_{\boldsymbol{k}_1+\boldsymbol{q}_1\sigma_1}$, see equation (2.73)—to the right, making two fermionic permutations. This is legal because the whole formula involving $H'_{\text{el-ph}}$ lies under a T_τ operator. At this stage, we use the fact that we can perform the phonon average separately and we make the following substitution in $H'_{\text{el-ph}}$:

$$B_{\boldsymbol{q}_1\lambda_1}(\tau) B_{\boldsymbol{q}_2\lambda_2}(\tau') \to \tfrac{1}{2}\langle T_\tau B_{\boldsymbol{q}_1\lambda_1}(\tau) B_{\boldsymbol{q}_2\lambda_2}(\tau')\rangle_0 = \tfrac{1}{2}\langle T_\tau B_{\boldsymbol{q}_1\lambda_1}(\tau) B^\dagger_{-\boldsymbol{q}_2\lambda_2}(\tau')\rangle_0$$
$$= -\tfrac{1}{2}\delta_{\lambda_1\lambda_2}\delta_{\boldsymbol{q}_1,-\boldsymbol{q}_2}\mathscr{D}^0_{\lambda_1}(\boldsymbol{q}_1, \tau - \tau'),$$

where we have used equations (5.49) and (5.50). The factor 1/2 corrects for a complication that we have overlooked in our derivation[18]. This substitution allows one to rewrite $\mathscr{G}(\boldsymbol{k}, \tau)$ in the form given by equation (5.53), where the effective phonon-mediated electron–electron interaction is given by equation (5.54). In performing this last step, we pull out the minus sign from $\langle BB^\dagger \rangle \propto -\mathscr{D}^0$—which appears m times in a term of order m—as a global factor $(-1)^m$, and we use the Kronecker deltas above to eliminate half of the $\boldsymbol{q}$ and λ sums.

[18] By replacing B_1B_2 by $\langle B_1B_2 \rangle_0$ in the expression of $H'_{\text{el-ph}}$, we retain only one out of the $(n-1) \times (n-3) \times \cdots \times 5 \times 3$ terms that are generated by Wick's theorem from the average of n phonon operators $\langle B_1 \cdots B_n \rangle_0$. On the other hand, when we rewrote the n sum of equation (5.52) as a sum on $m = 2n$, we did not correct accordingly for the factor $1/n!$ coming from the expansion of the evolution operator in equation (5.3), that was cancelled in equation (5.5) by retaining only the topologically different diagrams. In our case, the number of topologically different diagrams will be $m!$, not $(2m)!$, such that we must correct by a factor $m!/(2m)!$. It turns out that $(2m-1) \times (2m-3) \times \cdots \times 5 \times 3 \times m!/(2m)! = 1/2^m$, such that we can take this into account by inserting an additional 1/2 in front of $H'_{\text{el-ph}}$. See also Bruus & Flensberg (2004, p 315).

Doc A5.12 The electron–phonon self-energy at leading order

The first-order contribution of the electron–phonon interaction to the one-electron self-energy is

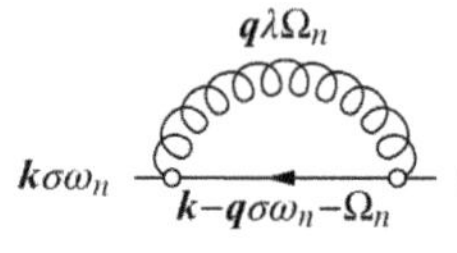

$$= -\sum_{q\lambda} |g_{q\lambda}|^2 \frac{1}{\beta} \sum_{i\Omega_n} \mathscr{D}^0_\lambda(\boldsymbol{q}, i\Omega_n)\mathscr{G}_0(\boldsymbol{k}-\boldsymbol{q}, i\omega_n - i\Omega_n)$$

$$= -\sum_{q\lambda} |g_{q\lambda}|^2 \frac{1}{\beta} \sum_{i\Omega_n} \left(\frac{1}{i\Omega_n - \hbar\omega_{q\lambda}} - \frac{1}{i\Omega_n + \hbar\omega_{q\lambda}} \right) \frac{1}{i\omega_n - i\Omega_n - \xi_{k-q}}.$$

Note that the result is intensive despite the $\boldsymbol{q}$ sum, owing to the $1/\mathcal{N}$ factor stemming from $|g_{q\lambda}|^2$ (equation (2.74)). The Matsubara sum gives, using equation (16):

$$\frac{-b(\hbar\omega_{q\lambda})}{i\omega_n - \hbar\omega_{q\lambda} - \xi_{k-q}} - \frac{-b(-\hbar\omega_{q\lambda})}{i\omega_n + \hbar\omega_{q\lambda} - \xi_{k-q}}$$

$$+ b(i\omega_n - \xi_{k-q}) \left(\frac{1}{i\omega_n - \xi_{k-q} - \hbar\omega_{q\lambda}} - \frac{1}{i\omega_n - \xi_{k-q} + \hbar\omega_{q\lambda}} \right),$$

which, together with the relations $b(-\varepsilon) = -1 - b(\varepsilon)$, $b(i\omega_n - \xi) = -f(-\xi)$, and $f(-\xi) = 1 - f(\xi)$, yields equation (5.58). See also Schrieffer (1964, p 200).

Doc A5.13 Particle–hole bubble

The zeroth-order term in the expansion of $\mathscr{C}^0_{\alpha\beta\gamma\delta}(\tau)$ in equation (5.59) is just

$$\begin{aligned}
\mathscr{C}^0_{\alpha\beta\gamma\delta}(\tau) &= -\langle T_\tau a^\dagger_\alpha(\tau) a_\beta(\tau) a^\dagger_\gamma(0) a_\delta(0)\rangle_0^{\text{con-diff}} \\
&= -\underbrace{\langle T_\tau a^\dagger_\alpha(\tau) a_\delta(0)\rangle_0}_{-\eta\mathscr{G}^0_{\delta\alpha}(-\tau)} \underbrace{\langle T_\tau a_\beta(\tau) a^\dagger_\gamma(0)\rangle_0}_{-\mathscr{G}^0_{\beta\gamma}(\tau)} \\
&= -\eta\, \mathscr{G}^0_{\beta\gamma}(\tau)\mathscr{G}^0_{\delta\alpha}(-\tau).
\end{aligned}$$

Graphically, this can be pictured like in equation (5.60): $\mathscr{G}^0_{\beta\gamma}(\tau)$ is represented by a line going from right to left (γ to β) as required by our convention for particles. $\mathscr{G}^0_{\delta\alpha}(-\tau)$ describes the hole propagating 'back in time' from left to right (α to δ). The η factor is taken care of by the diagrammatic rules because there is one particle loop in the diagram and the minus sign is explicitly kept in front of the diagram. The Fourier transform of this propagator is

$$\begin{aligned}
\mathscr{C}^0_{\alpha\beta\gamma\delta}(i\Omega_n) &= \int_0^\beta d\tau\, \mathscr{C}^0_{\alpha\beta\gamma\delta}(\tau) e^{i\Omega_n\tau} \\
&= -\eta \frac{1}{\beta^2} \sum_{i\bar{\nu}_1 i\bar{\nu}_2} \mathscr{G}^0_{\beta\gamma}(i\bar{\nu}_1)\mathscr{G}^0_{\delta\alpha}(i\bar{\nu}_2) \underbrace{\int_0^\beta d\tau\, e^{i\Omega_n\tau} e^{-i\bar{\nu}_1\tau} e^{-i\bar{\nu}_2(-\tau)}}_{\beta\delta_{\Omega_n+\bar{\nu}_2,\,\bar{\nu}_1}} \\
&= \frac{-\eta}{\beta} \sum_{i\nu_n} \mathscr{G}^0_{\delta\alpha}(i\nu_n)\mathscr{G}^0_{\beta\gamma}(i\nu_n + i\Omega_n).
\end{aligned}$$

In the representation where $\mathscr{G}^0$ is diagonal, we have using equation (16)

$$\mathscr{C}^0_{\alpha\beta\gamma\delta}(i\Omega_n) = \delta_{\alpha\delta}\delta_{\beta\gamma}\frac{-\eta}{\beta}\sum_{i\nu_n}\frac{1}{i\nu_n - \xi_\alpha}\frac{1}{i\nu_n + i\Omega_n - \xi_\beta}$$
$$= \delta_{\alpha\delta}\delta_{\beta\gamma}\left(\frac{d_{-\eta}(\xi_\alpha)}{\xi_\alpha + i\Omega_n - \xi_\beta} + \frac{d_{-\eta}(\xi_\beta)}{\xi_\beta - i\Omega_n - \xi_\alpha}\right),$$

which is just equation (5.61).

Along the same lines, we can derive the expression of the *renormalized bubble*, in which the free-particle propagators $\mathscr{G}^0$ are replaced by the exact propagators $\mathscr{G}$ expressed in terms of their spectral representation (equation (4.11)):

$$\mathscr{G}_{\alpha\beta}(i\nu_n) = \int_{-\infty}^{\infty} d\varepsilon \frac{\rho_{a_\alpha a_\beta^\dagger}(\varepsilon)}{i\nu_n - \varepsilon}.$$

We get

$$\mathscr{C}_{\alpha\beta\gamma\delta}(i\Omega_n) = \frac{-\eta}{\beta}\sum_{i\nu_n}\mathscr{G}_{\delta\alpha}(i\nu_n)\mathscr{G}_{\beta\gamma}(i\nu_n + i\Omega_n)$$
$$= \int_{-\infty}^{\infty} d\varepsilon_1 d\varepsilon_2\, \rho_{a_\delta a_\alpha^\dagger}(\varepsilon_1)\rho_{a_\beta a_\gamma^\dagger}(\varepsilon_2)\frac{-\eta}{\beta}\sum_{i\nu_n}\frac{1}{i\nu_n - \varepsilon_1}\frac{1}{i\nu_n + i\Omega_n - \varepsilon_2}$$
$$= \int_{-\infty}^{\infty} d\varepsilon_1 d\varepsilon_2\, \rho_{a_\delta a_\alpha^\dagger}(\varepsilon_1)\rho_{a_\beta a_\gamma^\dagger}(\varepsilon_2)\left(\frac{d_{-\eta}(\varepsilon_1)}{\varepsilon_1 + i\Omega_n - \varepsilon_2} + \frac{d_{-\eta}(\varepsilon_2)}{\varepsilon_2 - i\Omega_n - \varepsilon_1}\right),$$

which corresponds to equation (5.64).

Doc A5.14 Particle–hole bubble for free electrons at zero temperature

After performing the analytic continuation $i\Omega_n \to \varepsilon + i0^+$ and shifting the integration variable $\boldsymbol{k}$ by $-\boldsymbol{q}$ in the second term of equation (5.76), the electron–hole bubble becomes:

$$\chi^0_{nn}(\boldsymbol{q}, \varepsilon) = 2\sum_{\boldsymbol{k}}\left[\frac{f(\xi_{\boldsymbol{k}})}{\varepsilon + \xi_{\boldsymbol{k}} - \xi_{\boldsymbol{k}+\boldsymbol{q}} + i0^+} - \frac{f(\xi_{\boldsymbol{k}})}{\varepsilon + \xi_{\boldsymbol{k}-\boldsymbol{q}} - \xi_{\boldsymbol{k}} + i0^+}\right].$$

The free-electron dispersion is $\xi_{\boldsymbol{k}} = \frac{\hbar^2}{2m}(k^2 - k_F^2)$. In order to perform the $\boldsymbol{k}$ integration, we choose spherical coordinates with the k_z axis parallel to $\boldsymbol{q}$. We obtain $\xi_{\boldsymbol{k}} - \xi_{\boldsymbol{k}+\boldsymbol{q}} = -\frac{\hbar^2}{2m}(q^2 + 2qk\cos\vartheta)$. Introducing the dimensionless parameters $x = \frac{q}{2k_F}$, $z = \frac{2m\varepsilon}{\hbar^2 q^2}$ and the integration variable $u = k/k_F$, we then obtain

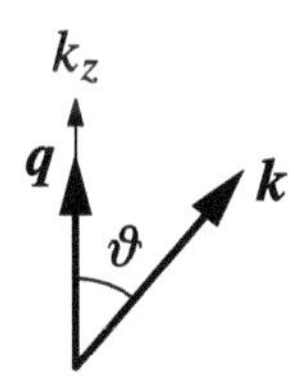

$$\chi_{nn}^0(\boldsymbol{q}, \varepsilon) = \frac{N_0^{\text{el}}(0)}{4x}\int_0^1 du\, u \int_0^\pi d\vartheta \left[\frac{\sin\vartheta}{\frac{x(z-1)}{u} - \cos\vartheta + i0^+} - \frac{\sin\vartheta}{\frac{x(z+1)}{u} - \cos\vartheta + i0^+}\right].$$

We have replaced the $\boldsymbol{k}$ sum by an integral according to $\sum_{\boldsymbol{k}} \to \frac{\mathcal{V}}{(2\pi)^3}\int d\boldsymbol{k}$, used the Fermi functions to restrict the k integration to $k < k_{\text{F}}$ ($u < 1$), and introduced the Fermi-level DOS of free electrons in 3D, $N_0^{\text{el}}(0) = mk_{\text{F}}\mathcal{V}/(\pi^2\hbar^2)$. We here make use of the formula (18) and get

$$\chi_{nn}^0(\boldsymbol{q}, \varepsilon) = \frac{N_0^{\text{el}}(0)}{4x}\int_0^1 du\, u\left\{\ln\left|\frac{u + x(z-1)}{u - x(z-1)}\right| - \ln\left|\frac{u + x(z+1)}{u - x(z+1)}\right| \right.$$
$$\left. - i\pi[\theta(u - |x(z-1)|) - \theta(u - |x(z+1)|)]\right\}.$$

It is not difficult to check that $\int_0^1 du\, u\, \theta(u - |a|) = \frac{1}{2}\max(0, 1 - a^2)$, such that the imaginary part is

$$\text{Im}\,\chi_{nn}^0(\boldsymbol{q}, \varepsilon) = -\frac{\pi N_0^{\text{el}}(0)}{8x}\{\max[0, 1 - x^2(z-1)^2] - \max[0, 1 - x^2(z+1)^2]\}.$$

We can see that this is an odd function of z, hence an odd function of the energy ε as required by the general property that the spectral function of $\chi_{nn}(\boldsymbol{q}, \varepsilon)$, $\rho_{n(\boldsymbol{q})n(-\boldsymbol{q})}(\varepsilon) = -\frac{1}{\pi}\text{Im}\,\chi_{nn}(\boldsymbol{q}, \varepsilon)$, must be odd. The real part follows by virtue of the formula (19):

$$\text{Re}\,\chi_{nn}^0(\boldsymbol{q}, \varepsilon) = \frac{N_0^{\text{el}}(0)}{4x}\left\{x(z-1) + \frac{1}{2}[1 - x^2(z-1)^2]\ln\left|\frac{1 + x(z-1)}{1 - x(z-1)}\right| \right.$$
$$\left. - x(z+1) - \frac{1}{2}[1 - x^2(z+1)^2]\ln\left|\frac{1 + x(z+1)}{1 - x(z+1)}\right|\right\}.$$

In order to study this function, we find it more convenient to introduce the variable $y = zx^2 = \varepsilon/(4\varepsilon_{\text{F}})$. We therefore rewrite

$$\text{Re}\,\chi_{nn}^0(\boldsymbol{q}, \varepsilon) = -N_0^{\text{el}}(0)\left\{\frac{1}{2} + \frac{1}{8x^3}[x^2 - (x^2 - y)^2]\ln\left|\frac{x + x^2 - y}{x - x^2 + y}\right| \right.$$
$$\left. + \frac{1}{8x^3}[x^2 - (x^2 + y)^2]\ln\left|\frac{x + x^2 + y}{x - x^2 - y}\right|\right\}$$

$$\text{Im}\,\chi_{nn}^0(\boldsymbol{q}, \varepsilon) = -\frac{\pi N_0^{\text{el}}(0)}{8x}\left\{\max\left[0, 1 - \frac{1}{x^2}(x^2 - y)^2\right] - \max\left[0, 1 - \frac{1}{x^2}(x^2 + y)^2\right]\right\}.$$

From these expressions, we find the zero-energy behavior by setting $y = 0$ and noting that the imaginary part vanishes:

$$\chi^0_{nn}(\boldsymbol{q}, 0) = -N^{\text{el}}_0(0)\left[\frac{1}{2} + \frac{1}{4}\left(\frac{1}{x} - x\right)\ln\left|\frac{1+x}{1-x}\right|\right].$$

This is known as the *Lindhard function* (figure 5.12). Since the function approaches unity as $x \to 0$, we find that in the static long-wavelength limit the free-electron density–density correlation function is just minus the Fermi-level DOS:

$$\chi^0_{nn}(0, 0) = -N^{\text{el}}_0(0).$$

This result is in fact valid for any system of independent fermions. Indeed, the static long-wavelength limit can be evaluated directly from the general expression (5.76) by expanding the Fermi factor $f(\xi_{\boldsymbol{k}+\boldsymbol{q}})$ around $\boldsymbol{q} = 0$ as $f(\xi_{\boldsymbol{k}}) + f'(\xi_{\boldsymbol{k}})(\xi_{\boldsymbol{k}+\boldsymbol{q}} - \xi_{\boldsymbol{k}})$. The imaginary part then involves a term like $x\delta(x)$ which vanishes and the real part becomes

$$\chi^0_{nn}(0, 0) = -\sum_{\boldsymbol{k}\sigma}[-f'(\xi_{\boldsymbol{k}})] = -\int_{-\infty}^{\infty} d\varepsilon\, N^{\text{el}}_0(\varepsilon)[-f'(\varepsilon)] \overset{T=0}{=} -N^{\text{el}}_0(0),$$

since $-f'(\varepsilon)$ becomes $\delta(\varepsilon)$ at zero temperature.

If we set $x = 0$ first instead of setting $y = 0$ first, we find $\chi^0_{nn}(0, \varepsilon) = 0$. The function $\chi^0_{nn}(\boldsymbol{q}, \varepsilon)$ is in fact discontinuous at $\boldsymbol{q} = \varepsilon = 0$, such that the value obtained for $\chi^0_{nn}(0, 0)$ depends on the order in which the two limits are taken. When $\boldsymbol{q}$ and ε are both small, we may develop the function along radial lines $y = \zeta x$. Doing so, we find that close to the origin and for $\zeta > 1$,

$$\chi^0_{nn}(\boldsymbol{q} \to 0, \varepsilon \propto q) = -N^{\text{el}}_0(0)\left(1 - \frac{\zeta}{2}\ln\left|\frac{1+\zeta}{1-\zeta}\right|\right).$$

Finally, at high frequency we obtain by expanding around $y^{-1} = 0$: $\chi^0_{nn}(\boldsymbol{q}, \varepsilon \to \infty) = -N^{\text{el}}_0(0)(-\frac{x^2}{3y^2}) = \frac{4}{3}N^{\text{el}}_0(0)\varepsilon_{\boldsymbol{q}}\varepsilon_{\text{F}}/\varepsilon^2$.

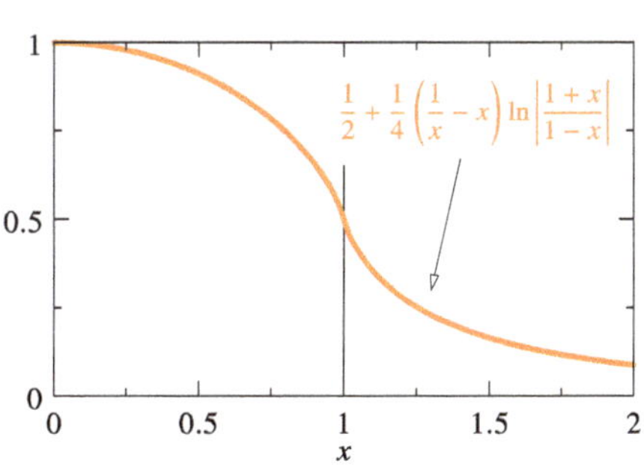

Figure 5.12. Lindhard function. It has a logarithmic singularity at $x = 1$ ($q = 2k_{\text{F}}$), corresponding to the largest momentum at which electron–hole pairs can be excited at $\varepsilon = 0$.

Doc A5.15 Plasmon dispersion in RPA

In the RPA approximation for the homogeneous electron gas, the equation (5.80) defining the plasmon excitation is, retaining the two terms of lowest order in $q/\varepsilon \equiv q/\hbar\omega$:

$$1 - \frac{e^2}{\epsilon_0 q^2}\frac{N_0^{\mathrm{el}}(0)}{\mathcal{V}}\left[\frac{1}{3}\left(\frac{\hbar k_{\mathrm{F}} q}{m\omega}\right)^2 + \frac{1}{5}\left(\frac{\hbar k_{\mathrm{F}} q}{m\omega}\right)^4\right] = 0.$$

The terms in the square brackets give the expansion of the function $-1 + \frac{\zeta}{2}\ln\left|\frac{1+\zeta}{1-\zeta}\right|$ in equation (5.77) around $\zeta^{-1} = 0$. Multiplying the equation by ω^4 and substituting the Fermi-level DOS $N_0^{\mathrm{el}}(0) = mk_{\mathrm{F}}\mathcal{V}/(\pi^2\hbar^2)$ gives

$$\omega^4 - \frac{e^2 n}{\epsilon_0 m}\left(\omega^2 + \frac{3}{5}v_{\mathrm{F}}^2 q^2\right) = 0$$

with $n = k_{\mathrm{F}}^3/(3\pi^2)$ the electron density and $v_{\mathrm{F}} = \hbar k_{\mathrm{F}}/m$ the Fermi velocity. At $q = 0$, we find the solution $\omega^2 = \frac{e^2 n}{\epsilon_0 m} \equiv \omega_p^2$, while the solution for all q is

$$\omega^2 = \omega_p^2\left(\frac{1}{2} + \frac{1}{2}\sqrt{1 + \frac{12}{5}\frac{v_{\mathrm{F}}^2}{\omega_p^2}q^2}\right).$$

Expanding $\omega(q)$ to order q^2, we obtain equation (5.84).

Doc A5.16 Longitudinal and transverse dielectric functions

In a linear and isotropic medium, the *macroscopic* electric and displacement fields $\boldsymbol{E}$ and $\boldsymbol{D}$ are colinear. In this case, the dielectric function $\epsilon(\boldsymbol{q}, \omega)$ relates the two fields according to

$$\boldsymbol{D}(\boldsymbol{q}, \omega) = \epsilon_0\epsilon(\boldsymbol{q}, \omega)\boldsymbol{E}(\boldsymbol{q}, \omega).$$

It is convenient to separate the fields in two components, one parallel and the other normal to the wave vector $\boldsymbol{q}$:

$$\boldsymbol{E}(\boldsymbol{q}, \omega) = \boldsymbol{E}_{\parallel}(\boldsymbol{q}, \omega) + \boldsymbol{E}_{\perp}(\boldsymbol{q}, \omega) \qquad \text{with} \qquad \boldsymbol{q}\cdot\boldsymbol{E}_{\perp}(\boldsymbol{q}, \omega) = 0,$$

and similarly for $\boldsymbol{D}(\boldsymbol{q}, \omega)$. Likewise, the dielectric function is split in longitudinal and transverse components $\epsilon_{\parallel}$ and $\epsilon_{\perp}$, such that

$$\boldsymbol{D}_{\parallel}(\boldsymbol{q}, \omega) = \epsilon_0\epsilon_{\parallel}(\boldsymbol{q}, \omega)\boldsymbol{E}_{\parallel}(\boldsymbol{q}, \omega) \qquad \text{and} \qquad \boldsymbol{D}_{\perp}(\boldsymbol{q}, \omega) = \epsilon_0\epsilon_{\perp}(\boldsymbol{q}, \omega)\boldsymbol{E}_{\perp}(\boldsymbol{q}, \omega).$$

In the macroscopic Maxwell's equations, the source of the displacement field $\boldsymbol{D}$ is the external or 'free' charge density, $\nabla\cdot\boldsymbol{D}(\boldsymbol{r}, t) = \rho^{\mathrm{ext}}(\boldsymbol{r}, t)$, while the source of the electric field $\boldsymbol{E}$ is the total or 'bound' charge density, $\epsilon_0\nabla\cdot\boldsymbol{E}(\boldsymbol{r}, t) = \rho^{\mathrm{tot}}(\boldsymbol{r}, t)$. Fourier transforming these two relations, we find (see equation (2*a*))

$$i\boldsymbol{q}\cdot\boldsymbol{D}(\boldsymbol{q}, \omega) = \rho^{\mathrm{ext}}(\boldsymbol{q}, \omega) \qquad \text{and} \qquad \epsilon_0 i\boldsymbol{q}\cdot\boldsymbol{E}(\boldsymbol{q}, \omega) = \rho^{\mathrm{tot}}(\boldsymbol{q}, \omega).$$

From the definitions of the longitudinal and transverse components, we deduce (omitting the $\boldsymbol{q}$ and ω arguments for briefness) $\rho^{\text{ext}} = i\boldsymbol{q}\cdot\boldsymbol{D} = iqD_{\|}$ and $\rho^{\text{tot}} = i\epsilon_0\boldsymbol{q}\cdot\boldsymbol{E} = i\epsilon_0 qE_{\|}$, while $D_{\|}$ and $E_{\|}$ are related by $D_{\|} = \epsilon_0\epsilon_{\|}E_{\|}$. Comparing these three relations, we find $\rho^{\text{ext}} = \epsilon_{\|}\rho^{\text{tot}}$ or

$$\epsilon_{\|}(\boldsymbol{q}, \omega) = \frac{\rho^{\text{ext}}(\boldsymbol{q}, \omega)}{\rho^{\text{tot}}(\boldsymbol{q}, \omega)} = \frac{V^{\text{ext}}(\boldsymbol{q}, \omega)}{V^{\text{tot}}(\boldsymbol{q}, \omega)}.$$

For the last equal sign, we have introduced the scalar potentials V^{ext} and V^{tot}, which are related to the corresponding charge densities by the Poisson equation $\epsilon_0\nabla^2 V(\boldsymbol{r}, t) = -\rho(\boldsymbol{r}, t)$ or, in Fourier space, $\epsilon_0 q^2 V(\boldsymbol{q}, \omega) = \rho(\boldsymbol{q}, \omega)$.

Doc A5.17 Longitudinal and transverse spin–spin correlation function

In order to separate the longitudinal and transverse components of the spin–spin correlation function defined in equation (5.100), we use the expressions of S^x and S^y in terms of S^+ and S^- (section 2.3.2) and write:

$$\begin{aligned}\boldsymbol{S}(\boldsymbol{q})\cdot\boldsymbol{S}(-\boldsymbol{q}) &= S_q^x S_{-q}^x + S_q^y S_{-q}^y + S_q^z S_{-q}^z \\ &= \frac{1}{4}\left(S_q^+ + S_q^-\right)\left(S_{-q}^+ + S_{-q}^-\right) - \frac{1}{4}\left(S_q^+ - S_q^-\right)\left(S_{-q}^+ - S_{-q}^-\right) + S_q^z S_{-q}^z \\ &= S_q^z S_{-q}^z + \frac{1}{2}\left(S_q^+ S_{-q}^- + S_q^- S_{-q}^+\right).\end{aligned}$$

Hence the longitudinal and transverse components are defined as

$$\chi_{ss}(\boldsymbol{q}, \tau) = \left(\frac{\hbar}{2}\right)^2\left[\chi_{\|}(\boldsymbol{q}, \tau) + \chi_{\perp}(\boldsymbol{q}, \tau)\right]$$

$$\left(\frac{\hbar}{2}\right)^2\chi_{\|}(\boldsymbol{q}, \tau) = -\langle T_\tau S^z(\boldsymbol{q}, \tau)S^z(-\boldsymbol{q}, 0)\rangle$$

$$\left(\frac{\hbar}{2}\right)^2\chi_{\perp}(\boldsymbol{q}, \tau) = \frac{1}{2}[-\langle T_\tau S^+(\boldsymbol{q}, \tau)S^-(-\boldsymbol{q}, 0)\rangle - \langle T_\tau S^-(\boldsymbol{q}, \tau)S^+(-\boldsymbol{q}, 0)\rangle].$$

The diagrammatic representation of these functions for electrons interacting via Coulomb forces is most easily written down in terms of the more general propagator

$$\chi_{\sigma_1\sigma_2\sigma_1'\sigma_2'}(\boldsymbol{q}, \tau) = -\sum_{\boldsymbol{k}\boldsymbol{k}'}\langle T_\tau c_{\boldsymbol{k}\sigma_1}^\dagger(\tau)c_{\boldsymbol{k}+\boldsymbol{q}\sigma_2}(\tau)c_{\boldsymbol{k}'\sigma_2'}^\dagger(0)c_{\boldsymbol{k}'-\boldsymbol{q}\sigma_1'}(0)\rangle,$$

which allows us to express the longitudinal and transverse functions as follows:

$$\chi_{\|} = \chi_{nn} - 2(\chi_{\uparrow\uparrow\downarrow\downarrow} + \chi_{\downarrow\downarrow\uparrow\uparrow}) \qquad \text{and} \qquad \chi_{\perp} = 2(\chi_{\uparrow\downarrow\uparrow\downarrow} + \chi_{\downarrow\uparrow\downarrow\uparrow}),$$

where we have removed the $(\boldsymbol{q}, \tau)$ arguments for briefness and χ_{nn} is the density–density correlation function discussed in section 5.1.4.5. The diagram representing

$\chi_{\sigma_1\sigma_2\sigma_1'\sigma_2'}$ is the same as equation (5.75), except that the spins indices at the vertices are free. A Dyson-like equation on the model of equation (5.69) may be formulated as:

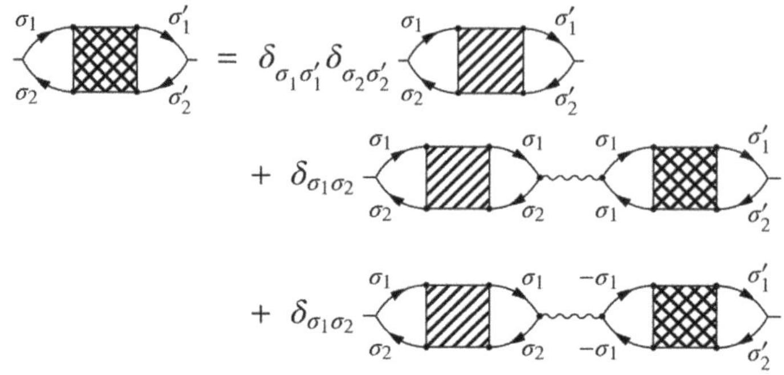

In writing this equation, we have taken into account that the spin must be conserved on the particle and hole lines in the polarization diagram, all interactions along these lines being spin-conserving. For the first term on the right, this gives rise to the product of delta functions. For the second and third terms, this fixes the value of the spin on two of the internal lines. As a result, a delta function appears due to the spin conservation at the vertex. One spin variable is left free on the second vertex. This spin can be either σ_1 or $-\sigma_1$, which explains the two terms. We can immediately translate this into the following formula:

$$-\chi_{\sigma_1\sigma_2\sigma_1'\sigma_2'} = \delta_{\sigma_1\sigma_1'}\delta_{\sigma_2\sigma_2'}\Pi_{\sigma_1\sigma_2} + \delta_{\sigma_1\sigma_2}\Pi_{\sigma_1\sigma_2}\frac{V}{\mathcal{V}}(\chi_{\sigma_1\sigma_1\sigma_1'\sigma_2'} + \chi_{-\sigma_1-\sigma_1\sigma_1'\sigma_2'}).$$

The definition of the various terms should be obvious by looking at the diagrams. This relation actually defines a linear system of 16 equations for the 16 correlators $\chi_{\sigma_1\sigma_2\sigma_1'\sigma_2'}$. It is immediately clear that $\chi_{\sigma-\sigma\sigma\sigma} = \chi_{\sigma-\sigma-\sigma\sigma} = \chi_{\sigma-\sigma-\sigma-\sigma} = 0$, since for these six correlators the right-hand side of the equation vanishes due to the delta functions. Furthermore, we also read from the equation that

$$\chi_{\sigma-\sigma\sigma-\sigma} = -\Pi_{\sigma-\sigma},$$

which is sufficient to determine $\chi_\perp$. We are then left with eight unknown correlators, namely $\chi_{\sigma\sigma\sigma\sigma}$, $\chi_{\sigma\sigma\sigma-\sigma}$, $\chi_{\sigma\sigma-\sigma\sigma}$, and $\chi_{\sigma\sigma-\sigma-\sigma}$. The diagrammatic relation leads to two decoupled systems of two equations, namely

$$-\chi_{\sigma\sigma\sigma\sigma} = \Pi_{\sigma\sigma} + \Pi_{\sigma\sigma}\frac{V}{\mathcal{V}}(\chi_{\sigma\sigma\sigma\sigma} + \chi_{-\sigma-\sigma\sigma\sigma})$$

$$-\chi_{-\sigma-\sigma\sigma\sigma} = \Pi_{-\sigma-\sigma}\frac{V}{\mathcal{V}}(\chi_{-\sigma-\sigma\sigma\sigma} + \chi_{\sigma\sigma\sigma\sigma})$$

with the solution

$$\chi_{\sigma\sigma\sigma\sigma} = -\frac{\Pi_{\sigma\sigma}\left(1 + \frac{V}{\mathcal{V}}\Pi_{-\sigma-\sigma}\right)}{1 + \frac{V}{\mathcal{V}}\Pi}, \quad \chi_{\sigma\sigma-\sigma-\sigma} = \frac{\frac{V}{\mathcal{V}}\Pi_{\uparrow\uparrow}\Pi_{\downarrow\downarrow}}{1 + \frac{V}{\mathcal{V}}\Pi}$$

and

$$
\begin{aligned}
-\chi_{\sigma\sigma\sigma-\sigma} &= \Pi_{\sigma\sigma}\frac{V}{\mathcal{V}}(\chi_{\sigma\sigma\sigma-\sigma} + \chi_{-\sigma-\sigma\sigma-\sigma}) \\
-\chi_{-\sigma-\sigma\sigma-\sigma} &= \Pi_{-\sigma-\sigma}\frac{V}{\mathcal{V}}(\chi_{-\sigma-\sigma\sigma-\sigma} + \chi_{\sigma\sigma\sigma-\sigma}),
\end{aligned}
$$

with the solution $\chi_{\sigma\sigma\sigma-\sigma} = \chi_{\sigma\sigma-\sigma\sigma} = 0$. Collecting the results and using equation (5.79), we find

$$
\chi_{\parallel} = -\frac{\Pi + 4\frac{V}{\mathcal{V}}\Pi_{\uparrow\uparrow}\Pi_{\downarrow\downarrow}}{1 + \frac{V}{\mathcal{V}}\Pi} \quad \text{and} \quad \chi_{\perp} = -2(\Pi_{\uparrow\downarrow} + \Pi_{\downarrow\uparrow}),
$$

as indicated in equation (5.101).

Doc A5.18 Equation of motion of the imaginary-time correlation functions

Writing explicitly the time ordering, $-\mathscr{C}_{AB}(\tau) = \theta(\tau)\langle A(\tau)B(0)\rangle + \eta\theta(-\tau)\langle B(0)A(\tau)\rangle$, and noting that $\partial_\tau\theta(\tau) = \delta(\tau)$, we find

$$
\begin{aligned}
-\partial_\tau\mathscr{C}_{AB}(\tau) = \delta(\tau)\langle A(\tau)B(0)\rangle - \eta\delta(\tau)\langle B(0)A(\tau)\rangle \\
+ \theta(\tau)\langle \partial_\tau A(\tau)B(0)\rangle + \eta\theta(-\tau)\langle B(0)\partial_\tau A(\tau)\rangle.
\end{aligned}
$$

In the first two terms, $A(\tau)$ can be replaced by $A(0)$ due to the delta function. This leads to $\delta(\tau)\langle A(0)B(0) - \eta B(0)A(0)\rangle$, i.e. the first term in the right-hand side of equation (5.108). The last two terms can be grouped as $\langle T_\tau\partial_\tau A(\tau)B(0)\rangle$. Finally, from the definition (4.3) we see that $\partial_\tau A(\tau) = [K, A]_\tau$, hence $\langle T_\tau\partial_\tau A(\tau)B(0)\rangle = -\langle T_\tau[A, K]_\tau B(0)\rangle = \mathscr{C}_{[A,\,K]B}(\tau)$, giving the second term in the right-hand side of equation (5.108).

Doc A5.19 Grand potential and one-particle Green's function

Consider the Hamiltonian $K = K_0 + \lambda V$, where K_0 contains no interaction term and can therefore be solved to yield the partition function $Z_0 = \mathrm{Tr}e^{-\beta K_0}$ and the grand potential $\Omega_0 = -\frac{1}{\beta}\ln Z_0$. Equation (2.5) gives the grand potential for $\lambda > 0$ as

$$
\Omega(\lambda) = -\frac{1}{\beta}\ln \mathrm{Tr}e^{-\beta(K_0+\lambda V)}.
$$

Taking the derivative with respect to λ, we obtain

$$
\frac{d\Omega(\lambda)}{d\lambda} = -\frac{1}{\beta}\frac{\mathrm{Tr}\,(-\beta V)e^{-\beta(K_0+\lambda V)}}{\mathrm{Tr}e^{-\beta(K_0+\lambda V)}} = \frac{\mathrm{Tr}e^{-\beta(K_0+\lambda V)}V}{\mathrm{Tr}e^{-\beta(K_0+\lambda V)}} = \langle V\rangle_\lambda,
$$

where $\langle \cdots \rangle_\lambda$ is the thermal average (2.2) with respect to the Hamiltonian $K_0 + \lambda V$. Integrating with the boundary condition $\Omega(\lambda = 0) = \Omega_0$ gives

$$\Omega = \Omega(1) = \Omega_0 + \int_0^1 d\lambda \, \langle V \rangle_\lambda .$$

We now express $\langle V \rangle_\lambda$ in terms of the correlation function that appears in the equation of motion of the Green's function, equation (5.114):

$$\begin{aligned}
\langle V \rangle_\lambda &= \frac{1}{2} \sum_{\alpha\gamma\mu_1\mu_2} V_{\alpha\gamma\mu_1\mu_2} \langle a_\alpha^\dagger a_\gamma^\dagger a_{\mu_2} a_{\mu_1} \rangle_\lambda \\
&= \frac{1}{2} \sum_{\alpha\gamma\mu_1\mu_2} V_{\alpha\gamma\mu_1\mu_2} \langle T_\tau a_\alpha^\dagger(0) a_\gamma^\dagger(0^-) a_{\mu_2}(0^-) a_{\mu_1}(0^-) \rangle_\lambda \\
&= \frac{\eta}{2} \sum_{\alpha\gamma\mu_1\mu_2} V_{\alpha\gamma\mu_1\mu_2} \underbrace{\langle T_\tau a_\gamma^\dagger(\tau) a_{\mu_2}(\tau) a_{\mu_1}(\tau) a_\alpha^\dagger(0) \rangle_\lambda}_{-\mathscr{C}^\lambda_{a_\gamma^\dagger a_{\mu_2} a_{\mu_1}, a_\alpha^\dagger}(\tau)} \Big|_{\tau=0^-} \\
&= -\frac{\eta}{2} \sum_{\alpha\gamma\mu_1\mu_2} V_{\alpha\gamma\mu_1\mu_2} \mathscr{C}^\lambda_{a_\gamma^\dagger a_{\mu_2} a_{\mu_1}, a_\alpha^\dagger}(\tau = 0^-).
\end{aligned}$$

On the other hand, equation (5.114) evaluated for $\beta = \alpha$ and the interaction λV and summed over α yields

$$\sum_\alpha \left[-\partial_\tau \mathscr{G}^\lambda_{\alpha\alpha}(\tau) - \sum_\gamma \xi_{\alpha\gamma} \mathscr{G}^\lambda_{\gamma\alpha}(\tau) - \delta(\tau) \right] = \lambda \sum_{\alpha\gamma\mu_1\mu_2} V_{\alpha\gamma\mu_1\mu_2} \mathscr{C}^\lambda_{a_\gamma^\dagger a_{\mu_2} a_{\mu_1}, a_\alpha^\dagger}(\tau),$$

such that

$$\langle V \rangle_\lambda = -\frac{\eta}{2\lambda} \sum_\alpha \left[-\partial_\tau \mathscr{G}^\lambda_{\alpha\alpha}(\tau) - \sum_\gamma \xi_{\alpha\gamma} \mathscr{G}^\lambda_{\gamma\alpha}(\tau) - \delta(\tau) \right]_{\tau=0^-} .$$

This, once inserted in the expression of Ω, proves equation (1.9). It is more convenient to write equation (1.9) in frequency space. Performing the Fourier transform, we have

$$\begin{aligned}
\langle V \rangle_\lambda &= -\frac{\eta}{2\lambda} \frac{1}{\beta} \sum_{i\nu_n} \sum_\alpha \left[i\nu_n \mathscr{G}^\lambda_{\alpha\alpha}(i\nu_n) - \sum_\gamma \xi_{\alpha\gamma} \mathscr{G}^\lambda_{\gamma\alpha}(i\nu_n) - 1 \right] e^{-i\nu_n 0^-} \\
&= -\frac{\eta}{2\lambda} \frac{1}{\beta} \sum_{i\nu_n} \sum_\alpha \left[\sum_\gamma (i\nu_n \delta_{\alpha\gamma} - \xi_{\alpha\gamma}) \mathscr{G}^\lambda_{\gamma\alpha}(i\nu_n) - 1 \right] e^{i\nu_n 0^+} \\
&= -\frac{\eta}{2\lambda} \frac{1}{\beta} \sum_{i\nu_n} \sum_\alpha \left\{ [\mathscr{G}_0^{-1}(i\nu_n) \mathscr{G}^\lambda(i\nu_n)]_{\alpha\alpha} - 1 \right\} e^{i\nu_n 0^+} \\
&= -\frac{\eta}{2\lambda} \frac{1}{\beta} \sum_{i\nu_n} \mathrm{Tr}[\mathscr{G}_0^{-1}(i\nu_n) \mathscr{G}^\lambda(i\nu_n) - \mathbb{1}] e^{i\nu_n 0^+}.
\end{aligned}$$

We have used equation (5.112) and adopted a matrix notation. Now we introduce the self-energy by means of Dyson's equation (5.31), which gives $\mathscr{G}_0^{-1}\mathscr{G}^\lambda = \mathbb{1} + \Sigma^\lambda\mathscr{G}^\lambda$ and consequently

$$\langle V\rangle_\lambda = -\frac{\eta}{2\lambda}\frac{1}{\beta}\sum_{i\nu_n}\mathrm{Tr}\,\Sigma^\lambda(i\nu_n)\mathscr{G}^\lambda(i\nu_n)e^{i\nu_n0^+}.$$

The resulting expression for the grand potential is (see also Abrikosov et al (1975, p 140))

$$\Omega = \Omega_0 - \frac{\eta}{2}\int_0^1\frac{d\lambda}{\lambda}\frac{1}{\beta}\sum_{i\nu_n}\mathrm{Tr}\,\Sigma^\lambda(i\nu_n)\mathscr{G}^\lambda(i\nu_n)e^{i\nu_n0^+}.$$

Remark that the calculations and results in the present document assume that $\Omega(\lambda)$ is a continuous function of λ with a well-defined derivative. This assumption reminds us of the hypothesis of adiabatic connection that underlies perturbation theory. In spite of its apparent generality, equation (1.9) breaks down if a phase transition occurs between $\lambda = 0$ and $\lambda = 1$ and induces a discontinuous derivative in $\Omega(\lambda)$.

Doc A5.20 Gor'kov equations

After performing the decoupling (5.115) and reshuffling the dummy indices, the last term in the equation of motion (5.114) of $\mathscr{G}_{\alpha\beta}(\tau)$ becomes

$$\begin{aligned}
&\sum_{\gamma\mu_1\mu_2} V_{\alpha\gamma\mu_1\mu_2}\Big[\langle a_\gamma^\dagger a_{\mu_2}\rangle\mathscr{G}_{\mu_1\beta}(\tau) + \eta\langle a_\gamma^\dagger a_{\mu_1}\rangle\mathscr{G}_{\mu_2\beta}(\tau) + \langle a_{\mu_2}a_{\mu_1}\rangle\mathscr{F}^\dagger_{\gamma\beta}(\tau)\Big]\\
&= \sum_\gamma\underbrace{\sum_{\mu_1\mu_2} V_{\alpha\mu_1\gamma\mu_2}\langle a_{\mu_1}^\dagger a_{\mu_2}\rangle}_{[V_\mathrm{H}]_{\alpha\gamma}}\mathscr{G}_{\gamma\beta}(\tau) + \sum_\gamma\eta\underbrace{\sum_{\mu_1\mu_2} V_{\alpha\mu_1\mu_2\gamma}\langle a_{\mu_1}^\dagger a_{\mu_2}\rangle}_{[V_\mathrm{x}]_{\alpha\gamma}}\mathscr{G}_{\gamma\beta}(\tau)\\
&+ \sum_\gamma\underbrace{\sum_{\mu_1\mu_2} V_{\alpha\gamma\mu_1\mu_2}\langle a_{\mu_2}a_{\mu_1}\rangle}_{\Delta_{\alpha\gamma}}\mathscr{F}^\dagger_{\gamma\beta}(\tau).
\end{aligned}$$

The mean fields V_H, V_x, and Δ are those reported in equation (5.119). With this, equation (5.114) becomes

$$\sum_\gamma(-\partial_\tau\delta_{\alpha\gamma} - \xi_{\alpha\gamma} - [V_\mathrm{H}]_{\alpha\gamma} - [V_\mathrm{x}]_{\alpha\gamma})\mathscr{G}_{\gamma\beta}(\tau) - \sum_\gamma\Delta_{\alpha\gamma}\mathscr{F}^\dagger_{\gamma\beta}(\tau) = \delta(\tau)\delta_{\alpha\beta}.$$

Moving to the frequency domain and using a matrix notation for the various quantities, we get

$$(i\nu_n\mathbb{1} - K_0 - V_\mathrm{H} - V_\mathrm{x})\mathscr{G}(i\nu_n) - \Delta\mathscr{F}^\dagger(i\nu_n) = \mathbb{1}.$$

This is the first Gor'kov equation displayed in equation (5.118). The notations V_H and V_x suggest that these quantities are the Hartree and exchange potentials, respectively. Let us see this in greater detail. From the definition of V_H, the representation of $a_{\mu_1}^\dagger$ in terms of the one-particle wave-function, $a_{\mu_1}^\dagger = \int dr\;\varphi_{\mu_1}(r)a_r^\dagger$ (see equation (2.91)), and the expression of the matrix element

$$V_{\alpha\mu_1\beta\mu_2} = \int d\mathbf{r}d\mathbf{r}' \; \varphi_\alpha^*(\mathbf{r})\varphi_{\mu_1}^*(\mathbf{r}')V(\mathbf{r}, \mathbf{r}')\varphi_\beta(\mathbf{r})\varphi_{\mu_2}(\mathbf{r}'),$$

(see equation (2.44)), we find

$$[V_\mathrm{H}]_{\alpha\beta} = \sum_{\mu_1\mu_2} \int d\mathbf{r}d\mathbf{r}'d\mathbf{r}''d\mathbf{r}''' \; \varphi_\alpha^*(\mathbf{r})\varphi_{\mu_1}^*(\mathbf{r}')V(\mathbf{r}, \mathbf{r}')\varphi_\beta(\mathbf{r})\varphi_{\mu_2}(\mathbf{r}') \; \varphi_{\mu_1}(\mathbf{r}'')\varphi_{\mu_2}^*(\mathbf{r}''')\langle a_{\mathbf{r}''}^\dagger a_{\mathbf{r}'''}\rangle.$$

The μ_1 and μ_2 sums yield closure relations: $\sum_{\mu_1}\varphi_{\mu_1}^*(\mathbf{r}')\varphi_{\mu_1}(\mathbf{r}'') = \delta(\mathbf{r}' - \mathbf{r}'')$ and similarly for μ_2 which gives $\delta(\mathbf{r}' - \mathbf{r}''')$. Hence we find

$$\begin{aligned}[V_\mathrm{H}]_{\alpha\beta} &= \int d\mathbf{r}d\mathbf{r}' \; \varphi_\alpha^*(\mathbf{r})V(\mathbf{r}, \mathbf{r}')\varphi_\beta(\mathbf{r})\underbrace{\langle a_{\mathbf{r}'}^\dagger a_{\mathbf{r}'}\rangle}_{n(\mathbf{r}')} \\ &= \int d\mathbf{r} \; \varphi_\alpha^*(\mathbf{r})\left[\int d\mathbf{r}' \; V(\mathbf{r}, \mathbf{r}')n(\mathbf{r}')\right]\varphi_\beta(\mathbf{r}) = \langle\alpha|V_\mathrm{H}|\beta\rangle.\end{aligned}$$

The quantity in brackets is just the classical potential produced at point $\mathbf{r}$ by a density of particles $n(\mathbf{r}')$ subject to the interaction $V(\mathbf{r}, \mathbf{r}')$. In the context of the Coulomb interaction, this is the Hartree potential. Following the same logic, one can check that V_x represents the exchange potential.

In order to complement the first Gor'kov equation, we must evaluate the equation of motion of the anomalous propagator $\mathscr{F}^\dagger_{\alpha\beta}(\tau)$. The latter is written in the main text, equation (5.117), as well as its decoupling by analogy with equation (5.115). Again, by reshuffling indices, we find that the last term takes the form

$$\begin{aligned}&-\sum_\gamma \underbrace{\sum_{\mu_1\mu_2} V^*_{\alpha\gamma\mu_1\mu_2}\langle a^\dagger_{\mu_1}a^\dagger_{\mu_2}\rangle}_{\Delta^*_{\alpha\gamma}} \mathscr{G}_{\gamma\beta}(\tau) - \sum_\gamma \underbrace{\eta\sum_{\mu_1\mu_2} V^*_{\alpha\mu_1\mu_2\gamma}\langle a^\dagger_{\mu_2}a_{\mu_1}\rangle}_{[V_\mathrm{x}]^*_{\alpha\gamma}} \mathscr{F}^\dagger_{\gamma\beta}(\tau) \\ &-\sum_\gamma \underbrace{\sum_{\mu_1\mu_2} V^*_{\alpha\mu_1\gamma\mu_2}\langle a^\dagger_{\mu_2}a_{\mu_1}\rangle}_{[V_\mathrm{H}]^*_{\alpha\gamma}} \mathscr{F}^\dagger_{\gamma\beta}(\tau).\end{aligned}$$

We thereby obtain in matrix form the second Gor'kov equation shown in equation (5.118):

$$\left(i\nu_n\mathbb{1} + K_0^* + V_\mathrm{H}^* + V_\mathrm{x}^*\right)\mathscr{F}^\dagger(i\nu_n) + \Delta^*\mathscr{G}(i\nu_n) = 0.$$

The mean fields can be related to the propagators $\mathscr{G}$ and $\mathscr{F}^\dagger$. Since

$$\langle a^\dagger_{\mu_1}a_{\mu_2}\rangle = \langle T_\tau a^\dagger_{\mu_1}(0)a_{\mu_2}(0^-)\rangle = \eta\langle T_\tau a_{\mu_2}(0^-)a^\dagger_{\mu_1}(0)\rangle = -\eta\mathscr{G}_{\mu_2\mu_1}(\tau = 0^-),$$

we obtain equations (5.119*a*) and (5.119*b*). On the other hand, we have

$$\langle a_{\mu_2}a_{\mu_1}\rangle = \langle a^\dagger_{\mu_1}a^\dagger_{\mu_2}\rangle^* = \langle T_\tau a^\dagger_{\mu_1}(0)a^\dagger_{\mu_2}(0^-)\rangle^* = \eta\langle T_\tau a^\dagger_{\mu_2}(0^-)a^\dagger_{\mu_1}(0)\rangle^* = -\eta\left[\mathscr{F}^\dagger_{\mu_2\mu_1}(\tau = 0^-)\right]^*,$$

which gives equation (5.119*c*).

Doc A5.21 Spin-singlet superconductor

In the real-space representation for fermions, the indices of the one-particle states are pairs of coordinates $(\boldsymbol{r}_i\sigma_i)$ for the position and spin of the electron and the wave functions are $\varphi_{\boldsymbol{r}_i\sigma_i}(\boldsymbol{r},\sigma)=\delta_{\sigma\sigma_i}\delta(\boldsymbol{r}-\boldsymbol{r}_i)$. The interaction matrix element therefore becomes, according to equation (2.44):

$$\begin{aligned}V_{\alpha\beta\gamma\delta}\to V_{\boldsymbol{r}_1\sigma_1\boldsymbol{r}_2\sigma_2\boldsymbol{r}_3\sigma_3\boldsymbol{r}_4\sigma_4}&=\int d\boldsymbol{r}\sum_{\sigma}\int d\boldsymbol{r}'\sum_{\sigma'}\\&\quad\times\delta_{\sigma\sigma_1}\delta(\boldsymbol{r}-\boldsymbol{r}_1)\delta_{\sigma'\sigma_2}\delta(\boldsymbol{r}'-\boldsymbol{r}_2)V(\boldsymbol{r},\boldsymbol{r}')\delta_{\sigma\sigma_3}\delta(\boldsymbol{r}-\boldsymbol{r}_3)\delta_{\sigma'\sigma_4}\delta(\boldsymbol{r}'-\boldsymbol{r}_4)\\&=\delta_{\sigma_1\sigma_3}\delta(\boldsymbol{r}_1-\boldsymbol{r}_3)\delta_{\sigma_2\sigma_4}\delta(\boldsymbol{r}_2-\boldsymbol{r}_4)V(\boldsymbol{r}_1,\boldsymbol{r}_2).\end{aligned}$$

The pairing field results from equation (5.119*c*):

$$\begin{aligned}\Delta_{\alpha\beta}\to\Delta_{\boldsymbol{r}\sigma\boldsymbol{r}'\sigma'}&=\int d\boldsymbol{r}_1\sum_{\sigma_1}\int d\boldsymbol{r}_2\sum_{\sigma_2}V_{\boldsymbol{r}\sigma\boldsymbol{r}'\sigma'\boldsymbol{r}_1\sigma_1\boldsymbol{r}_2\sigma_2}\langle c_{\boldsymbol{r}_2\sigma_2}c_{\boldsymbol{r}_1\sigma_1}\rangle\\&=V(\boldsymbol{r},\boldsymbol{r}')\langle c_{\boldsymbol{r}'\sigma'}c_{\boldsymbol{r}\sigma}\rangle=-V(\boldsymbol{r},\boldsymbol{r}')\langle c_{\boldsymbol{r}\sigma}c_{\boldsymbol{r}'\sigma'}\rangle.\end{aligned}$$

The assumption of spin-singlet pairing means that only electrons of opposite spins pair, in other words only $\Delta_{\uparrow\downarrow}$ and $\Delta_{\downarrow\uparrow}$ are nonzero. They are not independent though, since the anti-commutation rules of the fermions and the symmetry of the interaction $V(\boldsymbol{r},\boldsymbol{r}')=V(\boldsymbol{r}',\boldsymbol{r})$ imply

$$\Delta(\boldsymbol{r},\boldsymbol{r}')\equiv\Delta_{\boldsymbol{r}\uparrow\boldsymbol{r}'\downarrow}=-V(\boldsymbol{r},\boldsymbol{r}')\langle c_{\boldsymbol{r}\uparrow}c_{\boldsymbol{r}'\downarrow}\rangle=V(\boldsymbol{r}',\boldsymbol{r})\langle c_{\boldsymbol{r}'\downarrow}c_{\boldsymbol{r}\uparrow}\rangle=-\Delta_{\boldsymbol{r}'\downarrow\boldsymbol{r}\uparrow}.$$

For the same reason, only $\mathscr{F}^\dagger_{\uparrow\downarrow}$ and $\mathscr{F}^\dagger_{\downarrow\uparrow}$ are nonzero. The Gor'kov equations involve matrix products in the one-particle state indices. For instance

$$[\tilde{K}_0\mathscr{G}(i\omega_n)]_{\boldsymbol{r}\sigma\boldsymbol{r}'\sigma'}=\int d\boldsymbol{r}_1\sum_{\sigma_1}[\tilde{K}_0]_{\boldsymbol{r}\sigma\boldsymbol{r}_1\sigma_1}\mathscr{G}_{\boldsymbol{r}_1\sigma_1\boldsymbol{r}'\sigma'}(i\omega_n).$$

Here enters our assumption of a non-magnetic system: $[\tilde{K}_0]_{\boldsymbol{r}\sigma\boldsymbol{r}_1\sigma_1}=\delta_{\sigma\sigma_1}\tilde{K}_0$ and $\mathscr{G}_{\boldsymbol{r}_1\sigma_1\boldsymbol{r}'\sigma'}=\delta_{\sigma_1\sigma'}\mathscr{G}$. Hence

$$[\tilde{K}_0\mathscr{G}(i\omega_n)]_{\boldsymbol{r}\sigma\boldsymbol{r}'\sigma'}=\delta_{\sigma\sigma'}\int d\boldsymbol{r}_1\,\tilde{K}_0(\boldsymbol{r},\boldsymbol{r}_1)\mathscr{G}(\boldsymbol{r}_1,\boldsymbol{r}',i\omega_n).$$

Likewise

$$[\Delta\mathscr{F}^\dagger(i\omega_n)]_{\boldsymbol{r}\sigma\boldsymbol{r}'\sigma'}=\delta_{\sigma\sigma'}\int d\boldsymbol{r}_1\Delta_{\boldsymbol{r}\sigma\boldsymbol{r}_1-\sigma}\mathscr{F}^\dagger_{\boldsymbol{r}_1-\sigma\boldsymbol{r}'\sigma}(i\omega_n).$$

The first Gor'kov equation (5.118*a*) for $\sigma=\sigma'=\uparrow$ therefore becomes equation (5.133*a*). For the second we proceed similarly:

$$[\tilde{K}_0^* \mathscr{F}^\dagger(i\omega_n)]_{r\sigma r'\sigma'} = \delta_{\sigma,\,-\sigma'} \int dr_1\, \tilde{K}_0^*(\boldsymbol{r},\, \boldsymbol{r}_1) \mathscr{F}^\dagger_{r_1\sigma r'-\sigma}(i\omega_n)$$

$$\begin{aligned}[\Delta^* \mathscr{G}(i\omega_n)]_{r\sigma r'\sigma'} &= \delta_{\sigma,\,-\sigma'} \int dr_1 \Delta^*_{r\sigma r_1 -\sigma} \mathscr{G}(\boldsymbol{r}_1,\, \boldsymbol{r}',\, i\omega_n) \\ &= -\,\delta_{\sigma,\,-\sigma'} \int dr_1 \Delta^*_{r_1 -\sigma r\sigma} \mathscr{G}(\boldsymbol{r}_1,\, \boldsymbol{r}',\, i\omega_n),\end{aligned}$$

such that the second Gor'kov equation (5.118*b*) for $\sigma = -\sigma' = \downarrow$ reduces to equation (5.133*b*). Note the sign change in front of Δ^* with respect to equation (5.118). The gap equation is readily derived from the definitions:

$$\begin{aligned}\Delta(\boldsymbol{r},\, \boldsymbol{r}') &= -\,V(\boldsymbol{r},\, \boldsymbol{r}')\langle c^\dagger_{r'\downarrow} c^\dagger_{r\uparrow}\rangle^* = -V(\boldsymbol{r},\, \boldsymbol{r}')\langle T_\tau c^\dagger_{r'\downarrow}(0^+) c^\dagger_{r\uparrow}(0)\rangle^* \\ &= V(\boldsymbol{r},\, \boldsymbol{r}')\left[\mathscr{F}^\dagger_{\downarrow\uparrow}(\boldsymbol{r}',\, \boldsymbol{r},\, \tau = 0^+)\right]^* = V(\boldsymbol{r},\, \boldsymbol{r}')[\mathscr{F}^\dagger(\boldsymbol{r}',\, \boldsymbol{r},\, \tau = 0^+)]^*.\end{aligned}$$

In the Dyson-like formulation, the self-energy is diagonal in the spin indices and reads

$$\begin{aligned}\Sigma(\boldsymbol{r},\, \boldsymbol{r}',\, i\omega_n) &= \int dr_1 dr_2\; \Delta_{r\uparrow r_1\downarrow} \mathscr{G}_0(\boldsymbol{r}_2,\, \boldsymbol{r}_1,\, -i\omega_n) \Delta^*_{r_2\downarrow\, r'\uparrow} \\ &= -\int dr_1 dr_2\; \Delta(\boldsymbol{r},\, \boldsymbol{r}_1) \mathscr{G}_0(\boldsymbol{r}_2,\, \boldsymbol{r}_1,\, -i\omega_n) \Delta^*(\boldsymbol{r}',\, \boldsymbol{r}_2).\end{aligned}$$

This can be recast in matrix form as $\Sigma(i\omega_n) = -\Delta \mathscr{G}_0^T(-i\omega_n)\Delta^\dagger$, where $\mathscr{G}_0^T$ is the transpose of $\mathscr{G}_0$. In the Nambu formalism, we may write down the spin structure of the matrix explicitly:

$$\begin{pmatrix} i\omega_n - \tilde{K}_0 & 0 & 0 & -\Delta_{\uparrow\downarrow} \\ 0 & i\omega_n - \tilde{K}_0 & -\Delta_{\downarrow\uparrow} & 0 \\ 0 & \Delta^*_{\uparrow\downarrow} & i\omega_n + \tilde{K}_0^* & 0 \\ \Delta^*_{\downarrow\uparrow} & 0 & 0 & i\omega_n + \tilde{K}_0^* \end{pmatrix} \begin{pmatrix} \mathscr{G} & 0 & 0 & -\mathscr{F}_{\uparrow\downarrow} \\ 0 & \mathscr{G} & -\mathscr{F}_{\downarrow\uparrow} & 0 \\ 0 & \mathscr{F}^\dagger_{\uparrow\downarrow} & -\mathscr{G}^* & 0 \\ \mathscr{F}^\dagger_{\downarrow\uparrow} & 0 & 0 & -\mathscr{G}^* \end{pmatrix} = \mathbb{1},$$

and then reduce it again to a 'spinless' 2×2 system for $\mathscr{G}$, $\mathscr{G}^*$, $\mathscr{F}^\dagger \equiv \mathscr{F}^\dagger_{\downarrow\uparrow}$ and $\mathscr{F} \equiv -\mathscr{F}_{\uparrow\downarrow} = (\mathscr{F}^\dagger_{\downarrow\uparrow})^*$:

$$\begin{pmatrix} i\omega_n - \tilde{K}_0 & -\Delta \\ -\Delta^\dagger & i\omega_n + \tilde{K}_0^* \end{pmatrix} \begin{pmatrix} \mathscr{G} & \mathscr{F} \\ \mathscr{F}^\dagger & -\mathscr{G}^* \end{pmatrix} = \begin{pmatrix} 1 & 0 \\ 0 & 1 \end{pmatrix}.$$

This immediately leads us to the following Bogoliubov–de Gennes equations:

$$\begin{aligned}\int dr_1\left[\tilde{K}_0(\boldsymbol{r},\, \boldsymbol{r}_1) u_\gamma(\boldsymbol{r}_1) + \Delta(\boldsymbol{r},\, \boldsymbol{r}_1) v_\gamma(\boldsymbol{r}_1)\right] &= E_\gamma u_\gamma(\boldsymbol{r}) \\ \int dr_1\left[-\tilde{K}_0^*(\boldsymbol{r},\, \boldsymbol{r}_1) v_\gamma(\boldsymbol{r}_1) + \Delta^*(\boldsymbol{r}_1,\, \boldsymbol{r}) u_\gamma(\boldsymbol{r}_1)\right] &= E_\gamma v_\gamma(\boldsymbol{r}).\end{aligned}$$

Our last calculation here will be the derivation of the gap equation in the particular case where translation invariance is present. We must be careful to get the signs right. In case of translation invariance, the gap equation (5.134) is $\Delta(\boldsymbol{r}) = V(\boldsymbol{r})[\mathscr{F}^{\dagger}(-\boldsymbol{r}, \tau = 0^{+})]^{*}$. Fourier transforming, we get

$$\Delta_{\boldsymbol{k}} = \int d\boldsymbol{r}\ \Delta(\boldsymbol{r})e^{-i\boldsymbol{k}\cdot\boldsymbol{r}}$$

$$= \int d\boldsymbol{r} \underbrace{V(\boldsymbol{r})}_{\frac{1}{\mathscr{V}}\sum_{\boldsymbol{k}_1} V(\boldsymbol{k}_1)e^{i\boldsymbol{k}_1\cdot\boldsymbol{r}}} \Bigg[\underbrace{\mathscr{F}^{\dagger}(-\boldsymbol{r}, \tau = 0^{+})}_{\frac{1}{\mathscr{V}}\sum_{\boldsymbol{k}_2} \mathscr{F}^{\dagger}(\boldsymbol{k}_2, 0^{+})e^{-i\boldsymbol{k}_2\cdot\boldsymbol{r}}}\Bigg]^{*} e^{-i\boldsymbol{k}\cdot\boldsymbol{r}}.$$

The $\boldsymbol{r}$ integration yields $\mathscr{V}\delta_{\boldsymbol{k}_1+\boldsymbol{k}_2,\ \boldsymbol{k}}$. Therefore

$$\Delta_{\boldsymbol{k}} = \frac{1}{\mathscr{V}}\sum_{\boldsymbol{k}'} V(\boldsymbol{k} - \boldsymbol{k}')\Bigg[\underbrace{\mathscr{F}^{\dagger}(\boldsymbol{k}', \tau = 0^{+})}_{\frac{1}{\beta}\sum_{i\omega_n} \mathscr{F}^{\dagger}(\boldsymbol{k}', i\omega_n)e^{-i\omega_n 0^{+}}}\Bigg]^{*}$$

$$= \frac{1}{\mathscr{V}}\sum_{\boldsymbol{k}'} V(\boldsymbol{k} - \boldsymbol{k}')\frac{1}{\beta}\sum_{i\omega_n}[\mathscr{F}^{\dagger}(\boldsymbol{k}', i\omega_n)]^{*}e^{i\omega_n 0^{+}}.$$

The anomalous Green's function $\mathscr{F}^{\dagger}(\boldsymbol{k}, i\omega_n)$ is found by solving the algebraic equations (5.135) with the additional assumption $\xi_{-\boldsymbol{k}} = \xi_{\boldsymbol{k}}$:

$$\mathscr{F}^{\dagger}(\boldsymbol{k}, i\omega_n) = \frac{\Delta_{\boldsymbol{k}}^{*}}{i\omega_n + \xi_{\boldsymbol{k}}}\mathscr{G}(\boldsymbol{k}, i\omega_n) = \frac{\Delta_{\boldsymbol{k}}^{*}}{2E_{\boldsymbol{k}}}\left(\frac{1}{i\omega_n - E_{\boldsymbol{k}}} - \frac{1}{i\omega_n + E_{\boldsymbol{k}}}\right).$$

We get:

$$\Delta_{\boldsymbol{k}} = \frac{1}{\mathscr{V}}\sum_{\boldsymbol{k}'} V(\boldsymbol{k} - \boldsymbol{k}')\frac{\Delta_{\boldsymbol{k}'}}{2E_{\boldsymbol{k}'}}\frac{1}{\beta}\sum_{i\omega_n}\left(\frac{1}{i\omega_n - E_{\boldsymbol{k}'}} - \frac{1}{i\omega_n + E_{\boldsymbol{k}'}}\right)e^{-i\omega_n 0^{-}}.$$

The last step is to perform the frequency sums with the help of equation (17):

$$\frac{1}{\beta}\sum_{i\omega_n}\left(\frac{1}{i\omega_n - E_{\boldsymbol{k}'}} - \frac{1}{i\omega_n + E_{\boldsymbol{k}'}}\right)e^{-i\omega_n 0^{-}} = f(E_{\boldsymbol{k}'}) - f(-E_{\boldsymbol{k}'}) = -\tanh\left(\frac{\beta}{2}E_{\boldsymbol{k}'}\right).$$

This finally leads to equation (5.141).

Appendix B5

Bonus B5.1 One-dimensional electrons in a potential

The purpose of this document is to illustrate how the Green's function formalism can be put in action to painlessly solve the problem of non-interacting electrons in a local potential. Let us consider electrons in one dimension with the dispersion

$\xi_k = 2t\cos(ka) - \mu$. This corresponds to a tight-binding model with only nearest-neighbor hopping. We shall use the hopping energy as our unit of energy ($t \equiv 1$) and the lattice parameter as our unit of length ($a \equiv 1$). Furthermore, we shall assume that the system is initially half-filled (one electron per site), which is equivalent to setting $\mu = 0$ such that the chemical potential lies at the center of the band; the band extends from $-2t$ to $+2t$. The density of states (DOS) of such a system is easily deduced from equation (2.55) or equivalently equation (5.7), see also section 9.4.1.1:

$$N(\varepsilon) = \frac{1}{\pi t}\,\mathrm{Re}\left(\frac{1}{\sqrt{1-\left(\frac{\varepsilon+\mu}{2t}\right)^2}}\right).$$

To this uniform one-dimensional electron gas, we add a local potential $V(r)$. The translation invariance gets destroyed and the above uniform DOS becomes a spatially-varying local density of states (LDOS) $N(r, \varepsilon)$ given by equation (5.8). In order to calculate the LDOS, we must obtain the Green's function in real space using equation (5.18). The first step is to build the unperturbed retarded Green's function $G_0^R(r, r', \varepsilon)$, which is the analytic continuation of $\mathcal{G}_0(r, r', i\omega_n)$. For this we exploit the translation invariance of the system in the absence of potential, which means that

$$G_0^R(r, r', \varepsilon) = G_0^R(r - r', \varepsilon) = \frac{1}{\mathcal{V}}\sum_k \frac{e^{ik(r-r')}}{\varepsilon + i0^+ - \xi_k}.$$

Hence by Fourier transforming $1/(\varepsilon + i0^+ - \xi_k)$ we immediately get $G_0^R(r, \varepsilon)$, which we can then use to build the matrix $G_0^R(r, r', \varepsilon)$. The matrix $V_{rr'}$, on the other hand, is just the diagonal matrix with $V(r)$ on the diagonal: $V_{rr'} = \delta_{rr'}V(r)$. With this we have all ingredients needed in order to evaluate $N(r, \varepsilon)$.

The calculation is in fact very easily implemented, as shown in figure 5.13 (we used MATLAB® in this case). We consider a system of $N = 512$ sites closed with periodic boundary conditions. Line 1 defines the dispersion ξ_k as a vector of length N built from the discrete momenta $k_j = 2\pi j/N$. In MATLAB®, the syntax `[0:N-1]` returns the vector $(0, 1, \ldots, N-1)$. At line 2 we define the potential: we use a Gaussian of strength `V0=-3` and width `r0=0.05/N` with its minimum in the middle

```
1 N=512; xi=2*cos(2*pi*(0:N-1)/N);
2 V0=-3; r0=0.05; V=V0*exp(-((-N/2:N/2-1)/N/r0).^2/2);

3 w=(-5.1:0.01:2.1)'; for l=1:size(w)
4   g0=fft(1./(w(l)+i*12/N-xi))/N;
5   for r=1:N; for s=1:N; g(r,s)=g0(abs(r-s)+1); end; end
6   g=g*inv(eye(N)-diag(V)*g); LDOS(l,:)=-2*imag(diag(g))/pi;
7 end
8 for r=1:N; density(r)=sum(LDOS(:,r).*(w<=0)); end; density=0.01*density';

9 save LDOS.dat LDOS -ascii; save n.dat density -ascii
```

Figure 5.13. Calculation of the local density of states $N(r, \varepsilon)$ and electron density $n(r)$ for one-dimensional electrons in a local potential, using MATLAB®.

of the system. At line 3 we initialize a vector containing the energies at which the LDOS is to be calculated and we start the loop over the energies `w(l)`, a loop which is closed at line 7. Inside the loop, the first step is to calculate $G_0^R(r, \varepsilon)$ by Fourier transforming $1/(\varepsilon + i0^+ - \xi_k)$. In order to obtain a smooth LDOS (as a function of ε), the value used for 0^+ must be somewhat larger than the typical spacing of the levels ξ_k. In our case, we have $N/2$ levels (since $\xi_{-k} = \xi_k$) distributed over a bandwidth of $4t$, hence the typical inter-level spacing is $8t/N$. We use the value $12t/N$ for 0^+. On line 5 the matrix $G_0^R(r, r', \varepsilon)$ is built and stored in `g(r,s)`. On line 6 the Green's function is calculated following equation (5.18); instead of $G = (G_0^{-1} - V)^{-1}$, we use the equivalent form $G = G_0(\mathbb{1} - VG_0)^{-1}$ that requires only one matrix inversion. In MATLAB®, `inv(...)` is the function to invert a matrix and `eye(N)` is the $N \times N$ identity matrix. The second instruction on line 6 evaluates the LDOS following equation (5.8) and stores it in the array `LDOS`. On line 8 we also calculate the zero-temperature density using equation (3.38):

$$n(r) = \int_{-\infty}^{\infty} d\varepsilon\, \rho_{a_r a_r^\dagger}(\varepsilon) f(\varepsilon) = \int_{-\infty}^{\infty} d\varepsilon\, N(r, \varepsilon) f(\varepsilon).$$

In MATLAB®, the syntax `(w<=0)` returns a vector with the same length as the length of the vector `w` whose elements are either one or zero, depending on whether the corresponding element of `w` satisfies the inequality `w(l)<=0` or not: thus `(w<=0)` simply returns $f(\varepsilon)$ at zero temperature. Finally, on line 9 we save the LDOS and the density in the files `LDOS.dat` and `n.dat`. Running this code requires ~200 Mb of memory (including MATLAB's own memory), and takes a couple of minutes on a laptop.

The resulting LDOS is displayed in figure 5.14(a). The main effect of the potential is to shift locally the whole electronic spectrum by the value of the potential $V(r)$. This effect is the essence of the Thomas–Fermi approximation shown in figure 5.14(b):

$$N_{\mathrm{TF}}(r, \varepsilon) = N(\varepsilon - V(r)) = \frac{1}{\pi t} \mathrm{Re}\left(\frac{1}{\sqrt{1 - \left(\frac{\varepsilon + \mu - V(r)}{2t}\right)^2}}\right).$$

In addition to this 'macroscopic' effect, we see the formation of bound states in the potential well. As expected, the lowest bound state is nodeless with the maximum at the center of the well, the second has one node at the center, etc. We also see the quantum interferences between waves incoming and reflected, either outside or inside the well. Bound states and interferences are of course absent in the Thomas–Fermi approximation. The strength of the Green's function formalism appears very clearly here, as we could obtain these well-known quantum effects without using the wave-function machinery and solving the Schrödinger equation. Note also that the spectrum is 'less dense' inside the potential well due to the presence of bound states and the resulting electron density is lower than the value expected from the Thomas–Fermi approximation, namely $n(r) = 1 - (2/\pi)\mathrm{Re}\{\sin^{-1}[(V(r) - \mu)/(2t)]\}$.

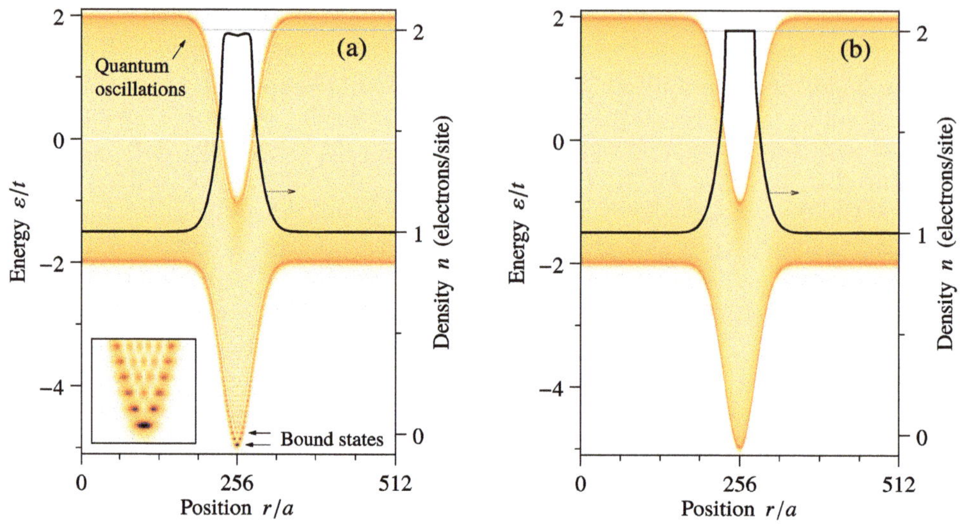

Figure 5.14. (a) Local density of states (LDOS) for one-dimensional electrons in a local potential (intensity-map, left scale). Darker regions denote larger values of the LDOS. Corresponding electron density (black line, right scale). (b) Same quantities calculated within the Thomas–Fermi approximation (see text).

Bonus B5.2 Density of states of BCS superconductors

Combining equations (5.7) and (5.136), we see that the DOS per unit volume of a BCS superconductor characterized by a normal-state dispersion $\xi_k = \xi_{-k}$ and a superconducting gap Δ_k is

$$N_{\mathrm{BCS}}(\varepsilon) = \frac{1}{\mathscr{V}}\sum_{k\sigma}\left(-\frac{1}{\pi}\right)\mathrm{Im}\left(\frac{1}{\varepsilon + i0^+ - \xi_k - \dfrac{|\Delta_k|^2}{\varepsilon + i0^+ + \xi_k}}\right).$$

Replacing 0^+ by a phenomenological constant scattering rate Γ leads to the so-called 'Dynes formula'[19]. For a numerical evaluation, one can directly use this equation, taking advantage of the symmetries in ξ_k and $|\Delta_k|$ for a better performance. Figure 5.15 gives a minimal implementation in MATLAB®—not exploiting these symmetries. The case considered is a two-dimensional tight-binding band, $\xi_k = 2t(\cos k_x + \cos k_y) - \mu$ with $t = -1$ and $\mu = 1$, the value of μ being arbitrarily chosen in order to break particle–hole symmetry. Line 2 builds the mesh of $\boldsymbol{k}$ points and the band ξ_k using the MATLAB® command `meshgrid`. We use a dense 2048 × 2048 mesh in order to achieve a good energy resolution. Lines 3–5 set up the array $|\Delta_k|^2$ for three possible gap symmetries: s symmetry with $\Delta_k = \Delta$; $d_{x^2-y^2}$ symmetry with $\Delta_k = \frac{\Delta}{2}(\cos k_x - \cos k_y)$; d_{xy} symmetry with $\Delta_k = \Delta \sin k_y \sin k_y$. Line 6 defines

[19] Dynes R C, Narayanamurti V and Garno J P 1978 *Phys. Rev. Lett.* **41** 1509.

```
1 t=-1; mu=1; D=0.75; N=2048;
2 [kx ky]=meshgrid(2*pi*(0:N-1)/N); xi=2*t*(cos(kx)+cos(ky))-mu;
3 D2=abs(D)^2;                          % use this for s-wave
4 D2=abs(D/2*(cos(kx)-cos(ky))).^2;     % use this for d-wave (x^2-y^2)
5 D2=abs(D*(sin(kx).*sin(ky))).^2;      % use this for d-wave (xy)

6 e=(-3*D:6*D/1000:3*D)'; i0=i*8/N;
7 for l=1:size(e,1);
8     e(l,2)=-sum(sum(imag(1./(e(l,1)+i0-xi-D2./(e(l,1)+i0+xi)))))/(pi*N*N);
9 end

10 save DOS.dat e -ascii
```

Figure 5.15. Calculation of the density of states (DOS) for two-dimensional BCS superconductors with s, $d_{x^2-y^2}$, and d_{xy} gap symmetries, using MATLAB®.

the vector of energies ε as well as the quantity `i0` representing $i0^+$. We use the value $8/N$ for 0^+: smaller values lead to spurious oscillations in the DOS while higher values broaden the DOS more than necessary. Lines 7–9 perform the loop over ε and the $\boldsymbol{k}$-sum for each ε. Finally, line 10 saves the data to file `DOS.dat`.

Results for the three gap symmetries are displayed in figure 5.16. For the s-wave symmetry there is a true gap of width 2Δ surrounded by two *coherence peaks*, which are square-root singularities (see below). The slight rounding at the gap edges reflects the finite value used for 0^+. The two additional peaks correspond to the Van Hove singularity of the normal-state DOS and are logarithmic singularities, weaker than the coherence peaks. The Van Hove singularity is due to the saddle point at $\boldsymbol{k} = (\pi/a,\, 0)$ and therefore appears at energy $\xi_{(\pi/a,\,0)} = -\mu$, i.e. -1 in our case. In the superconducting state, this is pushed to $-[\xi^2_{(\pi/a,\,0)} + \Delta^2]^{1/2} = -1.25$; furthermore, an 'echo' appears at $+1.25$ due to the particle–hole mixing of the Bogoliubov excitations (see section 5.2.2.3.5). The relative weights of the main Van Hove peak and the echo peak are controlled by the *coherence factors* $u_{\boldsymbol{k}}^2$ and $v_{\boldsymbol{k}}^2$ appearing in equation (5.139).

For d-wave pairing three main differences can be seen. First, there is no gap in the DOS but $N(\varepsilon)$ vanishes linearly like $|\varepsilon|$, showing that the gap has *nodes* on the Fermi surface, i.e. points where $\Delta_{\boldsymbol{k}_{\rm F}} = 0$; since, by definition of the Fermi surface $\xi_{\boldsymbol{k}_{\rm F}} = 0$, the excitation energy $E_{\boldsymbol{k}}$ vanishes at these points. As the insets show, the nodes are located along the zone diagonal for the $d_{x^2-y^2}$ symmetry and along the zone boundary for the d_{xy} symmetry. The linear increase of the DOS close to $\varepsilon = 0$ is due to the linear increase of $|\Delta_{\boldsymbol{k}}|$ near the nodes. Second, the coherence peaks—which are now weak logarithmic rather than strong square-root singularities—are not located at $\pm\Delta$. They appear at the energy corresponding to the largest value of $|\Delta_{\boldsymbol{k}}|$ along the Fermi surface. As illustrated in the insets, this maximum is always smaller than Δ. Third, the renormalization of the Van Hove singularity, which is controlled by the gap value at $\boldsymbol{k} = (\pi/a,\, 0)$, depends on the symmetry. For $d_{x^2-y^2}$ symmetry, the gap is maximum at this point and the renormalization is therefore the same as for the s symmetry. In contrast, the gap vanishes at $(\pi/a,\, 0)$ for the d_{xy} symmetry and there is no renormalization. For the same reason, there is no echo of the Van Hove peak in the d_{xy} case, as the coherence factor $v_{\boldsymbol{k}}^2$ vanishes at $(\pi/a,\, 0)$.

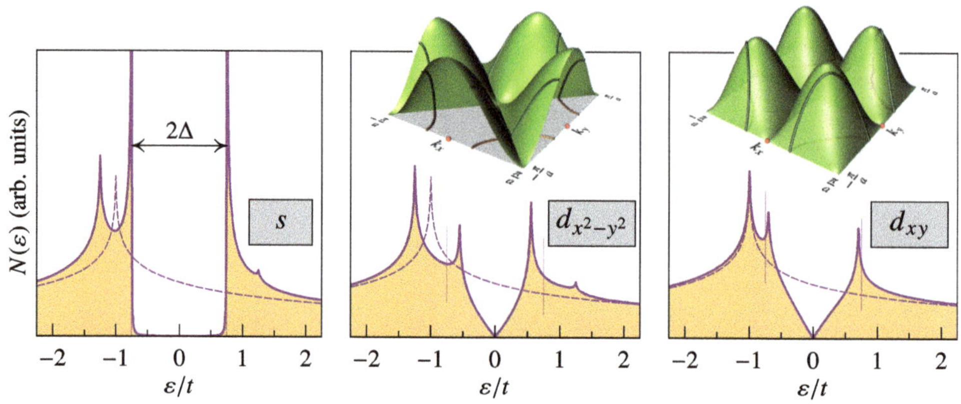

Figure 5.16. Density of states $N(\varepsilon)$ for two-dimensional superconductors with s, $d_{x^2-y^2}$, and d_{xy} pairing gaps. The dashed lines show the normal-state DOS with its Van Hove singularity at $\varepsilon/t = -\mu/t = -1$. The vertical lines indicate the maximum gap $\pm\Delta$ and do not coincide with the coherence peaks for d-wave pairing (see text). The insets show $|\Delta_{\boldsymbol{k}}|$ in the Brillouin zone (green surface) together with the Fermi surface (brown lines on the basal plane) and the value of $|\Delta_{\boldsymbol{k}}|$ along the Fermi surface (dark green lines). The red dots show $\boldsymbol{k} = (\pi/a, 0)$ and symmetry-equivalent points.

For s-wave pairing symmetry, $\Delta_{\boldsymbol{k}} \equiv \Delta$, the BCS DOS can be related exactly to the normal-state DOS $N_0^{\mathrm{el}}(\varepsilon)$. To show this, we define the even and odd parts $N_0^+(\varepsilon)$ and $N_0^-(\varepsilon)$ of the normal DOS,

$$N_0^{\pm}(\varepsilon) = \frac{1}{2}[N_0^{\mathrm{el}}(\varepsilon) \pm N_0^{\mathrm{el}}(-\varepsilon)]$$

and we use equation (5.140):

$$N_{\mathrm{BCS}}(\varepsilon) = \frac{1}{\mathcal{V}}\sum_{\boldsymbol{k}\sigma}\left[\frac{1}{2}\left(1 + \frac{\xi_{\boldsymbol{k}}}{E_{\boldsymbol{k}}}\right)\delta(\varepsilon - E_{\boldsymbol{k}}) + \frac{1}{2}\left(1 - \frac{\xi_{\boldsymbol{k}}}{E_{\boldsymbol{k}}}\right)\delta(\varepsilon + E_{\boldsymbol{k}})\right].$$

Since $E_{\boldsymbol{k}} = \sqrt{\xi_{\boldsymbol{k}}^2 + \Delta^2}$ is a function of $\xi_{\boldsymbol{k}}$ and not of $\boldsymbol{k}$ alone, we can rewrite the $\boldsymbol{k}$ sum as an energy integral:

$$N_{\mathrm{BCS}}^{(s\text{-wave})}(\varepsilon) = \int_{-\infty}^{\infty} d\xi\, N_0^{\mathrm{el}}(\xi)\left[\frac{1}{2}\left(1 + \frac{\xi}{E_\xi}\right)\delta(\varepsilon - E_\xi) + \frac{1}{2}\left(1 - \frac{\xi}{E_\xi}\right)\delta(\varepsilon + E_\xi)\right]$$

with $E_\xi = \sqrt{\xi^2 + \Delta^2}$. Now, since $N_0^{\mathrm{el}}(\xi) = N_0^+(\xi) + N_0^-(\xi)$, the integral can be split into

$$N_{\mathrm{BCS}}^{(s\text{-wave})}(\varepsilon) = \int_0^{\infty} d\xi N_0^+(\xi)\big[\delta(\varepsilon - E_\xi) + \delta(\varepsilon + E_\xi)\big] + \int_0^{\infty} d\xi N_0^-(\xi)\frac{\xi}{E_\xi}\big[\delta(\varepsilon - E_\xi) - \delta(\varepsilon + E_\xi)\big].$$

When evaluating the delta functions with the help of equation (11), we can use the fact that $\xi > 0$: $\delta(\varepsilon - E_\xi) = \delta(\xi - \xi_0)/|E'_{\xi_0}|$ with ξ_0 the positive solution of $\varepsilon - E_{\xi_0} = 0$, namely $\xi_0 = \sqrt{\varepsilon^2 - \Delta^2}$. Clearly $|E'_{\xi_0}| = \xi_0/|\varepsilon|$ and the delta function cannot be satisfied if $\varepsilon < \Delta$. Hence

$$\delta(\varepsilon - E_\xi) = \theta(\varepsilon - \Delta)\frac{|\varepsilon|}{\xi_0}\delta(\xi - \xi_0).$$

Similarly

$$\delta(\varepsilon + E_\xi) = \theta(-\varepsilon - \Delta)\frac{|\varepsilon|}{\xi_0}\delta(\xi - \xi_0).$$

Summing the two delta functions, we can write

$$\delta(\varepsilon - E_\xi) + \delta(\varepsilon + E_\xi) = \mathrm{Re}\left(\frac{|\varepsilon|}{\sqrt{\varepsilon^2 - \Delta^2}}\right)\delta(\xi - \xi_0)$$

because taking the real part has the same effect as the sum of the two step functions. The difference, on the other hand, reads

$$\begin{aligned}\frac{\xi}{E_\xi}\left[\delta(\varepsilon - E_\xi) - \delta(\varepsilon + E_\xi)\right] &= \theta(\varepsilon - \Delta)\frac{|\varepsilon|}{\varepsilon}\delta(\xi - \xi_0) - \theta(-\varepsilon - \Delta)\frac{|\varepsilon|}{-\varepsilon}\delta(\xi - \xi_0)\\ &= [\theta(\varepsilon - \Delta) - \theta(-\varepsilon - \Delta)]\delta(\xi - \xi_0).\end{aligned}$$

Collecting the particle–hole symmetric and anti-symmetric contributions, we find:

$$\begin{aligned}N_{\mathrm{BCS}}^{(s\text{-wave})}(\varepsilon) = &\, N_0^+\left(\sqrt{\varepsilon^2 - \Delta^2}\right)\mathrm{Re}\left(\frac{|\varepsilon|}{\sqrt{\varepsilon^2 - \Delta^2}}\right)\\ &+ N_0^-\left(\sqrt{\varepsilon^2 - \Delta^2}\right)[\theta(\varepsilon - \Delta) - \theta(-\varepsilon - \Delta)].\end{aligned} \tag{5.145}$$

The DOS vanishes for $|\varepsilon| < \Delta$ and diverges as $1/\sqrt{\varepsilon^2 - \Delta^2}$ at the gap edges. For a constant normal-state DOS $N_0^{\mathrm{el}}(\varepsilon) = N_0^{\mathrm{el}}(0)$, this reduces to equation (5.137) because $N_0^-(\varepsilon) = 0$ and $N_0^+(\varepsilon) = N_0^{\mathrm{el}}(\varepsilon) = N_0^{\mathrm{el}}(0)$.

For d-wave pairing symmetry, analytical calculations are less easy because $E_{\boldsymbol{k}}$ depends on $\boldsymbol{k}$ through both $\xi_{\boldsymbol{k}}$ and $\Delta_{\boldsymbol{k}}$. Progress is possible for example in two dimensions if one assumes cylindrical symmetry, i.e. $\xi_{\boldsymbol{k}} = \xi_k$ and $\Delta_{\boldsymbol{k}} = \Delta\cos 2\vartheta$, where $\boldsymbol{k} = k(\cos\vartheta, \sin\vartheta)$. This defines a generic model with $d_{x^2-y^2}$ gap symmetry. The DOS follows from equations (5.7) and (5.136) after rewriting the sum as an integral:

$$N_{\mathrm{BCS}}^{(d\text{-wave})}(\varepsilon) = \left(-\frac{1}{\pi}\right)\mathrm{Im}\frac{2}{(2\pi)^2}\int_0^\infty dk\, k\, \underbrace{\int_0^{2\pi} d\vartheta \frac{\varepsilon + i0^+ + \xi_k}{(\varepsilon + i0^+)^2 - \xi_k^2 - (\Delta\cos 2\vartheta)^2}}_{2\pi\frac{\varepsilon + i0^+ + \xi_k}{\sqrt{(\varepsilon + i0^+)^2 - \xi_k^2}\sqrt{(\varepsilon + i0^+)^2 - \xi_k^2 - \Delta^2}}}.$$

We see that after performing the angle integration we fall back to an expression that only depends on $\boldsymbol{k}$ through ξ_k, such that it is again possible to relate it exactly to the normal-state DOS as we did for s-wave symmetry:

$$N_{\mathrm{BCS}}^{(d\mathrm{-wave})}(\varepsilon) = \int_{-\infty}^{\infty} d\xi\, N_0^{\mathrm{el}}(\xi)\left(-\frac{1}{\pi}\right)\mathrm{Im}\frac{\varepsilon + i0^+ + \xi}{\sqrt{(\varepsilon + i0^+)^2 - \xi^2}\sqrt{(\varepsilon + i0^+)^2 - \xi^2 - \Delta^2}}.$$

For a flat DOS $N_0^{\mathrm{el}}(\xi) = N_0^{\mathrm{el}}(0)$, the term proportional to ξ in the integrand is odd and does not contribute. The remaining term may be evaluated thanks to the following identity, valid in the complex plane where $\mathrm{Im}\, z \neq 0$:

$$\int_{-\infty}^{\infty} d\xi \frac{z}{\sqrt{z^2 - \xi^2}\sqrt{z^2 - \xi^2 - \Delta^2}} = -2i\ \mathrm{sign}(\mathrm{Im}\, z)K\left(\frac{\Delta^2}{z^2}\right).$$

$K(z)$ is the complete elliptic integral of the fist kind. We deduce the DOS model first obtained by Won and Maki[20]:

$$N_{\mathrm{BCS}}^{(d\mathrm{-wave})}(\varepsilon) = N_0^{\mathrm{el}}(0)\frac{2}{\pi}\mathrm{Re}\, K\left(\frac{\Delta^2}{\varepsilon^2}\right). \tag{5.146}$$

This function is displayed in figure 5.11(c).

Bonus B5.3 Eliashberg equations

Perturbation about the BCS ground state The Eliashberg formalism is a self-consistent theory of pairing in superconductors which, unlike the BCS theory, can take into account retardation effects implied by the time dependence of the pairing interaction. It is well known that the phenomenon of pairing cannot be explained by a perturbation expansion around the non-interacting Fermi sea. This is manifested for instance in equation (5.143), which shows that the BCS gap cannot be expanded as a power series at the non-interacting point $V_0 = 0$. The reason is that the Fermi sea and the BCS ground state are not adiabatically connected. From the perspective of the low-lying excitations, it means that the electrons—or the Landau quasi-particles—are not well-defined excitations in a superconductor. In contrast, the pairing problem becomes a simple non-interacting problem once formulated in terms of the good excitations, which are the Bogoliubov quasi-particles of the superconductor. Therefore the latter formulation in terms of Bogoliubov quasi-particles is a good starting point in order to investigate retardation effects using perturbation theory. The Nambu formalism[21] is a neat way to implement the Bogoliubov transformation that brings out these new quasi-particles. The trick is to introduce *Nambu spinors*, which are pairs of an up-spin electron and a down-spin hole in the same orbital state $|\alpha\rangle$: $\gamma_\alpha^\dagger = \left(c_{\alpha\uparrow}^\dagger\ c_{\alpha\downarrow}\right)$. One then introduces the corresponding 2×2 Nambu matrix Green's function:

[20] Won H and Maki K 1994 *Phys. Rev.* **B 49** 1397.

[21] Nambu Y 1960 *Phys. Rev.* **117** 648.

$$\hat{\mathscr{G}}_{\alpha\beta}(\tau) = -\langle T_\tau \gamma_\alpha(\tau)\gamma_\beta^\dagger(0)\rangle = -\langle T_\tau \begin{pmatrix} c_{\alpha\uparrow}(\tau) \\ c_{\alpha\downarrow}^\dagger(\tau) \end{pmatrix} \begin{pmatrix} c_{\beta\uparrow}^\dagger(0) & c_{\beta\downarrow}(0) \end{pmatrix}\rangle$$
$$= \begin{pmatrix} -\langle T_\tau c_{\alpha\uparrow}(\tau)c_{\beta\uparrow}^\dagger(0)\rangle & -\langle T_\tau c_{\alpha\uparrow}(\tau)c_{\beta\downarrow}(0)\rangle \\ -\langle T_\tau c_{\alpha\downarrow}^\dagger(\tau)c_{\beta\uparrow}^\dagger(0)\rangle & -\langle T_\tau c_{\alpha\downarrow}^\dagger(\tau)c_{\beta\downarrow}(0)\rangle \end{pmatrix}. \tag{5.147}$$

Note that the matrix $\hat{\mathscr{G}}$ introduced here is the same object as the one introduced in equation (5.122), but specialized here to the case of a spin-singlet superconductor. The fact that in the usual BCS theory—often called 'weak-coupling BCS theory'—the excitations are independent Bogoliubov quasi-particles is manifested by the fact that the Nambu matrix $\hat{\mathscr{G}}$ has the same form as the Green's function of free fermions (see equation (5.124)), except for its 2×2 structure. Thus the 2×2 Nambu structure is the price to pay in order to move us from the part of the Hilbert space surrounding the Fermi sea to the part surrounding the BCS ground state.

A crucial property of the Nambu matrix $\hat{\mathscr{G}}$ is that it has a diagrammatic expansion identical to that of the usual electron Green's function $\mathscr{G}$, except that the electron propagators in the diagrams must be understood as 2×2 Nambu matrices and the interaction vertices carry additional Pauli matrices. One verifies this by extending the perturbation theory developed in section 5.1.3.4 in order to include diagrams involving propagators of the kind $\langle a^\dagger a^\dagger\rangle$ and $\langle aa\rangle$. This implies working with both the usual fermion propagator $\mathscr{G}$ and the anomalous propagator $\mathscr{F}^\dagger$ of equation (5.116), which are both nonzero in the superconducting state[22]. The two expansions for $\mathscr{G}$ and $\mathscr{F}^\dagger$ can then be recast in the 2×2 Nambu matrix form. Clearly, there are only two independent objects in this theory, $\mathscr{G}$ and $\mathscr{F}^\dagger$; therefore there are only two independent elements in the Nambu matrix $\hat{\mathscr{G}}$. One sees that the component 11 of the matrix equation (5.147) is just $\mathscr{G}_{\alpha\beta}(\tau)$ for up spins, while the component 21 is $\mathscr{F}^\dagger_{\alpha\beta}(\tau)$ as defined for instance in equation (5.132). In a paramagnetic state, the two remaining components are simply related to the two first. We indeed have

$$\left[\hat{\mathscr{G}}_{\alpha\beta}(\tau)\right]_{11} \equiv \mathscr{G}_{\alpha\beta}(\tau) = -\left[\hat{\mathscr{G}}_{\alpha\beta}(-\tau)\right]_{22}^{*} = -\left[\hat{\mathscr{G}}_{\alpha\beta}(-\tau)\right]_{22} \tag{5.148a}$$

$$\left[\hat{\mathscr{G}}_{\alpha\beta}(\tau)\right]_{21} \equiv \mathscr{F}^\dagger_{\alpha\beta}(\tau) = \left[\hat{\mathscr{G}}_{\beta\alpha}(\tau)\right]_{12}^{*} = \left[\hat{\mathscr{G}}_{\beta\alpha}(\tau)\right]_{12}. \tag{5.148b}$$

For checking these relations, one has to remember that $\hat{\mathscr{G}}_{\alpha\beta}(\tau)$ is real. This results directly from the definition of the correlation function and from the fact that creation and annihilation operators can be represented by real-valued matrices. Furthermore, one has to be careful that $[c_\alpha(\tau)]^\dagger = c_\alpha^\dagger(-\tau)$, as is clear from the definition of the imaginary-time evolution in equation (4.3). In frequency space, these relations become

[22] Gorkov L P 1958 *Sov. Phys. JETP* **7** 505.

$$\left[\hat{\mathscr{G}}_{\alpha\beta}(i\omega_n)\right]_{11} = -\left[\hat{\mathscr{G}}_{\alpha\beta}(i\omega_n)\right]^*_{22} = -\left[\hat{\mathscr{G}}_{\alpha\beta}(-i\omega_n)\right]_{22} \tag{5.149a}$$

$$\left[\hat{\mathscr{G}}_{\alpha\beta}(i\omega_n)\right]_{21} = \left[\hat{\mathscr{G}}_{\beta\alpha}(-i\omega_n)\right]^*_{12} = \left[\hat{\mathscr{G}}_{\beta\alpha}(i\omega_n)\right]_{12}. \tag{5.149b}$$

Consider now that the electrons interact via a generic retarded potential $V(\boldsymbol{q}, i\Omega_n)$. In conventional superconductors, this interaction is the sum of the screened Coulomb interaction (5.90) and the phonon-mediated interaction (5.54):

$$V(\boldsymbol{q}, i\Omega_n) = W(\boldsymbol{q}, i\Omega_n) + \sum_{\lambda} |g_{\boldsymbol{q}\lambda}|^2 \mathscr{D}^0_\lambda(\boldsymbol{q}, i\Omega_n). \tag{5.150}$$

Both contain retardation effects. In other situations, the interaction could be due to the exchange of spin-waves. This interaction leads to pairing, an effect that is automatically taken into account at the mean-field level via the Nambu formalism. The time dependence of the interaction furthermore leads to renormalization and damping of the Bogoliubov quasi-particles: this effect will be treated by perturbation theory. We specialize to a translation-invariant case, such that all matrices become diagonal in the $\alpha\beta$ indices, the latter being replaced by a wave vector $\boldsymbol{k}$. The matrix $\hat{\mathscr{G}}$ satisfies a Dyson equation in Nambu space:

$$\hat{\mathscr{G}}^{-1}(\boldsymbol{k}, i\omega_n) = \underbrace{\hat{\mathscr{G}}_0^{-1}(\boldsymbol{k}, i\omega_n)}_{\begin{pmatrix} i\omega_n - \xi_{\boldsymbol{k}} & 0 \\ 0 & i\omega_n + \xi_{\boldsymbol{k}} \end{pmatrix}} - \hat{\Sigma}(\boldsymbol{k}, i\omega_n). \tag{5.151}$$

In the Eliashberg approximation, the self-energy is given by the self-consistent exchange diagram[23]:

$$\hat{\Sigma}(\boldsymbol{k}, i\omega_n) =$$  $$= -\frac{1}{\beta\mathscr{V}} \sum_{\boldsymbol{q}i\Omega_n} V(\boldsymbol{q}, i\Omega_n)\, \tau^\mu \hat{\mathscr{G}}(\boldsymbol{k} - \boldsymbol{q}, i\omega_n - i\Omega_n)\tau^\mu. \tag{5.152}$$

The dashed line represents the effective pairing interaction $V(\boldsymbol{q}, i\Omega_n)$, the double-dashed line is the matrix $\hat{\mathscr{G}}$, and the gray circles represent Pauli matrices. For the Coulomb and phonon-mediated interactions, the appropriate Pauli matrices are τ^z ($\mu = z$), while for a spin-wave mediated interaction it is the identity matrix τ^0.

From the symmetry properties of the matrix $\hat{\mathscr{G}}$ and the fact that $V(\boldsymbol{q}, i\Omega_n)$ is real, we can deduce symmetry properties for the matrix $\hat{\Sigma}$:

$$\Sigma_{22}(\boldsymbol{k}, i\omega_n) = -\Sigma^*_{11}(\boldsymbol{k}, i\omega_n), \qquad \Sigma_{12}(\boldsymbol{k}, i\omega_n) = \Sigma_{21}(\boldsymbol{k}, i\omega_n). \tag{5.153}$$

The Dyson equation (5.151) can then be inverted to yield $\hat{\mathscr{G}}$, whose 11 and 12 components are

[23] Scalapino D J, Schrieffer J R and Wilkins J W 1966 *Phys. Rev.* **148** 263.

$$\mathscr{G}(\boldsymbol{k}, i\omega_n) = \frac{1}{i\omega_n - \xi_{\boldsymbol{k}} - \Sigma_{11} - \dfrac{\Sigma_{12}^2}{i\omega_n + \xi_{\boldsymbol{k}} + \Sigma_{11}^*}} \tag{5.154a}$$

$$\mathscr{F}(\boldsymbol{k}, i\omega_n) = \frac{\Sigma_{12}}{(i\omega_n - \xi_{\boldsymbol{k}} - \Sigma_{11})(i\omega_n + \xi_{\boldsymbol{k}} + \Sigma_{11}^*) - \Sigma_{12}^2}. \tag{5.154b}$$

Equations (5.152) and (5.154) form a set of self-consistent equations, known as the Eliashberg equations, that can be solved once $\xi_{\boldsymbol{k}}$ and $V(\boldsymbol{q}, i\Omega_n)$ are specified[24,25].

Example of solution In order to illustrate the content of the Eliashberg equations, we solve them in the simple case where the effective interaction between electrons is due to a single optical phonon branch and the Coulomb interaction is neglected:

$$V(\boldsymbol{q}, i\Omega_n) \equiv V(i\Omega_n) = g^2\left(\frac{1}{i\Omega_n - \hbar\Omega_0} - \frac{1}{i\Omega_n + \hbar\Omega_0}\right). \tag{5.155}$$

Ω_0 is the frequency of the optical phonon, assumed independent of $\boldsymbol{q}$ (see equation (5.51)) and g represents the electron–phonon coupling, also assumed independent of $\boldsymbol{q}$. As the interaction does not depend on momentum, the self-energy (5.152) becomes a function of energy only:

$$\hat{\Sigma}(i\omega_n) = -\frac{1}{\beta}\sum_{i\Omega_n} V(i\Omega_n)\hat{\mathscr{N}}(i\omega_n - i\Omega_n) = -\frac{1}{\beta}\sum_{i\omega_m} V(i\omega_n - i\omega_m)\hat{\mathscr{N}}(i\omega_m) \tag{5.156}$$

with $\mathscr{N}_{11}(i\omega_n) = \mathscr{V}^{-1}\sum_{\boldsymbol{k}}\mathscr{G}(\boldsymbol{k}, i\omega_n)$ and $\mathscr{N}_{12}(i\omega_n) = -\mathscr{V}^{-1}\sum_{\boldsymbol{k}}\mathscr{F}(\boldsymbol{k}, i\omega_n)$. Owing to the momentum independence of the self-energy, $\mathscr{G}(\boldsymbol{k}, i\omega_n)$ and $\mathscr{F}(\boldsymbol{k}, i\omega_n)$ in equations (5.154) are functions of $\xi_{\boldsymbol{k}}$ only and we can therefore replace the $\boldsymbol{k}$ sums by energy

---

[24] The BCS 'weak-coupling' limit is recovered when V is the time-independent BCS interaction: using the BCS result equation (5.136) in (5.152) we find that Σ_{11} becomes a momentum- and energy-independent constant (to be absorbed in the chemical potential) while the equation for Σ_{12} becomes the gap equation (5.141) with the solution $\Sigma_{12} = \Delta_{\boldsymbol{k}}$.

[25] Most often in the literature, the Eliashberg equations take a different form because the two unknown self-energies Σ_{11} and Σ_{12} are rewritten in terms of two real-valued functions Z and χ, and one complex function Δ or Φ. From the definition (5.152), we see that $\mathrm{Re}\,\Sigma_{11}(\boldsymbol{k}, i\omega_n) = \mathrm{Re}\,\Sigma_{11}(\boldsymbol{k}, -i\omega_n)$ and $\mathrm{Im}\,\Sigma_{11}(\boldsymbol{k}, i\omega_n) = -\mathrm{Im}\,\Sigma_{11}(\boldsymbol{k}, -i\omega_n)$, such that, without loss of generality, we can define

$$\Sigma_{11}(\boldsymbol{k}, i\omega_n) \equiv i\omega_n[1 - Z(\boldsymbol{k}, i\omega_n)] + \chi(\boldsymbol{k}, i\omega_n)$$

with $Z(\boldsymbol{k}, i\omega_n)$ and $\chi(\boldsymbol{k}, i\omega_n)$ two real functions that are even in $i\omega_n$. It is also customary to write

$$\Sigma_{12}(\boldsymbol{k}, i\omega_n) \equiv Z(\boldsymbol{k}, i\omega_n)\Delta(\boldsymbol{k}, i\omega_n) \equiv \Phi(\boldsymbol{k}, i\omega_n).$$

Expressed in terms of these new quantities, the Green's functions may be put in the form

$$\mathscr{G}(\boldsymbol{k}, i\omega_n) = \frac{i\omega_n Z(\boldsymbol{k}, i\omega_n) + \xi_{\boldsymbol{k}} + \chi(\boldsymbol{k}, i\omega_n)}{[i\omega_n Z(\boldsymbol{k}, i\omega_n)]^2 - [\xi_{\boldsymbol{k}} + \chi(\boldsymbol{k}, i\omega_n)]^2 - \Phi^2(\boldsymbol{k}, i\omega_n)}$$
$$\mathscr{F}(\boldsymbol{k}, i\omega_n) = \frac{\Phi(\boldsymbol{k}, i\omega_n)}{[i\omega_n Z(\boldsymbol{k}, i\omega_n)]^2 - [\xi_{\boldsymbol{k}} + \chi(\boldsymbol{k}, i\omega_n)]^2 - \Phi^2(\boldsymbol{k}, i\omega_n)}.$$

One sees that χ describes a renormalization of the dispersion $\xi_{\boldsymbol{k}}$ while Z renormalizes the energy. Δ, on the other hand, is known as the energy- and momentum-dependent pairing function.

integrals according to $\mathscr{V}^{-1}\sum_k(\cdots) = \int d\xi\, \tilde{N}_0^{\mathrm{el}}(\xi)(\cdots)$. $\tilde{N}_0^{\mathrm{el}}(\xi)$ is the density of states per spin direction for the band ξ_k. In order to be able to proceed analytically as far as possible, we make the simplification $\tilde{N}_0^{\mathrm{el}}(\xi) = \tilde{N}_0^{\mathrm{el}}(0)$ for $|\xi| < W$ and $\tilde{N}_0^{\mathrm{el}}(\xi) = 0$ for $|\xi| > W$. In other words, we assume that in the absence of pairing the density of states is flat over the bandwidth $2W$. The ξ integration can then be performed analytically. Separating out the real and imaginary parts of Σ_{11} through $\Sigma_{11} = \Sigma'_{11} + i\Sigma''_{11}$, we find:

$$\mathcal{N}_{11}(i\omega_n) = \tilde{N}_0^{\mathrm{el}}(0)\left[\left(1 + \frac{i\omega_n - i\Sigma''_{11}(i\omega_n)}{E(i\omega_n)}\right)\tanh^{-1}\left(\frac{W}{E(i\omega_n) - \Sigma'_{11}(i\omega_n)}\right)\right.$$
$$\left. -\left(1 - \frac{i\omega_n - i\Sigma''_{11}(i\omega_n)}{E(i\omega_n)}\right)\tanh^{-1}\left(\frac{W}{E(i\omega_n) + \Sigma'_{11}(i\omega_n)}\right)\right]$$
$$\mathcal{N}_{12}(i\omega_n) = -\tilde{N}_0^{\mathrm{el}}(0)\frac{\Sigma_{12}(i\omega_n)}{E(i\omega_n)}\left[\tanh^{-1}\left(\frac{W}{E(i\omega_n) - \Sigma'_{11}(i\omega_n)}\right)\right.$$
$$\left. + \tanh^{-1}\left(\frac{W}{E(i\omega_n) + \Sigma'_{11}(i\omega_n)}\right)\right]$$
$$E(i\omega_n) = \sqrt{[i\omega_n - i\Sigma''_{11}(i\omega_n)]^2 - \Sigma_{12}^2(i\omega_n)}\,.$$

Since $\hat{\mathcal{N}}$ is proportional to $\tilde{N}_0^{\mathrm{el}}(0)$, it is convenient to measure the interaction strength in units of $1/\tilde{N}_0^{\mathrm{el}}(0)$ by introducing the dimensionless parameter $\tilde{g}^2 \equiv 2g^2\tilde{N}_0^{\mathrm{el}}(0)/\hbar\Omega_0$. We are now in the position to perform the self-consistent loop numerically on the imaginary axis. Starting from the BCS solution $\Sigma_{11}(i\omega_n) = 0$ and $\Sigma_{12}(i\omega_n) = \hbar\Omega_0/\sinh(1/\tilde{g}^2)$, we compute $\mathcal{N}_{11}$ and $\mathcal{N}_{12}$ using the above relations. The latter quantities allow us to calculate new values for the self-energies $\Sigma_{11}(i\omega_n)$ and $\Sigma_{12}(i\omega_n)$ using equation (5.156) and the process is iterated to convergence. Since the self-energy is a convolution, it is most efficiently evaluated using a fast Fourier transform (FFT). This procedure works, but has one major drawback: the procedure yields numerical values for, e.g. $\mathcal{N}_{11}(i\omega_n)$, while we are primarily interested in the density of states on the real axis, which is formally given by $N(\varepsilon) = (-1/\pi)\,\mathrm{Im}\,\mathcal{N}_{11}(i\omega_n \to \varepsilon + i0^+)$. If we knew the function $\mathcal{N}_{11}(i\omega_n)$ *analytically*, the continuation to the real axis would be straightforward; but we happen to know this function only numerically for the discrete frequencies $i\omega_n$. The problem of the analytic continuation of numerical data is a very difficult one, because it is equivalent to inverting a matrix with exponentially small eigenvalues (see Bonus B5.4).

In order to avoid this problem, we solve the Eliashberg equations directly on the real axis. This can be done by means of the spectral representation of the Green's functions, which implies

$$\hat{\mathcal{N}}(i\omega_n) = \int_{-\infty}^{\infty} d\varepsilon \frac{\hat{N}(\varepsilon)}{i\omega_n - \varepsilon} \qquad \text{with} \qquad \hat{N}(\varepsilon) = -\frac{1}{\pi}\text{Im}\,\hat{\mathcal{N}}(i\omega_n \to \varepsilon + i0^+).$$

The real functions $N_{11}(\varepsilon)$ and $N_{12}(\varepsilon)$ will be our unknowns. Using the spectral representation and equation (16), we can perform the Matsubara sum for the self-energy:

$$\hat{\Sigma}(i\omega_n) = g^2 \int_{-\infty}^{\infty} d\varepsilon\ \hat{N}(\varepsilon)\left[\frac{f(\varepsilon) + b(\hbar\Omega_0)}{i\omega_n + \hbar\Omega_0 - \varepsilon} - \frac{f(\varepsilon) + b(-\hbar\Omega_0)}{i\omega_n - \hbar\Omega_0 - \varepsilon}\right]. \tag{5.157}$$

This expression is *analytic in* $i\omega_n$ and allows us to evaluate $\hat{\Sigma}(\varepsilon + i0^+)$ on the real axis. From this information, we deduce $\mathcal{N}_{11}(\varepsilon + i0^+)$ and $\mathcal{N}_{12}(\varepsilon + i0^+)$ using the formula derived previously and we can thus close the self-consistent loop by computing new values for the functions $N_{11}(\varepsilon)$ and $N_{12}(\varepsilon)$. For a practical implementation, we must discretize the ε-integral giving the self-energy. At high energy, we have $\hat{\Sigma}(\varepsilon + i0^+) \to 0$, which implies $N_{11}(|\varepsilon| \to \infty) = \tilde{N}_0^{\text{el}}(\varepsilon)$ and $N_{12}(|\varepsilon| \to \infty) = 0$. Hence the numerical evaluation of the self-energy Σ_{12} can be done by cutting the integral at some cutoff $\varepsilon_{\max}$, while for Σ_{11} we may perform analytically the integrations for $|\varepsilon| > \varepsilon_{\max}$, where the Fermi functions are either one or zero.

The complete solution is implemented in the code of figure 5.17. Line 1 defines the parameters. We measure energies in units of $\hbar\Omega_0$ and thus set `W0=1`; `W` is the bandwidth, `T` represents k_BT in units of $\hbar\Omega_0$, `g2` is the coupling $\tilde{g}^2$, `max` is the cutoff for the energy integral, and `n` is the number of discretization points. Line 2 builds the energy axis $[-\varepsilon_{\max}, \varepsilon_{\max}[$ in `e` and sets `i0`, which represents $i0^+$, to twice the discretization step. `a` is the *mixing factor*, which tells how much of the new solution is mixed with the old solution during the self-consistent loop; it is sometimes necessary to change this values in the course of the self-consistent loop. Lines 3 and 4 build the matrix `M` whose matrix elements are the expression in brackets in equation (5.157): one sees that after discretization of the integral, the calculation of the self-energy amounts to a matrix–vector multiplication; since the matrix does not depend

```
1  W0=1; W=1000; T=0.1; g2=2; max=15; n=3000;

2  de=2max/n; e=N[(Range[n]-1)de-max]; i0=I*2de; a=0.2;
3  f=Chop[1/(Exp[e/T]+1)]; b[x_]:=1/(Exp[x/T]-1);
4  M=g2*W0/2*de*Table[(f+b[W0])/(e[[k]]+i0+W0-e)-(f+b[-W0])/(e[[k]]+i0-W0-e),{k,1,n}];
5  dS11=g2*W0/2(2*b[W0](ArcTanh[W/(e+i0+W0)]+ArcTanh[W/(e+i0-W0)]-
6                        ArcTanh[max/(e+i0+W0)]-ArcTanh[max/(e+i0-W0)])+
7                        Log[1+W/(e+i0+W0)]-Log[1-W/(e+i0-W0)]-
8                        Log[1+max/(e+i0+W0)]+Log[1-max/(e+i0-W0)]);

9  Iteration:=Module[{u,EE,A1,A2,N12},
10         u=e+i0-I*Im[S11]; EE=Sqrt[u^2-S12^2]; u=u/EE;
11         A1=ArcTanh[W/(EE-Re[S11])]; A2=ArcTanh[W/(EE+Re[S11])];
12         N11=-Im[(1+u)A1-(1-u)A2]/Pi; N12=-Im[-S12/EE(A1+A2)]/Pi;
13         S11=(1-a)S11+a(M.N11+dS11); S12=(1-a)S12+a*M.N12;
14         ListLinePlot[Table[{e[[k]],N11[[k]]},{k,1,n}],PlotRange->{0,All}]
15 ];

16 S11=0; S12=W0/Sinh[1/g2]; Dynamic[Iteration]
```

Figure 5.17. Solution of the Eliashberg equations on the real-energy axis for an interaction mediated by a single optical phonon, using *Mathematica*.

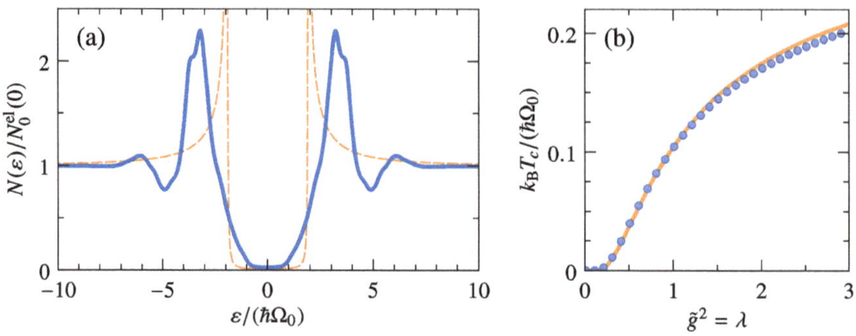

Figure 5.18. (a) Density of states $N(\varepsilon) = N_{11}(\varepsilon)$ calculated with the code of figure 5.17 (solid line) and corresponding weak-coupling BCS result (dashed line). (b) Evolution of the critical temperature T_c with the dimensionless coupling $\tilde{g}^2 = 2g^2\tilde{N}_0^{\rm el}(0)/\hbar\Omega_0$ as calculated with the code of figure 5.19 (dots). $\tilde{g}^2$ is equal to the electron–phonon coupling parameter λ (see text). The solid line is the McMillan formula (5.158) with $\langle\omega\rangle = \Omega_0$ and μ^*=0.

on $\hat{N}(\varepsilon)$, it can be set up outside the self-consistent loop. Lines 5–8 define the correction dS11 to the self-energy coming from the region outside $[-\varepsilon_{\rm max}, \varepsilon_{\rm max}[$. Lines 9–15 define a function that performs one self-consistent iteration. At line 12, the new values of $N_{11}(\varepsilon)$ and $N_{12}(\varepsilon)$ are computed and at line 13 the mixing of the new and old self-energies is performed. Line 14 displays a plot of $N_{11}(\varepsilon)$. Finally, at line 16 the loop is started by initializing the self-energies to their BCS values and performing the iterations. The *Mathematica* function Dynamic will run the iterations indefinitely; by looking at the plot, one can stop it at convergence. When using this code, it is better to move this last instruction in a new cell and launch it only after the other instructions have been executed. The calculation will take a few seconds.

Figure 5.18(a) shows the DOS $N_{11}(\varepsilon)$ obtained with the code of figure 5.17. The corresponding BCS solution (i.e. the first step of the self-consistent loop) is also displayed for comparison. One sees that the strong-coupling theory leads in this case to a gap wider than the weak-coupling BCS theory, but the gap edges are less sharp. One also notices features at $\hbar\Omega_0$, $2\hbar\Omega_0$, ..., that denote the threshold of the one-phonon, two-phonon, etc... absorption or emission processes. The self-consistency indeed includes these many-phonon processes through the subset of *rainbow diagrams*:

One also sees that the most prominent signature of strong coupling is a dip-hump at energies above the coherence peaks. This appears because the scattering rate is roughly proportional to the shifted DOS, e.g. $-{\rm Im}\,\Sigma_{11}(\varepsilon) \sim N_{11}(\varepsilon - \hbar\Omega_0)$ for $\varepsilon > 0$, and thus the scattering rate is peaked at $\varepsilon = \Delta + \hbar\Omega_0$ leading to a suppression of the DOS at this energy. Therefore, the higher the coherence peak in the DOS, the stronger the dip. It is clear that the analysis of the features in the DOS allows one in principle to determine the energy of the underlying phonon. This is the background of the celebrated McMillan and Rowell inversion procedure[26], by which the whole phonon spectrum is extracted from tunneling spectra assumed to be proportional to the DOS.

Linearized equations at T_c In order to determine T_c in this theory, we have to find the temperature above which the anomalous self-energy Σ_{12} vanishes. At T_c, where $\Sigma_{12} = 0$, we can expand the Eliashberg equations to lowest order in Σ_{12}. This means that we can replace in equation (5.154) Σ_{12}^2 by zero. Then $\mathscr{G}$ only depends on Σ_{11} and $\mathscr{F} = -\Sigma_{12}\mathscr{G}\mathscr{G}^*$. We thus obtain self-consistent equations for $\mathscr{G}$ and Σ_{11} that do not depend on $\mathscr{F}$ and Σ_{12}. Once they are solved for $\mathscr{G}$, we see that equation (5.152) for Σ_{12} has the form of an eigenvalue problem, namely (schematically) $\Sigma_{12} = V \otimes \mathscr{G}\mathscr{G}^*\Sigma_{12}$ where $\otimes$ stands for the convolution in Matsubara frequencies, in other words Σ_{12} is an eigenvector of $V \otimes \mathscr{G}\mathscr{G}^*$ with unity eigenvalue. The strategy to determine T_c is therefore to watch the eigenvalues of this matrix while varying T.

With our simple interaction (5.155), the self-consistent equation for Σ_{11} can be simplified and yields $\Sigma_{11}'(i\omega_n) = 0$ and

$$\mathscr{N}_{11}(i\omega_n) = -2i\tilde{N}_0^{\text{el}}(0)\tan^{-1}\left(\frac{W}{\omega_n - \Sigma_{11}''(i\omega_n)}\right).$$

We can check this by noting that $E(i\omega_n)$ is purely imaginary if $\Sigma_{12}^2 = 0$: $E(i\omega_n) = i|\omega_n - \Sigma_{11}'(i\omega_n)|$. The above formula follows, since $\tanh^{-1}(-ix) = -i\tan^{-1}(x)$. On the other hand, since $\mathscr{N}_{11}$ is purely imaginary and V is real, the real part of the self-energy Σ_{11} in equation (5.156) vanishes, consistently with our initial assumption. The equation for Σ_{12} can be recast in the form of an eigenvalue problem, indeed:

$$\Sigma_{12}(i\omega_n) = \sum_{i\omega_m}\left[-\frac{2\tilde{N}_0^{\text{el}}(0)}{\beta}\frac{V(i\omega_n - i\omega_m)}{\omega_m - \Sigma_{11}''(i\omega_m)}\tan^{-1}\left(\frac{W}{\omega_m - \Sigma_{11}''(i\omega_m)}\right)\right]\Sigma_{12}(i\omega_m),$$

with the term in brackets being the matrix of which we have to find the eigenvalues. At $T < T_c$, the largest eigenvalue is greater than one and consequently the self-consistent loop tends to increase the value of Σ_{12}. Inversely, at $T > T_c$ the self-consistency drives Σ_{12} to zero and all eigenvalues of the matrix are therefore smaller than one. At $T = T_c$, the largest eigenvalue is exactly one.

The code of figure 5.19 implements the calculation of T_c as a function of $\tilde{g}^2$. At line 3, we start the loop on temperature close to $T = 0$; for each T, the series of `2M` Matsubara frequencies ω_n is first built in `w`. Lines 4–7 solve for Σ_{11}. Since Σ_{11} is a convolution of V and $\mathscr{N}_{11}$ (equation (5.156)), we can evaluate it using Fourier transforms: schematically, $\Sigma_{11} = -(1/\beta)\text{FT}[\text{FT}^{-1}(V)*\text{FT}^{-1}(\mathscr{N}_{11})]$. The quantity $\text{FT}^{-1}(V)$ is prepared at line 4 and multiplied by the appropriate factors and

[26] McMillan W L and Rowell J M 1965 *Phys. Rev. Lett.* **14** 108.

```
1  W0=1; W=1000; M=200;

2  Tc[g2_]:=Module[{T,dT,w,V,S11,S0,dSA,max},
3    dT=0.1; T=0.001-dT; While[dT>10^-6, T=T+dT; w=(2Range[-M,M-1]+1)Pi*T;
4    V=-2T*g2*W0^2InverseFourier[RotateLeft[1/((2Range[-M,M-1]Pi*T)^2+W0^2),M]]Sqrt[2M];
5    S11=0; S0=0; dS=1; While[dS>10^-15,
6      S11=Fourier[V*InverseFourier[I*ArcTan[W/(w-Im[S11])]]];
7      dS=Abs[S11[[M]]-S0]; S0=S11[[M]]];
8    A=2T*g2*W0^2ArcTan[W/(w-Im[S11])]/(w-Im[S11]);
9    max=Max[Re[Eigenvalues[Table[A/((2(n+1)Pi*T-w)^2+W0^2),{n,-M,M-1}]]]];
10   If[max<1,T=T-dT; dT=dT/2]]; Max[T,0]
11 ]

12 Export["Tc.dat",Table[{g2,Tc[g2]},{g2,0.01,3,0.1}]];
```

Figure 5.19. Calculation of the critical temperature T_c in the strong-coupling Eliashberg theory, using *Mathematica.*

normalization; then the convolution is done at line 6. Lines 5 and 7 open and close the self-consistency loop for Σ_{11}. Once Σ_{11} is found, we build at lines 8 and 9 the matrix corresponding to the expression inside the brackets in the above equation and compute its largest eigenvalue `max`. If the latter is smaller than one it means that we have crossed T_c, such that we move one step back and decrease the temperature increment `dT` (line 10), continuing until the increment is sufficiently small (10^{-6} in our case).

The curve $T_c(\tilde{g}^2)$ obtained with this code is displayed in figure 5.18(b). Also shown is the behavior expected from the McMillan formula[27]

$$k_BT_c = \frac{\hbar\langle\omega\rangle}{1.2}\exp\left(-\frac{1.04(1+\lambda)}{\lambda-\mu^*(1+0.62\lambda)}\right). \tag{5.158}$$

This formula was derived by fitting numerical values of T_c for a model more elaborated than ours, but the results nevertheless match. McMillan considers a continuous phonon distribution rather than a single phonon and the electron–phonon coupling is represented by a dimensionless function $\alpha^2F(\omega)$. Our model corresponds to the simple case $\alpha^2F(\omega) = g^2\tilde{N}_0^{\text{el}}(0)\delta(\hbar\omega - \hbar\Omega_0)$. The overall electron–phonon coupling strength is measured by the parameter $\lambda = 2\int_0^\infty d\omega\, \alpha^2F(\omega)/\omega$, which becomes simply $2g^2\tilde{N}_0^{\text{el}}(0)/\hbar\Omega_0 = \tilde{g}^2$ in our case. μ^* is a parameter measuring the strength of the Coulomb repulsion, which we have neglected. Finally, $\langle\omega\rangle$ is a measure of the 'center of mass' of the electron–phonon coupling function, defined as $\langle\omega\rangle = (2/\lambda)\int_0^\infty d\omega\, \alpha^2F(\omega)$, i.e. simply Ω_0 for our model.

Bonus B5.4 The problem of numerical analytic continuation

The continuation of imaginary-time functions $\mathscr{C}(i\nu_n)$ from the imaginary to the real axis is a trivial task when the functional dependence of $\mathscr{C}$ on Matsubara frequencies

[27] McMillan W L 1968 *Phys. Rev.* **167** 331. Dynes R C 1972 *Solid. State Commun.* **10** 615. Allen P B and Dynes R C 1975 *Phys. Rev.* B **12** 905.

is known *analytically*. For instance, if $\mathscr{C}(i\nu_n) = 1/(i\nu_n - \xi)$, we immediately know from equation (4.12) that $C^R(\varepsilon) = 1/(\varepsilon + i0^+ - \xi)$. It is not quite that simple when the values $\mathscr{C}(i\nu_n)$ can only be obtained *numerically* for a finite—even if large—set of frequencies $i\nu_n$. In the latter case, we face a problem of general significance: given a finite number of values $C_n \equiv \mathscr{C}(i\nu_n)$, find the function $\rho(\varepsilon)$ such that equation (4.11) is satisfied. Rewriting the ε-integral as a discrete sum, we see that this is a linear problem of the kind $C_n = \sum_m A_{nm} R_m$, where $R_m \equiv \rho(\varepsilon_m)$ are the unknowns and the rectangular matrix A is given by $A_{nm} = d\varepsilon/(i\nu_n - \varepsilon_m)$. The formally simple solution $R = (A^T A)^{-1} A^T C$—where the superscript T means transposition and $(A^T A)^{-1} A^T$ is the left-inverse of A—falls apart when it comes to the practice, because it turns out that the square matrix $A^T A$ is badly conditioned and cannot be inverted numerically in any reliable way. This can be seen as follows. If we keep the whole series of Matsubara frequencies, the matrix $A^T A$ is given by

$$\begin{aligned}(A^T A)_{mm'} &= (d\varepsilon)^2 \sum_n \frac{1}{i\nu_n - \varepsilon_m} \frac{1}{i\nu_n - \varepsilon_{m'}} \\ &= -\eta \frac{(d\varepsilon)^2}{k_B T} \frac{d_{-\eta}(\varepsilon_m) - d_{-\eta}(\varepsilon_{m'})}{\varepsilon_m - \varepsilon_{m'}},\end{aligned}$$

where we have used equation (16). Writing $\varepsilon_m = m d\varepsilon$ with integer m and defining $\delta = d\varepsilon/(k_B T)$, this becomes

$$(A^T A)_{mm'} = \frac{\eta\delta}{(e^{m\delta} - \eta)(e^{m'\delta} - \eta)} \times \begin{cases} \delta e^{m\delta} & m = m' \\ \dfrac{e^{m\delta} - e^{m'\delta}}{m - m'} & m \neq m'. \end{cases}$$

The problem is most obvious for fermions at high temperature: expanding for small δ with $\eta = -1$, we see that all matrix elements are equal to $-\delta^2/4$ at leading order, such that the determinant of $A^T A$ is identically zero in this limit. It is not because just one eigenvalue happens to be zero: all but one eigenvalues are actually zero. In the opposite limit of low temperature, we see that the diagonal matrix elements approach $-\delta^2 e^{-|m|\delta}$. The off-diagonal matrix elements vanish even faster if m and m' have the same sign. We can assume this because the argument should not depend critically on which domain of energy we are considering. Hence we find that the determinant of $A^T A$ should typically vanish as $e^{-\delta M(M+1)/2}$ for a matrix of size M. The figure shows $\ln|\det(A^T A)|$ as a function of M for a grid of M energies ε_m taken symmetrically around $\varepsilon = 0$ and confirms that the matrix determinant plunges well below the accuracy of computers for just a few tenths of energies.

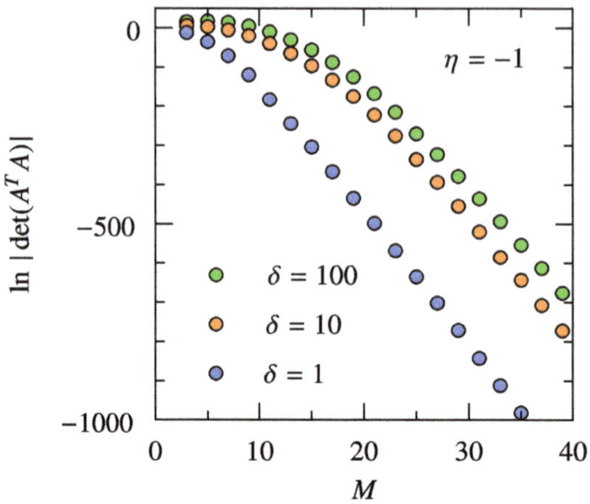

References

Abrikosov A A, Gorkov L P and Dzyaloshinski I E 1975 *Methods of Quantum Field Theory in Statistical Physics* (New York: Dover)

Bruus H and Flensberg K 2004 *Many-Body Quantum Theory in Condensed Matter Physics* (Oxford: Oxford University Press)

Fetter A L and Walecka J D 1971 *Quantum Theory of Many-Particle Systems* (New York: McGraw-Hill)

Schrieffer J R 1964 *Theory of Superconductivity* (New York: Benjamin)

Chapter 6

Response of matter to applied fields

Many condensed-matter experiments make use of an external field as a mean to probe a physical system. The method is in principle simple: turn on the external field and measure the change in some property of the system. This is called a *response* experiment, as opposed to *scattering* experiments which look at how a system scatters a beam of particles. From the theoretical viewpoint, the challenge is to understand how the external field affects the properties of the system and what is actually being measured in the experiment. If the intensity of the field is sufficiently weak and does not induce dramatic effects such as phase transitions, one can tackle the problem using perturbation theory. This chapter describes the response theory up to second order as well as the most common probing fields. This provides the basis for understanding response experiments (chapters 7, 8, and 9), while scattering experiments (chapter 10) are captured by the Fermi golden rule.

6.1 Linear and quadratic response

Before starting with the calculations, let us briefly outline the procedure. The system under investigation is initially represented by the Hamiltonian K, which contains all interactions needed to describe it (and is therefore usually intractable). The external field is adiabatically turned on and induces a perturbation V_t with the property $V_{-\infty} = 0$. The question is the following: what is the time evolution of the expectation value of some observable, say $\langle A(t)\rangle$?

In order to answer this question, we can either work within the Schrödinger or the Heisenberg picture. In the former, the time evolution is carried by the wave function or by the density matrix in a mixed state, while the operator A remains time independent; one therefore writes $\langle A(t)\rangle = \mathrm{Tr}\rho(t)A$ and one seeks an expansion of the density matrix $\rho(t)$ in powers of V. This route is followed, e.g. in Bruus & Flensberg (2004). We shall here use the Heisenberg and interaction pictures, in which case the density matrix remains time independent but the operators evolve in time according to $A(t) = \hat{U}^\dagger(t)\hat{A}(t)\hat{U}(t)$ (see section 2.2.1 and equation (2.20)).

doi:10.1088/978-0-7503-1741-2ch6

The reason for choosing this route is that we have already derived the perturbation expansion of $\hat{U}(t)$, equation (2.22), and therefore all the work is basically done. We indeed have

$$\begin{aligned}\langle A(t)\rangle_V &= \mathrm{Tr}\rho A(t) = \mathrm{Tr}\,\rho\ \hat{U}^\dagger(t)\hat{A}(t)\hat{U}(t)\\ &= \mathrm{Tr}\rho\left[\mathbb{1} + \sum_{n=1}^{\infty}\left(\frac{i}{\hbar}\right)^n \int_{-\infty}^{t} dt_1 \cdots \int_{-\infty}^{t_{n-1}} dt_n\ \hat{V}(t_n)\cdots\hat{V}(t_1)\right]\hat{A}(t)\\ &\quad\times\left[\mathbb{1} + \sum_{n=1}^{\infty}\left(-\frac{i}{\hbar}\right)^n \int_{-\infty}^{t} dt_1 \cdots \int_{-\infty}^{t_{n-1}} dt_n\ \hat{V}(t_1)\cdots\hat{V}(t_n)\right],\end{aligned} \tag{6.1}$$

where $\rho = e^{-\beta K}/(\mathrm{Tr}e^{-\beta K})$. In this chapter, we denote $\langle\cdots\rangle_V$ the thermal average in the presence of the perturbation V, while $\langle\cdots\rangle$ means the thermal average for $V = 0$.[1] At order zero in V, we have $\hat{U}(t) = \mathbb{1}$ such that the zeroth-order value is trivially

$$\langle A(t)\rangle_V^{(0)} = \mathrm{Tr}\rho\hat{A}(t) = \langle A(t)\rangle. \tag{6.2}$$

Remember that $\hat{A}(t) = A(t)$ in the absence of perturbation.

At order one, we get two terms. The first results from taking $\hat{U}^\dagger(t)$ at order one in V and $\hat{U}(t) = \mathbb{1}$ in equation (6.1) and the second from taking $\hat{U}^\dagger(t) = \mathbb{1}$ and $\hat{U}(t)$ at order one in V:

$$\begin{aligned}\langle A(t)\rangle_V^{(1)} &= \mathrm{Tr}\rho\left[\frac{i}{\hbar}\int_{-\infty}^{t} dt_1\ \hat{V}(t_1)\hat{A}(t) + \hat{A}(t)\left(-\frac{i}{\hbar}\right)\int_{-\infty}^{t} dt_1\ \hat{V}(t_1)\right]\\ &= -\frac{i}{\hbar}\int_{-\infty}^{t} dt_1\langle[A(t),\ V(t_1)]\rangle.\end{aligned} \tag{6.3}$$

Again, we could replace $\hat{A}(t)$ by $A(t)$ and $\hat{V}(t_1)$ by $V(t_1)$ because the Heisenberg and interaction pictures coincide at $V = 0$. $[\cdot,\ \cdot]$ is a short-hand for $[\cdot,\ \cdot]_-$ and denotes the commutator. Equation (6.3) is the foundation of the linear-response theory, which we shall elaborate more explicitly in the next section.

We will also need the quadratic response. At order two we get three terms, namely

$$\begin{aligned}\langle A(t)\rangle_V^{(2)} = \mathrm{Tr}\rho\Bigg[&\left(\frac{i}{\hbar}\right)^2 \int_{-\infty}^{t} dt_1 \int_{-\infty}^{t_1} dt_2\ \hat{V}(t_2)\hat{V}(t_1)\hat{A}(t)\\ &+ \frac{i}{\hbar}\int_{-\infty}^{t} dt_1\ \hat{V}(t_1)\hat{A}(t)\left(-\frac{i}{\hbar}\right)\int_{-\infty}^{t} dt_2\ \hat{V}(t_2)\\ &+ \hat{A}(t)\left(-\frac{i}{\hbar}\right)^2 \int_{-\infty}^{t} dt_1 \int_{-\infty}^{t_1} dt_2\ \hat{V}(t_1)\hat{V}(t_2)\Bigg].\end{aligned} \tag{6.4}$$

---

[1] Many authors prefer to use $\langle\cdots\rangle$ for $V \neq 0$ and $\langle\cdots\rangle_0$ for $V = 0$. This notation might lead speedy readers to understand that the thermal average $\langle\cdots\rangle_0$ is performed with *all* interactions turned off, i.e. using a quadratic Hamiltonian, which is incorrect since only the applied field must be discarded.

This expression can be rearranged in a more compact fashion by extending all integrations up to time t and using the fact that $\theta(t_1 - t_2) + \theta(t_2 - t_1) = 1$:

$$\langle A(t) \rangle_V^{(2)} = \left(-\frac{i}{\hbar}\right)^2 \int_{-\infty}^{t} dt_1 dt_2 \, \theta(t_1 - t_2) \langle [[A(t), V(t_1)], V(t_2)] \rangle. \quad (6.5)$$

The procedure can be carried on to arbitrary order. The nth order response involves $n!$ terms of the kind

$$\theta(t_1 - t_2) \cdots \theta(t_{n-1} - t_n) \langle [\cdots [[A(t), V(t_1)], V(t_2)], \ldots, V(t_n)] \rangle$$

obtained by taking all permutations of the time variables[2]. At order two, both permutations give the same contribution and we are thus left with just one term.

6.2 Response functions, susceptibilities

We have seen how the changes induced by an applied field to the value of an observable can be calculated using the response theory. To be more specific, we consider here the case where the applied field is a time-dependent classical field, say $F(t)$, that couples to the system via an observable B:

$$V_t = B * F(t). \quad (6.6)$$

The notation $B * F$ stands for a generalized scalar product. For instance, in the case where B and F are two vector fields, we would have $B * F(t) \equiv \int d\boldsymbol{r} \, \boldsymbol{B}(\boldsymbol{r}) \cdot \boldsymbol{F}(\boldsymbol{r}, t)$. In equation (6.6), we have written the perturbation V in the Schrödinger picture, in which the operator B is time independent. The perturbation has nevertheless an external (or 'explicit', see footnote in section 2.2.1) time dependence due to the external field. In order to use the response theory, we need $V_t(t)$ in the Heisenberg picture, i.e. we have to replace B by $B(t) = U^\dagger(t) B U(t)$ following equation (2.15). This adds an internal time dependence to $V_t(t)$ in addition to the external time dependence due to the applied field.

The effect of the field is to change certain properties of the system, in particular property B. For the sake of generality, we consider a property A which might or might not be the same as B. Using the response theory formulas (6.2) and (6.3), we can write

$$\begin{aligned} \langle A(t) \rangle_V - \langle A(t) \rangle &= \int_{-\infty}^{\infty} dt_1 \left(-\frac{i}{\hbar}\right) \theta(t - t_1) \langle [A(t), B(t_1)] \rangle * F(t_1) + \mathcal{O}(F^2) \\ &\equiv \int_{-\infty}^{\infty} dt_1 \, \chi_{AB}(t - t_1) * F(t_1) + \mathcal{O}(F^2). \end{aligned} \quad (6.7)$$

[2] Wehrum R P and Hermeking H 1974 *J. Phys. C: Solid State Phys.* **7** L107.

We have extended the time integration to $+\infty$ and corrected with the theta function. On the second line, we have defined the general linear susceptibility

$$\boxed{\chi_{AB}(t) = -\frac{i}{\hbar}\theta(t)\langle[A(t), B(0)]\rangle = C^{R}_{AB}(t).} \tag{6.8}$$

As indicated, the susceptibility $\chi_{AB}(t)$ is nothing but the retarded correlation function of the operators A and B (see equation (3.4)). Equation (6.7) is one of the pinnacles of many-body theory:

> *The first-order variation of the property A induced by a field that couples to the property B is proportional to the retarded correlation function of the operators A and B.*

The concept of susceptibility and its relation with the linear response of a system to applied fields shed light on the idea discussed in section 5.1.4. There, we saw that poles in the two-particle correlation functions correspond to collective modes (at finite energy) or to instabilities (at zero energy). Here, we find that the same two-particle correlation functions provide the susceptibility to applied fields. Therefore, in the present context, a pole in the correlation function should be regarded as a divergent susceptibility, i.e. the ability for the system to develop a finite response even in the absence of applied field. This is clearly the same physics as collective modes or instabilities.

Very often, the linear term dominates the response to an external field. However, there are situations like external photoemission, where the linear response vanishes and the dominant contribution therefore comes from second order. Using equation (6.5), we can define a second-order susceptibility and write

$$\langle A(t)\rangle_V^{(2)} = \int_{-\infty}^{\infty} dt_1 dt_2\, \chi_{AB}^{(2)}(t - t_1, t - t_2) \, * \, F_1(t_1) * F_2(t_2) \tag{6.9}$$

where

$$\begin{aligned}\chi_{AB}^{(2)}(t, t') = \left(-\frac{i}{\hbar}\right)^2 \frac{1}{2}\{&\theta(t)\theta(t' - t)\langle[[A(t), B_1(0)], B_2(t - t')]\rangle \\ &+ \theta(t')\theta(t - t')\langle[[A(t'), B_2(0)], B_1(t' - t)]\rangle\}.\end{aligned} \tag{6.10}$$

The indices in F_1, F_2, B_1, and B_2 represent the internal variables that may belong to these quantities (e.g. $B_1(0) * F_1(t_1) \equiv \int d\mathbf{r}_1\, B(\mathbf{r}_1)F(\mathbf{r}_1, t_1)$ and $B_2(t_2 - t_1) * F_2(t_2) \equiv \int d\mathbf{r}_2\, B(\mathbf{r}_2, t_2 - t_1)F(\mathbf{r}_2, t_2)$).

6.3 Examples of couplings

Until now, the discussion has been somewhat abstract. The purpose of this section is to briefly review typical examples of phenomena in condensed-matter physics that can be addressed using the linear-response theory.

6.3.1 Moving electrons with an electric potential

According to the basic laws of electrostatics, the total-energy variation of an electronic system subject to a scalar electric potential $\phi(\boldsymbol{r}, t)$ only depends on the electron density according to:

$$V_t = \int d\boldsymbol{r}\ (-e)n(\boldsymbol{r})\phi(\boldsymbol{r}, t). \tag{6.11}$$

This is an example of coupling of the form (6.6), where the observable B is $-en(\boldsymbol{r})$ and the field F is $\phi(\boldsymbol{r})$. The electric potential can displace charges and thus change the density in the system. According to equations (6.7) and (6.8), the first-order change in density is proportional to the dielectric susceptibility, in other words the retarded density–density correlation function $\chi_{nn}(\boldsymbol{r}, \boldsymbol{r}', t - t') = -(i/\hbar)\theta(t - t')\langle[n(\boldsymbol{r}, t), n(\boldsymbol{r}', t')]\rangle$:

$$\langle n(\boldsymbol{r}, t)\rangle_V - \langle n(\boldsymbol{r}, t)\rangle = -e \int d\boldsymbol{r}'dt'\ \chi_{nn}(\boldsymbol{r}, \boldsymbol{r}', t - t')\phi(\boldsymbol{r}', t') + \mathcal{O}(\phi^2). \tag{6.12}$$

In a translation-invariant system, $\chi_{nn}(\boldsymbol{r}, \boldsymbol{r}', t)$ becomes a function of $\boldsymbol{r} - \boldsymbol{r}'$ and its Fourier transform is just the analytic continuation of $\chi_{nn}(\boldsymbol{q}, \tau) = -\langle T_\tau n(\boldsymbol{q}, \tau)n(-\boldsymbol{q}, 0)\rangle$ discussed previously in section 5.1.4.5. An electric potential can also produce a current. The relevant response function in this case is the current–density correlation function $\chi_{jn}(\boldsymbol{q}, \tau) = -\langle T_\tau \boldsymbol{j}(\boldsymbol{q}, \tau)n(-\boldsymbol{q}, 0)\rangle$.

6.3.2 Animating spins with a magnetic field

In the presence of an applied magnetic field $\boldsymbol{H}(\boldsymbol{r}, t)$, the energy of a piece of matter changes according to

$$V_t = \int d\boldsymbol{r}\ (-\mu_0)\boldsymbol{M}(\boldsymbol{r}) \cdot \boldsymbol{H}(\boldsymbol{r}, t), \tag{6.13}$$

where $\boldsymbol{M}(\boldsymbol{r})$ is the magnetization of the material. This expression directly results from the facts that the electromagnetic energy is the volume integral of $\frac{1}{2\mu_0}B^2$ and that the total magnetic field is $\boldsymbol{B} = \mu_0(\boldsymbol{H} - \boldsymbol{M})$. There are several sources of magnetization. One is the polarization of the electron spins, which relates to the local spin density through $\boldsymbol{M}(\boldsymbol{r}) = g_e\mu_B\boldsymbol{S}(\boldsymbol{r})/\hbar$. For electrons, $g_e = -2.0023$. Other sources of magnetization include the orbital motion of the electrons in the field and the polarization of the nuclear spins. For this discussion, we disregard the magnetization of nuclear spins, which is typically smaller than that of the electron spins by a factor $\mu_N/\mu_B = m_e/m_p$. We also ignore the orbital magnetism as appropriate in insulators or metals in which the orbital moment is quenched by the crystal field.

Using the linear-response formula, we can express the first-order change in magnetization induced by the applied field in terms of the spin–spin correlation function:

$$\begin{aligned}\langle M_\mu(\boldsymbol{r}, t)\rangle_V - \langle M_\mu(\boldsymbol{r}, t)\rangle &= -\sum_\nu \int d\boldsymbol{r}'dt'\chi_{M_\mu(\boldsymbol{r})M_\nu(\boldsymbol{r}')}(t-t')\mu_0 H_\nu(\boldsymbol{r}', t') \\ &= -\left(\frac{g_e\mu_B}{\hbar}\right)^2 \sum_\nu \int d\boldsymbol{r}'dt'\,\chi_{ss}^{\mu\nu}(\boldsymbol{r}, \boldsymbol{r}', t-t')\mu_0 H_\nu(\boldsymbol{r}', t'). \quad (6.14)\end{aligned}$$

The susceptibility of interest here is $\chi_{ss}^{\mu\nu}(\boldsymbol{r}, \boldsymbol{r}', t) = -(i/\hbar)\theta(t)\langle[S_\mu(\boldsymbol{r}, t), S_\nu(\boldsymbol{r}', 0)]\rangle$ which, in a translation-invariant system, can be deduced from the corresponding imaginary-time correlator $\chi_{ss}^{\mu\nu}(\boldsymbol{q}, \tau) = -\langle T_\tau S_\mu(\boldsymbol{q}, \tau)S_\nu(-\boldsymbol{q}, 0)\rangle$. In the particular case of a static and uniform field $\boldsymbol{H}(\boldsymbol{r}, t) = H\boldsymbol{e}_z$, we have the change in magnetization

$$\delta M_\mu = -\left(\frac{g_e\mu_B}{\hbar}\right)^2 \chi_{ss}^{\mu z}(\boldsymbol{q}=0, \varepsilon=0)\mu_0 H. \quad (6.15)$$

Finally, for an isotropic system ($\chi_{ss}^{\mu\nu} \propto \delta_{\mu\nu}$) of independent electrons ($g_e \approx -2$ and $\chi_{ss}^{zz,0} = (\frac{\hbar}{2})^2\chi_{nn}^0$, see Doc A5.17), we recover using equation (5.77)

$$\delta \boldsymbol{M} = \mu_B^2 N_0^{\rm el}(0)\mu_0 \boldsymbol{H}. \quad (6.16)$$

This is the Pauli paramagnetism describing the response of independent-electron spins to magnetic fields and providing another measure of the Fermi-level DOS beside the specific heat (2.56). The contribution of the electronic orbital motion, neglected here, leads to Landau diamagnetism and in particular to de Haas–van Alphen oscillations in clean systems at low temperature.

6.3.3 Exciting electrons with photons

As we have seen in Doc A2.8, the energy resulting from the interaction of a material with an electromagnetic field is

$$V_t = \int d\boldsymbol{r}\, \boldsymbol{j}(\boldsymbol{r}) \cdot (-e)\boldsymbol{A}(\boldsymbol{r}, t). \quad (6.17)$$

$\boldsymbol{A}(\boldsymbol{r}, t)$ is the vector potential, the gauge is such that the scalar potential vanishes, and $\boldsymbol{j}$ is the *particle* current—not the *electric* current, the latter being $e\boldsymbol{j}$ with $e = -|e|$. Hence the light couples to the current operator and thereby puts the electrons in motion. The resulting current is, at leading order in $\boldsymbol{A}$:

$$\langle j_\mu(\boldsymbol{r}, t)\rangle_V - \langle j_\mu(\boldsymbol{r}, t)\rangle = -e\sum_\nu \int d\boldsymbol{r}'dt'\,\chi_{jj}^{\mu\nu}(\boldsymbol{r}, \boldsymbol{r}', t-t')A_\nu(\boldsymbol{r}', t'). \quad (6.18)$$

Here $\chi_{jj}^{\mu\nu}(\boldsymbol{r}, \boldsymbol{r}', t)$ is the retarded current–current correlation function, i.e. the analytic continuation of the corresponding imaginary-time function defined in real space as $\chi_{jj}^{\mu\nu}(\boldsymbol{r}, \boldsymbol{r}', \tau) = -\langle T_\tau j_\mu(\boldsymbol{r}, \tau)j_\nu(\boldsymbol{r}', 0)\rangle$. This function is closely related to the conductivity tensor $\sigma_{\mu\nu}(\boldsymbol{q}, \omega)$, as we shall see in chapter 8.

6.4 Response functions and imaginary-time functions

The response of a system to applied fields is given by the retarded correlation functions of that system evaluated in the absence of field. Therefore, in many situations, the comparison of experiment and theory requires us to calculate these retarded correlation functions. For independent particles, this is relatively straightforward once the single-particle energies are known—although the calculation may be technically difficult in practice. However, for interacting particles we will have to use an approximate method like e.g. perturbation theory in the interaction. This is where imaginary-time formalism is very useful. Indeed, there is no perturbative method for calculating retarded functions at finite temperature. Nevertheless, we have seen that they can be obtained by the analytic continuation[3] of the corresponding imaginary-time functions through $C^R_{AB}(\omega) = \mathscr{C}_{AB}(i\nu_n \to \hbar\omega + i0^+)$ and that, furthermore, $\mathscr{C}_{AB}(i\nu_n)$ does have a series expansion that can be evaluated using Feynman diagrams.

Introducing the time-Fourier transform of the applied field, $F(\omega) = \int_{-\infty}^{\infty} dt\, F(t)e^{i\omega t}$ and the short-hand notation $\hbar\omega^+ = \hbar\omega + i0^+$, we can rewrite the linear response in equation (6.7) in terms of the imaginary-time function as

$$\langle A(t)\rangle_V^{(1)} = \int_{-\infty}^{\infty} \frac{d\omega}{2\pi} e^{-i\omega t} \mathscr{C}_{AB}(i\nu_n \to \hbar\omega^+) * F(\omega). \tag{6.19}$$

Analogous expressions can be written at any order[4]. At order two, in particular, we have

$$\langle A(t)\rangle_V^{(2)} = \frac{1}{2}\int_{-\infty}^{\infty} \frac{d\omega_1}{2\pi}\frac{d\omega_2}{2\pi} e^{-i(\omega_1+\omega_2)t} \times \mathscr{C}_{AB_1B_2}(i\bar{\nu}_1 \to \hbar\omega_1^+,\, i\bar{\nu}_2 \to \hbar\omega_2^+) * F_1(\omega_1) * F_2(\omega_2) \tag{6.20}$$

where the appropriate imaginary-time function to consider is[5]

$$\mathscr{C}_{AB_1B_2}(\tau, \tau') = \langle T_\tau A(\tau)B_1(0)B_2(\tau - \tau')\rangle. \tag{6.21}$$

We will see an application of these formula when discussing external photoemission, which is a second-order process in the applied electromagnetic field.

Appendix B6

Bonus B6.1 Energy dissipation in an applied field

Imagine a system with rest Hamiltonian K_0 subject to an external field $F(t)$ that is switched on adiabatically. The field couples to observable B like in equation (6.6). We are interested in the total energy dissipated at leading order in the field. For any

[3] The analytical continuation is a hard problem when the imaginary-frequency function is known numerically rather than analytically (see Bonus B5.4).

[4] Wehrum R P and Hermeking H 1974 *J. Phys. C: Solid State Phys.* **7** L107.

[5] See appendix A of Berthod C, Köhl M and Giamarchi T 2015 *Phys. Rev.* A **92** 013626.

operator that has both the internal and an external time dependencies, the equation of motion is

$$i\hbar\frac{d}{dt}A_t(t) = U^\dagger(t)\left([A_t, K] + i\hbar\frac{\partial A_t}{\partial t}\right)U(t), \tag{6.22}$$

where the partial derivative is taken with respect to the external time. This results from equation (2.15) and the equation of motion of the evolution operator. In our case, $A_t \equiv K = K_0 + B * F(t)$ and it follows that $dK/dt = B(t) * dF(t)/dt$. The instantaneous power dissipated is $P(t) = d\langle K\rangle/dt = \langle B(t)\rangle * dF(t)/dt$. At first order in the field, $\langle B(t)\rangle$ is given by equation (6.19). Considering equations (4.12) and (6.8), introducing the Fourier transform of the field, and noting that $B = B^\dagger$, we have

$$\begin{aligned}P(t) &= \int_{-\infty}^{\infty}\frac{d\omega}{2\pi}e^{-i\omega t}\chi_{BB^\dagger}(\hbar\omega) * F(\omega) * \frac{d}{dt}\int_{-\infty}^{\infty}\frac{d\omega'}{2\pi}e^{-i\omega' t}F(\omega')\\ &= \int_{-\infty}^{\infty}\frac{d\omega\, d\omega'}{2\pi\ 2\pi}e^{-i(\omega+\omega')t}(-i\omega')\chi_{BB^\dagger}(\hbar\omega) * F(\omega) * F(\omega').\end{aligned}$$

The total energy dissipated is

$$\Delta E = \int_{-\infty}^{\infty} dt\ P(t) = i\int_{-\infty}^{\infty}\frac{d\omega}{2\pi}\ \omega\,\chi_{BB^\dagger}(\hbar\omega) * F(\omega) * F(-\omega).$$

As the field $F(t)$ is real, we have $F(-\omega) = F^*(\omega)$. ΔE is also real, therefore:

$$\begin{aligned}\Delta E &= i\int_{-\infty}^{\infty}\frac{d\omega}{2\pi}\ \omega\,\chi_{BB^\dagger}(\hbar\omega)**|F(\omega)|^2 = -i\int_{-\infty}^{\infty}\frac{d\omega}{2\pi}\ \omega\,\chi^*_{BB^\dagger}(\hbar\omega)**|F(\omega)|^2\\ &= \frac{i}{2}\int_{-\infty}^{\infty}\frac{d\omega}{2\pi}\ \omega\underbrace{\left[\chi_{BB^\dagger}(\hbar\omega) - \chi^*_{BB^\dagger}(\hbar\omega)\right]}_{2i\,\mathrm{Im}\,\chi_{BB^\dagger}(\hbar\omega)}\ **\ |F(\omega)|^2.\end{aligned}$$

The last line results by averaging the two expressions in the first line and the notation ** implies summation over all Cartesian and spatial indices associated with B and F. The final result is that the energy dissipated by a field coupling to observable B is controlled by the imaginary part of the retarded correlation function of B with itself:

$$\Delta E = \frac{1}{2}\int_{-\infty}^{\infty} d\omega\ \omega\left(-\frac{1}{\pi}\right)\mathrm{Im}\ C^R_{BB^\dagger}(\omega)\ **|F(\omega)|^2. \tag{6.23}$$

Owing to equation (3.47), the energy dissipation is related to the spectral function $\rho_{BB^\dagger}(\varepsilon)$, which expresses the fluctuation–dissipation theorem.

Reference

Bruus H and Flensberg K 2004 *Many-Body Quantum Theory in Condensed Matter Physics* (Oxford: Oxford University Press)

Part II

Spectroscopic probes

IOP Publishing

Spectroscopic Probes of Quantum Matter

Christophe Berthod

Chapter 7

External photoemission (XPS, PES, ARPES)

'*Photoemission presents a considerable challenge to the theorist since, for a complete solution, one must deal simultaneously with several complicated problems. A description is needed of both the equilibrium and excited electronic structure of a material and its interaction with the ion array and with the driving electromagnetic fields, whose form is modified by this interaction. Furthermore both problems must be solved in the vicinity of the surface, through which all system properties change dramatically. Not surprisingly, no complete solution of the problem has yet been developed.*'[1] This statement from 1978 remains up-to-date. The development of density-functional theory has enabled first-principle approaches that improve the description of the phenomenon by taking the microscopic details of the surface into account. Although promising, such developments are presently limited to materials for which the independent-electron approximations are appropriate. On the other hand, progress in the computational methods for strongly interacting electrons is still limited to bulk systems. Our understanding of what happens close to the surface of a material featuring correlated electrons is still at a very preliminary stage.

In 1964, Berglund and Spicer introduced the *three-step model*[2]. This model considers photoemission as a sequence of three independent processes: the primary excitation of the electron, its transport to the surface and its transmission through the surface. This phenomenological view was later replaced by *one-step* approaches that envision photoemission as a single quantum-mechanical process. These descriptions are based either on the response theory[3] (the photocurrent is viewed as the steady response of the system to an applied electromagnetic field) or on the

---

[1] Schaich W L 1978 , in *Photoemission in Solids* vol. 1 ed M Cardona and L Ley (Berlin: Springer-Verlag), p 105.

[2] Berglund C N and Spicer W E 1964 *Phys. Rev.* **136** A1030.

[3] Schaich W L and Ashcroft N W 1970 *Phys. Rev.* B **3** 2452. For a review, see Almbladh C-O 2006 *J. Phys.: Conf. Series* **35** 127.

doi:10.1088/978-0-7503-1741-2ch7

scattering theory[4] (the photoemission event is viewed as the inelastic scattering of a photon). Both approaches were shown to provide equivalent predictions in certain limits[5] and to reduce to the most popular 'golden-rule' expression where appropriate (see Bonus B7.1)[6].

In this chapter, we present the response theory of photoemission. We choose this approach firstly because it enters naturally within the flow of the book: the photocurrent is expressed in terms of a response function related to an imaginary-time correlation function that can be studied using diagrammatic techniques[7]. This approach, moreover, implements most naturally many-body effects and does not rest on uncontrolled assumptions like the assumption made in the golden-rule treatment that the final state is factorizable. Starting from the general expressions, we underline the various approximations needed in order to recover the common interpretation of the photoemission intensity in terms of the one-electron spectral function. The last two sections discuss the notion of quasi-particle and some phenomena that go beyond the spectral-function paradigm.

7.1 Response theory of external photoemission

A photoemission experiment uses light to extract electrons from a sample. Let us denote by $\boldsymbol{A}(\boldsymbol{r}, t)$ the vector potential describing the light. We work in the gauge where the scalar potential is zero. As we saw in Doc A2.8, the light couples to the electron current in the sample via

$$V_t = \int d\boldsymbol{r}\, \boldsymbol{j}(\boldsymbol{r}) \cdot (-e)\boldsymbol{A}(\boldsymbol{r}, t). \tag{7.1}$$

In response to the light, photo-electrons are emitted and collected in a detector at some point $\boldsymbol{R}$ outside the sample. The experiment measures the steady electron current $\boldsymbol{j}(\boldsymbol{R})$ at point $\boldsymbol{R}$. We can use the response theory for evaluating the expectation value of this current.

At zeroth order, equation (6.2) gives $\langle \boldsymbol{j}(\boldsymbol{R}, t)\rangle_V^{(0)} = \langle \boldsymbol{j}(\boldsymbol{R}, t)\rangle$. This contribution describes the current flowing out of the sample in the absence of applied electromagnetic field. Such a current can arise from thermally excited electrons that overcome the surface energy barrier and the effect is therefore called *thermionic emission*. This effect is completely negligible if the temperature is low compared with the surface barrier and $\boldsymbol{R}$ is far from the sample surface. The first-order contribution reads, according to equation (6.19),

$$\langle j_\mu(\boldsymbol{R}, t)\rangle_V^{(1)} = -e\int_{-\infty}^{\infty} \frac{d\omega}{2\pi} e^{-i\omega t} \sum_\nu \int d\boldsymbol{r}\, \mathscr{C}_{j_\mu(\boldsymbol{R}) j_\nu(\boldsymbol{r})}(i\Omega_n \to \hbar\omega^+) A_\nu(\boldsymbol{r}, \omega).$$

This term does not contribute to the dc photo-current. One indeed sees that its time average is proportional to $A_\nu(\boldsymbol{r}, 0) = 0$. Furthermore, the correlation function

[4] Mahan G D 1970 *Phys. Rev.* B **2** 4334.
[5] Schaich W L *op. cit.*
[6] Hermeking H and Wehrum R P 1975 *J. Phys. C: Solid State Phys.* **8** 3468.
[7] Keiter H 1978 *Z. Phys.* B **30** 167.

vanishes for $\boldsymbol{R}$ outside the sample: because $\mathscr{C}$ is to be evaluated in the absence of light, the trace is taken over a set of states $|a\rangle$ that have no electron at the point $\boldsymbol{R}$. All terms $\langle a|\,j_\mu(\boldsymbol{R},\tau)j_\nu(\boldsymbol{r},0)|a\rangle$ entering $\mathscr{C}_{j_\mu(\boldsymbol{R})j_\nu(\boldsymbol{r})}(\tau)$ simply vanish because $j_\mu(\boldsymbol{R})|a\rangle = 0$.

Hence, the external photoemission is of second order in the electromagnetic field. This is actually expected since the photo-current is known to be proportional to the intensity of the light, i.e. $\boldsymbol{A}^2$. The second-order photo-current is read directly from equation (6.20):

$$\langle j_\mu(\boldsymbol{R},t)\rangle_V^{(2)} = \frac{e^2}{2}\int_{-\infty}^{\infty}\frac{d\omega_1}{2\pi}\frac{d\omega_2}{2\pi}e^{-i(\omega_1+\omega_2)t}\sum_{\nu_1\nu_2}\int d\boldsymbol{r}_1 d\boldsymbol{r}_2$$
$$\times\,\mathscr{C}_{j_\mu^p(\boldsymbol{R})j_{\nu_1}^p(\boldsymbol{r}_1)j_{\nu_2}^p(\boldsymbol{r}_2)}(i\bar{\Omega}_1 \to \hbar\omega_1^+,\, i\bar{\Omega}_2 \to \hbar\omega_2^+)A_{\nu_1}(\boldsymbol{r}_1,\omega_1)A_{\nu_2}(\boldsymbol{r}_2,\omega_2) \quad (7.2)$$

with the three-current correlation function given by

$$\mathscr{C}_{j_\mu^p(\boldsymbol{R})j_{\nu_1}^p(\boldsymbol{r}_1)j_{\nu_2}^p(\boldsymbol{r}_2)}(\tau,\tau') = \langle T_\tau\, j_\mu^p(\boldsymbol{R},\tau)j_{\nu_1}^p(\boldsymbol{r}_1,0)j_{\nu_2}^p(\boldsymbol{r}_2,\tau-\tau')\rangle. \quad (7.3)$$

We could replace the total current $\boldsymbol{j}$ by the paramagnetic current $\boldsymbol{j}^p$ because the diamagnetic current (2.49*c*) is proportional to $\boldsymbol{A}$ and leads to contributions of order at least $\boldsymbol{A}^3$. Unlike the two-current correlator, the three-current correlator does not vanish because, owing to time ordering, it contains terms like $\langle a|\,j_{\nu_2}^p(\boldsymbol{r}_2)j_\mu^p(\boldsymbol{R})j_{\nu_1}^p(\boldsymbol{r}_1)|a\rangle$ in which the current $\boldsymbol{j}^p(\boldsymbol{R})$ does not act directly on the states $|a\rangle$, but on states like $j_{\nu_1}^p(\boldsymbol{r}_1)|a\rangle$ in which there is a finite amplitude that the operator $j_{\nu_1}^p(\boldsymbol{r}_1)$ has transferred an electron to the point $\boldsymbol{R}$.

We introduce three ingredients in order to proceed with the calculation. First, we specialize to a monochromatic light $\boldsymbol{A}(\boldsymbol{r},t) = \boldsymbol{A}(\boldsymbol{r})\cos(\omega_0 t)$, in other words

$$\boldsymbol{A}(\boldsymbol{r},\omega) = \pi\boldsymbol{A}(\boldsymbol{r})[\delta(\omega-\omega_0)+\delta(\omega+\omega_0)] \quad (7.4)$$

with $\omega_0 > 0$ for definiteness. Inserting this into equation (7.2) and performing the frequency integrations, we get four terms. Two of those terms evolve in time like $e^{2i\omega_0 t}$ and $e^{-2i\omega_0 t}$, respectively: such terms vanish on average and do not contribute to the dc photo-current. The two remaining terms are independent of time and obtained by taking one of the two combinations $\omega_1 = -\omega_2 = \pm\omega_0$. We reach the same conclusion by directly performing the time average in equation (7.2), which yields a factor $\delta(\omega_1+\omega_2)$. Second, we note that the experiment does not exactly measure the current density at point $\boldsymbol{R}$, but rather the total current through a solid angle $d\Omega$ around the direction $\boldsymbol{n} = \boldsymbol{R}/R$—or through a surface element $R^2 d\Omega$ in the direction $\boldsymbol{n}$. The quantity of interest is therefore $dJ(\boldsymbol{n}) = R^2 d\Omega\,\boldsymbol{n}\cdot\langle\boldsymbol{j}(\boldsymbol{R})\rangle_V$ in the limit $R\to\infty$. Third, we express the three-current correlator in terms of the real-space fermion operators $c_{\boldsymbol{r}\sigma}^\dagger$. For this purpose, we use the expression of the paramagnetic current (2.49*b*), recast for convenience in the following equivalent form:

$$\boldsymbol{j}^p(\boldsymbol{r}) = \frac{i\hbar}{2m}\sum_\sigma(\nabla_{\underline{\boldsymbol{r}}} - \nabla_{\boldsymbol{r}})c_{\underline{\boldsymbol{r}}\sigma}^\dagger c_{\boldsymbol{r}\sigma}\Big|_{\underline{\boldsymbol{r}}=\boldsymbol{r}}. \quad (7.5)$$

Upon insertion of these three ingredients in equation (7.2), the formula for the photo-current becomes somewhat hairy:

$$\frac{dJ(\boldsymbol{n})}{d\Omega} = \frac{-ie^2\hbar^3R^2}{64m^3}[\boldsymbol{n}\cdot(\nabla_{\underline{\boldsymbol{R}}} - \nabla_{\boldsymbol{R}})]\sum_{\sigma\sigma_1\sigma_2}\int d\boldsymbol{r}_1 d\boldsymbol{r}_2$$
$$\times\left[\boldsymbol{A}(\boldsymbol{r}_1)\cdot(\nabla_{\underline{\boldsymbol{r}}_1} - \nabla_{\boldsymbol{r}_1})\right]\left[\boldsymbol{A}(\boldsymbol{r}_2)\cdot(\nabla_{\underline{\boldsymbol{r}}_2} - \nabla_{\boldsymbol{r}_2})\right]$$
$$\times\mathscr{C}_{c^\dagger_{\underline{\boldsymbol{R}}\sigma}c_{\boldsymbol{R}\sigma}c^\dagger_{\underline{\boldsymbol{r}}_1\sigma_1}c_{\boldsymbol{r}_1\sigma_1}c^\dagger_{\underline{\boldsymbol{r}}_2\sigma_2}c_{\boldsymbol{r}_2\sigma_2}}(\hbar\omega_0^+, -\hbar\omega_0^+) + \{\omega_0 \to -\omega_0\}. \tag{7.6}$$

However, it is in a good shape for a systematic investigation using Feynman diagrams. The limits $\underline{\boldsymbol{R}} \to \boldsymbol{R}$, $\underline{\boldsymbol{r}}_1 \to \boldsymbol{r}_1$, and $\underline{\boldsymbol{r}}_2 \to \boldsymbol{r}_2$, as well as $R \to \infty$ are implicit in equation (7.6). The notation $\{\omega_0 \to -\omega_0\}$ represents a second term identical to the first, except that ω_0 is replaced by $-\omega_0$.

The correlation function that we have to consider is the following:

$$\mathscr{C}_{...}(\tau, \tau') = \langle T_\tau c^\dagger_{\underline{\boldsymbol{R}}\sigma}(\tau)c_{\boldsymbol{R}\sigma}(\tau)c^\dagger_{\underline{\boldsymbol{r}}_1\sigma_1}(0)c_{\boldsymbol{r}_1\sigma_1}(0)c^\dagger_{\underline{\boldsymbol{r}}_2\sigma_2}(\tau-\tau')c_{\boldsymbol{r}_2\sigma_2}(\tau-\tau')\rangle. \tag{7.7}$$

Its diagrammatic representation is depicted in figure 7.1(a). A second diagram (not shown) is obtained by reversing all arrows in figure 7.1(a) and is topologically different from the first. This second diagram gives the same contribution as figure 7.1(a), except that the roles of $(\boldsymbol{r}_1, \sigma_1)$ and $(\boldsymbol{r}_2, \sigma_2)$ are interchanged. This amounts to an exchange of τ and τ', which is the same as $\omega_0 \to -\omega_0$. Therefore, this second diagram is also generated by the second term in equation (7.6) and we can account for the two topologically different diagrams with an overall factor of two.

As usual, the Feynman diagrams permit an intuitive interpretation in terms of the physical processes taking place. The lower point at $\boldsymbol{r}_1$ in the diagram represents the excitation of the photo-electron by light: a photon is absorbed, an electron is created (line leaving to the right), and a hole is left behind (the photo-hole). This is like the particle–hole pair in the diagram (5.59). Before reaching the detector at point $\boldsymbol{R}$, the photo-electron experiences various interactions—interactions with the other electrons, impurities, phonons, with the sample surface, etc, attractive interaction with the photo-hole—represented by the shaded box. Finally, the photo-electron escapes and flies to the detector. Likewise, the photo-hole is affected by interactions as it

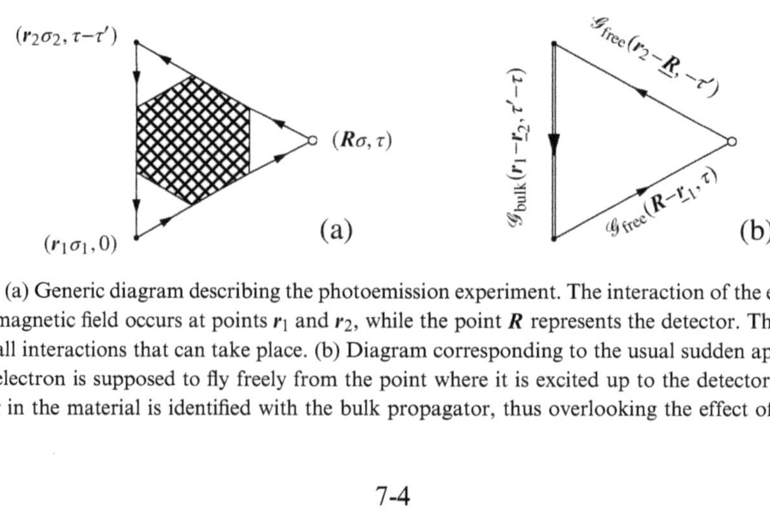

Figure 7.1. (a) Generic diagram describing the photoemission experiment. The interaction of the electrons with the electromagnetic field occurs at points $\boldsymbol{r}_1$ and $\boldsymbol{r}_2$, while the point $\boldsymbol{R}$ represents the detector. The shaded box stands for all interactions that can take place. (b) Diagram corresponding to the usual sudden approximation: the photo-electron is supposed to fly freely from the point where it is excited up to the detector and the hole propagator in the material is identified with the bulk propagator, thus overlooking the effect of the surface.

propagates 'back in time', i.e. vertically in the present case. The upper part of the diagram, which connects the detector at $\boldsymbol{R}$ with the point $\boldsymbol{r}_2$, represents the neutralization of the sample. As electrons are removed from the sample and absorbed by the detector, carriers must be supplied to keep the sample neutral. This establishes an electrical contact between detector and sample, for instance via the ground, and may be seen as electrons injected in the sample from the detector (upper line). These interact before recombining with the photo-holes at $\boldsymbol{r}_2$, each recombination leading to the emission of a photon. The sequence of events occurs in the 'logical' order if $0 < \tau < \tau - \tau'$ (first photoemission, then measurement and finally recombination). This is one of the time orderings that contribute to the correlation function, as discussed after equation (7.3).

In the next section, we present an approximation to equation (7.6) allowing us to recover the standard interpretation of the photo-current in terms of the one-electron spectral function. In the golden-rule approach, this same interpretation results from the so-called *sudden approximation* (see Bonus B7.1). Our approach provides a different and somewhat clearer perspective on the nature of the sudden approximation. A qualitative discussion of a few effects that go beyond the sudden approximation is given in section 7.4.

7.2 Sudden approximation and spectral function

In order to get a fruitful intuition on what the photoemission experiment actually measures, we certainly need a formula simpler than the general expression (7.6). Three assumptions will be needed before we can arrive at such a simple formula. The first assumption is that the photo-electron is directly emitted into the vacuum. In other words, it behaves like a free electron as soon as it is excited. The possible energy losses and momentum changes experienced by the photo-electron during its travel inside the material and its jump over the surface barrier are therefore neglected. Since the experiment actually probes a very thin layer below the surface, it is not unreasonable to think that at least some of the photo-electrons reach the vacuum without substantial losses. Our approximation might therefore be valid for this part of the photoemission signal (the so-called *no-loss* part). Formally, this approximation means that the propagations from point $\boldsymbol{r}_1$ to point $\boldsymbol{R}$ and from point $\boldsymbol{R}$ to point $\boldsymbol{r}_2$ in the diagram of figure 7.1(a) are replaced by free-particle propagations in the vacuum. They are described by the free propagator $\mathscr{G}_{\text{free}}$ discussed in Doc A7.1.

The second assumption is that the remaining propagation from the point $\boldsymbol{r}_2$ to the point $\boldsymbol{r}_1$ can be identified with the bulk propagation described by the Green's function $\mathscr{G}_{\text{bulk}}$, despite the presence of the nearby surface. In practice, this means that upon performing the $\boldsymbol{r}_1$ and $\boldsymbol{r}_2$ integrations we shall ignore the presence of the surface and pretend we are in the bulk.

The third assumption is that the electromagnetic field to which the electrons couple at the points $\boldsymbol{r}_1$ and $\boldsymbol{r}_2$ is the same as the electromagnetic field in the vacuum. In other words, the possibly complicated modifications of the field due to the presence of the surface as well as the screening of the field inside the material are not taken into account. The latter approximation is quite unphysical for metals because in reality the penetration depth of the light inside the material is no more than a few nanometers. If, as we assume, the light penetrates the whole sample and the photo-electrons are directly excited into the vacuum, then the whole bulk of the material contributes to the photo-current. Hence our final result should be proportional to the sample volume. This problem will be discussed further below.

The diagram implementing our three assumptions is displayed in figure 7.1(b). From this point, we can proceed with the calculation without further approximation and completely evaluate the diagram to get (see Doc A7.2)

$$\frac{dJ^{\mathrm{SA}}(\boldsymbol{n})}{d\Omega} = \frac{e^2\mathcal{V}}{8\pi^2 m\hbar}|\boldsymbol{A}\cdot\boldsymbol{n}|^2\int_{-\infty}^{\infty} d\varepsilon\, \kappa^3(\varepsilon)A(\kappa\boldsymbol{n},\, \varepsilon)f(\varepsilon). \tag{7.8}$$

Equation (7.8) is valid provided that the momentum of the photons is negligible compared with the typical momenta of the electrons, which is true in the case of photons with energies $\lesssim 20$ eV. In the opposite situation, a slightly more complicated formula replaces equation (7.8) (see Doc A7.2). $A(\boldsymbol{k}, \varepsilon) \equiv \rho_{c_{\boldsymbol{k}\sigma}c^\dagger_{\boldsymbol{k}\sigma}}(\varepsilon)$ is the spectral function of the retarded one-electron Green's function. It can be deduced from the corresponding imaginary-time Green's function according to equations (3.47) and (4.12):

$$A(\boldsymbol{k}, \varepsilon) = -\frac{1}{\pi}\operatorname{Im}\mathcal{G}(\boldsymbol{k}, i\omega_n \to \varepsilon + i0^+). \tag{7.9}$$

The quantity κ entering equation (7.8) is the magnitude of the free photo-electron wave vector, $\kappa = \sqrt{2m(\varepsilon + \hbar\omega_0 - \phi)/\hbar^2} = \sqrt{2mE_{\mathrm{kin}}/\hbar^2}$, with E_{kin} the energy of the photo-electron measured above the vacuum level and ϕ the surface work function. The presence of a κ^3 factor is a direct consequence of the photo-current being a three-current correlation function, equation (7.2). κ has a weak dependence on ε, because the typical energies ε of interest for valence-band photoemission ($\lesssim 0.5$ eV) are small compared with the photon energy $\hbar\omega_0$ ($\gtrsim 10$ eV)[8].

We see that the photo-current (7.8) is proportional to the sample volume as anticipated. In order to cure this problem, one could phenomenologically introduce the finite penetration depth ℓ_p of the photons and replace the sample volume $\mathcal{V}$ by the volume of a thin layer of thickness ℓ_p below the surface. Alternatively, one could introduce the finite escape depth ℓ_e of the photo-electrons due to their various interactions inside the material, such that $\mathcal{V}$ would be replaced by the volume of a layer of thickness ℓ_e. Which one of these two phenomena is the actual limiting factor depends on which one of ℓ_p and ℓ_e is the shortest: most likely ℓ_e, which is usually just a few atomic layers. In section 7.4, we will see that both phenomena can in principle be taken into account by including the appropriate diagrams in the modeling.

Our model gives an emission that vanishes in the directions normal to the vector potential $\boldsymbol{A}$ and is largest when $\boldsymbol{A}\|\boldsymbol{n}$. This is consistent with the experimental

[8] Modern laser-ARPES setups use lower-energy photons in the range 5–10 eV.

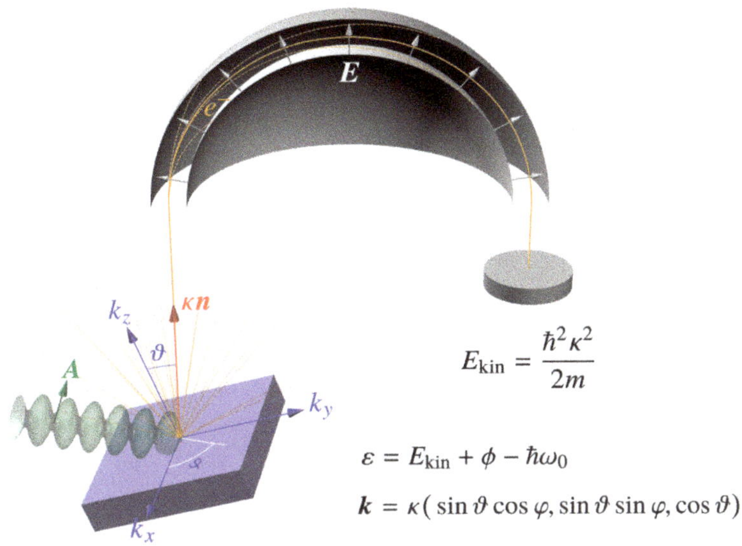

Figure 7.2. Angle-resolved photoemission spectroscopy (ARPES). Photo-electrons are collected along the direction $\boldsymbol{n}$ while the sample is rotated by the angles ϑ and φ. A tunable radial electric field $\boldsymbol{E}$ is used to discriminate photo-electrons according to their kinetic energy $E_{\rm kin}$ in a hemispherical detector. The knowledge of $E_{\rm kin}$, ϑ, and φ allows one to deduce the energy ε and momentum $\boldsymbol{k}$ of the electron before its excitation. See also figure 7.9.

observation that the efficiency is better with light polarized in the plane of incidence (so-called p-polarization) as compared to light polarized parallel to the sample surface (s-polarization). The form $\boldsymbol{A}\cdot\boldsymbol{n}$ results from taking a linearly-polarized light and a single plane wave for the final state of the photo-electron. Other light polarizations and a more detailed modeling of the final state can lead to less caricatural forms (see Bonus B7.2).

Equation (7.8) gives the total—energy-integrated—electron count in the direction $\boldsymbol{n}$ through the solid angle $d\Omega$ ($dJ^{\rm SA}/d\Omega$ has the dimension s^{-1}). For energy-resolved experiments, one uses a hemispherical analyzer in which a radial electric field curves the trajectories in order to select electrons with one particular energy $E_{\rm kin}$ in the vacuum, as sketched in figure 7.2. This corresponds to selecting electrons that have in the material an energy $\varepsilon = E_{\rm kin} + \phi - \hbar\omega_0$ measured from the chemical potential (see figure 7.9). Apart from corrections due to the finite angular and energy resolutions, the angle *and* energy resolved photoemission intensity is therefore given in the sudden approximation by

$$\boxed{\frac{d^2J^{\rm SA}(\boldsymbol{n}, \varepsilon)}{d\Omega d\varepsilon} = M(\boldsymbol{n}, \hbar\omega_0 - \phi)A(\boldsymbol{k}, \varepsilon)f(\varepsilon),} \tag{7.10}$$

where the relations between $(\boldsymbol{k}, \varepsilon)$ and $(\vartheta, \varphi, E_{\rm kin})$ are recalled in figure 7.2. We have neglected the ε-dependence of κ and introduced the 'matrix element' $M(\boldsymbol{n}, \hbar\omega_0 - \phi) = e^2|\boldsymbol{A}\cdot\boldsymbol{n}|^2\kappa^3\mathcal{V}/(8\pi^2 m\hbar)$ which depends upon the light intensity, polarization, and frequency. The very crude form of this matrix element, in

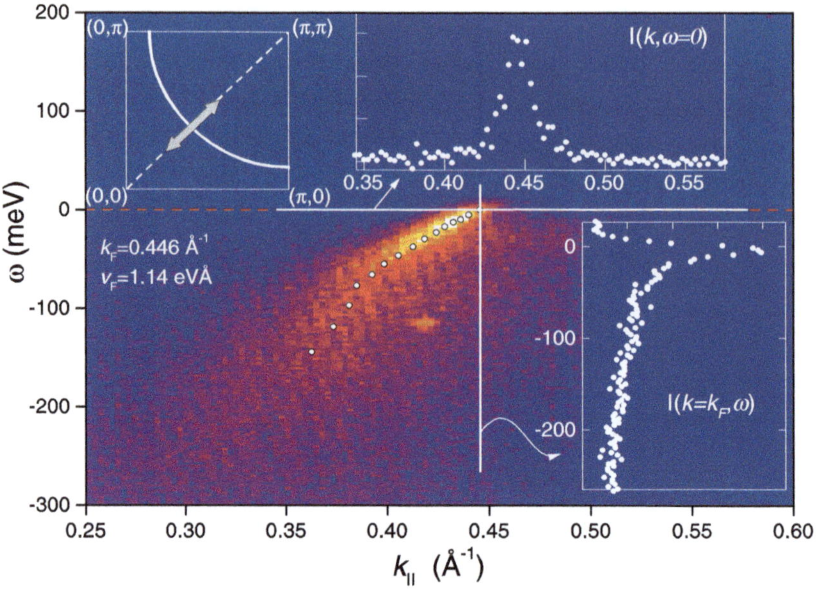

Figure 7.3. Illustration of the EDC and MDC photoemission spectra. The data are shown for the high-T_c superconductor $Bi_2Sr_2CaCu_2O_{8+\delta}$, a quasi-two dimensional material with a nearly cylindrical Fermi surface centered at $\boldsymbol{k}_{\|} = (\pi/a, \pi/a)$ (inset). The color map shows $J(\boldsymbol{k}, \hbar\omega)$ for $\boldsymbol{k}$ along the segment represented by the double arrow in the inset and $\hbar\omega$ between −300 and +200 meV. The temperature is $T = 48$ K (~4 meV), such that the signal is suppressed above $\hbar\omega \sim 4$ meV. The EDCs are obtained by plotting $J(\boldsymbol{k}, \hbar\omega)$ at fixed $\boldsymbol{k}$ as a function of $\hbar\omega$ (vertical cut), while the MDCs are obtained by plotting at fixed $\hbar\omega$ as a function of $\boldsymbol{k}$ (horizontal cut). Reprinted figure with permission from Valla *et al* 1999 *Science* **285** 2110. Copyright (1999) by the American Association for the Advancement of Science.

particular the fact that it does not depend on the electronic wave functions in the sample, reflects our assumption of translation invariance for the sample Green's function. A generalization is proposed in Bonus B7.2, where some matrix-element effects are also discussed[9].

Equation (7.10) provides the foundation for interpreting most angle-resolved photoemission spectroscopy (ARPES) experiments nowadays. The procedure is to fix the light polarization ($\boldsymbol{A}$), frequency (ω_0), and the direction of observation ($\boldsymbol{n}$)—i.e. to fix the matrix element M—and then to rotate the sample in order to vary the internal momentum $\boldsymbol{k}$ at which the spectral function is measured (see figure 7.2). Intensity spectra recorded at constant $\boldsymbol{k}$ as a function of ε are called *energy-distribution curves* (EDC), while intensity spectra recorded at constant ε as a function of $\boldsymbol{k}$ are called *momentum-distribution curves* (MDC).

Figure 7.3 illustrates the difference between EDC and MDC. The MDCs often have a symmetric Lorentzian line-shape, while the EDCs are more asymmetric and show a background at high binding energy. This is a consequence of the fact that the

[9] For a more in-depth discussion of the photoemission matrix element, in particular regarding the role of the surface, see Moser S 2017 *J. Elec. Spectrosc. Rel. Phenom.* **214** 29.

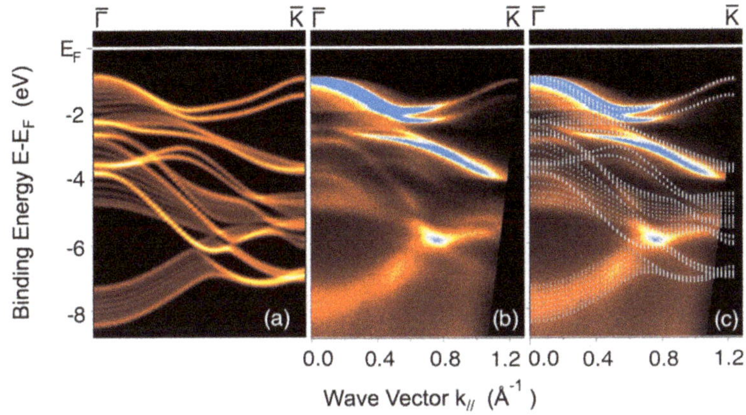

Figure 7.4. (a) Calculated valence band structure of a ten-layer slab of the metal dichalcogenide WSe_2, plotted along the $\bar{\Gamma}$–$\bar{K}$ line of the surface Brillouin zone. The Fermi level was adjusted and the band energies scaled by 6% to match the experimental bandwidth. The color scale shows the spectral weight of states that are even relative to the plane of the figure. (b) Spectral intensity measured by ARPES using the p-polarization ($\boldsymbol{A}$ in the plane of the figure). (c) Comparison of the measured and calculated band dispersions. Reprinted figure with permission from Tanabe *et al* 2016 *J. Phys.: Condens. Matter* **28** 345503. Copyright (2016) by IOP Publishing.

energy dependence of the self-energy is generally pronounced, while its momentum dependence is weak. The main features of figure 7.3, in particular the difference between the EDC and MDC line-shapes, can be captured in a very simple model with a phenomenological self-energy (see Bonus B7.4).

The scientific literature on photoemission is vast[10]. We pick just a few random examples in order to illustrate the capabilities of the technique. ARPES can be used as a band-mapping tool. Thanks to steady progress in energy and momentum resolutions, it has become possible to resolve fine details such as spin–orbit splitting in the valence band of materials. For the Shockley state on the Cu(111) surface, a spin–orbit splitting of only 0.006 $Å^{-1}$ was resolved using a 6 eV laser source, while the He lamp at 21 eV showed only one broad electronic band[11]. Another example of band mapping is shown in figure 7.4. In this study, the light polarization was varied in order to distinguish states that are even or odd relative to a mirror plane of the crystal.

Beside the band-mapping activity which focuses on the one-particle physics of the photoemission matrix element and dispersion $\xi_{\boldsymbol{k}}$, the ARPES technique is also used to gain information about the many-particle physics going on in the self-energy $\Sigma(\boldsymbol{k}, \varepsilon)$. Following a pioneering work on Mo(110),[12] where the effects of impurities

[10] For review articles, see for instance Damascelli A, Hussain Z and Shen Z-X 2003 *Rev. Mod. Phys.* **75** 473. Reinert F and Hüfner S 2005 *New J. Phys.* **7** 97. Lu D, Vishik I M, Yi M, Chen Y, Moore R G and Shen Z-X 2012 *Annu. Rev. Condens. Matter Phys.* **3** 129. Plumb N C and Radović M 2017 *J. Phys.: Condens. Matter* **29** 433005.

[11] Tamai A, Meevasana W, King P D C, Nicholson C W, de la Torre A, Rozbicki E and Baumberger F 2003 *Phys. Rev.* B **87** 075113.

[12] Valla T, Fedorov A V, Johnson P D and Hulbert S L 1999 *Phys. Rev. Lett.* **83** 2085.

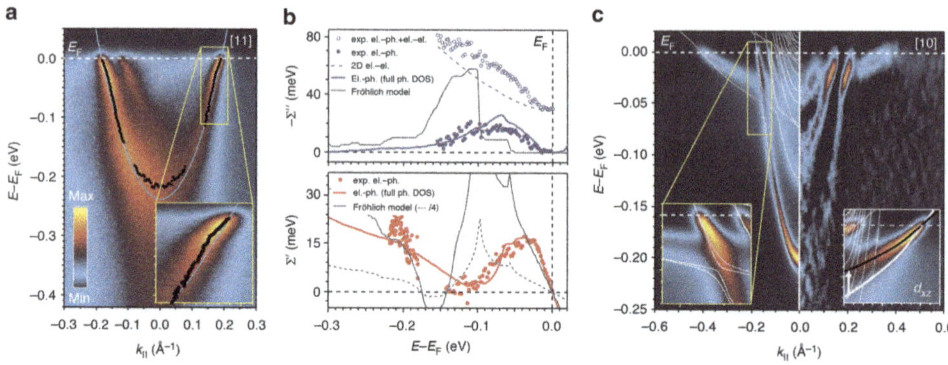

Figure 7.5. (a) ARPES data for electrons confined to the $SrTiO_3$ (100) surface. The peak positions and a model dispersion are indicated in black and blue, respectively. (b) Extracted self-energy compared with different models. The real part has a broad maximum between −20 and −60 meV, pointing to several phonon modes (red dots). (c) Comparison of theory (left and insets) and measurement (right, second-derivative data) showing the renormalized heavy band in addition to the light bands seen in (a). Reprinted figure with permission from King *et al* 2014 *Nat. Commun.* **4** 3414. Copyright (2014) by Springer Nature Limited. All rights reserved.

and electron–electron interactions could be investigated, many studies have tried to determine the self-energy from the photoemission spectral line-shape. In graphite[13], alongside the characteristic linear hour-glass dispersion above and below E_F, a strong damping due to electron–phonon scattering was observed with a linear energy dependence at high energy. In $1T$-$TiTe_2$, a quasi-two-dimensional metal, the effects of impurity, electron–phonon, and electron–electron scattering were tentatively distinguished by a detailed fitting of the quasi-particle peak[14]. A subsequent study of the same system revealed deviations from the sudden approximation[15]. Figure 7.5 shows data recorded from a two-dimensional electron gas at the surface of $SrTiO_3$. This data set displays, among other things, a dispersion kink and concomitant line-width broadening typical of electron–phonon interaction. The document Bonus B7.5 presents an analytical expression for the temperature-dependent self-energy of electrons coupled to dispersionless optical phonons, as well as a practical application to analyze EDC data taken on the (0001) surface of beryllium.

7.3 The notion of quasi-particle

For independent particles, we have seen that the spectral function is $A(\boldsymbol{k}, \varepsilon) = \delta(\varepsilon - \xi_{\boldsymbol{k}})$ with $\xi_{\boldsymbol{k}}$ the particle energy measured from the chemical potential. In a hypothetical system of independent particles, the photoemission signal would therefore display a sharp peak, allowing for an unambiguous determination of the particle energy as a function of $\boldsymbol{k}$ (for occupied states).

[13] Sugawara K, Sato T, Souma S, Takahashi T and Suematsu H 2007 *Phys. Rev. Lett.* **98** 036801.

[14] Perfetti L, Rojas C, Reginelli A, Gavioli L, Berger H, Margaritondo G, Grioni M, Gaál R, Forró L and Albenque F R 2001 *Phys. Rev.* B **64** 115102. Nicolay G, Eltner B, Hüfner S, Reinert F, Probst U and Bucher E 2006 *Phys. Rev.* B **73** 045116.

[15] Krasovskii E E, Rossnagel K, Fedorov A, Schattke W and Kipp L 2007 *Phys. Rev. Lett.* **98** 217604.

The vanishing width of this peak is the signature of the long-lived nature of the one-particle excitations in such systems. Moreover, the spectral weight of the peak is unity: the excited particle leaves nothing behind. Knowing the dispersion ξ_k, one could then evaluate the group velocity $\boldsymbol{v}_k = \hbar^{-1}\nabla\xi_k$, as well as the mass tensor $m_{\mu\nu}^{-1} = \hbar^{-2}\partial_{k_\mu}\partial_{k_\nu}\xi_k$.

Of course, independent particles do not exist in Nature. Nevertheless, for weakly interacting particles this picture is not expected to change dramatically: one still expects to observe a peak in the spectral function at some energy $E_k \approx \xi_k$. This peak has to be broadened due to interactions to a width Γ_k and its spectral weight might be smaller than unity. In such a case, we can consider that the one-particle excitations are *like* independent particles: they have a well-defined energy E_k, a reasonably long life-time $\tau_k \propto \Gamma_k^{-1}$, and a spectral weight $Z_k \lesssim 1$. When these conditions are fulfilled, one speaks about *quasi-particles* with a quasi-particle energy and life-time E_k and τ_k, respectively, and a quasi-particle *residue* Z_k. Such excitations are also often called *dressed particles* and a system exhibiting such excitations is known as a *Fermi liquid.*

Although the above description is not quite rigorous, it corresponds to the phenomenology that is commonly associated with quasi-particle physics in the context of photoemission. We shall give below a precise meaning to the concepts of life-time, residue, etc. Before moving on, we stress that the very existence of quasi-particles in many materials in Nature is the reason for the success of the conventional—mostly perturbative—descriptions of the low-energy properties in these systems, as formalized in the Landau theory of Fermi liquids. The existence of quasi-particles is not ineluctable, however, and research is nowadays largely dedicated to the study of materials in which the low-energy excitations most likely are *not* quasi-particles, such as fractional quantum Hall systems or high-T_c superconductors[16].

In the remainder of this section, we denote the real and imaginary parts of the retarded self-energy by Σ' and Σ'', respectively, a convention that is popular in the photoemission community:

$$\Sigma^R(\boldsymbol{k}, \varepsilon) = \Sigma'(\boldsymbol{k}, \varepsilon) + i\Sigma''(\boldsymbol{k}, \varepsilon). \tag{7.11}$$

From the general definition (7.9), we see that the spectral function takes the form

$$A(\boldsymbol{k}, \varepsilon) = \frac{-\Sigma''(\boldsymbol{k}, \varepsilon)/\pi}{[\varepsilon - \xi_k - \Sigma'(\boldsymbol{k}, \varepsilon)]^2 + [\Sigma''(\boldsymbol{k}, \varepsilon)]^2}. \tag{7.12}$$

There are constraints on $A(\boldsymbol{k}, \varepsilon)$ and $\Sigma^R(\boldsymbol{k}, \varepsilon)$ that result from the analytic properties of the Green's function. As we saw in section 3.4, $A(\boldsymbol{k}, \varepsilon)$ integrates to unity. From the general spectral representation (3.12) and (3.14), we further see that $A(\boldsymbol{k}, \varepsilon) > 0$, because it is the spectral function of the correlation function of two adjoint operators (i.e. $c_{k\sigma}$ and $c_{k\sigma}^\dagger$). From equation (7.12), it is obvious that the positivity of $A(\boldsymbol{k}, \varepsilon)$

[16] For a perspective on this subject, see Jain J K and Anderson P W 2009 *PNAS (USA)* **106** 9131.

requires the negativity of $\Sigma''(\boldsymbol{k}, \varepsilon)$. Some other useful analytic properties of the self-energy are discussed in Bonus B7.3.

Most often, one is interested in the single-particle excitations at low energy and close to the Fermi surface. But the very notion of Fermi surface might lose its meaning in the presence of interactions: for independent particles, the Fermi surface is defined by the discontinuity of the distribution function $\langle n_{\boldsymbol{k}} \rangle = f(\xi_{\boldsymbol{k}})$ (equations (3.39) and (3.25)) at zero temperature, which marks the frontier between occupied and empty states. In the presence of interactions—or at finite temperature—it is natural to define the Fermi 'surface' as the locus of zero-energy excitations in momentum space, in other words the maximum of $A(\boldsymbol{k}, 0)$. From equation (7.12), one sees that, up to small corrections involving the momentum derivative of $\Sigma''(\boldsymbol{k}, 0)$, this maximum is given by the equation

$$\boxed{\xi_{\boldsymbol{k}_\mathrm{F}} + \Sigma'(\boldsymbol{k}_\mathrm{F}, 0) = 0.} \tag{7.13}$$

This shows that, unless the real part of Σ vanishes at zero energy, the effects encoded in the self-energy displace the 'bare' Fermi surface defined by $\xi_{\boldsymbol{k}_\mathrm{F}^0} = 0$.

For studying the low-energy one-particle excitations in weakly-interacting systems, the next step is to expand the Green's function $G^R(\boldsymbol{k}, \varepsilon)$ close to $\boldsymbol{k} = \boldsymbol{k}_\mathrm{F}$ and $\varepsilon = 0$. Because G^R has a nearly singular structure near $\boldsymbol{k} = \boldsymbol{k}_\mathrm{F}$ and $\varepsilon = 0$, it is more convenient to expand $1/G^R$:

$$\begin{aligned}\frac{1}{G^R(\boldsymbol{k}, \varepsilon)} &= \varepsilon - \xi_{\boldsymbol{k}} - \Sigma'(\boldsymbol{k}, \varepsilon) - i\Sigma''(\boldsymbol{k}, \varepsilon) \\ &= \varepsilon - [\xi_{\boldsymbol{k}_\mathrm{F}} + (\boldsymbol{k} - \boldsymbol{k}_\mathrm{F}) \cdot \nabla \xi_{\boldsymbol{k}}|_{\boldsymbol{k}_\mathrm{F}}] \\ &\quad - [\Sigma'(\boldsymbol{k}_\mathrm{F}, 0) + (\boldsymbol{k} - \boldsymbol{k}_\mathrm{F}) \cdot \nabla \Sigma'(\boldsymbol{k}, 0)|_{\boldsymbol{k}_\mathrm{F}} + \varepsilon \partial_\varepsilon \Sigma'(\boldsymbol{k}_\mathrm{F}, \varepsilon)|_{\varepsilon=0}] \\ &\quad - i\Sigma''(\boldsymbol{k}, \varepsilon) + \mathcal{O}[(\boldsymbol{k} - \boldsymbol{k}_\mathrm{F})^2, \varepsilon^2].\end{aligned} \tag{7.14}$$

Thanks to equation (7.13), we can eliminate the terms $-\xi_{\boldsymbol{k}_\mathrm{F}} - \Sigma'(\boldsymbol{k}_\mathrm{F}, 0)$. The remaining terms can be recast in the form

$$\frac{1}{G^R(\boldsymbol{k}, \varepsilon)} = \frac{1}{Z(\boldsymbol{k}_\mathrm{F})}\left[\varepsilon - E_{\boldsymbol{k}} + \frac{i\hbar}{2\tau(\boldsymbol{k}, \varepsilon)}\right] + \mathcal{O}[(\boldsymbol{k} - \boldsymbol{k}_\mathrm{F})^2, \varepsilon^2] \tag{7.15}$$

with the help of the definitions

$$\frac{1}{Z(\boldsymbol{k}_\mathrm{F})} \equiv 1 - \partial_\varepsilon \Sigma'(\boldsymbol{k}_\mathrm{F}, \varepsilon)|_{\varepsilon=0} \tag{7.16a}$$

$$E_{\boldsymbol{k}} \equiv Z(\boldsymbol{k}_\mathrm{F})\,(\boldsymbol{k} - \boldsymbol{k}_\mathrm{F}) \cdot \nabla[\xi_{\boldsymbol{k}} + \Sigma'(\boldsymbol{k}, 0)]_{\boldsymbol{k}_\mathrm{F}} \tag{7.16b}$$

$$\frac{\hbar}{\tau(\boldsymbol{k}, \varepsilon)} \equiv -2Z(\boldsymbol{k}_\mathrm{F})\Sigma''(\boldsymbol{k}, \varepsilon). \tag{7.16c}$$

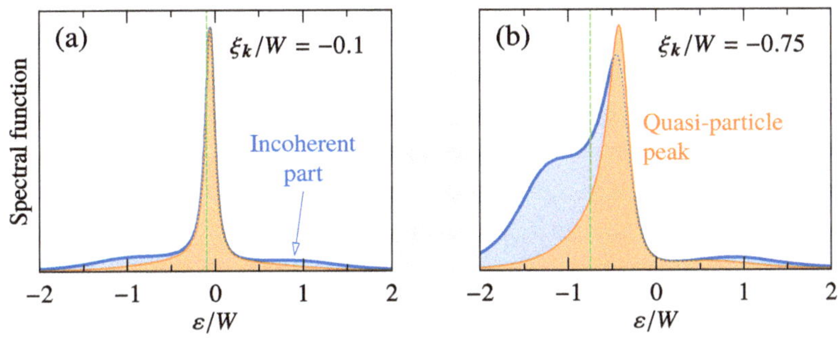

Figure 7.6. Spectral function (blue) and coherent quasi-particle peak (orange) for the model self-energy (7.31) with parameters $\Gamma = 0.1$, $\alpha = 1$, and $W = 1$. The dashed green lines indicate the bare dispersion ξ_k, which differ from the quasi-particle peak energy E_k. (a) When $|\xi_k|$ is small, the quasi-particle peak captures well the main feature of the spectral function. (b) At higher value of $|\xi_k|$, the incoherent part increases and the quasi-particle description progressively breaks down. The vertical scales are different in (a) and (b) as indicated by the gray shade.

These various quantities correspond to the concepts of quasi-particle residue, energy, and life-time introduced in the previous qualitative discussion, as can be seen by rewriting the Green's function (7.15) in the more suggestive form

$$G^R(\boldsymbol{k}, \varepsilon) = \frac{Z(\boldsymbol{k}_F)}{\varepsilon - E_k + \dfrac{i\hbar}{2\tau(\boldsymbol{k}, \varepsilon)}} + G_{\text{inc}}(\boldsymbol{k}, \varepsilon). \tag{7.17}$$

The first term describes a peak with the maximum at $\varepsilon = E_k$ and a width $\hbar/\tau$ (the so-called 'coherent' quasi-particle peak). The weight of the peak is not one but $Z(\boldsymbol{k}_F)$. If the life-time τ does not depend on $\boldsymbol{k}$ and ε, the quasi-particle peak has a Lorentzian shape and a width $\hbar/\tau$ (full width at half maximum). G_{inc} contains 'the rest', i.e. the part not contained in the coherent quasi-particle peak; it is not expected *a priori* to have a pole structure. Figure 7.6 illustrates the concept of quasi-particle peak using the model self-energy introduced in Bonus B7.4.

In many systems—but there are important exceptions—the self-energy depends on momentum much less than it depends on energy and neglecting the momentum dependence altogether can be a reasonable approximation. There are plenty of theoretical models in which the self-energy is momentum independent, e.g. the second-order Coulomb interaction diagram (section 8.4), the Kondo problem (section 8.5), or the case of electrons interacting with optical phonons (Bonus B7.5). Momentum independence of the self-energy occurs when the interaction itself is local (i.e. momentum independent). For the Coulomb repulsion, the potential can be considered momentum independent for electrons close to the Fermi surface; more precisely, the screening transforms the long-range Coulomb potential into a screened potential which is almost local (see section 5.1.4.6). In the Kondo problem, the electrons interact with localized spins, leading to a momentum-independent potential. Finally, for electrons coupled to optical phonons, the momentum independence stems from neglecting the weak dispersion of the optical phonons. If we develop the real part of a momentum-independent self-energy for low energies,

$\Sigma'(\boldsymbol{k}, \varepsilon) \approx \Sigma'(0) - \lambda\varepsilon + \mathcal{O}(\varepsilon^2)$, we see that the Green's function takes the quasi-particle form (7.17) at low energy:

$$G^R(\boldsymbol{k}, \varepsilon) \approx \frac{1}{\varepsilon - \xi_k - [\Sigma'(0) - \lambda\varepsilon] - i\Sigma''(\varepsilon)} = \frac{Z}{\varepsilon - E_k + \dfrac{i\hbar}{2\tau(\varepsilon)}}, \tag{7.18}$$

with $Z = 1/(1 + \lambda)$, $E_k = (\varepsilon_k - \mu')/(1 + \lambda)$, $\mu' = \mu - \Sigma'(0)$, and $\hbar\tau^{-1}(\varepsilon) = 2Z|\Sigma''(\varepsilon)|$. This Green's function describes low-energy quasi-particles with a life-time $\tau(\varepsilon)$, a renormalized dispersion E_k, i.e. a renormalized velocity

$$v_k \longrightarrow v_k^* = \frac{v_k}{1 + \lambda}, \tag{7.19}$$

a reduced spectral weight

$$1 \longrightarrow Z = \frac{1}{1 + \lambda}, \tag{7.20}$$

and a correspondingly enhanced mass and density of states at the Fermi level,

$$m \longrightarrow m^* = (1 + \lambda)m, \qquad N_0^{\mathrm{el}}(0) \longrightarrow N^{\mathrm{el}}(0) = (1 + \lambda)N_0^{\mathrm{el}}(0). \tag{7.21}$$

The dimensionless parameter λ, related to the zero-energy derivative of the real part of the self-energy,

$$\lambda = -\left.\frac{\partial\Sigma'(\varepsilon)}{\partial\varepsilon}\right|_{\varepsilon=0}, \tag{7.22}$$

plays a central role in describing the properties of quasi-particles. It is often called the *renormalization factor*. Since it is possible to measure v_k^* by photoemission, the renormalization factor can in principle be determined as $\lambda = v_k/v_k^* - 1$ if the noninteracting velocity v_k is known.

More generally, a momentum-independent self-energy can in principle be completely determined (at negative energies) by photoemission, provided that the noninteracting dispersion can be assumed to be linear with un-renormalized Fermi surface, i.e. $\xi_k \approx v_{\mathrm{F}}(k - k_{\mathrm{F}})$. This is possible even if v_{F} is unknown, by exploiting the causal nature of the self-energy. For a momentum-independent self-energy, the spectral function reads

$$A(\boldsymbol{k}, \varepsilon) = \frac{-\Sigma''(\varepsilon)/\pi}{[\varepsilon - \xi_k - \Sigma'(\varepsilon)]^2 + [\Sigma''(\varepsilon)]^2}. \tag{7.23}$$

For a linear dispersion, a measurement of the MDC, i.e. $A(\boldsymbol{k}, \varepsilon_0)$ versus $\boldsymbol{k}$ at fixed energy ε_0 must yield a Lorentzian peaked at the wave vector $\boldsymbol{k}_0$ that is the solution of $\varepsilon_0 - \xi_{k_0} - \Sigma'(\varepsilon_0) = 0$ (figure 7.7(a)). By varying ε_0 and plotting $\boldsymbol{k}_0$ as a function of ε_0, one obtains the quasi-particle dispersion as illustrated in figure 7.7(b). The width $2\Delta k$ of the Lorentzian gives the imaginary part according to $\Sigma''(\varepsilon_0) = -v_{\mathrm{F}}\Delta k$. One can thus construct the curve $-\Sigma''(\varepsilon)$ (figure 7.7(c)), up to the unknown factor v_{F}. This

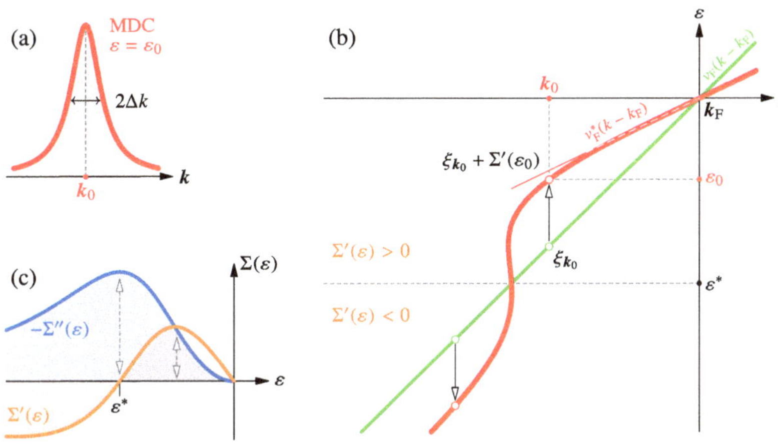

Figure 7.7. Determination of the self-energy by photoemission for a linear dispersion. (a) The peak position $\boldsymbol{k}_0$ in the Lorentzian-shaped MDC at energy ε_0 gives the quasi-particle dispersion (b) as ε_0 is varied. The width of the MDC gives $-\Sigma''$ as shown in (c). Once the noninteracting dispersion is known, Σ' is given by the difference between the quasi-particle and noninteracting dispersions. As $\Sigma(\varepsilon)$ is causal, a maximum in Σ' corresponds to a change of curvature in $-\Sigma''$ and a zero in Σ' to a maximum in $-\Sigma''$.

curve should present a maximum at some energy ε^*, because the self-energy must vanish at sufficiently high energy; Kramers–Kronig consistency then implies that the real part $\Sigma'(\varepsilon)$ must vanish close to ε^*. This allows one to fix v_F and the noninteracting dispersion, since the quasi-particle and noninteracting dispersions are equal at ε^*. Once v_F is fixed, the curve $\Sigma'(\varepsilon)$ can be constructed by measuring the vertical difference between the quasi-particle and noninteracting dispersions. Another check of Kramers–Kronig consistency is that a maximum in $\Sigma'(\varepsilon)$ must coincide with a change of curvature in $-\Sigma''(\varepsilon)$. This procedure is often not directly applicable in practice because the scale ε^* corresponds to energies where the curvature of the dispersion can no longer be neglected.

7.4 Beyond the sudden approximation

The main virtue of the response theory of photoemission is to pinpoint directions where the theory can be improved beyond the mainstream paradigm. In this section, we sketch a few of them, suggesting the road to follow rather than doing actual calculations.

7.4.1 Surface barrier

Unlike we assumed in section 7.2, the photo-electrons are not excited directly into the vacuum, but within the material. In order to reach the vacuum they must overcome the surface barrier, which in a first approximation is a potential step of height ϕ (see figure 7.9). In order to understand the effect of this barrier, we may keep using a diagram of the type displayed in figure 7.1(b) but replace the propagators $\mathscr{G}_{\text{free}}$ by a better approximation, closer to the exact propagator $\mathscr{G}$ that should actually be used in evaluating the diagram, e.g.

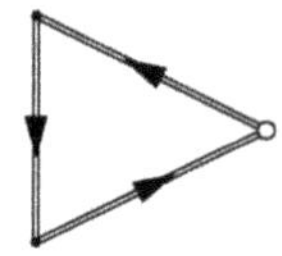

(7.24)

In the calculation of Doc A7.2, this amounts to replacing the product

$$G^R_{\text{free}}(\boldsymbol{R} - \underline{\boldsymbol{r}}_1, \varepsilon + \hbar\omega_0)G^A_{\text{free}}(\boldsymbol{r}_2 - \underline{\boldsymbol{R}}, \varepsilon + \hbar\omega_0)$$

by

$$\begin{aligned} &G^R(\boldsymbol{R}, \underline{\boldsymbol{r}}_1, \varepsilon + \hbar\omega_0)G^A(\boldsymbol{r}_2, \underline{\boldsymbol{R}}, \varepsilon + \hbar\omega_0) = \\ &\quad \mathcal{G}(\boldsymbol{R}, \underline{\boldsymbol{r}}_1, i\omega_n \to \varepsilon + \hbar\omega_0 + i0^+)\mathcal{G}(\boldsymbol{r}_2, \underline{\boldsymbol{R}}, i\omega_n \to \varepsilon + \hbar\omega_0 - i0^+). \end{aligned} \tag{7.25}$$

$\mathcal{G}$ describes the actual propagation from $\boldsymbol{r}_1$ to $\boldsymbol{R}$ and $\boldsymbol{R}$ to $\boldsymbol{r}_2$. In the presence of a potential $V(\boldsymbol{r})$—describing the surface barrier and/or the lattice potential—this propagator $\mathcal{G}$ becomes, according to equation (5.18*a*),

$$\mathcal{G}(\boldsymbol{r}, \boldsymbol{r}', i\omega_n) = \mathcal{G}_{\text{free}}(\boldsymbol{r} - \boldsymbol{r}', i\omega_n) + \int d\boldsymbol{r}''\mathcal{G}_{\text{free}}(\boldsymbol{r} - \boldsymbol{r}'', i\omega_n)V(\boldsymbol{r}'')\mathcal{G}(\boldsymbol{r}'', \boldsymbol{r}', i\omega_n). \tag{7.26}$$

In order to describe the effect of the surface barrier, we would therefore have to solve this equation for $\mathcal{G}$ and insert the result in the calculation of Doc A7.2. At lowest order in V, we could replace $\mathcal{G}(\boldsymbol{r}'', \boldsymbol{r}', i\omega_n)$ by $\mathcal{G}_{\text{free}}(\boldsymbol{r}'' - \boldsymbol{r}', i\omega_n)$ in the second term, reducing the calculation to the evaluation of a single integral.

For a proper treatment of the surface barrier, we should also replace $\mathcal{G}_{\text{bulk}}$ in figure 7.1(b) by the true propagator in the material, which is modified by the presence of the surface. We would in this way take into account the modifications of the hole spectral function due to the presence of the surface as well as the interference between the photo-electron and electrons of the material that are reflected by the surface.

7.4.2 Intrinsic losses, damping

Damping effects are essential for a correct interpretation of photoemission data. On one hand, they determine the self-energy which enters in the spectral function of equation (7.10); on the other hand, they are responsible for the finite escape depth of the photo-electrons. Much of these damping effects can already be investigated within the approximation given in equation (7.24) by improving the calculation of the three propagators involved in the diagram.

The damping of the excitations can have various origins: scattering on impurities (section 5.1.3.6), electron–electron interaction (section 5.1.3.7), electron–phonon interaction (section 5.1.3.8), spin–orbit interaction, etc. The crudest approach is to consider that these phenomena lead to a purely imaginary self-energy $\Sigma(\boldsymbol{k}, \varepsilon) \approx -i\Gamma$ for the photo-electron. If this correction is introduced in the free propagator of Doc A7.1 (replacing $i\nu_n$ by $i\nu_n + i\Gamma$), the propagator vanishes exponentially with distance like $e^{-|\text{Im}\,\kappa|r}$. This defines the electron mean-free path as $1/\ell_e = 2|\text{Im}\,\kappa|$. As a result, the photo-electrons created at a depth exceeding ℓ_e in the material have no chance to reach and cross the surface.

7.4.3 Extrinsic losses

Beside the diagram (7.24), the three-current correlator involves many terms that account for the interactions of the photo-electron with the material. These terms can represent impurity scattering, Coulomb interaction, or electron–phonon interaction, like in the following diagrams:

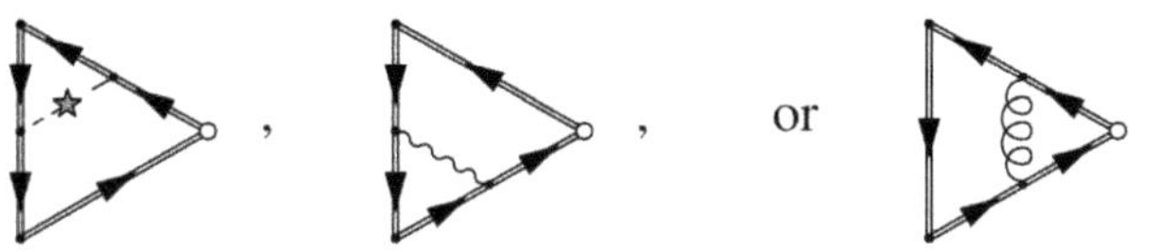

The case of the Coulomb interaction was treated by Chang and Langreth[17]. In this analysis, they describe a phenomenon of great importance for the XPS (core-electron) spectroscopy, namely the losses due to the excitation of one or more plasmons by the photo-electron. These losses are captured by the diagrams

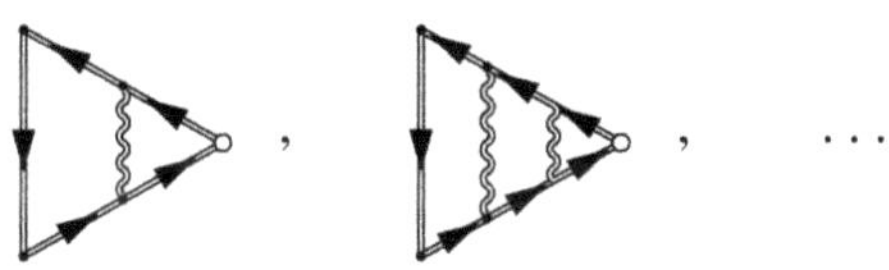

and they lead to satellite peaks in the core-level photoemission spectrum of solids beside the main quasi-particle (or 'no-loss') peak (see figure 7.8 for an example).

The diagrams above with the wavy lines replaced by phonon lines describe losses due to the photo-electron exciting phonons. The phonons typically have energies below 50–100 meV in solids, two orders of magnitude lower than the typical plasmon energies, such that their effect on the core-electron photoemission lineshape cannot be seen due to insufficient energy resolution. The various effects of electron–phonon scattering in photoemission were discussed by Caroli *et al*[18].

7.4.4 Screening of the electromagnetic field

In the vicinity of the surface, the electromagnetic field may change dramatically due to the screening by the material. In metals, in particular, the field vanishes beyond a short distance ℓ_p inversely proportional to the conductivity (skin effect). One can approach this problem with the macroscopic Maxwell equations and a suitable dielectric function in order to determine the vector potential in the vicinity of the surface. From our more microscopic perspective, we recognize that the diagrams

[17] Chang J-J and Langreth D C 1973 *Phys. Rev.* B **8** 4638.

[18] Caroli C, Lederer-Rozenblatt D, Roulet B and Saint-James D 1973 *Phys. Rev.* B **8** 4552.

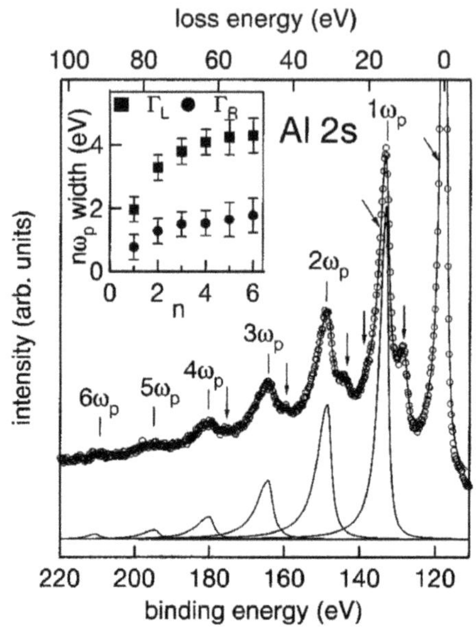

Figure 7.8. XPS photoemission spectrum of the 2*s* core level in elemental Al (circles). Beside the main peak, there are several equidistant satellite peaks at lower energies corresponding to 2*s* photo-electrons having excited one or more plasmons before escaping. The Al bulk plasmon has an energy $\hbar\omega_p \approx 15$ eV. The photo-electrons can also excite surface plasmons, which in Al have a lower energy near 10 eV. These losses lead to smaller satellite peaks indicated by the vertical arrows. Reprinted figure with permission from Biswas *et al* 2003 *Phys. Rev.* B **67** 165416. Copyright (2003) by the American Physical Society.

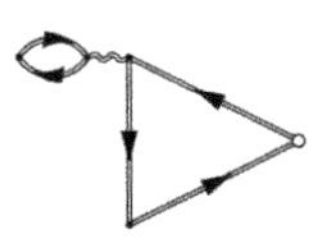 and 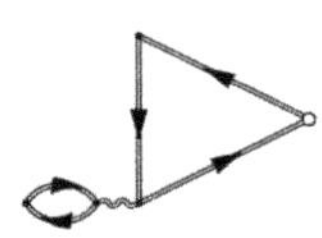

contain this screening effect treated within the self-consistent RPA approximation (see sections 5.1.4.6 and 5.1.4.4).

Turning the few ideas sketched here into concrete calculations is a formidable task. Fully analytical calculations are generally impossible, except for simple models that are designed to exhibit this or that particular aspect of the phenomenology. As an illustration of this difficulty, and in order to close this chapter as we opened it, let us quote W L Schaich once again: '*(...) the possible utility of such calculations in revealing new physics as opposed to simply providing a better fit to experiment does not seem compelling when weighted against the extra effort.*'[19]

[19] Schaich W L *op. cit.*, p 117.

Appendix A7

Doc A7.1 Free-particle propagator in the vacuum

The Green's function of a free particle in the vacuum is given in momentum space by $\mathscr{G}_{\text{free}}(\boldsymbol{k}, i\nu_n) = (i\nu_n - \xi_{\boldsymbol{k}})^{-1}$, where $\xi_{\boldsymbol{k}} = \hbar^2 k^2/(2m) - E_0$ and E_0 sets the origin on the energy axis. We wish to derive the expression of $\mathscr{G}_{\text{free}}$ in real space:

$$\mathscr{G}_{\text{free}}(\boldsymbol{r}, i\nu_n) = \frac{1}{\mathscr{V}} \sum_{\boldsymbol{k}} \frac{e^{i\boldsymbol{k}\cdot\boldsymbol{r}}}{i\nu_n - \xi_{\boldsymbol{k}}}.$$

In order to perform the integration, we choose the coordinate system like in Doc A2:

$$\begin{aligned}\mathscr{G}_{\text{free}}(\boldsymbol{r}, i\nu_n) &= \frac{1}{(2\pi)^2} \int_0^\infty dk \int_0^\pi d\vartheta \frac{k^2 \sin\vartheta \, e^{ikr\cos\vartheta}}{i\nu_n + E_0 - \hbar^2 k^2/(2m)} \\ &= \frac{m}{\pi^2\hbar^2 r} \int_0^\infty dk \frac{k \sin kr}{\kappa^2 - k^2} = \frac{m}{2\pi^2\hbar^2 r} \frac{1}{i} \int_{-\infty}^\infty dk \frac{k e^{ikr}}{\kappa^2 - k^2}.\end{aligned}$$

We have introduced $\kappa^2 = \frac{2m}{\hbar^2}(i\nu_n + E_0)$. The remaining wave vector integration is performed by means of the residue theorem (7), by closing the integration contour in the upper half of the complex plane, where the integral vanishes due to the factor e^{ikr}. The integrand has two poles at $k = \pm\kappa$. If $\text{Im}\,\kappa > 0$, the pole at $+\kappa$ in enclosed in the contour, while if $\text{Im}\,\kappa < 0$ it is the pole at $-\kappa$ that is enclosed. Hence we obtain

$$\mathscr{G}_{\text{free}}(\boldsymbol{r}, i\nu_n) = -\frac{m}{2\pi\hbar^2 r} e^{i\,\text{sign}(\text{Im}\,\kappa)\kappa r}, \qquad \kappa = \sqrt{\frac{2m}{\hbar^2}(i\nu_n + E_0)}.$$

This propagator is often used to describe particles that are asymptotically free, i.e. become free at a distance R large compared with other typical length scales in the problem. In this situation, the free propagation from a point $\boldsymbol{r}$ in the system to a distant point $\boldsymbol{R}$, which only depends on the distance $|\boldsymbol{r} - \boldsymbol{R}|$, can be written as

$$\mathscr{G}_{\text{free}}(\boldsymbol{r} - \boldsymbol{R}, i\nu_n) = -\frac{m}{2\pi\hbar^2 R} e^{i\,\text{sign}(\text{Im}\,\kappa)\kappa(R - \boldsymbol{n}\cdot\boldsymbol{r})} \qquad (R \gg r),$$

where we have used the fact that $|\boldsymbol{r} - \boldsymbol{R}| = R - \boldsymbol{n} \cdot \boldsymbol{r} + \mathcal{O}[(r/R)^2]$ with $\boldsymbol{n} = \boldsymbol{R}/R$.

Doc A7.2 Photoemission intensity and spectral function

The no-loss contribution to photoemission intensity is given by the diagram displayed in figure 7.1(b). The various approximations involved and the relation with the sudden approximation are discussed in section 7.2 and illustrated in figure 7.9. Translating the diagram with the help of the standard diagrammatic rules and inserting in equation (7.6), we arrive at

$$\begin{aligned}\frac{dJ^{\text{SA}}(\boldsymbol{n})}{d\Omega} &= \frac{-ie^2\hbar^3 R^2}{32m^3}[\boldsymbol{n} \cdot (\nabla_{\underline{\boldsymbol{R}}} - \nabla_{\boldsymbol{R}})] \sum_{\sigma\sigma_1\sigma_2} \int d\boldsymbol{r}_1 d\boldsymbol{r}_2 [\boldsymbol{A}(\boldsymbol{r}_1) \cdot (\nabla_{\underline{\boldsymbol{r}}_1} - \nabla_{\boldsymbol{r}_1})] \\ &[\boldsymbol{A}(\boldsymbol{r}_2) \cdot (\nabla_{\underline{\boldsymbol{r}}_2} - \nabla_{\boldsymbol{r}_2})] \int_0^\beta d\tau d\tau' \, e^{i\bar{\Omega}_1\tau} e^{i\bar{\Omega}_2\tau'} \delta_{\sigma_1\sigma_2} \mathscr{G}_{\text{bulk}}(\boldsymbol{r}_1 - \underline{\boldsymbol{r}}_2, \tau' - \tau) \\ &\delta_{\sigma\sigma_1} \mathscr{G}_{\text{free}}(\boldsymbol{R} - \underline{\boldsymbol{r}}_1, \tau) \delta_{\sigma\sigma_2} \mathscr{G}_{\text{free}}(\boldsymbol{r}_2 - \underline{\boldsymbol{R}}, -\tau')|_{\substack{i\bar{\Omega}_1 \to \hbar\omega_0^+ \\ i\bar{\Omega}_2 \to -\hbar\omega_0^+}} + \{\omega_0 \to -\omega_0\}.\end{aligned}$$

A factor of 2 for the two topologically different diagrams has been added as discussed near the end of section 7.1 and the factor −1 due to the fermion loop in the diagram is cancelled by the fact that the correlation function is actually minus the diagram, like in equation (5.59); one can check this by directly applying Wick's theorem to equation (7.7). In order to be able to perform the time integrations and the analytic continuation, we introduce the spectral representations (4.11) of $\mathscr{G}_{\text{bulk}}$ and $\mathscr{G}_{\text{free}}$:

$$\mathscr{G}_{\text{bulk}}(\boldsymbol{r},\tau) = \frac{1}{\mathscr{V}}\sum_{\boldsymbol{k}}\frac{1}{\beta}\sum_{i\omega_n}\underbrace{\mathscr{G}_{\text{bulk}}(\boldsymbol{k},i\omega_n)}_{\int_{-\infty}^{\infty}d\varepsilon\frac{A(\boldsymbol{k},\varepsilon)}{i\omega_n-\varepsilon}}\, e^{i(\boldsymbol{k}\cdot\boldsymbol{r}-\omega_n\tau)}$$

$$\mathscr{G}_{\text{free}}(\boldsymbol{r},\tau) = \frac{1}{\beta}\sum_{i\omega_n}\underbrace{\mathscr{G}_{\text{free}}(\boldsymbol{r},i\omega_n)}_{\int_{-\infty}^{\infty}d\varepsilon\frac{A_{\text{free}}(\boldsymbol{r},\varepsilon)}{i\omega_n-\varepsilon}}\, e^{-i\omega_n\tau}.$$

$A(\boldsymbol{k},\varepsilon)$ and $A_{\text{free}}(\boldsymbol{r},\varepsilon)$ are the total spectral functions of the bulk and free Green's functions, respectively. We shall always denote the vector potential by a bold $\boldsymbol{A}$ and the spectral function by a roman A, which should preclude any confusion. The time integrations can now be evaluated:

$$\int_0^\beta d\tau d\tau'\, e^{i\bar{\Omega}_1\tau}e^{i\bar{\Omega}_2\tau'}e^{-i\omega_n(\tau'-\tau)}e^{-i\bar{\omega}_1\tau}e^{-i\bar{\omega}_2(-\tau')} = \beta^2\delta_{i\bar{\omega}_1,\, i\omega_n+i\bar{\Omega}_1}\delta_{i\bar{\omega}_2,\, i\omega_n-i\bar{\Omega}_2}.$$

One Matsubara sum remains to be performed before the analytic continuation:

$$\frac{1}{\beta}\sum_{i\omega_n}\frac{1}{i\omega_n-\varepsilon}\frac{1}{i\omega_n+i\bar{\Omega}_1-\varepsilon_1}\frac{1}{i\omega_n-i\bar{\Omega}_2-\varepsilon_2} = \frac{f(\varepsilon)}{(\varepsilon+i\bar{\Omega}_1-\varepsilon_1)(\varepsilon-i\bar{\Omega}_2-\varepsilon_2)}.$$

We have discarded two terms proportional to $f(\varepsilon_1)$ and $f(\varepsilon_2)$. The reason is that ε_1 and ε_2 enter the spectral functions $A_{\text{free}}(\boldsymbol{r},\varepsilon_1)$ and $A_{\text{free}}(\boldsymbol{r},\varepsilon_2)$. Since $A_{\text{free}}(\boldsymbol{k},\varepsilon) = \delta(\varepsilon - E)$ with $E = \frac{\hbar^2\kappa^2}{2m} + \phi$ the energy of the free photo-electron (see figure 7.9), the spectral weight is concentrated at an energy of the order of $\hbar\omega_0$ which justifies the approximation $f(\varepsilon_1) = f(\varepsilon_2) = 0$. These two terms correspond to the two time orderings in equation (7.3) that are such that $\boldsymbol{j}^p(\boldsymbol{R})$ acts directly on the system eigenstates. We can now perform the analytic continuation $i\bar{\Omega}_1 \to \hbar\omega_0^+$ and $i\bar{\Omega}_2 \to -\hbar\omega_0^+$ and reintroduce the retarded and advanced free correlation functions (see equations (3.18) and (3.19)):

$$\frac{dJ^{\text{SA}}(\boldsymbol{n})}{d\Omega} = \frac{-ie^2\hbar^3R^2}{16m^3}[\boldsymbol{n}\cdot(\nabla_{\underline{R}}-\nabla_{R})]\int d\boldsymbol{r}_1 d\boldsymbol{r}_2[\boldsymbol{A}(\boldsymbol{r}_1)\cdot(\nabla_{\underline{r}_1}-\nabla_{r_1})]$$
$$[\boldsymbol{A}(\boldsymbol{r}_2)\cdot(\nabla_{\underline{r}_2}-\nabla_{r_2})]\frac{1}{\mathscr{V}}\sum_{\boldsymbol{k}}e^{i\boldsymbol{k}\cdot(\boldsymbol{r}_1-\underline{\boldsymbol{r}}_2)}\int_{-\infty}^{\infty}d\varepsilon\, A(\boldsymbol{k},\varepsilon)f(\varepsilon)$$
$$\underbrace{\int_{-\infty}^{\infty}d\varepsilon_1\frac{A_{\text{free}}(\boldsymbol{R}-\underline{\boldsymbol{r}}_1,\varepsilon_1)}{\varepsilon+\hbar\omega_0-\varepsilon_1+i0^+}}_{G^R_{\text{free}}(\boldsymbol{R}-\underline{\boldsymbol{r}}_1,\,\varepsilon+\hbar\omega_0)}\underbrace{\int_{-\infty}^{\infty}d\varepsilon_2\frac{A_{\text{free}}(\boldsymbol{r}_2-\underline{\boldsymbol{R}},\varepsilon_2)}{\varepsilon+\hbar\omega_0-\varepsilon_2-i0^+}}_{G^A_{\text{free}}(\boldsymbol{r}_2-\underline{\boldsymbol{R}},\,\varepsilon+\hbar\omega_0)}.$$

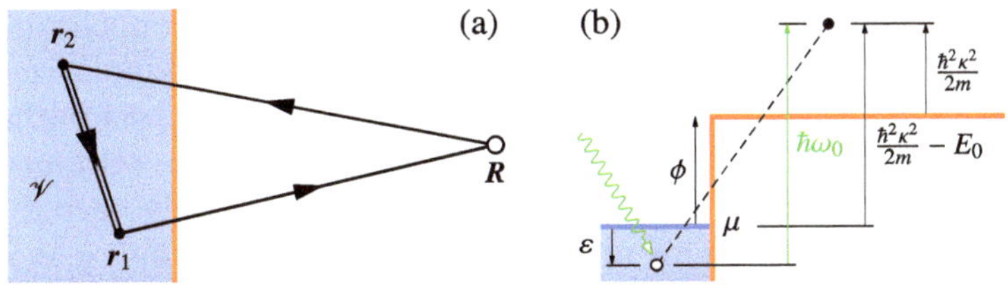

Figure 7.9. The sudden approximation for photoemission. (a) The sample is treated as a bulk material of volume $\mathscr{V}$, overlooking the presence of the surface, and the propagation from within the sample to the detector at point $\boldsymbol{R}$ is assumed to be free. (b) Energy diagram. Within the sample, the energies ε are measured from the chemical potential μ. ϕ is the surface work function and $\hbar\omega_0$ is the photon energy. The photo-electron is directly excited into the vacuum and κ measures its kinetic energy above the vacuum level. In order to refer this kinetic energy to the chemical potential μ, we must set $E_0 = -\phi$.

We have dropped the second term with ω_0 replaced by $-\omega_0$. This term involves the free propagators at energy $\varepsilon - \hbar\omega_0 < 0$ (remember that $\hbar\omega_0 > 0$ and that $\varepsilon \lesssim 0$ due to the product $A(\boldsymbol{k}, \varepsilon)f(\varepsilon)$). At negative energy, the free propagators $G^{R,A}_{\text{free}}$ decay exponentially like $e^{-|\text{Im}(\kappa)|r}$ because κ is imaginary (see Doc A7.1): such contributions disappear as $R \to \infty$. This represents the physically obvious fact that only electrons can propagate to the detector, not holes. Using the relations $G^R_{\text{free}}(\boldsymbol{r}, \varepsilon) = \mathscr{G}_{\text{free}}(\boldsymbol{r}, \varepsilon + i0^+)$, $G^A_{\text{free}}(\boldsymbol{r}, \varepsilon) = \mathscr{G}_{\text{free}}(\boldsymbol{r}, \varepsilon - i0^+)$, and the expansion of $\mathscr{G}_{\text{free}}$ at large distances (see Doc A7.1), we are now ready to evaluate the gradients involving $\boldsymbol{R}$ and $\underline{\boldsymbol{R}}$. As we measure the energies relative to the chemical potential and since the photo-electrons are at rest in the vacuum if their excitation energy above the chemical potential equals the surface work function (see the energy diagram in figure 7.9), we conclude that the energy E_0 must be set to $E_0 = -\phi$ and that κ must be taken as $\kappa = \sqrt{2m(\varepsilon + \hbar\omega_0 - \phi)/\hbar^2}$ when we use the free propagator of Doc A7.1. After checking the identity (the limit $\underline{\boldsymbol{R}} \to \boldsymbol{R}$ is tacit)

$$R^2[\boldsymbol{n} \cdot (\nabla_{\underline{\boldsymbol{R}}} - \nabla_{\boldsymbol{R}})]G^R_{\text{free}}(\boldsymbol{R} - \underline{\boldsymbol{r}}_1, \varepsilon + \hbar\omega_0)G^A_{\text{free}}(\boldsymbol{r}_2 - \underline{\boldsymbol{R}}, \varepsilon + \hbar\omega_0) = -\frac{im^2\kappa}{2\pi^2\hbar^4}e^{i\kappa\boldsymbol{n}\cdot(\boldsymbol{r}_2 - \underline{\boldsymbol{r}}_1)},$$

inserting it in the expression of the photo-current, and noting that the reality of the vector potential implies $\boldsymbol{A}(-\boldsymbol{q}) = \boldsymbol{A}^*(\boldsymbol{q})$, we obtain

$$\begin{aligned}\frac{dJ^{\text{SA}}(\boldsymbol{n})}{d\Omega} &= -\frac{e^2}{32\pi^2 m\hbar}\frac{1}{\mathscr{V}}\sum_{\boldsymbol{k}}\int_{-\infty}^{\infty} d\varepsilon\, A(\boldsymbol{k}, \varepsilon)f(\varepsilon)\kappa(\varepsilon)\\ &\times \underbrace{\int d\boldsymbol{r}_1\, \boldsymbol{A}(\boldsymbol{r}_1)\cdot(\nabla_{\underline{\boldsymbol{r}}_1} - \nabla_{\boldsymbol{r}_1})e^{i\boldsymbol{k}\cdot\boldsymbol{r}_1}e^{-i\kappa\boldsymbol{n}\cdot\underline{\boldsymbol{r}}_1}}_{-i(\kappa\boldsymbol{n} + \boldsymbol{k})\cdot\boldsymbol{A}(\kappa\boldsymbol{n} - \boldsymbol{k})}\underbrace{\int d\boldsymbol{r}_2\, \boldsymbol{A}(\boldsymbol{r}_2)\cdot(\nabla_{\underline{\boldsymbol{r}}_2} - \nabla_{\boldsymbol{r}_2})e^{-i\boldsymbol{k}\cdot\underline{\boldsymbol{r}}_2}e^{i\kappa\boldsymbol{n}\cdot\boldsymbol{r}_2}}_{-i(\boldsymbol{k} + \kappa\boldsymbol{n})\cdot\boldsymbol{A}(\boldsymbol{k} - \kappa\boldsymbol{n})}\\ &= \frac{e^2}{32\pi^2 m\hbar}\frac{1}{\mathscr{V}}\sum_{\boldsymbol{k}}\int_{-\infty}^{\infty} d\varepsilon\, A(\boldsymbol{k}, \varepsilon)f(\varepsilon)\kappa(\varepsilon)|(\kappa\boldsymbol{n} + \boldsymbol{k})\cdot\boldsymbol{A}(\kappa\boldsymbol{n} - \boldsymbol{k})|^2.\end{aligned}$$

The last step is to remember that we have been considering a monochromatic light of frequency ω_0; we furthermore assume that the light is linearly polarized and takes the simple form $\boldsymbol{A}(\boldsymbol{r}, t) = \boldsymbol{A}\cos(\boldsymbol{q}_0 \cdot \boldsymbol{r} - \omega_0 t)$ with $q_0 = \omega_0/c$. The Fourier transform is $\boldsymbol{A}(\boldsymbol{q}, \omega) = \pi\mathcal{V}\boldsymbol{A}[\delta_{\boldsymbol{q}\boldsymbol{q}_0}\delta(\omega - \omega_0) + \delta_{\boldsymbol{q}, -\boldsymbol{q}_0}\delta(\omega + \omega_0)]$. The term with $\omega = -\omega_0$ gives no contribution as we have seen above. We therefore need only keep the first term, i.e. we take $\boldsymbol{A}(\boldsymbol{q}) = \mathcal{V}\boldsymbol{A}\delta_{\boldsymbol{q}\boldsymbol{q}_0}$. This implies that $\kappa\boldsymbol{n} - \boldsymbol{k} = \boldsymbol{q}_0$ or $\boldsymbol{k} + \boldsymbol{q}_0 = \kappa\boldsymbol{n}$, which is the expected momentum-conservation rule if we interpret $\boldsymbol{k}$ as the wave vector of the electron before it is excited and $\kappa\boldsymbol{n}$ as the wave vector of the photo-electron. This leads us to our final result:

$$\frac{dJ^{\mathrm{SA}}(\boldsymbol{n})}{d\Omega} = \frac{e^2\mathcal{V}}{8\pi^2 m\hbar}\int_{-\infty}^{\infty} d\varepsilon \left| \boldsymbol{A} \cdot \left(\boldsymbol{n} - \frac{\boldsymbol{q}_0}{2\kappa}\right)\right|^2 \kappa^3(\varepsilon)\, A(\kappa\boldsymbol{n} - \boldsymbol{q}_0, \varepsilon) f(\varepsilon).$$

For a light of energy 1 keV (x-rays), the wave vector $q_0 = 0.51$ Å^{-1} is similar to the typical electron wave vectors in a crystal. Such photon energies are usually used to extract core electrons. However, for the lights of energy $\lesssim 20$ eV that are generally used in valence-electron photoemission, the wave vector $q_0 \lesssim 10^{-2}$ Å^{-1} is negligible compared with the electron wave vectors and we can assume that the light induces purely $\boldsymbol{q} = 0$ excitations. In this latter case, taking $\boldsymbol{q}_0 = 0$, we arrive at the expression given in equation (7.8).

Appendix B7

Bonus B7.1 Golden-rule calculation of the photoemission intensity

The momentum-resolved photoemission process may be viewed as an absorption of a photon that leaves the system in an excited state containing one free electron of wave vector $\boldsymbol{k}$. More precisely, the state of the photo-electron is a so-called time-reversed LEED state, which is built by matching an outgoing plane wave outside the material with a high-energy excited state inside the material; such LEED states with ingoing plane waves are used for the description of low-energy electron diffraction. To evaluate the absorption rate, we use the following version of the Fermi golden rule:

$$\Gamma = \frac{2\pi}{\hbar}\sum_{ab}\frac{e^{-\beta K_b}}{Z}|\langle a|V|b\rangle|^2\delta(\hbar\omega_0 + K_b - K_a).$$

Three modifications were brought to equation (2.27). First, we have replaced the t-matrix by its first-order value, namely V. Second, instead of a single initial state, we perform a thermal average over all possible initial states here denoted $|b\rangle$, which are eigenstates of K, with Boltzmann weight $e^{-\beta K_b}/Z$; we also sum over all possible final states $|a\rangle$. Third, we have generalized the golden rule to the case of an oscillating perturbation as discussed after equation (2.27) and kept only the term corresponding to the photon absorption. In the expression of Γ, the final state is unconstrained if not for the energy conservation $K_a = K_b + \hbar\omega_0$. As V conserves the number of particles, the final state has the same number of electrons as the initial state. In the photoemission experiment, there is one additional constraint since the only final states measured are those containing the photo-electron. In the sudden approximation, this is made explicit by replacing in the expression of the absorption rate $|a\rangle$ by $c_k^\dagger|a\rangle$ and K_a by $E_a + \varepsilon_k - \mu(N_a + 1) = K_a + \xi_k$, where $c_k^\dagger$ creates the time-reversed LEED state and ε_k is the energy of the photo electron. In other words, it is assumed that the photo-excitation from the state $|b\rangle$ leaves all of a sudden the system in an eigenstate $|a\rangle$ of energy $E_a = E_b + \hbar\omega_0 - \varepsilon_k$ with one electron less than $|b\rangle$. This is an approximation because the photo-excitation in fact produces a superposition of eigenstates determined by all the processes leading to the relaxation of the photo-hole. The photoemission current $dJ(\boldsymbol{n})/d\Omega = e\Gamma$ in the sudden approximation is therefore

$$\frac{dJ^{\mathrm{SA}}(\boldsymbol{n})}{d\Omega} = \frac{2\pi e}{\hbar}\sum_{ab}\frac{e^{-\beta K_b}}{Z}|\langle a|c_k\Delta|b\rangle|^2\delta(\hbar\omega_0 + K_b - K_a - \xi_k),$$

where $\boldsymbol{n}$ is the direction of $\boldsymbol{k}$ and for V we used the dipole operator (see Doc A2.8)

$$\Delta = \frac{-e}{2m}\sum_{\alpha\beta}\langle\alpha|\boldsymbol{p}\cdot\boldsymbol{A} + \boldsymbol{A}\cdot\boldsymbol{p}|\beta\rangle c_\alpha^\dagger c_\beta \equiv \sum_{\alpha\beta}\Delta_{\alpha\beta}c_\alpha^\dagger c_\beta.$$

We take the vector potential $\boldsymbol{A}$ as constant throughout space and $|\alpha\rangle$, $|\beta\rangle$ are single-electron states with $c_\alpha^\dagger$ and $c_\beta^\dagger$ the corresponding operators, see equation (2.43).

As the LEED orbital created by $c_k^\dagger$ is not occupied in the initial state $|b\rangle$—rather, since we sum over *all* states $|b\rangle$, the states such that $c_k|b\rangle \neq 0$ have exponentially small Boltzmann weight of order $e^{-\hbar\omega_0/k_\mathrm{B}T}$ in the sum—the only way for the matrix element $\langle a|c_k c_\alpha^\dagger c_\beta|b\rangle$ to be nonzero is to have $c_\alpha^\dagger = c_k^\dagger$. We thus get

$$\langle a|c_k\Delta|b\rangle = \sum_\beta \Delta_{k\beta}\langle a|c_k c_k^\dagger c_\beta|b\rangle = \sum_\beta \Delta_{k\beta}\langle a|c_\beta|b\rangle,$$

where we have used $c_k|b\rangle = 0$. More formally, this can be established by means of the commutation rules (2.41), which give $c_k c_\alpha^\dagger c_\beta = \delta_{k\alpha}c_\beta + c_\alpha^\dagger c_\beta c_k$. The photocurrent follows as

$$\frac{dJ^{\mathrm{SA}}(\boldsymbol{n})}{d\Omega} = \frac{2\pi e}{\hbar}\sum_{\alpha\beta}\Delta_{k\alpha}\Delta_{k\beta}^*\sum_{ab}\frac{e^{-\beta K_b}}{Z}\langle a|c_\alpha|b\rangle\langle b|c_\beta^\dagger|a\rangle\delta(\hbar\omega_0 + K_b - K_a - \xi_k).$$

The relation with the spectral function is now easily seen. The occupied part of the single-electron spectral function is (see equations (3.32) and (3.14))

$$A_{\alpha\beta}^<(\varepsilon) \equiv \rho_{c_\alpha c_\beta^\dagger}^<(\varepsilon) = \rho_{c_\alpha c_\beta^\dagger}(\varepsilon)f(\varepsilon) = \sum_{ab}\frac{e^{-\beta K_b}}{Z}\langle a|c_\alpha|b\rangle\langle b|c_\beta^\dagger|a\rangle\delta(\varepsilon + K_a - K_b).$$

A direct comparison leads to

$$\frac{dJ^{\mathrm{SA}}(\boldsymbol{n})}{d\Omega} = \frac{2\pi e}{\hbar}\sum_{\alpha\beta}\Delta_{k\alpha}\Delta_{k\beta}^* A_{\alpha\beta}^<(\xi_k - \hbar\omega_0).$$

$\xi_k - \hbar\omega_0$ is the energy of the photo-hole measured from the chemical potential, which corresponds to ε in equation (7.10), as seen in figure 7.9. In a plane-wave basis, and taking the LEED state as a plane wave of wave vector $\boldsymbol{k}$ for simplicity, the matrix element of the dipole operator is $\Delta_{kk'} = -e\hbar/m\ (\boldsymbol{A} \cdot \boldsymbol{k})\delta_{kk'}$ and we get a result very similar to equation (7.10):

$$\frac{dJ^{\mathrm{SA}}(\boldsymbol{n})}{d\Omega} \propto |\boldsymbol{A} \cdot \boldsymbol{k}|^2 A(\boldsymbol{k}, \varepsilon) f(\varepsilon).$$

While the derivation based on the Fermi golden rule is much simpler than the response theory of section 7.1, it relies heavily on the specific form adopted for the final state. Unlike in the response theory, there is no clear strategy for a systematic improvement of the simplest approximation.

Bonus B7.2 Photoemission matrix element and selection rules

To understand photoemission in real materials, it is necessary to go beyond the idealizations of a translation-invariant sample and plane-wave final state that have been adopted in order to arrive at the simple result (7.10). Still based on the diagram in figure 7.1(b), here we provide a more general expression that takes into account the wave functions of the initial and final states. The calculation proceeds exactly like in Doc A7.2, but with the following replacements:

$$\mathscr{G}(\boldsymbol{r}_1, \underline{\boldsymbol{r}}_2, i\omega_n) \to \sum_\alpha \varphi_\alpha(\boldsymbol{r}_1)\varphi_\alpha^*(\underline{\boldsymbol{r}}_2) \int_{-\infty}^{\infty} d\varepsilon \frac{A_\alpha(\varepsilon)}{i\omega_n - \varepsilon}$$

$$G_{\mathrm{free}}^{R}(\boldsymbol{R} - \underline{\boldsymbol{r}}_1, \varepsilon + \hbar\omega_0) \to -\frac{m\sqrt{\mathscr{V}}}{2\pi\hbar^2}\psi_{\kappa n}^*(\underline{\boldsymbol{r}}_1)\frac{e^{i\kappa R}}{R}$$

$$G_{\mathrm{free}}^{A}(\boldsymbol{r}_2 - \underline{\boldsymbol{R}}, \varepsilon + \hbar\omega_0) \to -\frac{m\sqrt{\mathscr{V}}}{2\pi\hbar^2}\psi_{\kappa n}(\boldsymbol{r}_2)\frac{e^{-i\kappa \underline{R}}}{\underline{R}}.$$

The first expression extends equation (3.27) to the case of interacting particles with the assumption that the interacting Green's function remains diagonal in the basis $\varphi_\alpha(\boldsymbol{r})$ of the non-interacting problem[20]. The free retarded and advanced Green's functions are replaced by expressions in which the plane wave $\mathscr{V}^{-1/2}e^{i\boldsymbol{k}\cdot\boldsymbol{r}}$ at positions inside the sample is replaced by the wave function $\psi_{\boldsymbol{k}}(\boldsymbol{r})$ of the time-reversed LEED state (see Bonus B7.1). This leads us to the following total photo-electron current:

$$\frac{dJ^{\mathrm{SA}}(\boldsymbol{n})}{d\Omega} = \frac{e^2\mathscr{V}}{8\pi^2 m\hbar}\int_{-\infty}^{\infty} d\varepsilon \sum_\alpha |M_\alpha(\kappa\boldsymbol{n})|^2 A_\alpha(\varepsilon) f(\varepsilon) \tag{7.27a}$$

$$M_\alpha(\kappa\boldsymbol{n}) = \frac{i}{2}\sqrt{\kappa(\varepsilon)}\int d\boldsymbol{r}[\boldsymbol{A}(\boldsymbol{r}) \cdot (\nabla_{\underline{\boldsymbol{r}}} - \nabla_{\boldsymbol{r}})]\varphi_\alpha(\boldsymbol{r})\psi_{\kappa n}^*(\underline{\boldsymbol{r}}). \tag{7.27b}$$

The result has the same general form as equation (7.8), but with a more involved expression for the matrix element. In the $\boldsymbol{q} \to 0$ limit for the electromagnetic field,

[20] The generalization is not difficult but involves a non-diagonal spectral function $A_{\alpha\beta}(\varepsilon)$.

where $A(r) \equiv A$, we can integrate by parts the term involving the gradient of $\varphi_\alpha(r)$ and thus transfer the gradient to the final state. The matrix element simplifies to

$$M_\alpha(\kappa n) = i\sqrt{\kappa(\varepsilon)} \int dr\ \varphi_\alpha(r)(A \cdot \nabla_r)\psi^*_{\kappa n}(r) \qquad (7.27c)$$
$$\sim \kappa^{3/2}(A \cdot n)\varphi_\alpha(\kappa n).$$

Because the final state is not far from the plane wave $\psi^*_{\kappa n}(r) \sim e^{-i\kappa n\cdot r}$, we can remember as a rule of thumb that the matrix element is proportional to the Fourier component of the initial wave function at the wave vector κn of the photo-electron, corrected by a geometric factor depending on the light polarization. If the initial and final states are expanded on a basis of angular-momentum eigenstates $|n\ell m\rangle$, the matrix element becomes a superposition of matrix elements of the form $\langle n\ell m|A \cdot p|n'\ell' m'\rangle$ which obey the standard dipole selection rules.

As a case study, let us consider a Bloch crystal and ignore final-state and surface effects. In order to describe this situation, we make the substitutions

$$\varphi_\alpha(r) \to \varphi_{k\nu}(r) = \frac{1}{\sqrt{\mathcal{V}}}\sum_G u_{k\nu}(G)e^{i(k+G)\cdot r} \quad \text{and} \quad \psi_{\kappa n}(r) \to \frac{1}{\sqrt{\mathcal{V}}}e^{i\kappa n\cdot r},$$

where k is in the first Brillouin zone and ν is a band index. We furthermore restrict to a constant vector potential and arrive at

$$\frac{d^2 J^{\rm SA}(n, \varepsilon)}{d\Omega d\varepsilon} = \frac{e^2\mathcal{V}}{8\pi^2 m\hbar}\kappa^3|A \cdot n|^2|u_{k\nu}(\kappa n - k)|^2 A_{k\nu}(\varepsilon)f(\varepsilon).$$

The wave vector k is the irreducible value of the photo-electron wave vector κn in the first Brillouin zone. This result resembles equation (7.10), except that the photoemission intensity is modulated by the Fourier component of the initial-state wave function in the zone $G = \kappa n - k$. For electrons in tight orbitals, the wave function has components $u_{k\nu}(G)$ up to large values of G and the photoemission signal spreads over several Brillouin zones. The more extended the Bloch states are, the more concentrated $u_{k\nu}(G)$ is close to $G = 0$ and the more the signal tends to be confined to a single Brillouin zone, the one closest to κn. Because κ increases like the square root of the photon energy, the signal may shift from one zone to another as $\hbar\omega_0$ varies.

Another approach is to expand the Bloch waves on localized Wannier orbitals. The Wannier functions are defined as $|W_{R_n\nu}\rangle = \mathcal{N}^{-1/2}\sum_k e^{-ik\cdot R_n}|\varphi_{k\nu}\rangle$ with the sum extending over the first Brillouin zone. They are localized around the cell R_n and have the symmetry of the orbital(s) composing the band ν. The Bloch states are expanded as $|\varphi_{k\nu}\rangle = \mathcal{N}^{-1/2}\sum_{R_n} e^{ik\cdot R_n}|W_{R_n\nu}\rangle$. The matrix element (7.27c) becomes

$$M_{k\nu}(\kappa n) = \sqrt{\kappa}\int dr\ \frac{1}{\sqrt{\mathcal{N}}}\sum_{R_n} e^{ik\cdot R_n}W_{R_n\nu}(r)\ \underbrace{(A \cdot i\nabla_r)\psi^*_{\kappa n}(r)}_{\frac{1}{\sqrt{\mathcal{V}}}\kappa(A\cdot n)e^{-i\kappa n\cdot r}}$$
$$= \kappa^{3/2}(A \cdot n)\frac{1}{\sqrt{\mathcal{N}\mathcal{V}}}\sum_{R_n} e^{ik\cdot R_n}\int dr'\ \underbrace{W_{R_n\nu}(r' + R_n)}_{=W_{0\nu}(r')}e^{-i\kappa n\cdot(r'+R_n)}$$
$$= \kappa^{3/2}(A \cdot n)\frac{\delta_{k+G,\,\kappa n}}{\sqrt{\mathcal{V}_{\rm cell}}}\int dr\ W_{0\nu}(r)e^{-i\kappa n\cdot r}.$$

The momentum conservation expresses that $\kappa\boldsymbol{n} = \boldsymbol{k} + \boldsymbol{G}$ with $\boldsymbol{k}$ in the first Brillouin zone and $\boldsymbol{G}$ some reciprocal-space vector. We find that the matrix element is proportional to the Fourier component of the Wannier function at wave vector $\kappa\boldsymbol{n}$. We can study this Fourier component by expanding the integrand on angular-momentum eigenstates. It is a well-known result of scattering theory that the plane wave has a representation in terms of spherical waves,

$$e^{-i\boldsymbol{k}\cdot\boldsymbol{r}} = 4\pi\sum_{\ell m}(-i)^{\ell}j_{\ell}(kr)Y_{\ell}^{m}(\vartheta_{\boldsymbol{k}}, \varphi_{\boldsymbol{k}})Y_{\ell}^{m}(\vartheta_{\boldsymbol{r}}, \varphi_{\boldsymbol{r}})^{*},$$

where $j_{\ell}(x)$ are the spherical Bessel functions, $Y_{\ell}^{m}(\vartheta, \varphi)$ are the spherical harmonics, $\boldsymbol{k} = k(\sin\vartheta_{\boldsymbol{k}}\cos\varphi_{\boldsymbol{k}}, \sin\vartheta_{\boldsymbol{k}}\sin\varphi_{\boldsymbol{k}}, \cos\vartheta_{\boldsymbol{k}})$, and similarly for $\boldsymbol{r}$. In general, the Wannier function involves series of spherical waves centered on different atoms in the unit cell. For the sake of argument, we consider a single component:

$$W_{0\nu}(\boldsymbol{r}) \propto R_{n\ell}(r)Y_{\ell}^{m}(\vartheta_{\boldsymbol{r}}, \varphi_{\boldsymbol{r}}).$$

The orthogonality of the spherical harmonics yields the Fourier transform

$$W_{0\nu}(\kappa\boldsymbol{n}) \propto Y_{\ell}^{m}(\vartheta_{\boldsymbol{n}}, \varphi_{\boldsymbol{n}})\int_{0}^{\infty} dr\, r^{2}R_{n\ell}(r)j_{\ell}(\kappa r). \tag{7.28}$$

The key observation here is that the dependence of the matrix element on sample orientation follows the angular dependence of the wave function and, consequently, the photoemission intensity in principle allows one to determine the symmetry of the initial state. The radial integral introduces a dependence on κ—hence on $\hbar\omega_0$—that is non-monotonic and can lead to extinction of the intensity at particular photon energies. Figure 7.10 depicts the sample-orientation and photon-energy dependencies of the matrix element for the case of a $5d_{x^2-y^2}$ initial state ($n = 5$, $\ell = 2$, $m = \pm 2$). The angular dependence is given by $\frac{1}{2}|Y_2^2(\vartheta, \varphi) + Y_2^{-2}(\vartheta, \varphi)|^2 = \frac{15}{16\pi}\sin^4(\vartheta)\cos^2(2\varphi)$, shown in the figure as a stereographic projection. For the radial wave function, we used the hydrogen-like form

$$R_{n\ell}(r) = \sqrt{\left(\frac{2Z}{na_0}\right)^3\frac{(n-\ell-1)!}{2n(n+\ell)!}}\, e^{-\frac{Zr}{na_0}}\left(\frac{2Zr}{na_0}\right)^{\ell}L_{n-\ell-1}^{2\ell+1}\left(\frac{2Zr}{na_0}\right),$$

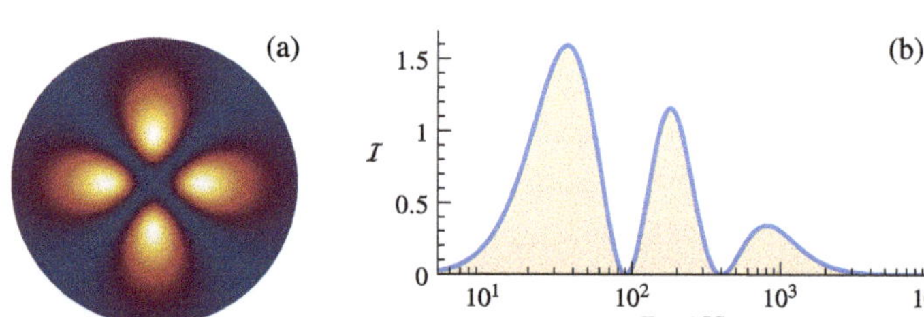

Figure 7.10. (a) Stereographic view of the photoemission matrix element for an initial state of $d_{x^2-y^2}$ symmetry. The factor $|\boldsymbol{A}\cdot\boldsymbol{n}|^2$ is not included. (b) Dependence of the matrix element on the photo-electron kinetic energy for a $5d$ state with effective nuclear charge $Z = 18.7$.

where L are the generalized Laguerre polynomials, a_0 is the Bohr radius, and we took the value $Z = 18.7$, which is the effective nuclear charge for the $5d$ shell of iridium. The radial integral is known analytically and yields the result

$$\mathcal{I} = \kappa^3 \left| \int_0^\infty dr\, r^2 R_{5,2}(r) j_2(\kappa r) \right|^2 = \frac{2^{11} 7\, x^{7/2}}{(1+x)^{12}} \left(1 - \frac{18}{7}x + x^2\right)^2, \quad x = \frac{E_{\text{kin}}}{(Z/5)^2 \text{Ry}},$$

where we used $\kappa = \sqrt{2mE_{\text{kin}}/\hbar^2}$ and $\text{Ry} = \hbar^2/(2ma_0^2) = 13.6$ eV is the Rydberg. Figure 7.10(b) shows that this function defines three energy ranges favorable for photoemission separated by blind windows where the intensity is suppressed.

Bonus B7.3 Analytic properties of the self-energy

The self-energy is a key quantity since it encodes all effects that cannot be understood on the basis of mean-field approximations. Obviously, this is also the reason why the self-energy is most often impossible to calculate exactly. To check the validity of approximations or the consistency of self-energies derived from experiments, it is useful to notice a few analytic properties that the self-energy must obey on general grounds[21]. We restrict here to translation-invariant systems and we regard the self-energy $\Sigma(\boldsymbol{k}, z)$ as a function of the complex energy z.

A first property is that the self-energy is analytic everywhere in the complex plane, except possibly on the real axis $\text{Im}\, z = 0$. This follows directly from Dyson's equation $\Sigma(\boldsymbol{k}, z) = \mathcal{G}_0^{-1}(\boldsymbol{k}, z) - \mathcal{G}^{-1}(\boldsymbol{k}, z)$ using the facts that both $\mathcal{G}_0$ and $\mathcal{G}$ are analytic everywhere except on the real axis and that neither $\mathcal{G}_0$ nor $\mathcal{G}$ can vanish for $\text{Im}\, z \neq 0$. The analyticity of $\mathcal{G}$ is obvious from the spectral representation (see equation (4.13))

$$\mathcal{G}(\boldsymbol{k}, z) = \int_{-\infty}^{\infty} d\varepsilon \frac{A(\boldsymbol{k}, \varepsilon)}{z - \varepsilon} = \int_{-\infty}^{\infty} d\varepsilon\, (z^* - \varepsilon) \frac{A(\boldsymbol{k}, \varepsilon)}{|z - \varepsilon|^2}.$$

Clearly $\mathcal{G}(\boldsymbol{k}, z)$ cannot be singular if $\text{Im}(z) \neq 0$. From the second equality, since $A(\boldsymbol{k}, \varepsilon)$ is real and non-negative, we find that

$$\text{Im}\, \mathcal{G}(\boldsymbol{k}, z) = -\text{Im}(z) \underbrace{\int_{-\infty}^{\infty} d\varepsilon \frac{A(\boldsymbol{k}, \varepsilon)}{|z - \varepsilon|^2}}_{>0}.$$

The inequality is strict, because for the integrand to vanish $A(\boldsymbol{k}, \varepsilon)$ must be identically zero, which contradicts the sum rule $\int_{-\infty}^{\infty} d\varepsilon\, A(\boldsymbol{k}, \varepsilon) = 1$. Hence $\text{Im}\, \mathcal{G}(\boldsymbol{k}, z)$ and *a fortiori* $\mathcal{G}(\boldsymbol{k}, z)$ cannot vanish for $\text{Im}(z) \neq 0$.

The second property is $\Sigma(\boldsymbol{k}, |z| \to \infty) = \Sigma_{\text{HF}}(\boldsymbol{k})$, where $\Sigma_{\text{HF}}(\boldsymbol{k})$ is the real and energy-independent contribution coming from the mean-field decoupling of the interaction. In the case of the Coulomb interaction, this is the sum of the Hartree and exchange terms calculated in section 5.1.3.7. This property can be established quite generally using the equation-of-motion method. The trick is to note that the

[21] Luttinger J M 1961 *Phys. Rev.* **121** 942.

equation of motion of the Green's function equation (5.114) can be put in the form $-\partial_\tau \mathcal{G}_{\alpha\beta}(\tau) = \delta(\tau)\delta_{\alpha\beta} + [K_0\mathcal{G}(\tau)]_{\alpha\beta} + \Lambda_{\alpha\beta}(\tau)$, with $\Lambda_{\alpha\beta}(\tau) = -\langle T_\tau[a_\alpha, V](\tau)a_\beta^\dagger(0)\rangle$. In the frequency domain and in matrix notation, this reads $(i\omega_n\mathbb{1} - K_0)\mathcal{G}(i\omega_n) = \mathbb{1} + \Lambda(i\omega_n)$. Since $i\omega_n\mathbb{1} - K_0 = \mathcal{G}_0^{-1}(i\omega_n)$ (see equation (5.112)), this is nothing but Dyson's equation provided that we *define* the self-energy by $\Lambda \equiv \Sigma\mathcal{G}$. The relation $\Sigma(z) = \Lambda(z)\mathcal{G}^{-1}(z)$ allows us to investigate the asymptotic properties of $\Sigma(z)$ by studying the moment expansion of $\Lambda(z)$. For $|z| \to \infty$ we have (see section 3.4.5) $\Lambda(z) = M_0/z$, where $[M_0]_{\alpha\beta} = \langle[[a_\alpha, V]_-, a_\beta^\dagger]_{-\eta}\rangle$ is the first moment. Evaluating the matrix element, we find $M_0 = V_\text{H} + V_\text{x}$, with V_H the generic Hartree potential defined in equation (5.119*a*) and V_x the corresponding exchange potential equation (5.119*b*). Therefore, since $\mathcal{G}^{-1}(z) = z$ for $|z| \to \infty$ (see equation (3.45)), we find $\lim_{|z|\to\infty} \Sigma(z) = V_\text{H} + V_\text{x}$.

The third property results from the first two: since $\Sigma(\boldsymbol{k}, z) - \Sigma_\text{HF}(\boldsymbol{k})$ vanishes at infinity and is analytic in the upper half of the complex plane, we can deduce the Kramers–Kronig relations for the retarded self-energy (see Bonus B3.2):

$$\text{Re}\,\Sigma(\boldsymbol{k}, \varepsilon + i0^+) = \Sigma_\text{HF}(\boldsymbol{k}) - \frac{1}{\pi}⨍_{-\infty}^{\infty} d\varepsilon' \frac{\text{Im}\,\Sigma(\boldsymbol{k}, \varepsilon' + i0^+)}{\varepsilon - \varepsilon'} \tag{7.29a}$$

$$\text{Im}\,\Sigma(\boldsymbol{k}, \varepsilon + i0^+) = \frac{1}{\pi}⨍_{-\infty}^{\infty} d\varepsilon' \frac{\text{Re}\,\Sigma(\boldsymbol{k}, \varepsilon' + i0^+) - \Sigma_\text{HF}(\boldsymbol{k})}{\varepsilon - \varepsilon'}. \tag{7.29b}$$

The easiest way to check this is to integrate the function $[\Sigma(\boldsymbol{k}, z) - \Sigma_\text{HF}(\boldsymbol{k})]/(\varepsilon - z)$ along a contour C passing immediately above the real axis, avoiding the pole at ε and closing at infinity in the upper half. The contour contains no pole, such that the integral vanishes. The contribution of the real axis avoiding ε yields a principal value integral and the infinitesimal half-circle around the pole at ε gives $-i\pi$ (we turn counter-clockwise) times the residue which is $-[\Sigma(\boldsymbol{k}, \varepsilon + i0^+) - \Sigma_\text{HF}(\boldsymbol{k})]$:

$$\begin{aligned} 0 &= \oint_C dz \frac{\Sigma(\boldsymbol{k}, z) - \Sigma_\text{HF}(\boldsymbol{k})}{\varepsilon - z} \\ &= ⨍_{-\infty}^{\infty} d\varepsilon' \frac{\Sigma(\boldsymbol{k}, \varepsilon' + i0^+) - \Sigma_\text{HF}(\boldsymbol{k})}{\varepsilon - \varepsilon'} + i\pi[\Sigma(\boldsymbol{k}, \varepsilon + i0^+) - \Sigma_\text{HF}(\boldsymbol{k})]. \end{aligned}$$

The Kramers–Kronig relations result by taking the real and imaginary parts of the above relation.

Bonus B7.4 Modeling the photoemission spectrum of $Bi_2Sr_2CaCu_2O_{8+\delta}$

The photoemission spectrum displayed in figure 7.3 presents several characteristics that are common to many systems: a linear dispersion close to the Fermi energy, symmetric Lorentzian-like momentum distribution curves (MDC), and asymmetric energy distribution curves (EDC) with a long tail. Our purpose here is to show how a simple model can be constructed to reproduce these three features. Simple models of this kind are useful for understanding how such spectra must be understood and help to distinguish what is expected from what is surprising. We assume that the sudden

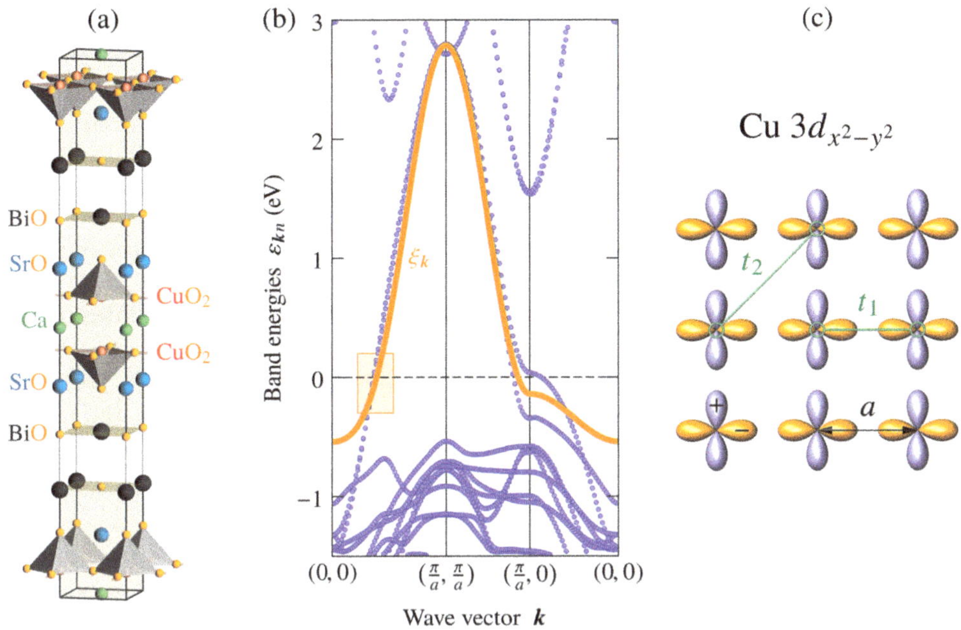

Figure 7.11. (a) Crystal structure of $Bi_2Sr_2CaCu_2O_{8+\delta}$. (b) Band structure plotted along high-symmetry lines in the two-dimensional Brillouin zone of the square CuO_2 lattice. The orange curve shows the model dispersion ξ_k. The orange rectangle indicates the range of the measurement in figure 7.3. (c) Representation of the Cu $3d_{x^2-y^2}$ Wannier functions localized on the Cu atoms and hopping amplitudes t_1 and t_2.

approximation is appropriate here, such that the quantity plotted in figure 7.3 is proportional to $A(\boldsymbol{k}, \varepsilon)f(\varepsilon)$ as we have obtained in equation (7.10). Two ingredients are needed in order to model the spectral function (7.12): the dispersion $\xi_{\boldsymbol{k}}$ and the self-energy $\Sigma(\boldsymbol{k}, \varepsilon)$. We start with a tight-binding model for $\xi_{\boldsymbol{k}} = \varepsilon_{\boldsymbol{k}} - \mu$.

The high-T_c cuprate superconductor $Bi_2Sr_2CaCu_2O_{8+\delta}$ has a layered atomic structure illustrated in figure 7.11(a). Planes of bismuth oxide (BiO), strontium oxide (SrO), copper oxide (CuO_2), and calcium are superposed in a periodic sequence. A band structure calculation[22] yields the result shown in figure 7.11(b): only two bands cross the Fermi energy and contribute to the photoemission signal in the narrow energy range considered in the experiment. It turns out that these two bands originate mostly from the $3d_{x^2-y^2}$ orbitals of the Cu atoms with some admixture of the $2p_x$ and $2p_y$ orbitals of the oxygen atoms located in the CuO_2 planes. The other Cu $3d$ orbitals are pushed to lower energies by the crystal field and are full because Cu has nine electrons in the $3d$ shell. The atomic levels of Sr and Ca are quite far from the Fermi energy, such that these atoms play no role close to ε_F. Finally, the Bi

[22] Lin H, Sahrakorpi S, Markiewicz R S and Bansil A 2006 *Phys. Rev. Lett.* **96** 097001.

$6p$ states are slightly above ε_F and hybridize with the oxygen $2p$ states in the same planes to form the band with the minimum at $(\pi/a, 0)$ near 1.5 eV, well above the ARPES measurement window. The pair of bands crossing ε_F is the bonding–antibonding pair formed by the two sets of $3d_{x^2-y^2}$ orbitals in the two CuO_2 planes of the unit cell, which are weakly coupled across the Ca layer. This coupling lifts the degeneracy of the two bands, except along the $(0, 0)$–$(\pi/a, \pi/a)$ line, precisely the line on which the measurement in figure 7.3 was performed. Therefore, we do not need to take this *bilayer splitting* into account and we can work with a one-band model.

Let $|W_{R_n}\rangle$ be the Wannier functions built from the electronic states in this band, i.e. a collection of orbitals localized at the Cu lattice sites $\boldsymbol{R}_n$ and having $d_{x^2-y^2}$ symmetry. These functions are schematically shown in figure 7.11(c). They are all identical if not for a shift of their origin, such that $W_{\boldsymbol{R}_n}(\boldsymbol{r}) = W_0(\boldsymbol{r} - \boldsymbol{R}_n)$. Let $c^\dagger_{\boldsymbol{R}_n}$ be the operator creating an electron in the state $|W_{\boldsymbol{R}_n}\rangle$. Following equation (2.43), the one-electron Hamiltonian—of which $\varepsilon_{\boldsymbol{k}}$ are the eigenvalues—takes the form

$$H_0 = \sum_{nm} \left\langle W_{\boldsymbol{R}_n} \middle| \frac{p^2}{2m} + V \middle| W_{\boldsymbol{R}_m} \right\rangle c^\dagger_{\boldsymbol{R}_n} c_{\boldsymbol{R}_m} = \sum_{nm} \left\langle W_0 \middle| \frac{p^2}{2m} + V \middle| W_{\boldsymbol{R}_m} \right\rangle c^\dagger_{\boldsymbol{R}_n} c_{\boldsymbol{R}_n+\boldsymbol{R}_m},$$

where $V(\boldsymbol{r})$ is the crystal potential and the discrete translation symmetry of the lattice has been used. Because the Wannier orbitals are localized, the matrix element falls off rapidly as $|\boldsymbol{R}_m|$ increases. For our simple model we shall assume that the matrix element is negligible for $|\boldsymbol{R}_m| > \sqrt{2}\,a$, with a the Cu–Cu distance. Hence we retain only matrix elements between Wannier functions on nearest neighbor and next-nearest neighbor lattice sites. The on-site matrix element $\langle W_0|p^2/2m + V|W_0\rangle$ sets the zero of energy and can be absorbed in the chemical potential. We are left with two distinct matrix elements—or *hopping amplitudes*—namely $t_1 = \langle W_0|p^2/2m + V|W_{\boldsymbol{\delta}_1}\rangle$ with $\boldsymbol{\delta}_1 = \pm a\hat{\boldsymbol{x}}, \pm a\hat{\boldsymbol{y}}$ for nearest-neighbor sites and $t_2 = \langle W_0|p^2/2m + V|W_{\boldsymbol{\delta}_2}\rangle$ with $\boldsymbol{\delta}_2 = a(\pm\hat{\boldsymbol{x}} \pm \hat{\boldsymbol{y}})$ for next-nearest neighbor sites. Explicitly, the Hamiltonian becomes

$$\begin{aligned} H_0 = \sum_n c^\dagger_{\boldsymbol{R}_n}\Big[& t_1(c_{\boldsymbol{R}_n+a\hat{\boldsymbol{x}}} + c_{\boldsymbol{R}_n-a\hat{\boldsymbol{x}}} + c_{\boldsymbol{R}_n+a\hat{\boldsymbol{y}}} + c_{\boldsymbol{R}_n-a\hat{\boldsymbol{y}}}) \\ & + t_2(c_{\boldsymbol{R}_n+a(\hat{\boldsymbol{x}}+\hat{\boldsymbol{y}})} + c_{\boldsymbol{R}_n+a(\hat{\boldsymbol{x}}-\hat{\boldsymbol{y}})} + c_{\boldsymbol{R}_n+a(-\hat{\boldsymbol{x}}+\hat{\boldsymbol{y}})} + c_{\boldsymbol{R}_n+a(-\hat{\boldsymbol{x}}-\hat{\boldsymbol{y}})})\Big]. \end{aligned}$$

Thanks to translation symmetry, we can now diagonalize H_0 by moving to momentum space: we express the real-space operators in terms of $c^\dagger_{\boldsymbol{k}}$ using equation (2.90): $c^\dagger_{\boldsymbol{R}_n} = \mathcal{N}^{-\frac{1}{2}} \sum_{\boldsymbol{k}} e^{-i\boldsymbol{k}\cdot\boldsymbol{R}_n} c^\dagger_{\boldsymbol{k}}$ and $c_{\boldsymbol{R}_m} = \mathcal{N}^{-\frac{1}{2}} \sum_{\boldsymbol{k}'} e^{i\boldsymbol{k}'\cdot\boldsymbol{R}_m} c_{\boldsymbol{k}'}$ ($\mathcal{V}$ has been replaced by $\mathcal{N}$ because here the real space is discrete). This gives

$$
\begin{aligned}
H_0 = \sum_n \frac{1}{\mathcal{N}} \sum_{kk'} e^{-i\boldsymbol{k}\cdot\boldsymbol{R}_n} c_{\boldsymbol{k}}^\dagger \Big\{ & t_1 e^{i\boldsymbol{k}'\cdot\boldsymbol{R}_n} \Big(e^{i\boldsymbol{k}'\cdot a\hat{x}} + e^{-i\boldsymbol{k}'\cdot a\hat{x}} + e^{i\boldsymbol{k}'\cdot a\hat{y}} + e^{-i\boldsymbol{k}'\cdot a\hat{y}} \Big) \\
& + t_2 e^{i\boldsymbol{k}'\cdot\boldsymbol{R}_n} \big[e^{i\boldsymbol{k}'\cdot a(\hat{x}+\hat{y})} + e^{i\boldsymbol{k}'\cdot a(\hat{x}-\hat{y})} + e^{i\boldsymbol{k}'\cdot a(-\hat{x}+\hat{y})} + e^{i\boldsymbol{k}'\cdot a(-\hat{x}-\hat{y})} \big] \Big\} c_{\boldsymbol{k}'} \\
= \sum_{kk'} c_{\boldsymbol{k}}^\dagger c_{\boldsymbol{k}'} \Big\{ & t_1 \Big(e^{ik'_x a} + e^{-ik'_x a} + e^{ik'_y a} + e^{-ik'_y a} \Big) + t_2 \Big[e^{i(k'_x+k'_y)a} \\
& + e^{i(k'_x-k'_y)a} + e^{i(-k'_x+k'_y)a} + e^{i(-k'_x-k'_y)a} \Big] \Big\} \underbrace{\frac{1}{\mathcal{N}} \sum_n e^{i(\boldsymbol{k}'-\boldsymbol{k})\cdot\boldsymbol{R}_n}}_{\delta_{kk'}} \\
= \sum_{k} \varepsilon_{\boldsymbol{k}} c_{\boldsymbol{k}}^\dagger c_{\boldsymbol{k}} \quad & \text{with} \quad \varepsilon_{\boldsymbol{k}} = 2t_1(\cos k_x a + \cos k_y a) + 4t_2 \cos k_x a \cos k_y a.
\end{aligned}
$$

Therefore our model for the dispersion $\xi_{\boldsymbol{k}}$ has three parameters and reads

$$
\xi_{\boldsymbol{k}} = 2t_1(\cos k_x a + \cos k_y a) + 4t_2 \cos k_x a \cos k_y a - \mu. \tag{7.30}
$$

To determine the parameters, we require that the model reproduces the measured Fermi surface shown in figure 7.3. The Fermi surface is defined by the condition $\xi_{\boldsymbol{k}_F} = 0$: it cannot fix all parameters but only two ratios, for instance t_2/t_1 and μ/t_1. The result is $t_2/t_1 = -0.38$ and $\mu/t_1 = 1.18$. With $a = 3.885$ Å, this gives $k_F = 0.446$ Å^{-1} along the line $(0, 0)$–$(\pi/a, \pi/a)$ and $\boldsymbol{k}_F = (\pi/a, 0.115\ \text{Å}^{-1})$ along the line $(\pi/a, 0)$–$(\pi/a, \pi/a)$. To fix t_1, we require that the model reproduces as much as possible the calculated band shown in figure 7.11(b): specifically, we take $t_1 = -416$ meV such that both agree at $(\pi/a, \pi/a)$. The resulting dispersion is shown on top of the calculated band structure in figure 7.11(b) and the corresponding Fermi surface is displayed in figure 7.12(b). Improving the model by adding more hopping amplitudes is straightforward: the dispersion is given in general by $\varepsilon_{\boldsymbol{k}} = \sum_n \langle W_0 | p^2/2m + V | W_{\boldsymbol{R}_n} \rangle e^{i\boldsymbol{k}\cdot\boldsymbol{R}_n}$. However, this increases the number of adjustable parameters, a complication which is unnecessary for the purpose followed here.

We turn now to the self-energy. Most often the self-energy depends much more on energy than it depends on momentum. Our phenomenological model is a momentum-independent self-energy with the following properties. (i) The scattering on impurities and the finite temperature give rise to a constant (energy-independent) scattering rate, such that $\operatorname{Im}\Sigma(\varepsilon = 0) = -\Gamma$. (ii) The electron–electron interaction leads to a ε^2 increase of the scattering rate at low energy. (iii) The imaginary part of Σ vanishes as ε^{-2} at high energy, where the effects of interactions disappear. (iv) The real part of Σ is 'Kramers–Kronig consistent' with the imaginary part, as required by the general analytic properties of the self-energy (see Bonus B7.3). This leads us to the following model:

$$
\Sigma(\varepsilon) = -i\frac{\Gamma + \alpha\varepsilon^2}{1 + (\varepsilon/W)^4} + \frac{1}{\sqrt{2}}\frac{\varepsilon}{W}\frac{\Gamma[1 + (\varepsilon/W)^2] - \alpha W^2[1 - (\varepsilon/W)^2]}{1 + (\varepsilon/W)^4}. \tag{7.31}
$$

The parameter α sets the magnitude of the ε^2 term and the energy scale W controls the crossover from ε^2 to ε^{-2}. The imaginary part is even as a function of ε and the real part

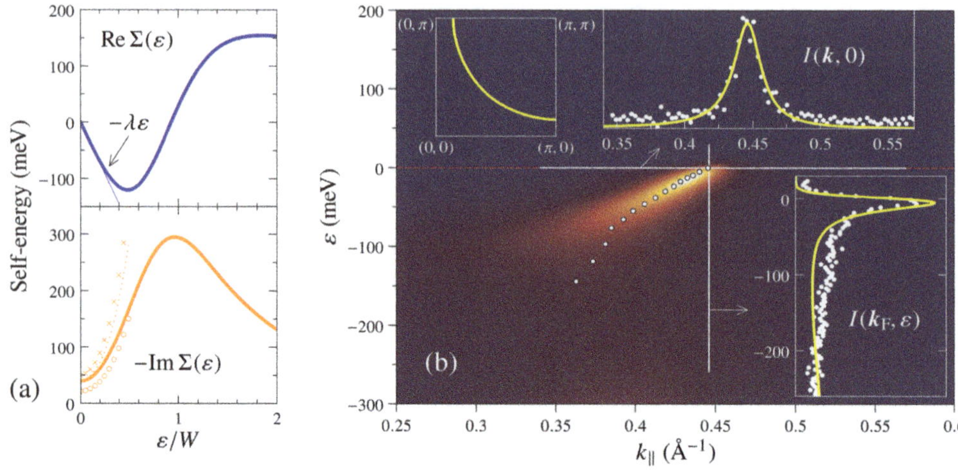

Figure 7.12. (a) Kramers–Kronig consistent model self-energy. The slope of Re Σ at $\varepsilon = 0$ gives the renormalization factor λ. The circles, dotted line, and crosses show the value of $-\mathrm{Im}\,\Sigma$ obtained by fitting the MDC to a Lorentzian, assuming a linear dispersion with $\lambda = 0.5$, 1.8, and 2.5, respectively. (b) Calculated Fermi surface and photoemission intensity. The dots are the experimental results of figure 7.3 and the dotted line is the bare dispersion.

is odd. As Re Σ vanishes at $\varepsilon = 0$, there is no renormalization of the Fermi surface: the maximum of $A(\boldsymbol{k}, 0)$ in the Brillouin zone coincides with the bare Fermi surface. There is a renormalization of the Fermi velocity, though (see also section 7.3):

$$v_{\mathrm{F}} = |\nabla\xi_{\boldsymbol{k}}|_{\boldsymbol{k}_{\mathrm{F}}} \longrightarrow v_{\mathrm{F}}^{*} = |\nabla E_{\boldsymbol{k}}|_{\boldsymbol{k}_{\mathrm{F}}} = \frac{v_{\mathrm{F}}}{1+\lambda} \quad \text{with} \quad \lambda = -\partial_{\varepsilon}\mathrm{Re}\,\Sigma(\varepsilon)|_{\varepsilon=0}.$$

The model gives $\lambda = (\alpha W^2 - \Gamma)/(\sqrt{2}\,W)$ and, with our dispersion parameters, we have $v_{\mathrm{F}} = 3.19$ eV Å at the Fermi crossing along $(0, 0)$–$(\pi/a, \pi/a)$. The measured velocity $v_{\mathrm{F}}^{*} = 1.14$ eV Å implies $\lambda = 1.8$. We substitute this value in the model by eliminating the parameter $\alpha = (\sqrt{2}\,W\lambda + \Gamma)/W^2$. The parameter Γ can be determined by fitting the MDC of figure 7.3, because the spectral function only depends on Γ at $\varepsilon = 0$. The result is $\Gamma = 39$ meV. Finally, fitting the EDC at k_{F} yields $W \approx 200$ meV.

The outcome of the model is compared with the experimental data in figure 7.12. Figure 7.12(a) shows the self-energy. The imaginary part increases like ε^2 at low energy, peaks at $\varepsilon \approx W$ and then decreases. The real part goes like $-\lambda\varepsilon$ at low energy, has a minimum where the imaginary part changes curvature and a zero where the imaginary part is maximal. Figure 7.12(b) displays the calculated ARPES intensity map and the MDC and EDC cuts to be compared with the data in figure 7.3. One sees that the overall behavior is crudely reproduced. In particular, the contrast between a symmetric Lorentzian-like MDC and an asymmetric EDC is a direct consequence of the fact that the self-energy is independent of momentum[23].

[23] A strictly Lorentzian line-shape requires a linear dispersion in addition to the momentum-independent self-energy. The dispersion is nonlinear in our model and the MDC is not exactly Lorentzian.

This feature has an important implication: if the experimental self-energy were indeed momentum independent and *if the bare dispersion were known*, the imaginary part of the self-energy could be extracted exactly from the ARPES data by fitting the MDC at different energies. Unfortunately, the bare dispersion is not *a priori* known. One can start with the linearized form $\xi_k = v_F^*(1+\lambda)(k-k_F^*)$: v_F^* and k_F^* are experimentally measurable quantities, but λ is not. If one uses this form, the extracted self-energy depends on the unknown λ. This is illustrated is figure 7.12(a): the model MDC for energies between 0 and −100 meV were fitted assuming the linearized λ-dependent bare dispersion and the fitted self-energy is shown for $\lambda = 0.5$, 1.8, and 2.5. One sees that the procedure can lead to significant errors on $\text{Im}\,\Sigma$, even if the correct value of λ is used. Extracting reliable self-energies from ARPES data still requires a great deal of expertise…

The model is of course too crude to capture the fine details of the $Bi_2Sr_2CaCu_2O_{8+\delta}$ photoemission data. One can see, in particular, that the shape of the EDC tail is not properly reproduced. The change of slope seen experimentally in the momentum dependence of E_k near −70 meV also goes beyond our model: it is believed to be the signature of the coupling to a collective mode, either a phonon or a spin excitation. In fact, the excitations in $Bi_2Sr_2CaCu_2O_{8+\delta}$ are far from being completely understood.

Bonus B7.5 Electrons coupled to Einstein phonons

At lowest order in the electron–phonon coupling, the self-energy due to the interaction of electrons with phonons is given by equation (5.58). Here we consider the simplest case of a nondispersive optical-phonon branch—in other words Einstein phonons—whose frequency is independent of wave vector: $\omega_{q\lambda} \equiv \Omega_0$. Likewise for the electron–phonon coupling: $g_{q\lambda} \equiv g$. The self-energy is independent of momentum and can be rewritten as an energy integral because the integrand only depends on $\boldsymbol{q}$ through ξ_{k-q}:

$$\Sigma(i\omega_n) = g^2 \int_{-\infty}^{\infty} d\xi\, \tilde{N}_0^{\text{el}}(\xi) \left[\frac{1 - f(\xi) + b(\hbar\Omega_0)}{i\omega_n - \hbar\Omega_0 - \xi} + \frac{f(\xi) + b(\hbar\Omega_0)}{i\omega_n + \hbar\Omega_0 - \xi} \right]. \tag{7.32}$$

$\tilde{N}_0^{\text{el}}(\xi)$ is the density of states per spin direction for the dispersion ξ_k. We can readily deduce the scattering rate, related to the imaginary part of Σ on the real-energy axis, by making the analytic continuation $i\omega_n \to \varepsilon + i0^+$ and using equation (10):

$$\begin{aligned} -\text{Im}\,\Sigma(\varepsilon) = \pi g^2 \Big\{ &[1 - f(\varepsilon - \hbar\Omega_0) + b(\hbar\Omega_0)]\tilde{N}_0^{\text{el}}(\varepsilon - \hbar\Omega_0) + \\ &[f(\varepsilon + \hbar\Omega_0) + b(\hbar\Omega_0)]\tilde{N}_0^{\text{el}}(\varepsilon + \hbar\Omega_0) \Big\}. \end{aligned} \tag{7.33}$$

$-\text{Im}\,\Sigma(\varepsilon)$ gives the scattering rate for an electron or a hole of energy ε. We see two terms, one corresponding to relaxation of electrons above the Fermi energy and one for holes below the Fermi energy. An electron of energy $\varepsilon > \hbar\Omega_0 > 0$ can relax to a lower energy $\varepsilon - \hbar\Omega_0$ by emitting a phonon of frequency Ω_0 at a rate proportional to the number of final states available at the final energy—hence a factor $\tilde{N}_0^{\text{el}}(\varepsilon - \hbar\Omega_0)$—and provided that the final state is not already occupied—hence a factor $1 - f(\varepsilon - \hbar\Omega_0)$. This relaxation rate $\propto [1 - f(\varepsilon - \hbar\Omega_0)]\tilde{N}_0^{\text{el}}(\varepsilon - \hbar\Omega_0)$ is enhanced by $b(\hbar\Omega_0)$: this is the phenomenon of *stimulated emission*; the probability for an electron to decay by emitting a phonon is larger if a phonon is already present. The stimulated emission contributes only for temperatures similar to, or higher than the phonon energy, such that thermally excited phonons are present. The second term in the scattering rate corresponds to the relaxation of a hole at energy $\varepsilon < -\hbar\Omega_0 < 0$. Such a hole can relax to a lower *binding* energy $\varepsilon + \hbar\Omega_0$ by 'absorbing' a phonon[24], provided that there are final states available—factor $\tilde{N}_0^{\text{el}}(\varepsilon + \hbar\Omega_0)$—and that the state at $\varepsilon + \hbar\Omega_0$ is occupied by an electron—factor $f(\varepsilon + \hbar\Omega_0)$.

In order to discuss the spectral function, we also need the real part of the self-energy. One possible approach is to use the Kramers–Kronig relations (see Bonus B7.3) and the known imaginary part. We follow another route by calculating both the real and imaginary parts together. We take the simplest possible model for the density of states: $\tilde{N}_0^{\text{el}}(\varepsilon) = \tilde{N}_0^{\text{el}}(0)$ for $|\varepsilon| < W$ and $\tilde{N}_0^{\text{el}}(\varepsilon) = 0$ for $|\varepsilon| > W$. We assume that the bandwidth $2W$ is much larger than $\hbar\Omega_0$ and k_BT and also much larger than the energies ε of interest (i.e. a few times $\hbar\Omega_0$). This will allow us to obtain an expression for the self-energy that in independent of the bandwidth and only depends on three parameters: the product $g^2\tilde{N}_0^{\text{el}}(0)$ controlling the overall strength of the coupling, the phonon frequency Ω_0, and the temperature. The term proportional to $b(\hbar\Omega_0)$ is readily evaluated and yields

$$\begin{aligned}\Sigma_b(\varepsilon) &= g^2\tilde{N}_0^{\text{el}}(0)b(\hbar\Omega_0)\int_{-W}^{W} d\xi\left(\frac{1}{\varepsilon - \hbar\Omega_0 - \xi + i0^+} + \frac{1}{\varepsilon + \hbar\Omega_0 - \xi + i0^+}\right)\\ &= g^2\tilde{N}_0^{\text{el}}(0)b(\hbar\Omega_0)\left\{\ln\left|\frac{(W+\varepsilon)^2 - (\hbar\Omega_0)^2}{(W-\varepsilon)^2 - (\hbar\Omega_0)^2}\right|\right.\\ &\quad\left. - i\pi[\theta(W - |\varepsilon - \hbar\Omega_0|) + \theta(W - |\varepsilon + \hbar\Omega_0|)]\right\}. \end{aligned} \tag{7.34}$$

The logarithm in the real part behaves as $4\varepsilon/W$ at low energy and drops in the limit $W \gg |\varepsilon|$. The imaginary part survives in that limit and gives $-2\pi g^2\tilde{N}_0^{\text{el}}(0)b(\hbar\Omega_0)$. This constant term is only relevant if $k_BT \gtrsim \hbar\Omega_0$ due to the occupation $b(\hbar\Omega_0)$: we drop it as well in the following since we are interested in the opposite regime $k_BT \ll \hbar\Omega_0$.

[24] The actual physical process is that an electron initially at energy $\varepsilon + \hbar\Omega_0$ below the Fermi level recombines with the hole at ε by emitting a phonon, thus leaving a hole at $\varepsilon + \hbar\Omega_0$; there is in fact no phonon absorption which would require a real phonon to be present (see section 5.1.3.8).

The contribution involving the Fermi factors is convergent at high energy and does not require the introduction of a bandwidth. Using $1 - f(\xi) = f(-\xi)$ and changing variable from ξ to $-\xi$ in the first term, we see that the integral to evaluate is

$$\begin{aligned}\mathcal{I} &= \int_{-\infty}^{\infty} f(\xi)\left[\frac{1}{\xi - (-i\omega_n + \hbar\Omega_0)} - \frac{1}{\xi - (i\omega_n + \hbar\Omega_0)}\right]\\ &= -\frac{1}{\beta}\sum_{i\omega_{n'}}\int_{-\infty}^{\infty}\frac{1}{\xi - i\omega_{n'}}\left[\frac{1}{\xi - (-i\omega_n + \hbar\Omega_0)} - \frac{1}{\xi - (i\omega_n + \hbar\Omega_0)}\right].\end{aligned}$$

The first line shows why this contribution is independent of the bandwidth: the integrand is cut by the Fermi function at $\xi \to \infty$ and behaves as $-2i\omega_n/\xi^2$ for $\xi \to -\infty$. At the second line, we have used equation (16) and expressed the Fermi function as a sum over a set of Matsubara frequencies $i\omega_{n'}$. In this form, the integral is readily performed using the residue theorem by closing the contour in the upper or lower half of the complex plane, where the integrand vanishes. If $\omega_{n'} < 0$, we close the contour in the upper half: the pole at $\xi = i\omega_{n'}$ lies outside the contour and does not contribute, the pole at $\xi = -i\omega_n + \hbar\Omega_0$ contributes if $\omega_n < 0$, and the pole at $\xi = i\omega_n + \hbar\Omega_0$ contributes if $\omega_n > 0$. If $\omega_{n'} > 0$, we close the contour in the lower half instead. This leads to (see equation (7))

$$\begin{aligned}\mathcal{I} &= -\frac{2\pi i}{\beta}\sum_{i\omega_{n'}}\theta(-\omega_{n'})\left[\frac{\theta(-\omega_n)}{-i\omega_n + \hbar\Omega_0 - i\omega_{n'}} - \frac{\theta(\omega_n)}{i\omega_n + \hbar\Omega_0 - i\omega_{n'}}\right]\\ &\quad + \frac{2\pi i}{\beta}\sum_{i\omega_{n'}}\theta(\omega_{n'})\left[\frac{\theta(\omega_n)}{-i\omega_n + \hbar\Omega_0 - i\omega_{n'}} - \frac{\theta(-\omega_n)}{i\omega_n + \hbar\Omega_0 - i\omega_{n'}}\right]\\ &= -\frac{2\pi i}{\beta}\sum_{n'=0}^{\infty}\left[\frac{1}{i\omega_{n'} + \mathrm{sign}(\omega_n)(i\omega_n - \hbar\Omega_0)} - \frac{1}{i\omega_{n'} + \mathrm{sign}(\omega_n)(i\omega_n + \hbar\Omega_0)}\right].\end{aligned}$$

The last line results after changing variable from $\omega_{n'}$ to $-\omega_{n'}$ in the term involving $\theta(-\omega_{n'})$. This kind of sum over one half of the Matsubara frequencies is typically a case for the digamma function, which is defined for $z \in \mathbb{C}$ as

$$\psi(z) = \lim_{M\to\infty}\left(\ln M - \sum_{n=0}^{M}\frac{1}{n + z}\right). \tag{7.35}$$

The connection becomes clear if we substitute for $i\omega_{n'}$ its expression (4.9) and then rearrange our integral as:

$$\mathcal{I} = \sum_{n'=0}^{\infty}\left\{-\frac{1}{n' + \left[\frac{1}{2} - i\mathrm{sign}(\omega_n)\frac{i\omega_n - \hbar\Omega_0}{2\pi k_\mathrm{B}T}\right]} + \frac{1}{n' + \left[\frac{1}{2} + i\mathrm{sign}(\omega_n)\frac{i\omega_n + \hbar\Omega_0}{2\pi k_\mathrm{B}T}\right]}\right\}.$$

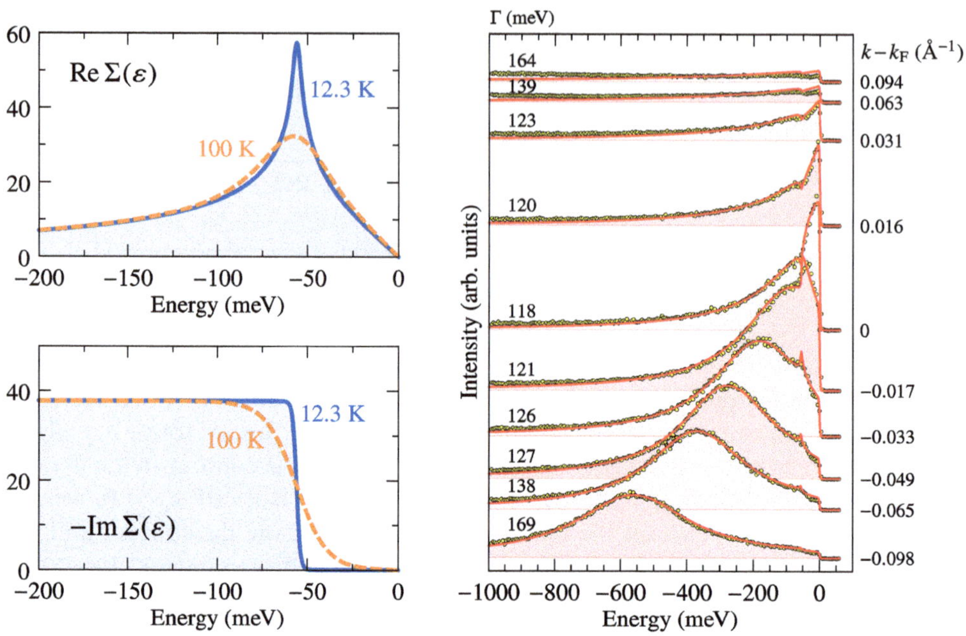

Figure 7.13. Left: self-energy (7.36) due to coupling of electrons with Einstein phonons. The parameters are $g^2\tilde{N}_0^{\text{el}}(0) = 12$ meV, $\hbar\Omega_0 = 56$ meV, and the two indicated temperatures. Right: photoemission data for Be(0001) (circles) and fit (lines). Experimental data from Hengsberger *et al* 1999 *Phys. Rev. Lett.* **83** 592.

We recognize here two sums identical to those entering the digamma function, while the logarithmically divergent terms cancel each other. Our final formula for the self-energy follows after analytic continuation:

$$\Sigma(\varepsilon) = g^2\tilde{N}_0^{\text{el}}(0)\left[\psi\left(\frac{1}{2} - i\,\text{sign}(\varepsilon)\frac{\varepsilon - \hbar\Omega_0}{2\pi k_\text{B}T}\right) - \psi\left(\frac{1}{2} + i\,\text{sign}(\varepsilon)\frac{\varepsilon + \hbar\Omega_0}{2\pi k_\text{B}T}\right)\right]. \quad (7.36)$$

As $T \to 0$, $\Sigma(\varepsilon)$ approaches $g^2\tilde{N}_0^{\text{el}}(0)\left(\ln\frac{\varepsilon - \hbar\Omega_0}{\varepsilon + \hbar\Omega_0} - i\pi\right)$. It is customary to quantify the strength of the electron–phonon coupling by a number λ corresponding to $-\partial_\varepsilon \text{Re}\,\Sigma|_{\varepsilon=0}$ (see section 7.3). In our case, $\lambda = 2g^2\tilde{N}_0^{\text{el}}(0)/(\hbar\Omega_0)$ at $T = 0$. The real part of $\Sigma(\varepsilon)$ is odd with a negative slope $-\lambda$ at $\varepsilon = 0$ and peaks at $|\varepsilon| = \hbar\Omega_0$. The imaginary part is even and negative with a jump at $|\varepsilon| = \hbar\Omega_0$. The peak in Re Σ and the jump in Im Σ are strongly temperature dependent (see figure 7.13).

In spite of its simplicity, this model can be useful for interpreting real data. As an illustration, we consider measurements performed on the Be(0001) surface[25].

[25] Hengsberger M, Purdie D, Segovia P, Garnier M and Baer Y 1999 *Phys. Rev. Lett.* **83** 592. Hengsberger M, Frésard R, Purdie D, Segovia P and Baer Y 1999 *Phys. Rev* B **60** 10796.

The photoemission EDC curves shown in figure 7.13 can be nicely approximated by the function

$$I(k, \varepsilon) \propto -\frac{1}{\pi}\mathrm{Im}\left(\frac{1}{\varepsilon - v_{\mathrm{F}}(k - k_{\mathrm{F}}) - \Sigma(\varepsilon) + i\Gamma}\right)\frac{1}{e^{\frac{\varepsilon}{k_{\mathrm{B}}T}} + 1}.$$

We have assumed a linear dispersion $\xi_k = v_{\mathrm{F}}(k - k_{\mathrm{F}})$ using the values of $k - k_{\mathrm{F}}$ given with the experimental data and a Fermi velocity of 5756 meV Å = 8.8×10^7 cm s^{-1}. The strength of the electron–phonon coupling is $g^2\tilde{N}_0^{\mathrm{el}}(0) = 12$ meV and the phonon energy is $\hbar\Omega_0 = 56$ meV. The temperature is set to $k_{\mathrm{B}}T = 1.06$ meV = 12.3 K like in the experiment. We have added to the model a phenomenological scattering rate Γ, which can account for effects not described by our simple self-energy and was left free to adjust independently for each EDC curve. The values of Γ are indicated on the figure. Clearly, the model captures the main trends seen in the experiment—in particular the peculiar double-peak line shape close to the Fermi point—and allows one to conclude that the electrons on the Be(0001) surface interact with phonons in the 60 meV range. We find an electron–phonon coupling strength $\lambda = 2g^2\tilde{N}_0^{\mathrm{el}}(0)/(\hbar\Omega_0) \approx 0.42$. Hengsberger *et al* report a value of 1.18 based on a different modeling that takes into account, in particular, the finite momentum resolution of the experiment. This relatively large difference in the values of λ shows that extracting reliable physical information from photoemission line shapes is delicate and that a good fit is not always the end of the story. A better modeling of a second self-energy component is probably needed here: one sees that the Γ values are larger than $-\mathrm{Im}\,\Sigma(\varepsilon)$ and increase like $(k - k_{\mathrm{F}})^2$ above a minimum at $k = k_{\mathrm{F}}$, suggesting a superposition of impurity scattering and electron–electron interaction as assumed in the work of Hengsberger *et al* (the finite energy resolution may also contribute to Γ). This second component should ideally be Kramers–Kronig consistent as well and involve a real part which also contributes to the renormalization of the dispersion, like in the model equation (7.31).

IOP Publishing

Spectroscopic Probes of Quantum Matter

Christophe Berthod

Chapter 8

Electrical resistivity

The electrical characterization of new materials or new samples often begins with a measurement of the dc resistivity versus temperature. Such a measurement would normally not be considered as a spectroscopic probe. Yet it does provide access to the electronic self-energy: the present chapter emphasizes this relationship. It is not aimed at a general presentation of resistivity in solids, but at providing a few streetlights along the route that the many-body theory exposed in the previous chapters provides for describing the phenomenon of resistivity. Despite its importance, resistivity is neither straightforward to measure precisely nor easy to calculate theoretically. More often than not, the theoretical description of resistivity is phenomenological rather than first-principles. One difficulty is that several mechanisms contribute to resistivity and a complete theory seems to be out of reach. In a given temperature range, though, the resistivity is often dominated by a single mechanism which can thus be distinguished and studied separately. In the following, we first show how resistivity is connected with the microscopic self-energy via Ohm's law and linear-response theory. We then discuss impurity scattering and electron–electron interaction and their roles in the resistivity of metals. More advanced topics are then addressed, like the Kondo effect and the phenomenon of weak localization.

8.1 Kubo formula for the conductivity

Ohm's law expresses a linear relationship between the electric field E and the current j (i.e. the *electric* current, which is e times the *particle* current) in the form $j = \sigma E$. In this expression, the conductivity σ characterizes the linear response of the system to the electric field E, the response being an induced current j. It is therefore natural to try and express the conductivity as a response function. Like any response theory, Ohm's law must be envisioned in general as a non-local (in space and time) relation between the electric field and the current, which at first order has to take the form

doi:10.1088/978-0-7503-1741-2ch8

$$e\langle j_\mu(\boldsymbol{r},\, t)\rangle_V = \sum_\nu \int d\boldsymbol{r}' \int_{-\infty}^{\infty} dt'\; \sigma_{\mu\nu}(\boldsymbol{r},\, \boldsymbol{r}',\, t - t')E_\nu(\boldsymbol{r}',\, t'). \tag{8.1}$$

In this definition, $j_\mu(\boldsymbol{r},\, t)$ is the particle-current operator defined in equation (2.49), $\boldsymbol{E}(\boldsymbol{r},\, t)$ is the electric field, and $\sigma_{\mu\nu}$ is the *conductivity tensor*. Equation (8.1) has the same form as the general linear-response relation (6.7). Therefore, in order to determine $\sigma_{\mu\nu}$ one has to derive a relation similar to (8.1) from the response theory and deduce the conductivity by identification. The perturbation V due to the electric field is, like in equations (6.17) and (7.1), $V_t = \int d\boldsymbol{r}\, \boldsymbol{j}(\boldsymbol{r}) \cdot (-e)\boldsymbol{A}(\boldsymbol{r},\, t)$. Since equation (8.1) involves the electric field $\boldsymbol{E}$ rather than the vector potential $\boldsymbol{A}$, it is better to work in the gauge where the scalar potential $\phi(\boldsymbol{r},\, t) \equiv 0$ and to use the relation $\boldsymbol{E}(\boldsymbol{r},\, t) = -\partial_t \boldsymbol{A}(\boldsymbol{r},\, t)$ in the form $\boldsymbol{E}(\boldsymbol{r},\, \omega) = i\omega \boldsymbol{A}(\boldsymbol{r},\, \omega)$. Then, from the general principle expressed in equation (6.8) and the fact that the electric field couples to the current operator, one foresees that the conductivity tensor must be related to a current–current correlation function. The explicit derivation is performed in Doc A8.1 and yields

$$\sigma_{\mu\nu}(\boldsymbol{r},\, \boldsymbol{r}',\, \omega) = \frac{ie^2}{\omega}\left[\mathscr{C}_{j_\mu^p(\boldsymbol{r})j_\nu^p(\boldsymbol{r}')}(i\Omega_n \to \hbar\omega + i0^+) + \delta_{\mu\nu}\delta(\boldsymbol{r} - \boldsymbol{r}')\frac{\langle n(\boldsymbol{r})\rangle}{m}\right]. \tag{8.2}$$

The factor $1/\omega$ reflects the fact that the current is not proportional to the vector potential $\boldsymbol{A}$ but to its time derivative $\boldsymbol{E}$. The current–current correlation function involves only the paramagnetic current (2.49*b*), while the contribution of the diamagnetic current (2.49*c*) is given by the second term in the square brackets.

When translation invariance is assumed, the conductivity takes a simpler form in reciprocal space (see Doc A8.1):

$$\boxed{\sigma_{\mu\nu}(\boldsymbol{q},\, \omega) = \frac{ie^2}{\omega}\left[\frac{1}{\mathscr{V}}\chi_{jj}^{\mu\nu}(\boldsymbol{q},\, i\Omega_n \to \hbar\omega + i0^+) + \delta_{\mu\nu}\frac{\langle n\rangle}{m}\right],} \tag{8.3}$$

where we have introduced the current–current correlation function in momentum space

$$\chi_{jj}^{\mu\nu}(\boldsymbol{q},\, \tau) = -\langle T_\tau j_\mu^p(\boldsymbol{q},\, \tau) j_\nu^p(-\boldsymbol{q},\, 0)\rangle. \tag{8.4}$$

The latter two relations provide the starting point for calculating the conductivity from first-principles. The non-translation invariant case is very similar (see Bonus B8.1).

Looking at the expression (2.50) of the paramagnetic current, we see that the correlation function $\chi_{jj}^{\mu\nu}$ can be represented by the generic diagram

$$\chi_{jj}^{\mu\nu}(\boldsymbol{q}, \tau) = - \tag{8.5}$$

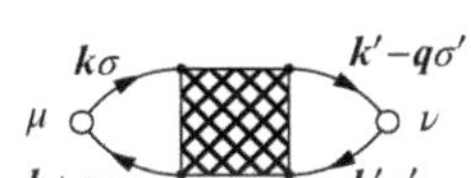

Each circle represents a *bare current vertex*, i.e. $\frac{\hbar}{m}(k_\mu + q_\mu/2)$ on the left side and $\frac{\hbar}{m}(k'_\nu - q_\nu/2)$ on the right side of the diagram if we work in the plane-wave basis.

Apart from this current vertex, the expression of the $\chi_{jj}^{\mu\nu}$ correlator is very similar to the density–density correlator in equation (5.75). In effect, it is possible to derive an explicit relation between the density–density and current–current correlation functions by using the continuity equation (see Doc A2.8), which expresses the conservation of particle number and links the density and current operators. This relation is called a Ward identity[1].

As for the density–density correlation function, we can distinguish two types of diagrams: those corresponding to renormalizations of the propagators and those corresponding to renormalizations of the current vertex. We thus have

$$\chi_{jj}^{\mu\nu}(\boldsymbol{q}, i\Omega_n) = -$$

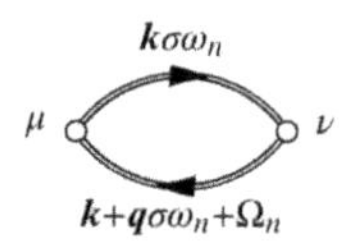

$$+ \quad \text{vertex corrections}$$

$$= \left(\frac{\hbar}{m}\right)^2 \sum_{\boldsymbol{k}\sigma} \left(k_\mu + \frac{q_\mu}{2}\right)\left(k_\nu + \frac{q_\nu}{2}\right) \int_{-\infty}^{\infty} d\varepsilon_1 d\varepsilon_2\, A(\boldsymbol{k}, \varepsilon_1)$$
$$\times A(\boldsymbol{k} + \boldsymbol{q}, \varepsilon_2) \frac{f(\varepsilon_1) - f(\varepsilon_2)}{i\Omega_n + \varepsilon_1 - \varepsilon_2} \quad + \text{ vertex corrections.} \tag{8.6}$$

The calculation leading to the second line proceeds exactly like for the renormalized bubble in Doc A5.13: we use the one-electron spectral function $A(\boldsymbol{k}, \varepsilon) \equiv \rho_{c_{\boldsymbol{k}\sigma}c^\dagger_{\boldsymbol{k}\sigma}}(\varepsilon)$ to replace the product of Green's functions $\mathscr{G}(\boldsymbol{k}, i\omega_n)\mathscr{G}(\boldsymbol{k} + \boldsymbol{q}, i\omega_n + i\Omega_n)$ appearing in the expression of the diagram and we evaluate the Matsubara sum over $i\omega_n$ using equation (16).

We close this section by a derivation of the dc conductivity tensor ($\boldsymbol{q} = 0$ and $\omega = 0$) when the vertex corrections are neglected. Starting from equation (8.3) and using equations (8.6) and (10), we can write the conductivity as

$$\begin{aligned}\operatorname{Re} \sigma_{\mu\nu}(\boldsymbol{q} = 0, \omega) &= -\frac{e^2}{\omega}\frac{1}{\mathscr{V}} \operatorname{Im} \chi_{jj}^{\mu\nu}(\boldsymbol{0}, i\Omega_n \to \hbar\omega + i0^+) \\ &= \frac{\pi e^2\hbar^2}{m^2\omega}\frac{1}{\mathscr{V}} \sum_{\boldsymbol{k}\sigma} k_\mu k_\nu \int_{-\infty}^{\infty} d\varepsilon\, A(\boldsymbol{k}, \varepsilon) A(\boldsymbol{k}, \varepsilon + \hbar\omega) \\ &\quad \times [f(\varepsilon) - f(\varepsilon + \hbar\omega)],\end{aligned} \tag{8.7}$$

since the diamagnetic contribution is purely imaginary. In the dc limit $\omega \to 0$, this expression simplifies using $f(\varepsilon) - f(\varepsilon + \hbar\omega) = -f'(\varepsilon)\hbar\omega + \mathcal{O}(\omega^2)$:

[1] The Ward identity relating the density–density and current–current correlation functions reads

$$\Omega_n^2 \chi_{nn}(\boldsymbol{q}, i\Omega_n) + \hbar^2 \sum_{\mu\nu} q_\mu q_\nu \chi_{jj}^{\mu\nu}(\boldsymbol{q}, i\Omega_n) = -N\frac{\hbar^2 q^2}{m}.$$

It is an instructive exercise to check that this relation is obeyed by free electrons. See also section 8.6.

$$\mathrm{Re}\,\sigma_{\mu\nu}(\mathbf{0},\,0) = \frac{\pi e^2\hbar^3}{m^2}\frac{1}{\mathcal{V}}\sum_{k\sigma} k_\mu k_\nu \int_{-\infty}^{\infty} d\varepsilon\,[-f'(\varepsilon)]A^2(\boldsymbol{k},\,\varepsilon). \tag{8.8}$$

This is the exact expression of the dc conductivity tensor for a translation-invariant system when all vertex corrections are neglected. At $T = 0$, the conductivity is determined by $A^2(\boldsymbol{k}, 0)$; since $A(\boldsymbol{k}, 0)$ gives the zero-energy distribution of spectral weight for single-electron excitations, in other words the region of $\boldsymbol{k}$-space where zero-energy excitations are present, we recognize in equation (8.8) the idea that there must be zero-energy single-electron excitations to produce conductivity (unless vertex corrections introduce another mechanism of conduction). The presence of the square of the spectral function shows that electrical conductivity is produced by *scattering* phenomena close to the Fermi surface, which imply contributions from initial and final states. This view contrasts with the phenomenological interpretation of charge transport by electrons following semiclassical orbits along the Fermi surface. Lastly, we see that equation (8.8) correctly gives an infinite conductivity for systems of independent particles where $A(\boldsymbol{k}, \varepsilon) = \delta(\varepsilon - \xi_{\boldsymbol{k}})$. This is actually expected, since for independent particles there are no vertex corrections and equation (8.8) is exact.

8.2 Derivation of the Drude formula

In this section, we derive the Drude formula starting from equation (8.8). If the material is isotropic, we can obtain the dc conductivity by averaging the three directions:

$$\sigma_{\mathrm{dc}} = \frac{1}{3}\,\mathrm{Re}\big[\sigma_{xx}(\mathbf{0},\,0) + \sigma_{yy}(\mathbf{0},\,0) + \sigma_{zz}(\mathbf{0},\,0)\big]. \tag{8.9}$$

We proceed with a simplifying assumption, namely that the electron self-energy $\Sigma(\boldsymbol{k}, \varepsilon)$ has a weak momentum dependence which can be neglected. This is often a good approximation for three-dimensional metals. It does not need to be true for all momenta, but only in the vicinity of the Fermi surface because momenta far from there do not contribute to equation (8.8). It can also be shown[2] that a momentum-independent self-energy implies that all vertex corrections vanish: equation (8.8) therefore gives the exact conductivity in the framework of this approximation. We then get for σ_{dc} the result (see Doc A8.2):

$$\boxed{\sigma_{\mathrm{dc}} = \frac{ne^2}{m}\int_{-\infty}^{\infty} d\varepsilon\,[-f'(\varepsilon)]\tau_{\mathrm{tr}}(\varepsilon),} \tag{8.10}$$

where we have defined the *transport life-time*

[2] Khurana A 1990 *Phys. Rev. Lett.* **64** 1990.

$$\tau_{\mathrm{tr}}(\varepsilon) = \frac{\hbar}{2|\mathrm{Im}\,\Sigma(\boldsymbol{k}_{\mathrm{F}}, \varepsilon)|}. \tag{8.11}$$

The transport life-time is shorter than the quasi-particle life-time $\tau(\varepsilon)$ defined in equation (7.16*c*) by a factor $Z(\boldsymbol{k}_{\mathrm{F}})$: $\tau_{\mathrm{tr}}(\varepsilon) = Z(\boldsymbol{k}_{\mathrm{F}})\tau(\varepsilon)$. For a momentum-independent self-energy we have $Z(\boldsymbol{k}_{\mathrm{F}}) = m/m^* < 1$. Note also that m in equation (8.10) is the bare electron mass (or the band mass, see Bonus B8.1), not the effective mass m^*. The usual Drude formula $\sigma_{\mathrm{dc}} = ne^2\tau_{\mathrm{tr}}(0)/m$ is recovered at $T=0$ where $-f'(\varepsilon) = \delta(\varepsilon)$ or if $\tau_{\mathrm{tr}}(\varepsilon) \equiv \tau_{\mathrm{tr}}(0)$, since $\int_{-\infty}^{\infty} d\varepsilon\,[-f'(\varepsilon)] = 1$. Equation (8.10) shows that, at finite temperature, the life-time entering the dc conductivity is an average of the transport life-time in a small region of width $\sim k_{\mathrm{B}}T$ around the Fermi surface. This eventually leads to the insightful notion that *the temperature dependence of the resistivity resembles the energy dependence of the self-energy*. The resistivity $\rho = 1/\sigma_{\mathrm{dc}}$ is proportional to the inverse of the thermally-averaged inverse self-energy. This double inversion reduces to the identity at $T=0$ and remains close to an identity at low enough temperatures, provided that the self-energy varies slowly on the scale of $k_{\mathrm{B}}T$. In other words, if we can make sure that

$$\left\{\int_{-\infty}^{\infty} d\varepsilon \frac{[-f'(\varepsilon)]}{|\mathrm{Im}\,\Sigma(\boldsymbol{k}_{\mathrm{F}}, \varepsilon)|}\right\}^{-1} \approx \int_{-\infty}^{\infty} d\varepsilon\,[-f'(\varepsilon)]|\mathrm{Im}\,\Sigma(\boldsymbol{k}_{\mathrm{F}}, \varepsilon)|,$$

then the resistivity is given in terms of the thermally-averaged self-energy. Imagine now that the self-energy is regular close to $\varepsilon = 0$, as it is usually the case, and has a Taylor expansion $|\mathrm{Im}\,\Sigma(\boldsymbol{k}_{\mathrm{F}}, \varepsilon)| = \sum_\ell \alpha_\ell |\varepsilon|^\ell$. Because $-f'(\varepsilon)$ is a function of the form $\frac{1}{T}F(\varepsilon/T)$, we can write

$$\rho \approx \frac{2m}{ne^2\hbar}\sum_\ell \alpha_\ell \underbrace{\int_{-\infty}^{\infty}\frac{d\varepsilon}{T}F\left(\frac{\varepsilon}{T}\right)|\varepsilon|^\ell}_{T^\ell \int_{-\infty}^{\infty} dx\,F(x)|x|^\ell} = \sum_\ell A_\ell T^\ell \tag{8.12}$$

with $A_\ell = \frac{2m}{ne^2\hbar}\alpha_\ell \int_{-\infty}^{\infty} dx\,F(x)|x|^\ell$. This shows that the temperature dependence of the dc resistivity informs us about the energy dependence of the self-energy. For instance, the explanation of a T-linear resistivity is often sought in microscopic models that produce a linear-in-energy self-energy. Of course, these considerations assume that the vertex corrections can be neglected, which cannot always be true. It is known, in particular, that the exact transport life-time (i.e. the value of $\tau_{\mathrm{tr}}(\varepsilon)$ extracted from measurements of the conductivity) generally differs from equation (8.11) and from the quasi-particle life-time measured e.g. by photoemission. Equation (8.10) is nevertheless a good starting point for investigating the effects of various scattering mechanisms on the conductivity, as will be done in the next three sections.

8.3 Residual resistivity of metals and impurity scattering

In ordinary metals, the resistivity is dominated at low temperature by scattering off defects and impurities. The amount of defects (vacancies, dislocations, grain boundaries, etc) can in principle be reduced by specific treatments that improve the crystalline quality of the materials. Impurities, on the contrary, are often introduced on purpose, either to dope or to alloy a material and thus change its electrical and mechanical properties. In this section, we consider the case of a defect-free metal in which substitutional impurities are introduced. If the number of impurities is not too large, one observes an increase of the residual resistivity ρ_0 (the resistivity at $T = 0$) and this increase is proportional to the impurity concentration. This is usually written as $\rho_0 = \alpha x$ with x the impurity concentration in atomic percent. The parameter α depends weakly on the host material—materials with higher electron densities tend to have lower α values—but increases strongly with increasing the valence difference Z between the host and impurity atoms (see figure 8.1). In this section, we use the results of section 5.1.3.6 on impurity scattering and the expression (8.10) of the dc conductivity and find that they allow one to explain semi-quantitatively the data in figure 8.1.

The self-energy due to scattering on impurities at low impurity concentration is given in the first Born approximation by equation (5.41). For a heterovalent impurity with a difference of valence Z in a metallic host, we can approximate the impurity potential by a Yukawa potential:

$$v(\boldsymbol{q}) = \frac{Ze^2}{\epsilon_0} \frac{1}{q^2 + k_{\mathrm{TF}}^2}. \tag{8.13}$$

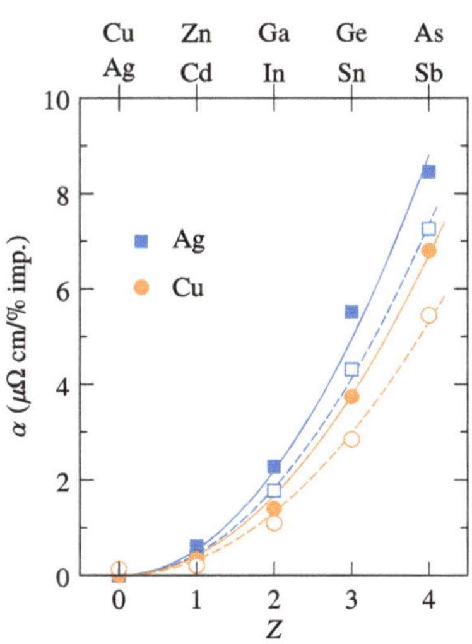

Figure 8.1. Coefficient α describing the linear increase of the residual resistivity as a function of the valence difference Z between the host and impurity atoms. Filled squares (circles) correspond to impurities Cu, Zn, Ga, Ge, As introduced in silver (copper), respectively, while open squares and circles correspond to the series Ag, Cd, In, Sn, Sb. The lines are quadratic fits $\alpha \propto Z^2$. Experimental data from Blatt 1957 *Phys. Rev.* **108** 285.

k_{TF} is the Thomas–Fermi wave vector introduced in section 5.1.4.6, which gives the leading contribution to the static screening in metals. For the host metal, we use a nearly-free electron description with a dispersion $\xi_k = \hbar^2k^2/(2m_b) - \varepsilon_{\mathrm{F}}$, where m_b is the band mass (see Bonus B8.1). The calculation of the self-energy can then be performed exactly (see Doc A8.3),

$$\Sigma^{1\mathrm{BA}}(\boldsymbol{k}, \varepsilon) = \underbrace{\frac{\pi e^2}{8\epsilon_0 a_0}}_{134\ \mathrm{eV}} \frac{n_i a_0^3}{(k_{\mathrm{F}} a_0)^2} Z^2 \times \frac{k_{\mathrm{F}}}{k}\left(\frac{1}{\frac{k_{\mathrm{F}}}{k_{\mathrm{TF}}}\sqrt{1+\frac{\varepsilon}{\varepsilon_{\mathrm{F}}}} - \frac{k}{k_{\mathrm{TF}}} + i} - \frac{1}{\frac{k_{\mathrm{F}}}{k_{\mathrm{TF}}}\sqrt{1+\frac{\varepsilon}{\varepsilon_{\mathrm{F}}}} + \frac{k}{k_{\mathrm{TF}}} + i}\right), \quad (8.14)$$

where $a_0 = 4\pi\epsilon_0\hbar^2/(me^2) = 0.53$ Å is the Bohr radius. In spite of the large prefactor of 134 eV, the self-energy remains in the 10 meV range because $n_i a_0^3$, the number of impurities in a volume a_0^3, is of the order 10^{-4}–10^{-5}.

The energy and momentum dependencies of $\Sigma^{1\mathrm{BA}}$ are displayed in figure 8.2. The first important thing to remark is that $\Sigma^{1\mathrm{BA}}$ has an imaginary part. The latter vanishes for energies below the bottom of the band ($\varepsilon/\varepsilon_{\mathrm{F}} < -1$) and decreases as $\varepsilon^{-3/2}$ at high energy. The presence of an imaginary part means that impurity scattering induces damping in the first Born approximation and thus produces resistivity. The damping results from the interference due to scattering twice on the same impurity—this is the only process contained in equation (5.41). We have seen in section 5.1.3.6 that the term corresponding to a single scattering produces no damping and is just $n_i v(\boldsymbol{q} = 0)$. The *double* scattering on *one* impurity is the reason for the behavior proportional to $n_i Z^2$: scattering on several impurities leads to corrections of order at least n_i^2, while scattering more that twice leads to corrections of order at least Z^3.

Another very useful observation is that the self-energy is almost constant in the vicinity of the Fermi surface. In many practical applications (low energy and wave vectors close to k_{F}), one can therefore take into account the effect of impurities by introducing a phenomenological self-energy that is purely imaginary:

$$\Sigma_{\mathrm{imp}}(\boldsymbol{k}, \varepsilon) \approx -i\Gamma. \quad (8.15)$$

The almost constant real part can be lumped in the chemical potential. In the expression of the retarded Green's function, this simply amounts to replacing $\varepsilon + i0^+$ by $\varepsilon + i\Gamma$. When performing numerical calculations, one *has* to replace 0^+ by a small (but not infinitesimal) positive number; we see that this can be justified as a phenomenological way to take into account impurity scattering. A slightly more elaborate model would incorporate the linear term in the real part by taking $\Sigma_{\mathrm{imp}}(\varepsilon) \approx (1 - Z^{-1})\varepsilon - i\Gamma$ with Z the quasi-particle residue (see equation (7.16*a*)).

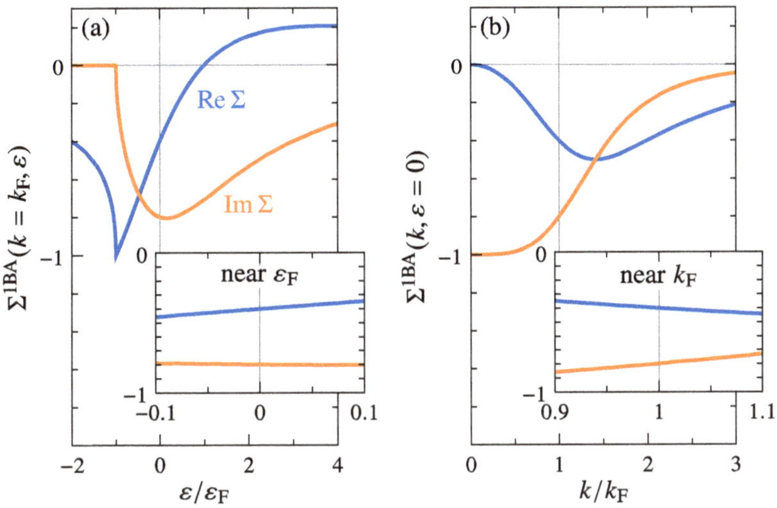

Figure 8.2. Self-energy in the first Born approximation (a) as a function of energy at $k = k_F$ and (b) as a function of momentum at $\varepsilon = 0$. The quantity plotted is the self-energy (8.14) divided by the constant prefactor in the case $k_{TF} = k_F$. The insets show the behavior at low energy close to the Fermi surface.

In Doc A8.3, we estimate the value of ρ_0 obtained from equation (8.14) using equation (8.10), and we get the numbers $\alpha_{Cu}/Z^2 = 0.37$ μΩ cm/%imp. and $\alpha_{Ag}/Z^2 = 0.44$ $\mu\Omega$ cm/% imp. for copper and silver, respectively, if the free-electron value of k_{TF} is increased by a factor two in order to account for the screening by d electrons. These values reproduce the trends observed in figure 8.1, except the different α values found for impurities with the same valence but different masses: heavier impurity atoms increase resistivity less than light atoms. In order to understand this effect, one would have to improve our model (8.13) for the impurity potential. Since d electrons tend to be more localized in heavier atoms, one could argue that k_{TF} in equation (8.13) increases with the atomic mass, which would reduce the scattering. Crystal-field effects, which lead to a non-spherical potential, can also play a role.

8.4 T^2 law and electron–electron interaction

8.4.1 Phenomenology

At low temperature, the resistivity of many metals increases like T^2. This is considered a signature of Fermi-liquid behavior because, as we shall see, it corresponds to the expected effect that the Coulomb interaction has on the resistivity in a fluid of weakly-interacting electrons. This behavior is not only observed in simple metals, but in a large variety of materials. Among those, the famous heavy-fermion compound $CeAl_3$ is often cited as an example: this material has a well-defined Fermi surface showing de

Haas van Alphen oscillations in a magnetic field and displays T^2 resistivity at very low T,[3] two characteristics of a Fermi liquid. Another famous heavy-fermion material is $LaRu_2Si_2$, which also displays T^2 resistivity[4]. More exotic materials such as sodium cobaltate $Na_{0.7}CoO_2$,[5] or iron pnictide $CaFe_4As_3$,[6] also exhibit a T^2 law. Often materials enter an ordered phase, for instance superconductivity at low temperature. In such a situation, the T^2 law, if present, is hidden by the specific signature of the ordered phase. Suppressing the ordered phase by applying pressure or magnetic field, one can then recover a T^2 law like in the heavy-fermion U_2Pt_2In,[7] in the pyrochlore $Hg_2Ru_2O_7$,[8] or even in the high-T_c cuprate superconductor $Tl_2Ba_2CuO_{6+x}$ and other similar copper oxides[9].

The T^2 behavior can also appear in conjunction with other mechanisms. For example, it has been proposed that the low-temperature resistivity of the high-T_c superconductor $La_{2-x}Sr_xCuO_4$ is the superposition of a T-linear and a T^2 term[10]. The same phenomenology was reported for several classes of superconducting materials (organic conductors, iron pnictides, cuprates), and it was found that the amplitude of the T-linear term is proportional to the critical temperature[11]. These observations suggest that these materials are Fermi liquids in which another scattering mechanism develops at low temperature, responsible for both the T-linear resistivity and the formation of the superconducting state.

Our goal in this section is to derive the T^2 law from perturbation theory using the diagrammatic technique for the Coulomb interaction. We will also discuss the Kadowaki–Woods scaling, which relates the coefficient of the T^2 resistivity to the coefficient of the T-linear specific heat and was recently generalized and corrected to include 2D and 1D materials.

8.4.2 Life-time at second order

We saw in section 5.1.3.7 that the Coulomb interaction induces electron damping at second order because the self-energy terms (5.48) are complex. In order to calculate the electron life-time at low energy and/or low temperature, we need the

[3] Andres K, Graebner J E and Ott H R 1975 *Phys. Rev. Lett.* **35** 1779.
[4] Marumoto K, Takeuchi T, Miyako Y, Ocio M, Pari P and Hammann J 2001 *Solid. State Commun.* **117** 245.
[5] Li S Y, Taillefer L, Hawthorn D G, Tanatar M A, Paglione J, Sutherland M, Hill R W, Wang C H and Chen X H 2004 *Phys. Rev. Lett.* **93** 056401.
[6] Zhao L L, Yi T, Fettinger J C, Kauzlarich S M and Morosan E 2009 *Phys. Rev.* B **80** 020404(R).
[7] Estrela P, de Visser A, Naka T, de Boer F R and Pereira L C J 2002 *Physica B* **312** 482.
[8] Takeshita N, Terakura C, Tokura Y, Yamamoto A and Takagi H 2007 *J. Phys. Soc. Jpn.* **76** 063707.
[9] Shibauchi T, Krusin-Elbaum L, Hasegawa M, Kasahara Y, Okazaki R and Matsuda Y 2008 *Proc. Nat. Acad. Sci. U.S.A.* **105** 7120. Barišić N, Chan M K, Li Y, Yu G, Zhao X, Dressel M, Smontara A and Greven M 2013 *Proc. Natl. Acad. Sci. U.S.A.* **110** 12235.
[10] Cooper R A *et al* 2009 *Science* **323** 603.
[11] Doiron-Leyraud N, Auban-Senzier P, René de Cotret S, Bourbonnais C, Jérome D, Bechgaard K and Taillefer L 2009 *Phys. Rev.* B **80** 214531.

imaginary part of these diagrams near $\varepsilon = 0$. The calculation cannot be easily done analytically, even for an isotropic free-electron dispersion, due to the kinematic constraint between momenta appearing both in the Coulomb potential and in the dispersion. In three dimensions, as the constraint is easily satisfied it is a reasonable approximation to relax it and furthermore approximate the Coulomb potential by a constant value V_0. With these simplifications, we obtain (see Doc A8.4)

$-\mathrm{Im}\Big($  $+$ 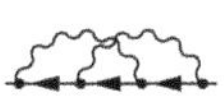 $\Big) \approx \mathrm{cste} \times \left[\varepsilon^2 + (\pi k_\mathrm{B} T)^2\right],$ (8.16)

where the constant does not depend on energy or temperature. This result is very important. It provides the microscopic justification for the phenomenological Landau theory of Fermi liquids. The latter theory postulates the existence at $T = 0$ of quasi-particles close to the Fermi surface, i.e. long-lived single-particle excitations with the same quantum numbers as the electrons. Equation (8.16) confirms that the quasi-particle life-time diverges as $1/\varepsilon^2$ at zero temperature, in other words the damping due to electron–electron interactions disappears on the Fermi surface. The fact that the scattering rate is proportional to ε^2 means that the width of the quasi-particle peak in the electron spectral function (see section 7.3) is smaller than the energy $E_k \approx \varepsilon$ of the peak when $\varepsilon \to 0$. We see in equation (8.16) the appearance of two regimes: $\varepsilon \gg k_\mathrm{B}T$ corresponds to the low-temperature regime just discussed, where the scattering rate goes like ε^2. When $\varepsilon \ll k_\mathrm{B}T$, on the other hand, the scattering rate becomes independent of energy and goes like T^2. The latter regime corresponds to the measurement conditions of the dc conductivity, as well as the low-frequency optical conductivity where a Drude-like response is expected due to the energy-independent scattering rate. We also see that the energy and the temperature play symmetric roles in the self-energy, another generic feature of Fermi liquids. The correspondence $\varepsilon \leftrightarrow \pi k_\mathrm{B}T$ can be traced back to the specific relationship between time and temperature or between real and imaginary times, as illustrated by the fact that the Matsubara frequencies $i\nu_n \propto \pi k_\mathrm{B}T$ (equation (4.9)). Let us finally note that, in spite of the fact that equation (8.16) gives the result at second order, it is possible to show that this behavior is robust and remains the same at all orders in perturbation theory[12]. This is illustrated in Bonus B8.2 with the example of the RPA (random phase approximation), which sums an infinite subset of diagrams and recovers the generic energy and temperature dependence given in equation (8.16). A simple interpretation of the ε^2 behavior is also proposed there: a direct consequence of

[12] Luttinger J M 1961 *Phys. Rev.* **121** 942.

the fact that the number of electron–hole excitations increases linearly with energy[13].

8.4.3 T^2 resistivity and Kadowaki–Woods scaling

Using equation (8.11), the Drude formula (8.10) for a band mass m_b can be rewritten in terms of the self-energy as

$$\sigma_{\rm dc} = \frac{ne^2\hbar}{2m_b}\int_{-\infty}^{\infty} d\varepsilon \frac{[-f'(\varepsilon)]}{|\text{Im}\,\Sigma(\varepsilon)|}. \tag{8.17}$$

With the expression (8.16), $|\text{Im}\,\Sigma(\varepsilon)| = \alpha[\varepsilon^2 + (\pi k_{\rm B}T)^2]$, we have

$$\int_{-\infty}^{\infty} d\varepsilon \frac{[-f'(\varepsilon)]}{|\text{Im}\,\Sigma(\varepsilon)|} = \frac{1}{\alpha(k_{\rm B}T)^2}\underbrace{\int_{-\infty}^{\infty} dx \frac{e^x/(e^x+1)^2}{x^2+\pi^2}}_{\frac{1}{12}} = \frac{1}{12\alpha(k_{\rm B}T)^2}, \tag{8.18}$$

and we thus see that the resistivity increases like T^2:

$$\rho(T) = AT^2, \qquad A = \frac{24m_b k_{\rm B}^2}{e^2\hbar}\frac{\alpha}{n}. \tag{8.19}$$

In order to check whether the numbers make sense, we may use for the value of α the RPA result given in Bonus B8.2: $\alpha \approx (\pi/16)^{3/2}/(\varepsilon_{\rm F}\sqrt{k_{\rm F}a_0})$. It is thus seen that the coefficient A varies like the square of the mass, which explains the famous empirical Kadowaki–Woods graph[14]. In this graph, the experimental value of A is plotted as a function of γ^2, where γ is the coefficient of the T-linear term in the low-temperature electronic specific heat. For weakly-interacting electrons with quadratic dispersion,

---

[13] This objection is often raised: The Coulomb interaction cannot induce resistivity because it conserves momentum. The argument goes as follows. Resistivity essentially measures the relaxation time of the current: in the Kubo formula (8.3), this is represented by the current–current correlation function, which is the propagator of the current. If the current is a conserved quantity in the long-wavelength limit, the resistivity vanishes. Now, according to equation (2.50), the operator giving the current at $\boldsymbol{q} = 0$ is $\hbar\boldsymbol{K}/m$, where $\hbar\boldsymbol{K}$ is the total momentum of the electrons. The Coulomb interaction allows momentum exchanges among the electrons but does not change the total momentum. One is lead to the conclusion that $\boldsymbol{K}$ is a conserved quantity and therefore the Coulomb interaction does not induce any resistivity… More formally, working in the plane-wave basis and using equations (2.50), (2.76), and (5.43), one indeed verifies that $[\boldsymbol{j}^p(\boldsymbol{q}=0),\ V_{\rm Cb}] = 0$. This operator identity is true in any basis. Equations (2.50) and (2.45) furthermore lead to $[\boldsymbol{j}^p(\boldsymbol{0}),\ K_0] = (\hbar/m)\sum_{\boldsymbol{k}\boldsymbol{k}'}(\boldsymbol{k}-\boldsymbol{k}')\xi_{\boldsymbol{k}\boldsymbol{k}'}c^\dagger_{\boldsymbol{k}}c_{\boldsymbol{k}'}$ in the plane-wave basis. For a translation-invariant system, $\xi_{\boldsymbol{k}\boldsymbol{k}'} \propto \delta_{\boldsymbol{k}\boldsymbol{k}'}$ and the current is conserved if the Hamiltonian is $K_0 + V_{\rm Cb}$. Such a system has infinite conductivity. The translation invariance is broken in real materials, though. The periodic lattice potential implies that $\xi_{\boldsymbol{k}\boldsymbol{k}'}$ contains terms $V(\boldsymbol{G})\delta_{\boldsymbol{k},\,\boldsymbol{k}'+\boldsymbol{G}}$: the electrons can experience Umklapp processes and transfer a momentum $\hbar\boldsymbol{G}$ to the lattice. When phonons are excited at finite temperature, even the discrete translation symmetry of the lattice is broken. Defects and impurities disrupt it even further. Hence the current is not conserved. The global coherence among all momenta needed for the Coulomb interaction to conserve the current is lost and the initial current decay is governed by the fastest process which, at low temperature, may be the quasi-particle decay due to electron–electron interaction.

[14] Rice M J 1968 *Phys. Rev. Lett.* **20** 1439. Kadowaki K and Woods S B 1986 *Solid. St. Commun.* **58** 507.

we know (equation (2.56)) that $\gamma = (\pi^2/3)k_B^2N^{el}(0) = k_B^2m^*k_F/(3\hbar^2)\mathscr{V}$, such that the so-called *Kadowaki–Woods ratio* becomes

$$\frac{A}{\gamma^2} = \frac{8\pi}{3}\frac{\hbar a_0^7}{e^2k_B^2}\frac{(m_b/m^*)^2}{\mathscr{V}^2}r_s^{15/2}. \tag{8.20}$$

We have used the electron gas formula $n = k_F^3/(3\pi^2)$ as well as the density parameter r_s given by $1/n = (4/3)\pi(r_sa_0)^3$ with a_0 the Bohr radius. Empirically, it is found that this ratio takes a nearly constant value for transition metals ($\approx 0.4\,\mu\Omega$ cm mol^2 K^2 J^{-2}) and another nearly constant value for heavy-fermion materials ($\approx 10\,\mu\Omega$ cm mol^2 K^2 J^{-2}). This suggests that the material-dependent factor $(m_b/m^*)^2r_s^{15/2}$ does not vary much within one class of materials. This material-dependent factor changes if a more accurate model is used instead of the isotropic electron gas within RPA. We may nevertheless check that the expression found gives the correct order of magnitude for simple metals, where the dynamical mass renormalization is small. Introducing the molar volume $\mathscr{V}_{mol}$, we rewrite

$$\frac{A}{\gamma^2} = 0.021\frac{(m_b/m^*)^2r_s^{15/2}}{(\mathscr{V}_{mol}/\text{cm}^3)^2}\mu\Omega\ \text{cm mol}^2\ \text{K}^2\ \text{J}^{-2}. \tag{8.21}$$

The numbers for gold ($m^*/m_b = 1.1$, $\mathscr{V}_{mol} = 10.21$ cm^3, $r_s = 3.01$) yield the very reasonable value 0.65 $\mu\Omega$ cm mol^2 K^2 J^{-2}, while an average over the 27 simple metals listed in Ashcroft & Mermin (1976) gives an average of 0.6.

There are classes of materials displaying T^2 resistivity and T-linear specific heat without following the Kadowaki–Woods scaling: these are, for example, organic conductors and many oxides. A new scaling was recently proposed[15], which also takes into account the real part of the self-energy and the dimensionality of the Fermi surface. With this new scaling, the four classes of materials (transition metals, heavy fermions, organic conductors, and oxides) all fall onto the same straight line in spite of the fact that A varies by ten orders of magnitude (see figure 8.3). The new scaling law reads

$$\frac{A}{\gamma^2} \times [N^{el}(0)]^2\langle v_{0x}^2\rangle n\xi^2 = \frac{81}{4\pi}\frac{1}{e^2\hbar k_B^2} \tag{8.22}$$

where $\langle v_{0x}^2\rangle$ is the Fermi-surface average of the unrenormalized squared Fermi velocity and ξ is a pure number of order one.

8.5 Magnetic impurities and Kondo effect

The scattering on dilute impurities leads to a temperature-independent residual resistivity as $T \to 0$ (section 8.3). In many magnetic alloys such as AuFe, however, one observes a qualitatively different behavior with a minimum in the $\rho(T)$ curve at low temperature (see figure 8.4). For illustrations of the same behavior in other materials, see e.g. the electron-doped cuprate Ce_2CuO_4,[16] or the cobaltate

[15] Jacko A C, Fjærestad J O and Powell B J 2009 *Nat. Phys.* **5** 422.
[16] Sekitani T, Naito M and Miura N 2003 *Phys. Rev.* B **67** 174503.

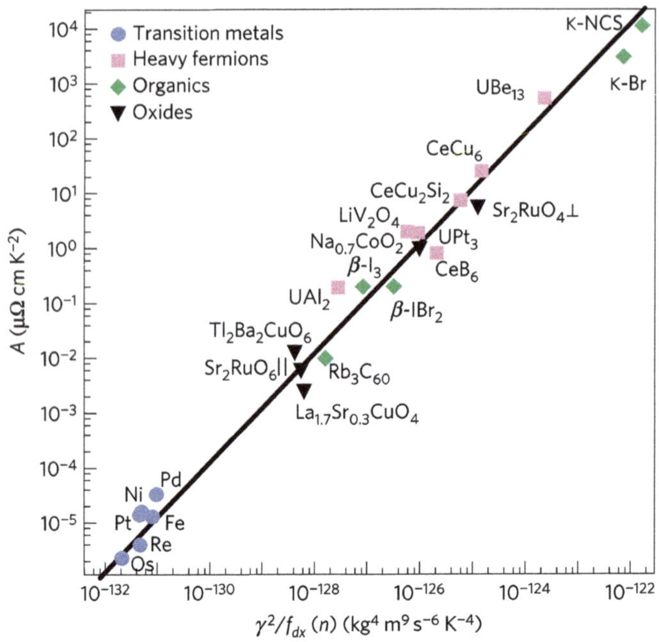

Figure 8.3. The Kadowaki–Woods plot revisited. The coefficient A of the T^2 term in the resistivity is plotted as a function of $\gamma^2/f_{dx}(n)$, with γ the coefficient of the T-linear specific heat and $f_{dx}(n)$ a material-dependent factor (see equation (8.22)). One sees that many materials with largely different characteristics and different dimensionalities all fall onto the same straight line. Reprinted figure with permission from Jacko *et al* 2009 *Nat. Phys.* **5** 422. Copyright (2009) by Springer Nature Limited. All rights reserved.

$Na_{0.7}CoO_2$[17]. Although this phenomenology has been known since the 1930s, the effect remained a mystery until 1964 with the work of Jun Kondo[18]. Kondo showed that scattering on the impurity *spins* leads within perturbation theory to a resistivity correction that diverges logarithmically as temperature goes to zero. This divergent result pointed to a failure of perturbation theory as $T \to 0$, a failure which became known as the Kondo problem. The search for a solution to the Kondo problem initiated the era of the renormalization-group theory in condensed-matter physics.

8.5.1 Models for electron scattering on magnetic impurities

An impurity atom in a host metal not only leads to a change in the local potential—this effect was considered in section 8.3—but also introduces new energy levels associated with the foreign atom. A famous model that describes one such foreign level (e.g. a d-electron level in a gas of s electrons) is the Anderson impurity model

[17] Zhang Z, Zhang J, Xu Y, Jing C, Cao S and Zhao Y 2006 *Phys. Rev.* **B 74** 045108.

[18] Kondo J 1964 *Prog. Theor. Phys.* **32** 37. See also the 40th birthday special issue: Kondo J 2005 *J. Phys. Soc. Jpn.* **74** 1.

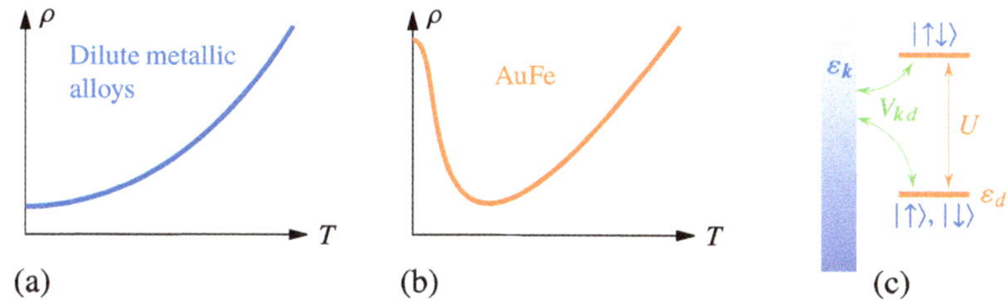

Figure 8.4. Typical temperature dependencies of the resistivity at low temperature for (a) non-magnetic dilute metallic alloys and (b) metals containing magnetic impurities. (c) Schematic representation of the Anderson impurity model (8.23): the two impurity levels ε_d and $\varepsilon_d + U$ are degenerate with the conduction band ε_k. If the chemical potential lies between ε_d and $\varepsilon_d + U$, the impurity is occupied by only one electron and carries a spin 1/2. Otherwise the impurity is occupied by either zero or two electrons and is non-magnetic.

$$H_{\rm A} = \sum_{k\sigma} \xi_k c^\dagger_{k\sigma} c_{k\sigma} + \xi_d \sum_{\sigma} c^\dagger_{d\sigma} c_{d\sigma} + \sum_{k\sigma} \Big(V_{kd} c^\dagger_{k\sigma} c_{d\sigma} + \text{h.c.} \Big) + U n_{d\uparrow} n_{d\downarrow}. \qquad (8.23)$$

The first term represents the unperturbed conduction electrons (see equation (2.53)) in a one-band scheme, the second term represents the foreign level, the third term describes processes in which electrons hop from the d level to the conduction band and inversely, and the last term accounts for the Coulomb repulsion when two electrons occupy the impurity level. The model is depicted in figure 8.4(c). Two different situations can arise depending upon the position of the chemical potential with respect to the impurity levels. If μ lies below ε_d or above $\varepsilon_d + U$, the impurity level is either unoccupied or occupied by two electrons: in both cases it carries no spin and the impurity behaves like a non-magnetic impurity. If the chemical potential is between ε_d and $\varepsilon_d + U$, the impurity is occupied by only one electron and carries a spin 1/2.

The two-body interaction term in equation (8.23) prevents any closed analytical solution. The problem with $U = 0$ is exactly solvable, but its physical properties are different from those of the original model because the doubly-degenerate impurity level has no magnetic degree of freedom. The net result at $U = 0$ is that the level ε_d gets broadened and slightly shifted due to the hybridization V_{kd}. One can also attempt a mean-field treatment, replacing $U n_{d\uparrow} n_{d\downarrow}$ by $U[n_{d\uparrow}\langle n_{d\downarrow}\rangle + n_{d\downarrow}\langle n_{d\uparrow}\rangle - \langle n_{d\uparrow}\rangle\langle n_{d\downarrow}\rangle]$. In this case, beside the non-magnetic solution $\langle n_{d\uparrow}\rangle = \langle n_{d\downarrow}\rangle$ that always exists, there are magnetic solutions corresponding to the formation of a magnetic moment on the impurity when U exceeds a critical value. These solutions are unphysical artifacts of the mean-field approximation, however, since there is nothing in the initial Hamiltonian breaking the spin-rotation invariance.

Another approach known as the Schrieffer–Wolff transformation[19] simplifies equation (8.23) and reduces it to a model in which the conduction electrons only interact with the impurity spin. In a spin-only model, the occupation of the impurity level must be fixed and therefore no term like the third one in equation (8.23) can

[19] Schrieffer J R and Wolff P A 1966 *Phys. Rev.* **149** 491.

appear. The idea is therefore to set up a unitary transformation S of the Hamiltonian such that the transformed Hamiltonian $\tilde{H}_A = e^{iS}H_A e^{-iS}$ does not depend on V_{kd}, at least approximately for low energies and wave vectors close to the Fermi surface. The derivation is performed in Bruus & Flensberg (2004, p 169) and leads to the so-called *sd* Hamiltonian, or Kondo Hamiltonian, which was originally considered by Kondo:

$$H_{sd} = \sum_{k\sigma} \xi_k c^\dagger_{k\sigma} c_{k\sigma} + J \int d\boldsymbol{r}\, \boldsymbol{S}_d(\boldsymbol{r}) \cdot \boldsymbol{S}(\boldsymbol{r}). \tag{8.24}$$

We have written the model for a situation where there are N_i magnetic impurities at random positions $\boldsymbol{R}_\ell$ rather that just one impurity like in equation (8.23). Similar to equation (5.32), $\boldsymbol{S}_d(\boldsymbol{r})$ is the spin density due to the N_i impurities,

$$\boldsymbol{S}_d(\boldsymbol{r}) = \sum_{\ell=1}^{N_i} \boldsymbol{s}_d\, \delta(\boldsymbol{r} - \boldsymbol{R}_\ell), \tag{8.25}$$

with $\boldsymbol{s}_d = \frac{\hbar}{2}\boldsymbol{\tau}$ the spin carried by each impurity and $\boldsymbol{\tau}$ the vector of Pauli matrices given in equation (22). $\boldsymbol{S}(\boldsymbol{r})$ is the spin density of the conduction electrons given by equation (2.47). The coupling parameter J resulting from the Schrieffer–Wolff transformation is

$$J = \frac{2|V_{k_F d}|^2 U}{-\xi_d(\xi_d + U)}. \tag{8.26}$$

We see that the coupling is anti-ferromagnetic ($J > 0$)—i.e. the energy is lower when $\boldsymbol{S}_d$ and $\boldsymbol{S}$ are anti-parallel—if $\xi_d < 0$ and $\xi_d + U > 0$;[20] this is the situation where the chemical potential is between ε_d and $\varepsilon_d + U$.

8.5.2 Third-order perturbation calculation

Following Kondo, we study the Hamiltonian (8.24) by means of the perturbation theory in J. We perform an impurity average like in section 5.1.3.6 and thus restore the translation invariance. Expressed in terms of the electron operators, the perturbation is

$$V = \frac{\hbar J}{2} \int d\boldsymbol{r} \sum_{\mu\sigma\sigma'} \tau^\mu_{\sigma\sigma'} S^\mu_d(\boldsymbol{r}) c^\dagger_{\boldsymbol{r}\sigma} c_{\boldsymbol{r}\sigma'}. \tag{8.27}$$

This form is similar to the generic one-body perturbation (5.9) with two differences: as we are coupling the vector fields $\boldsymbol{S}_d(\boldsymbol{r})$ and $\boldsymbol{S}(\boldsymbol{r})$, there is an additional μ sum for the scalar product; more importantly, in the 'matrix element' we have $S^\mu_d(\boldsymbol{r})$ which is an operator rather than a scalar. This complicates the perturbation theory because there is no equivalent of Wick's theorem for spin operators.

[20] The opposite convention is often taken in the literature for the sign of J, with an extra minus sign in equations (8.24) and (8.26).

Applying the general perturbation expansion[21] (5.5) to the case of the Green's function $\mathscr{G}_{\sigma\sigma'}(\boldsymbol{k}, \tau) = \mathscr{C}_{c_{\boldsymbol{k}\sigma}c^{\dagger}_{\boldsymbol{k}\sigma}}(\tau)$, inserting the expression of the perturbation (8.27), and performing the impurity average, we obtain the expansion

$$\begin{aligned}\mathscr{G}_{\sigma\sigma'}(\boldsymbol{k}, \tau) = &-\sum_{n=0}^{\infty}\left(-\frac{\hbar J}{2}\right)^n \int_0^{\beta} d\tau_1 \cdots d\tau_n \int d\boldsymbol{r}_1 \sum_{\mu_1\sigma_1\sigma_1'} \cdots \int d\boldsymbol{r}_n \sum_{\mu_n\sigma_n\sigma_n'} \\ &\times \tau^{\mu_1}_{\sigma_1\sigma_1'} \cdots \tau^{\mu_n}_{\sigma_n\sigma_n'} \langle\langle T_\tau S_d^{\mu_1}(\boldsymbol{r}_1, \tau_1) \cdots S_d^{\mu_n}(\boldsymbol{r}_n, \tau_n)\rangle_0\rangle_{\mathrm{imp}} \\ &\times \langle T_\tau c^{\dagger}_{\boldsymbol{r}_1\sigma_1}(\tau_1) c_{\boldsymbol{r}_1\sigma_1'}(\tau_1) \cdots c^{\dagger}_{\boldsymbol{r}_n\sigma_n}(\tau_n) c_{\boldsymbol{r}_n\sigma_n'}(\tau_n) c_{\boldsymbol{k}\sigma}(\tau) c^{\dagger}_{\boldsymbol{k}\sigma'}(0)\rangle_0^{\mathrm{con-diff}}. \end{aligned} \tag{8.28}$$

The thermal average in equation (5.5) could be split into one part for the impurity spins and one part for the conduction electrons, because in the absence of interaction these two sub-systems are decoupled. The impurity average denoted by $\langle\cdots\rangle_{\mathrm{imp}}$ like in equation (5.35) only concerns the correlation function of the impurity spins and has no effect on the conduction electrons.

The calculation of the terms up to order $n = 3$ is performed in Doc A8.5. The Green's function turns out to be diagonal in the spin indices, as required by the spin-rotation invariance of the problem: $\mathscr{G}_{\sigma\sigma'} \propto \delta_{\sigma\sigma'}$. The first term ($n = 1$) vanishes. For non-magnetic impurities, the first-order term (5.37) is a constant shift of the chemical potential compensating the average potential induced by the impurity charges. As spins are neutral, there is no such correction in the present case[22]. At second order, the self-energy is (see Doc A8.5)

$$\Sigma^{(2)}(\boldsymbol{k}, i\omega_n) = n_i \frac{3\hbar^4 J^2}{8} \frac{1}{\mathscr{V}} \sum_{\boldsymbol{q}} \frac{1}{i\omega_n - \xi_{\boldsymbol{q}}}. \tag{8.29}$$

This is the same expression as we obtained for scattering on non-magnetic impurities within the first Born approximation, equation (5.41), applied to the case of point-like defects with a potential $v(\boldsymbol{q}) = (3/8)^{\frac{1}{2}}\hbar^2 J$. Therefore, at second order the scattering on impurities carrying a spin is similar to the scattering on non-magnetic impurities. In particular, it induces a temperature-independent residual resistivity (section 8.3) and is unable to explain the resistivity upturn depicted in figure 8.4(b). Looking back at the calculation in Doc A8.5, one sees that the contribution of the impurity spins to the scattering is independent of time, meaning that the internal spin state of the impurity is not changed during the second-order scattering process. This is the reason for the deceptive result (8.29): at second order, the impurity spins show no dynamical behavior but instead behave like static objects.

---

[21] See Bruus & Flensberg (2004, p 241) for a proof that equation (5.5) remains valid in the absence of Wick's theorem for the spin operators.

[22] The situation would be different in the presence of a magnetic field which aligns the impurity spins and leads to a finite magnetization: in this case, the first-order correction is finite and accounts for the difference in chemical potential for up and down electrons.

Something new happens at third order: here the scattering involves an intermediate state in which the spin of the impurity is changed and the intrinsic dynamics of the impurity starts to play a role. The resulting scattering rate is (see Doc A8.5)

$$-\mathrm{Im}\,\Sigma^{(3)}(\boldsymbol{k},\,\varepsilon) = \frac{3\pi n_i \hbar^6 J^3}{32} N_0^{\mathrm{el}}(\varepsilon) \int_{-\infty}^{\infty} d\xi\; N_0^{\mathrm{el}}(\xi) \frac{\tanh(\beta\xi/2)}{\xi - \varepsilon}. \tag{8.30}$$

This scattering rate diverges as $\ln(1/T)$ as $T \to 0$. To fix ideas, let us take a density of states equal to the constant N_0 over the bandwidth extending from $-W/2$ to $W/2$ and let us look at $\varepsilon = 0$: then, using equation (20),

$$\begin{aligned} -\mathrm{Im}\Sigma^{(3)}(\boldsymbol{k},\,0) &\approx \frac{3\pi n_i \hbar^6 (JN_0)^2}{32} J \int_{-W/2}^{W/2} d\xi \frac{\tanh(\beta\xi/2)}{\xi} \\ &\approx \frac{3\pi n_i \hbar^6 (JN_0)^2}{16} J \ln\left(0.567 \frac{W}{k_{\mathrm{B}}T}\right). \end{aligned} \tag{8.31}$$

The first thing to remark is that this contribution is *positive*—and therefore increases the scattering rate—if the coupling is anti-ferromagnetic ($J > 0$). Therefore, this correction is able to produce a resistivity upturn like in figure 8.4(b).

The singular nature of this correction points to a failure of perturbation theory as $T\to 0$. The fact that the third-order correction diverges, while the second-order correction is finite, suggests that one has to sum all terms in the perturbation series in order to get a finite result at $T = 0$. This is not the case, though: summing logarithmic contributions from higher-order terms leads to an even more singular result, which diverges at a finite temperature T_{K} known as the Kondo temperature[23]. We must conclude that the perturbation theory is unreliable at temperatures below T_{K}. Above T_{K}, the perturbation theory works because the third-order correction is smaller than the second-order one. Collecting equations (8.29) and (8.31), we see that the scattering rate on the Fermi surface is

$$-\mathrm{Im}\,\Sigma(\boldsymbol{k},\,0) = \frac{3\pi n_i \hbar^2 J}{16}\left[(\hbar^2 JN_0) + (\hbar^2 JN_0)^2 \ln\left(0.567\frac{W}{k_{\mathrm{B}}T}\right) + \cdots\right], \tag{8.32}$$

where the ellipsis indicates higher-order terms. We can estimate the Kondo temperature by equating the first two terms in the expansion, with the result

$$k_{\mathrm{B}}T_{\mathrm{K}} = 0.567\; We^{-1/(\hbar^2 JN_0)}. \tag{8.33}$$

Note the similarity of this expression with other non-perturbative results such as equations (5.88) and (5.144).

8.5.3 Beyond perturbation theory

The breakdown of perturbation theory for scattering on local magnetic moments has become known as the *Kondo problem*. In fact, similar phenomena may occur

[23] Abrikosov A A 1965 *Physics* **2** 5.

whenever the scattering centers have internal degrees of freedom, i.e. a proper dynamics which contributes to the scattering process. Another well-known example exhibiting Kondo physics is the transport through the localized levels of a quantum dot or through individual atoms on surfaces probed by scanning tunneling spectroscopy. Theoretical progress has been achieved in these problems thanks to the renormalization-group method. In this section, we provide a birds'-eye view of the renormalization-group approach with the aim of introducing the basic idea and jargon. Interested readers should read introductions by experts[24].

In order to understand why perturbation theory fails, we must take one step back and remember the very first assumptions on which perturbation theory relies. The first assumption is that the observable quantities that we are calculating *do have* a regular expansion in powers of the perturbation. In practice, this assumption is already implicitly made when we write the series (2.23) for the evolution operator. The second assumption is that of *adiabatic connection*, stating that the actual ground (or thermodynamic) state in the presence of interactions can be reached by switching on the interaction adiabatically starting from the non-interacting state. If any of these conditions are not met, we must expect the perturbative approach to fail, even if the perturbation series can be completely summed up to infinite order. As a simple analogy, consider a function $f(x)$ that we try to evaluate at $x = x_0$ near $x = 0$. If this function is singular exactly *at* $x = 0$, like $1/x$, $\ln x$, or $e^{-1/x}$, the Taylor expansion is simply ill-defined and we get singular values if we try and evaluate the terms of the expansion. The second possibility is that the function is regular at $x = 0$ but has a singularity between $x = 0$ and $x = x_0$, like $\tan x$ if $x_0 = \pi$: in this situation, the expansion around $x = 0$ is of no use to evaluate $f(x_0)$.

The quest for a solution to the Kondo problem has led to the advent of the renormalization group in condensed-matter physics. The goal of this methodology is to find the effective Hamiltonian (or the effective action) that determines the properties of the system at low energy and/or low temperature: when $T \to 0$, only the few lowest levels of the original Hamiltonian contribute significantly to the thermal average. The idea is therefore to build a *low-energy effective Hamiltonian* by excluding high-energy levels, with the hope that the effective theory will be easier to solve than the original problem. The process of eliminating the high-energy degrees of freedom changes the parameters of the original Hamiltonian. In some cases, the low-energy theory turns out to be just a Fermi-liquid theory with renormalized parameters (velocities, masses, etc). In this situation, one can generally rely on perturbation theory. In other words, one can take the $J \to 0$ limit *first*—if J is the parameter controlling the interactions—and the limit $T \to 0$ *after*: theorists say 'J is the smallest energy scale in the problem' or 'the renormalization group flows to weak coupling'. In other cases, the low-energy theory has a different character reflecting the fact that the ground-state wave function is not adiabatically connected with the non-interacting ground state. In this situation, one should take the $T \to 0$ limit first, in other words solve the problem at finite J: now T is the smallest energy scale in the

[24] Maris H J and Kadanoff L P 1971 *Am. J. Phys.* **46** 652. Delamotte B 2004 *Am. J. Phys.* **72** 170; Delamotte B 2012 *Lecture Notes in Physics* **852**.

problem and the renormalization group 'flows to strong coupling' (i.e. $J \gg T$ as $T \to 0$).

To be slightly more explicit, we consider the Kondo Hamiltonian characterized by the partition function $Z(T, J)$. The renormalization is a transformation $\mathscr{R}$—to be chosen wisely according to the problem at hand—which delivers a new partition function Z' and new temperature and interaction parameters T' and J':

$$Z(T, J) = \mathrm{Tr} e^{-\beta K} \overset{\mathscr{R}}{\longmapsto} Z'(T', J') = \mathrm{Tr} e^{-\beta' K'} = \mathscr{R} Z(T, J). \tag{8.34}$$

The transformation is useful if the new problem has fewer degrees of freedom than the original problem. If the *function* Z' is the same function of its arguments than the function Z, the problem is called *exactly renormalizable*. If this is not the case, one has to choose a transformation $\mathscr{R}$ such that the terms that make the function Z' differ from Z are small. If $Z \equiv Z'$, we can iterate:

$$Z(T, J) \overset{\mathscr{R}}{\longmapsto} Z(T', J') \overset{\mathscr{R}}{\longmapsto} Z(T'', J'') \overset{\mathscr{R}}{\longmapsto} \dots \tag{8.35}$$

In this iterative process, the parameters T and J 'flow' toward *fixed points* such that $\mathscr{R} Z(T^*, J^*) = Z(T^*, J^*)$. We can envision three possibilities: $(T^* = 0, J^* = \infty)$ corresponds to the strong-coupling fixed point, where only the second term in equation (8.24) matters; $(T^* = \infty, J^* = 0)$ corresponds to the weak-coupling fixed point, where the perturbation theory is reliable; and $(T^* = T_c, J^* = J_c)$ would describe a phase transition between two different regimes at some particular temperature and coupling.

In the case of the Kondo problem, it turns out that J flows to the strong-coupling fixed point $J^* = \infty$. This was demonstrated in a famous work by Kenneth Wilson[25]. Wilson's solution starts by rewriting H_{sd} in terms of new operators $f^\dagger_{n\sigma}$ corresponding to spherical wave functions centered on the impurity and localized in a shell whose radius increases exponentially with the index n. At low values of n, the wave function is localized on the impurity and is built from states $c^\dagger_{k\sigma}$ covering a large part of the band, while high values of n correspond to extended states built from states $c^\dagger_{k\sigma}$ closer and closer to the Fermi energy. With this rewriting, the Hamiltonian takes the form of a one-dimensional chain that couples state n to state $n-1$ and state $n = 0$ to the impurity spin. By increasing n and eliminating high-energy states, Wilson was able to set up a renormalization scheme and show that the Kondo Hamiltonian (8.24) flows to strong coupling.

The physical interpretation is that the effective anti-ferromagnetic interaction between the conduction electrons and the spin of the impurities increases as $T \to 0$ and the electrons therefore tend to form singlet states together with the impurity spins. The formation of a bound state between conduction electrons and impurity spins is the non-perturbative process that disconnects the ground state from the Fermi sea, very much like the formation of Cooper pairs disconnects the BCS ground state from the Fermi sea. Once the singlets are formed, the impurity spins are

[25] Wilson K G 1975 *Rev. Mod. Phys.* **47** 773.

screened such that the impurities become effectively non-magnetic at $T = 0$. Since non-magnetic impurities give a non-singular residual resistivity at $T = 0$, the saturation of the resistivity in figure 8.4(b) can finally be understood.

8.6 Effects beyond quasi-particle scattering

In the previous sections, we have considered applications that neglected the vertex corrections in equation (8.5). As a result, the resistivity was entirely controlled by the self-energy Σ, as expressed in the Drude formula (8.10). In this section, we discuss qualitatively some effects corresponding to vertex corrections that therefore go beyond this version of the Drude formula.

8.6.1 Vertex functions, Ward identities, conserving approximations

Separating the contributions with and without vertex corrections in equation (8.5), we have

$\chi_{jj}^{\mu\nu} = -$ 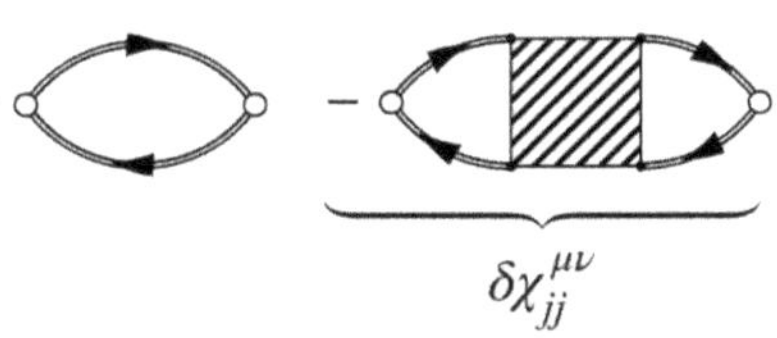 (8.36)

where the second term gives the vertex corrections, i.e. all diagrams in which the electron and hole lines are connected by the interaction. The shaded box is a *vertex function*; it depends on three energy–momentum variables and we shall denote it Λ:

$$\equiv \Lambda(K, K', Q). \tag{8.37}$$

In order to lighten the formula in this section, we use the 'four-vector' notation:

$$K \equiv (\boldsymbol{k}, \sigma, i\omega_n), \qquad Q \equiv (\boldsymbol{q}, i\Omega_n), \qquad \sum_K \equiv \sum_{\boldsymbol{k}\sigma} \frac{1}{\beta} \sum_{i\omega_n}. \tag{8.38}$$

Translating the second diagram in equation (8.36), the vertex corrections to the current–current correlation function can be formally written as[26]

$$\delta\chi_{jj}^{\mu\nu}(Q) = \sum_{KK'} \Gamma_0^{\mu}(K, Q)\mathscr{G}(K)\mathscr{G}(K+Q)\Lambda(K, K', Q) \\ \times \mathscr{G}(K')\mathscr{G}(K'-Q)\Gamma_0^{\nu}(K', -Q), \tag{8.39}$$

where $\boldsymbol{\Gamma}_0(K, Q) = \frac{\hbar}{m}(\boldsymbol{k} + \frac{\boldsymbol{q}}{2})$ is the bare current vertex and $\mathscr{G}$ is the full Green's function containing all self-energy corrections. The goal is to find good

[26] There is no consensus in the literature as to what is the best convention for labeling the diagram and naming the arguments of Λ. Here we adopt the following convention (see equation (8.37)): K and K' for input legs, $K + Q$ and $K' - Q$ for output legs, K, K', and Q for the arguments of the function Λ.

approximation schemes for the vertex function Λ. In another equivalent formulation, one introduces the *renormalized current vertex* $\mathbf{\Gamma}(K, Q)$ by writing

$\chi_{jj}^{\mu\nu}(Q) = -$ 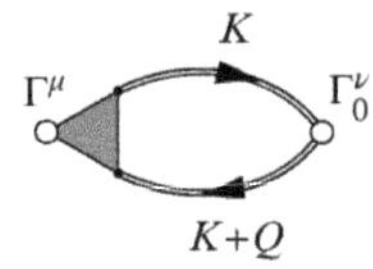

$$= \sum_K \Gamma^{\mu}(K, Q)\mathscr{G}(K)\mathscr{G}(K+Q)\Gamma_0^{\nu}(K+Q, -Q). \qquad (8.40)$$

Comparing with equation (8.36) or equation (8.39), we can deduce the relation between the renormalized vertex $\mathbf{\Gamma}$ and the vertex function Λ:

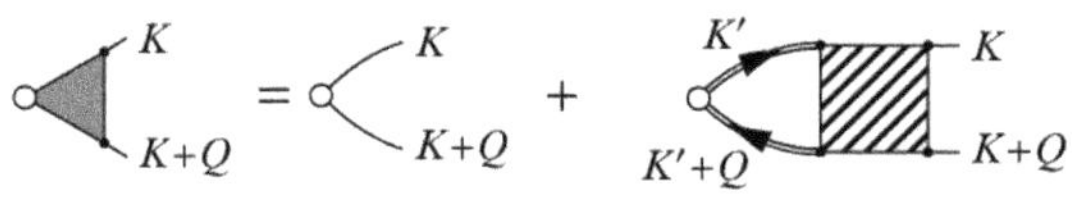

$$\mathbf{\Gamma}(K, Q) = \mathbf{\Gamma}_0(K, Q) + \sum_{K'} \mathbf{\Gamma}_0(K', Q)\mathscr{G}(K')\mathscr{G}(K'+Q)\Lambda(K', K+Q, Q). \qquad (8.41)$$

The expressions (8.39) and (8.40) formally provide exact representations of the current–current correlation function in which the vertex corrections have been encapsulated in the functions Λ and $\mathbf{\Gamma}$. Such formulas are useful for establishing exact relations between correlation functions known as Ward identities, which arise from conservation laws.

The conservation of electric charge results in the continuity equation discussed in Doc A2.8, a relation between the density and current operators: $i\Omega_n n - \hbar \boldsymbol{q} \cdot \boldsymbol{j} = 0$. Not surprisingly, this relation which binds together density and current also has implications for the density–density and current–current correlation functions. By analogy with equation (8.40), the density–density correlation $\chi_{nn}(Q)$ defined in equation (5.75) can be formally written in terms of a bare density vertex, which is just unity, and a renormalized density vertex $\Gamma_c(K, Q)$ according to

$$\chi_{nn}(Q) = \sum_K \Gamma_c(K, Q)\mathscr{G}(K)\mathscr{G}(K+Q). \qquad (8.42)$$

The Ward identity corresponding to the conservation of charge takes a simple form when expressed in terms of the renormalized vertices and Green's function[27]:

$$\boxed{i\Omega_n\Gamma_c(K, Q) - \hbar \boldsymbol{q} \cdot \mathbf{\Gamma}(K, Q) = \mathscr{G}^{-1}(K+Q) - \mathscr{G}^{-1}(K).} \qquad (8.43)$$

This important formula should be satisfied by any approximation scheme that claims to be *conserving*, i.e. that satisfies the conservation of charge (an analogous relation between the spin–spin and spin–current correlation functions expresses the conservation of spin). We can readily check that the Ward identity (8.43) is satisfied by free electrons: take $\Gamma_c(K, Q) \equiv \Gamma_{c0} = 1$, $\mathbf{\Gamma}(K, Q) \equiv \mathbf{\Gamma}_0(K, Q) = \frac{\hbar}{m}(\boldsymbol{k} + \frac{\boldsymbol{q}}{2})$,

[27] For a derivation, see Schrieffer (1964, p 228) or Metzner W and Di Castro C 1993 *Phys. Rev.* **B 47** 16107.

$\mathscr{G}^{-1}(K+Q) = i\omega_n + i\Omega_n - \xi_{\boldsymbol{k}+\boldsymbol{q}}$, $\mathscr{G}^{-1}(K) = i\omega_n - \xi_{\boldsymbol{k}}$, and $\xi_{\boldsymbol{k}} = \hbar^2 k^2/(2m) - \mu$. The approximation behind the Drude formula is not conserving, because it assumes $\Gamma_c \equiv \Gamma_{c0}$ and $\boldsymbol{\Gamma} \equiv \boldsymbol{\Gamma}_0$, which according to equation (8.43) constrains the self-energy to obey

$$\Sigma(\boldsymbol{k}+\boldsymbol{q}, i\omega_n + i\Omega_n) - \Sigma(\boldsymbol{k}, i\omega_n) = \xi_{\boldsymbol{k}} + \frac{\hbar^2}{2m}(2\boldsymbol{k}\cdot\boldsymbol{q} + q^2) - \xi_{\boldsymbol{k}+\boldsymbol{q}},$$

a relation generally violated by the approximations used for the self-energy in the Drude formula.

8.6.2 Ladder approximation, diffuson, transport life-time

The first term in the right-hand side of equation (8.36) describes the propagation of uncorrelated electron–hole pairs: it takes into account all scattering effects for the electron and the hole separately, but neglects all interactions between electron and hole. Hence the second term accounts for the corrections due to these interactions between the electrons and the holes. When the electron–hole pair density is low, it is likely that the dominant terms are those where the same electron repeatedly interacts with the same hole. The corresponding set of diagrams in the vertex function Λ is known as the *diffuson* Λ^{D}. The diffuson is given by a *ladder sum* (for definiteness, we draw the diagrams for the case of the Coulomb interaction):

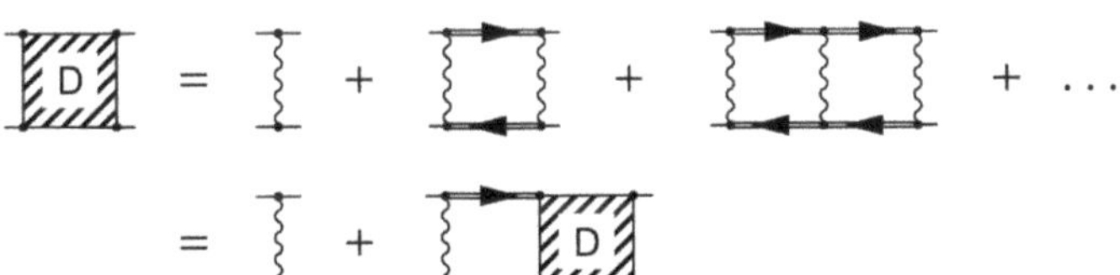

(8.44)

The second line corresponds to a closed integral equation which can in principle be solved for $\Lambda^{\mathrm{D}}(K, K', Q)$. The solution is particularly simple in the case of scattering on point-like impurities (i.e. such that $v(\boldsymbol{q}) \equiv v_0$ in equation (5.41)), since in this case Λ^{D} does not depend on K and K':

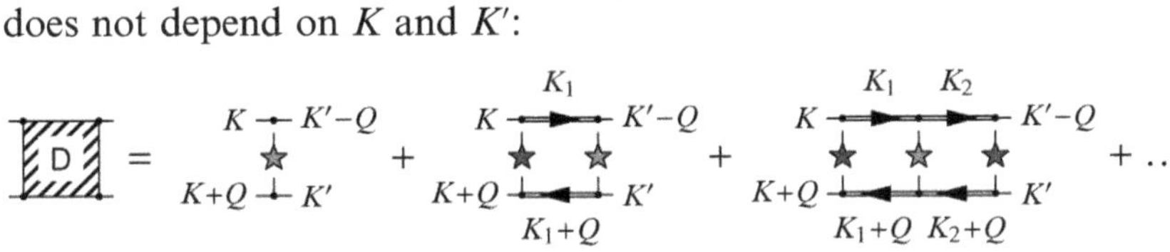

$$
\begin{aligned}
&= n_i v_0^2 + \left(n_i v_0^2\right)^2 \sum_{K_1} \mathscr{G}(K_1)\mathscr{G}(K_1+Q) \\
&\quad + \left(n_i v_0^2\right)^3 \sum_{K_1} \mathscr{G}(K_1)\mathscr{G}(K_1+Q) \sum_{K_2} \mathscr{G}(K_2)\mathscr{G}(K_2+Q) + \ldots \\
&= n_i v_0^2 \left\{1 + n_i v_0^2 \zeta(Q) + \left[n_i v_0^2 \zeta(Q)\right]^2 + \ldots\right\} = \frac{n_i v_0^2}{1 - n_i v_0^2 \zeta(Q)}
\end{aligned}
\tag{8.45}
$$

with

$$\zeta(Q) = \sum_K \mathcal{G}(K)\mathcal{G}(K+Q). \tag{8.46}$$

Introducing this form in equation (8.39), one gets a contribution to the conductivity that corrects the Drude formula (8.10), replacing the simple transport life-time defined in equation (8.11) by a shorter time (see Bruus & Flensberg (2004, p 291)).

8.6.3 Cooperon, weak localization

Another class of diagrams appearing in the vertex function Λ is the series of so-called *maximally-crossed diagrams*:

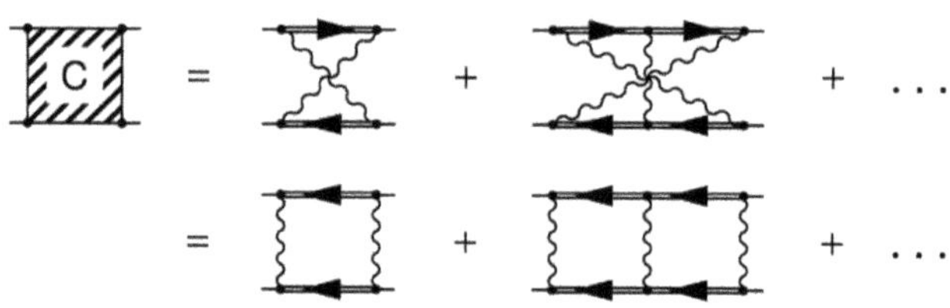

(8.47)

At the second line, we have twisted the upper part of the diagrams, thus showing that this series again gives rise to a ladder sum like in the expression of the diffuson, however with the important difference that the ladder is now in the *particle–particle channel* rather than the particle–hole channel. This set of terms describes the propagation of two electrons repeatedly interacting with each other. If the interaction is attractive, these two electrons eventually form a bound state which is nothing but a Cooper pair. This series of diagrams has therefore become known as the *cooperon*. In Bonus B8.3, we show explicitly the relationship between the cooperon and superconductivity by introducing the *pairing susceptibility* and using the Thouless criterion for determining the critical temperature.

In the context of impurity scattering, the cooperon describes processes in which electrons follow the same closed path in real space in opposite directions (time-reversed paths) and interfere constructively. The constructive interference for this particular type of trajectories enhances the probability for an electron to follow a closed path and thus be more localized than purely itinerant electrons: the phenomenon is known as *weak localization* and gives rise to a logarithmic upturn in the resistivity at very low temperature.

In order to link this qualitative description with the cooperon diagrams, let us consider the first of these diagrams. Since we have twisted the upper part, the momenta on the left legs are now $K + Q$ and $K' - Q$:

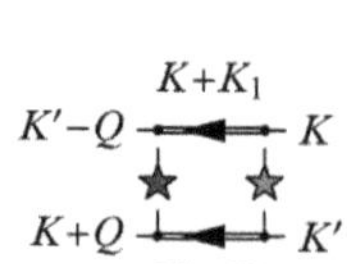

The value of the diagram is $(n_i v_0^2)^2 \tilde{\zeta}(K+K')$, where $\tilde{\zeta}(K+K') = \sum_{K_1} \mathcal{G}(K+K_1) \mathcal{G}(K'-K_1)$. It is independent of Q, in contrast to the diffuson diagrams which depend on Q but not on K and K'. Hence we can evaluate the diagram at $Q = 0$ without loss of generality; visualizing the diagram in imaginary time rather than imaginary frequency, we see two electrons evolving, one starting at time 0 with momentum $\boldsymbol{k}'$ and ending at time τ with momentum $\boldsymbol{k}$ (lower part of the diagram), the other starting at time τ with momentum $\boldsymbol{k}$ (recall that we have twisted the upper part, thus also exchanging times) and ending at time 0 with momentum $\boldsymbol{k}'$. This is the momentum-space analog of the real-space time-reversed trajectories discussed above.

The constructive interference can be understood as follows. The complete set of maximally-crossed diagrams form a geometric series and can be summed giving the value of the cooperon as $(n_i v_0^2)^2 \tilde{\zeta}(P)/[1 - n_i v_0^2 \tilde{\zeta}(P)]$ with $P \equiv K + K'$. It turns out that within the first Born approximation, $n_i v_0^2 \tilde{\zeta}(0) = 1$ (Bruus & Flensberg, 2004, p 304), meaning that the cooperon diverges at $P = 0$ and thus the corresponding correction to the conductivity comes from the region of four-momenta $K' \approx -K$. Let us now consider the case of two scattering events as in the diagram shown above. If the impurities are at positions $\boldsymbol{R}_1$ and $\boldsymbol{R}_2$ and the electron initially at position $\boldsymbol{r}$, the phase accumulated along the first trajectory (lower part of the diagram) is $e^{i\boldsymbol{k}'\cdot(\boldsymbol{R}_1-\boldsymbol{r})} e^{i(\boldsymbol{k}'-\boldsymbol{k}_1)\cdot(\boldsymbol{R}_2-\boldsymbol{R}_1)} e^{i\boldsymbol{k}\cdot(\boldsymbol{r}-\boldsymbol{R}_2)}$. Along the time-reversed trajectory (upper part of the diagram) it is $e^{i\boldsymbol{k}\cdot(\boldsymbol{R}_1-\boldsymbol{r})} e^{i(\boldsymbol{k}+\boldsymbol{k}_1)\cdot(\boldsymbol{R}_2-\boldsymbol{R}_1)} e^{i\boldsymbol{k}'\cdot(\boldsymbol{r}-\boldsymbol{R}_2)}$. The diagram is proportional to the product of these two phases, which in general depends on the impurity positions $\boldsymbol{R}_1$ and $\boldsymbol{R}_2$. But a phase that depends on the impurity positions vanishes when the impurity average is performed and such contributions disappear. However, if we take $\boldsymbol{k}' = -\boldsymbol{k}$ ($P = 0$) the two phases exactly cancel, such that the relative phase no longer depends on the impurity positions and survives the impurity average.

The detailed calculation of the correction $\delta\chi_{jj}^{\mu\nu}$ to the conductivity is somewhat tedious (Bruus & Flensberg, 2004) and yields a logarithmically divergent result. The divergence arises because time-reversed paths of arbitrary lengths are retained in the calculation, while physically the electrons do not maintain phase coherence over arbitrarily long times. Hence there should be an upper cutoff given by $v_F \tau_\phi$ in the length of the paths, where τ_ϕ is the *phase coherence time*. This new length scale, in addition to the usual mean free path $\ell_0 = v_F \tau_0$ due to conventional impurity scattering, regularizes the divergence. One then finds that the correction is largest in two dimensions, where it is given by

$$\frac{\delta\sigma_{\mathrm{dc}}^{\mathrm{WL}}}{\sigma_{\mathrm{dc}}} = -\frac{1}{\pi k_F \ell_0} \ln\left(\frac{\tau_\phi}{\tau_0}\right). \tag{8.48}$$

Since τ_ϕ decreases rapidly in the presence of a magnetic field, systems in the weak-localization regime show a characteristic *negative* magneto-resistance, i.e. the conductivity increases in the field due to the suppression of the weak localization effect (see figure 8.5).

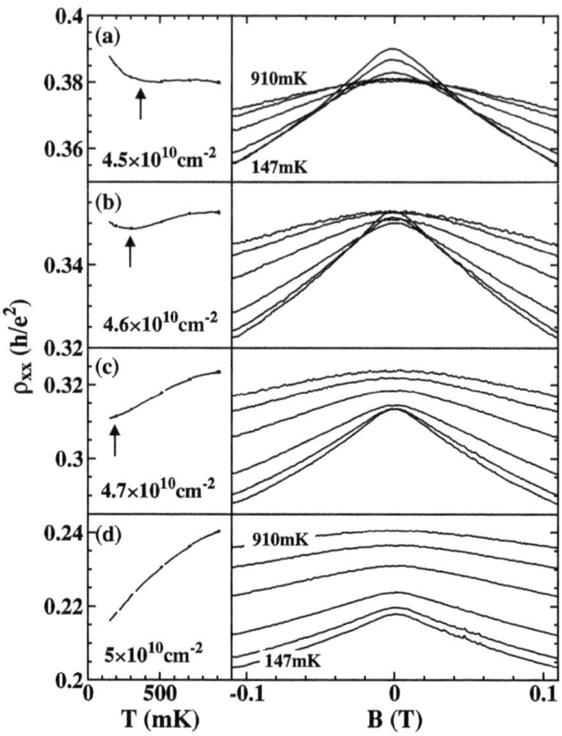

Figure 8.5. The phenomenon of weak localization observed in a two-dimensional electron gas at the interface between GaAs and AlGaAs. When the carrier density is low, the resistivity increases for $T \to 0$. In addition, a negative magneto-resistance appears, which is strongest at the lowest temperatures. Reprinted figure with permission from Simmons *et al* 2000 *Phys. Rev. Lett.* **84** 2489. Copyright (2000) by the American Physical Society.

Appendix A8

Doc A8.1 Kubo formula

Introducing the Fourier transforms of the conductivity tensor and of the electric field with equation (2*a*) and performing the t' time integration with equation (3*b*), we rewrite equation (8.1) in the frequency domain as

$$e\langle j_\mu(\boldsymbol{r}, t)\rangle_V = \int_{-\infty}^{\infty} \frac{d\omega}{2\pi} e^{-i\omega t} \sum_\nu \int d\boldsymbol{r}' \, \sigma_{\mu\nu}(\boldsymbol{r}, \boldsymbol{r}', \omega) E_\nu(\boldsymbol{r}', \omega). \tag{8.49}$$

On the other hand, the expansion of the current in powers of the vector potential reads

$$\langle j_\mu(\boldsymbol{r}, t)\rangle_V = \langle j_\mu^p(\boldsymbol{r}, t)\rangle + \langle j_\mu^p(\boldsymbol{r}, t)\rangle_V^{(1)} + \langle j_\mu^d(\boldsymbol{r}, t)\rangle_V^{(1)} + \mathcal{O}(A^2). \tag{8.50}$$

We assume there is no current flowing in the absence of field, such that the first term on the right-hand side vanishes. The second term is the paramagnetic current induced by the vector potential which is, according to equation (6.19),

$$\langle j_\mu^p(\boldsymbol{r},t)\rangle_V^{(1)} = -e\int_{-\infty}^{\infty}\frac{d\omega}{2\pi}e^{-i\omega t}\sum_\nu\int d\boldsymbol{r}'\mathscr{C}_{j_\mu^p(\boldsymbol{r})j_\nu^p(\boldsymbol{r}')}(i\Omega_n\to\hbar\omega^+)\underbrace{A_\nu(\boldsymbol{r}',\omega)}_{\frac{1}{i\omega}E_\nu(\boldsymbol{r}',\omega)}.$$

The third term in equation (8.50) is just the average of the diamagnetic current, computed directly from equation (2.49*c*) because $\boldsymbol{j}^d$ itself is first-order in $\boldsymbol{A}$:

$$\begin{aligned}\langle j_\mu^d(\boldsymbol{r},t)\rangle_V^{(1)} &= -\frac{e}{m}A_\mu(\boldsymbol{r},t)\langle n(\boldsymbol{r},t)\rangle\\ &= -\frac{e}{m}\langle n(\boldsymbol{r})\rangle\int_{-\infty}^{\infty}\frac{d\omega}{2\pi}e^{-i\omega t}\underbrace{A_\mu(\boldsymbol{r},\omega)}_{\frac{1}{i\omega}E_\mu(\boldsymbol{r},\omega)}.\end{aligned}$$

Here we have used the fact that the average $\langle n(\boldsymbol{r},t)\rangle$ must be evaluated in the absence of field and is therefore independent of time. Collecting the two terms, we find

$$\begin{aligned}e\langle j_\mu(\boldsymbol{r},t)\rangle_V = \int_{-\infty}^{\infty}\frac{d\omega}{2\pi}e^{-i\omega t}\sum_\nu\int d\boldsymbol{r}'\bigg\{&-\frac{e^2}{i\omega}\mathscr{C}_{j_\mu^p(\boldsymbol{r})j_\nu^p(\boldsymbol{r}')}(i\Omega_n\to\hbar\omega^+)\\ &-\delta_{\mu\nu}\delta(\boldsymbol{r}-\boldsymbol{r}')\frac{e^2}{i\omega m}\langle n(\boldsymbol{r})\rangle\bigg\}E_\nu(\boldsymbol{r}',\omega).\end{aligned}$$

Comparing with the Ohm's law (8.49), we deduce the expression of the conductivity tensor reported in equation (8.2).

In case of translational invariance, we have $\langle n(\boldsymbol{r})\rangle \equiv \langle n\rangle$ and $\sigma_{\mu\nu}(\boldsymbol{r},\boldsymbol{r}',\omega) = \sigma_{\mu\nu}(\boldsymbol{r}-\boldsymbol{r}',\boldsymbol{0},\omega)$, such that

$$\begin{aligned}\sigma_{\mu\nu}(\boldsymbol{q},\omega) &= \int d\boldsymbol{r}e^{-i\boldsymbol{q}\cdot\boldsymbol{r}}\sigma_{\mu\nu}(\boldsymbol{r},\boldsymbol{0},\omega) = \frac{1}{\mathscr{V}}\int d\boldsymbol{r}d\boldsymbol{r}'e^{-i\boldsymbol{q}\cdot\boldsymbol{r}}\sigma_{\mu\nu}(\boldsymbol{r}+\boldsymbol{r}',\boldsymbol{r}',\omega)\\ &= \frac{ie^2}{\omega}\frac{1}{\mathscr{V}}\int d\boldsymbol{r}d\boldsymbol{r}'e^{-i\boldsymbol{q}\cdot\boldsymbol{r}}\left[\mathscr{C}_{j_\mu^p(\boldsymbol{r}+\boldsymbol{r}')j_\nu^p(\boldsymbol{r}')}(i\Omega_n\to\hbar\omega^+) + \delta_{\mu\nu}\delta(\boldsymbol{r})\frac{\langle n\rangle}{m}\right]\\ &= \frac{ie^2}{\omega}\left\{\frac{1}{\mathscr{V}^2}\sum_{\boldsymbol{q}_1\boldsymbol{q}_2}\mathscr{C}_{j_\mu^p(\boldsymbol{q}_1)j_\nu^p(\boldsymbol{q}_2)}(\hbar\omega^+)\underbrace{\frac{1}{\mathscr{V}}\int d\boldsymbol{r}d\boldsymbol{r}'e^{-i\boldsymbol{q}\cdot\boldsymbol{r}}e^{i\boldsymbol{q}_1\cdot(\boldsymbol{r}+\boldsymbol{r}')}e^{i\boldsymbol{q}_2\cdot\boldsymbol{r}'}}_{\mathscr{V}\delta_{\boldsymbol{q}\boldsymbol{q}_1}\delta_{\boldsymbol{q}_1,-\boldsymbol{q}_2}} + \delta_{\mu\nu}\frac{\langle n\rangle}{m}\right\}\\ &= \frac{ie^2}{\omega}\left[\frac{1}{\mathscr{V}}\mathscr{C}_{j_\mu^p(\boldsymbol{q})j_\nu^p(-\boldsymbol{q})}(i\Omega_n\to\hbar\omega^+) + \delta_{\mu\nu}\frac{\langle n\rangle}{m}\right],\end{aligned}$$

consistently with equation (8.3).

Doc A8.2 Drude formula

For a momentum-independent self-energy $\Sigma(\varepsilon)$, the spectral function entering equation (8.8) is given by (see equation (7.12)):

$$A(\boldsymbol{k}, \varepsilon) = \frac{-\mathrm{Im}\,\Sigma(\varepsilon)/\pi}{[\varepsilon - \xi_{\boldsymbol{k}} - \mathrm{Re}\,\Sigma(\varepsilon)]^2 + [\mathrm{Im}\,\Sigma(\varepsilon)]^2}.$$

The dc conductivity resulting from equations (8.9) and (8.8) can be recast in the form

$$\sigma_{\mathrm{dc}} = \int_{-\infty}^{\infty} d\varepsilon[-f'(\varepsilon)][\mathrm{Im}\,\Sigma(\varepsilon)]^2 \frac{1}{\pi}\int_{-\infty}^{\infty} dE \frac{\Phi(E)}{\{[\varepsilon - \mathrm{Re}\,\Sigma(\varepsilon) - E]^2 + [\mathrm{Im}\,\Sigma(\varepsilon)]^2\}^2}.$$

We have used $\sum_\mu k_\mu^2 = k^2$ and introduced the material-specific *transport function*

$$\Phi(E) = \frac{e^2\hbar}{3}\frac{1}{\mathcal{V}}\sum_{\boldsymbol{k}\sigma} v_{\boldsymbol{k}}^2 \delta(\xi_{\boldsymbol{k}} - E).$$

$\boldsymbol{v}_{\boldsymbol{k}} = \hbar^{-1}\nabla\xi_{\boldsymbol{k}}$ is the group velocity, which is $\hbar\boldsymbol{k}/m$ in the plane-wave basis. We see that the transport function is proportional to the average squared velocity for all states having energy E. $\Phi(E)$ will vary from one material to the next depending on the details of the dispersion relation $\xi_{\boldsymbol{k}}$, but it is expected to be a slow function of E. In order to illustrate this, we assume a free-electron dispersion $\xi_{\boldsymbol{k}} = \hbar^2 k^2/(2m) - \varepsilon_{\mathrm{F}}$. Using $(\hbar k/m)^2 = 2(\xi_{\boldsymbol{k}} + \varepsilon_{\mathrm{F}})/m$, $\Phi(E)$ can be related to the free-electron DOS:

$$\begin{aligned}\Phi(E) &= \frac{2e^2\hbar}{3m}\frac{1}{\mathcal{V}}\sum_{\boldsymbol{k}\sigma}(\xi_{\boldsymbol{k}} + \varepsilon_{\mathrm{F}})\delta(\xi_{\boldsymbol{k}} - E)\\ &= \frac{2e^2\hbar}{3m}\int_{-\infty}^{\infty} d\xi\, N_0^{\mathrm{el}}(\xi)(\xi + \varepsilon_{\mathrm{F}})\delta(\xi - E)\\ &= \frac{2e^2\hbar}{3m} N_0^{\mathrm{el}}(E)(E + \varepsilon_{\mathrm{F}}).\end{aligned}$$

This is indeed a slow function of E that varies on the scale of ε_{F}. In contrast, the denominator in the expression of σ_{dc} varies over a much smaller scale $\sim\mathrm{Im}\,\Sigma(\varepsilon)$, with ε constrained to remain of the order $k_{\mathrm{B}}T$. We can therefore take $\Phi(E) \approx \Phi(0)$. Substituting the free-electron expressions, we find:

$$\Phi(0) = \frac{2e^2\hbar}{3m}N_0^{\mathrm{el}}(0)\varepsilon_{\mathrm{F}} = \frac{2e^2\hbar}{3m}\frac{mk_{\mathrm{F}}}{\pi^2\hbar^2}\frac{\hbar^2 k_{\mathrm{F}}^2}{2m} = \frac{e^2\hbar}{m}\frac{k_{\mathrm{F}}^3}{3\pi^2} = \frac{ne^2\hbar}{m}.$$

Replacing $\Phi(E)$ by $\Phi(0)$ in the expression of σ_{dc}, we are left with an integral that can be evaluated analytically:

$$\frac{1}{\pi}\int_{-\infty}^{\infty}\frac{dE}{[(a-E)^2 + \Gamma^2]^2} = \frac{1}{\pi|\Gamma|^3}\int_{-\infty}^{\infty}\frac{du}{(u^2+1)^2} = \frac{1}{2|\Gamma|^3}.$$

Finally we get

$$\sigma_{\mathrm{dc}} = \frac{ne^2\hbar}{2m}\int_{-\infty}^{\infty} d\varepsilon \frac{[-f'(\varepsilon)]}{|\mathrm{Im}\,\Sigma(\varepsilon)|}.$$

It is interesting to note that the real part of the self-energy disappears completely, leaving only the imaginary part in equation (8.10). The real part of Σ encodes information about the renormalization of the dispersion, like e.g. the effective mass. The observation that the real part of the self-energy disappears underlines the fact that the *bare* electron mass, not the effective mass, should be used in the Drude formula (8.10). For a more detailed discussion of this point, see Bonus B8.1.

Doc A8.3 Self-energy in the first Born approximation

For real energies, the self-energy (5.41) reads

$$\Sigma^{1\mathrm{BA}}(\boldsymbol{k}, \varepsilon) = n_i \frac{1}{\mathcal{V}} \sum_{\boldsymbol{q}} \frac{|v(\boldsymbol{q})|^2}{\varepsilon - \xi_{\boldsymbol{k}-\boldsymbol{q}} + i0^+} = n_i \frac{1}{\mathcal{V}} \sum_{\boldsymbol{q}} |v(\boldsymbol{q})|^2 G_0^R(\boldsymbol{k} - \boldsymbol{q}, \varepsilon)$$

with G_0^R the free retarded Green's function (see equation (3.26)). $\Sigma^{1\mathrm{BA}}$ is a convolution in reciprocal space with the generic form $f(\boldsymbol{k}) = \mathcal{V}^{-1} \sum_{\boldsymbol{q}} g(\boldsymbol{q}) h(\boldsymbol{k} - \boldsymbol{q})$. In real space, such convolutions become simple products: $f(\boldsymbol{r}) = g(\boldsymbol{r}) h(\boldsymbol{r})$. Hence we can write $\Sigma^{1\mathrm{BA}}$ in real space as

$$\Sigma^{1\mathrm{BA}}(\boldsymbol{r}, \varepsilon) = n_i \left(\frac{Ze^2}{\epsilon_0} \right)^2 \frac{e^{-k_{\mathrm{TF}} r}}{8\pi k_{\mathrm{TF}}} G_0^R(\boldsymbol{r}, \varepsilon).$$

We have used equation (8.13) for $v(\boldsymbol{q})$ and equation (15) for the Fourier transform of $|v(\boldsymbol{q})|^2$. For an isotropic dispersion $\xi_{\boldsymbol{k}} = \hbar^2 k^2/(2m_b) - \varepsilon_{\mathrm{F}}$, the real-space Green's function was calculated in Doc A7.1:

$$G_0^R(\boldsymbol{r}, \varepsilon) = -\frac{m_b}{2\pi\hbar^2 r} e^{i\kappa r} \qquad \text{with} \qquad \kappa = \sqrt{\frac{2m_b}{\hbar^2}(\varepsilon + \varepsilon_{\mathrm{F}})} = k_{\mathrm{F}} \sqrt{1 + \varepsilon/\varepsilon_{\mathrm{F}}}\,.$$

The last step is to Fourier transform back to momentum space, using spherical coordinates like in Doc A2 and Doc A5.14:

$$\Sigma^{1\mathrm{BA}}(\boldsymbol{k}, \varepsilon) = -\left(\frac{Ze^2}{\epsilon_0} \right)^2 \frac{m_b n_i}{16\pi^2 \hbar^2 k_{\mathrm{TF}}} 2\pi \int_0^\infty dr \int_0^\pi d\vartheta\, r^2 \sin\vartheta \frac{e^{-k_{\mathrm{TF}} r} e^{i\kappa r}}{r} e^{-ikr\cos\vartheta}.$$

The ϑ-integration is elementary and yields $2\sin(kr)/kr$. The r integration is easily done as well:

$$\begin{aligned}\Sigma^{1\mathrm{BA}}(\boldsymbol{k}, \varepsilon) &= -\left(\frac{Ze^2}{\epsilon_0} \right)^2 \frac{m_b n_i}{4\pi\hbar^2 k_{\mathrm{TF}} k} \underbrace{\int_0^\infty dr \sin kr e^{-k_{\mathrm{TF}} r} e^{i\kappa r}}_{-\frac{1}{2}\left(\frac{1}{\kappa - k + ik_{\mathrm{TF}}} - \frac{1}{\kappa + k + ik_{\mathrm{TF}}} \right)} \\ &= \left(\frac{Ze^2}{\epsilon_0} \right)^2 \frac{m_b n_i}{8\pi\hbar^2 k_{\mathrm{TF}} k} \left(\frac{1}{\kappa - k + ik_{\mathrm{TF}}} - \frac{1}{\kappa + k + ik_{\mathrm{TF}}} \right).\end{aligned}$$

Equation (8.14) results by introducing the expression of κ and using the isotropic-electron gas expression $k_{\mathrm{TF}}^2 = (e^2/\epsilon_0)N^{\mathrm{el}}(0) = (e^2/\epsilon_0)m_b k_{\mathrm{F}}/(\pi^2\hbar^2)$ to rewrite the prefactor.

Let us estimate the order of magnitude of the residual resistivity in order to check whether it can explain the data of figure 8.1. The residual resistivity is deduced from equations (8.10) and (8.11) evaluated at $T = 0$:

$$\rho_0 = \frac{m_b}{ne^2}\frac{2}{\hbar}|\mathrm{Im}\,\Sigma^{1\mathrm{BA}}(\boldsymbol{k}_{\mathrm{F}}, 0)|.$$

We write the impurity density as $n_i = (x/100) \times 4/a^3$, where x is the impurity concentration in atomic percent introduced in the first paragraph of section 8.3 and $4/a^3$ is the number of atoms per unit volume for fcc crystals like copper and silver, with a the lattice parameter. The electronic density is expressed as $n = 4N_v/a^3$ with N_v the number of valence electrons per atom. Denoting $s = k_{\mathrm{TF}}/k_{\mathrm{F}}$, we arrive at

$$\alpha = \frac{\rho_0}{x} = \frac{1}{3(10\pi)^2}\frac{ma_0^2}{\epsilon_0\hbar}\frac{(a/a_0)^3(m_b/m)^2}{N_v^2}\frac{Z^2}{s^2(4+s^2)}$$
$$= 9.23 \times 10^{-2}\mu\Omega\,\mathrm{cm}\frac{(a/a_0)^3(m_b/m)^2}{N_v^2}\frac{Z^2}{s^2(4+s^2)}.$$

Copper and silver have a band mass of the order of the electron mass and one valence electron in the $4s$ and $5s$ shell, respectively. The cubic lattice parameters are $a = 6.83a_0$ for Cu and $a = 7.72a_0$ for Ag. The parameter s measures the efficiency of the screening and is more difficult to estimate. The free-electron expression with $N_v = 1$ electron per atom underestimates s because it ignores the screening effect of occupied d levels. The free-electron formula gives

$$s^2 = \frac{k_{\mathrm{TF}}^2}{k_{\mathrm{F}}^2} = \frac{2^{4/3}}{3^{1/3}\pi^{5/3}}\frac{(a/a_0)(m_b/m)}{N_v^{1/3}},$$

which leads to values of $s \approx 1.5$ for Cu and Ag and to somewhat too large α coefficients $\alpha_{\mathrm{Cu}}/Z^2 = 2.9\ \mu\Omega\,\mathrm{cm}/\%$ imp. and $\alpha_{\mathrm{Ag}}/Z^2 = 3.5\ \mu\Omega\,\mathrm{cm}/\%$ imp. Increasing s by a factor two relative to the free-electron result gives α coefficients in very good agreement with the measurements: $\alpha_{\mathrm{Cu}}/Z^2 = 0.37\ \mu\Omega\,\mathrm{cm}/\%$ imp. and $\alpha_{\mathrm{Ag}}/Z^2 = 0.44\ \mu\Omega\,\mathrm{cm}/\%$ imp.

Doc A8.4 Second-order Coulomb self-energy

Performing the analytic continuation $i\omega_n \to \varepsilon + i0^+$ in equation (5.48), taking the imaginary part with equation (10), and using the notations $\boldsymbol{k}_3 \equiv \boldsymbol{k} + \boldsymbol{k}_1 - \boldsymbol{k}_2$ and $\xi_i \equiv \xi_{\boldsymbol{k}_i}$, we get

$-\mathrm{Im}\Big($ 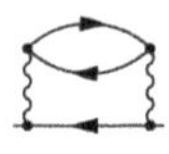 $+$ 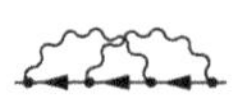 $\Big)$

$$= \frac{\pi}{\mathcal{V}^2}\sum_{k_1 k_2}\left[2|V(\boldsymbol{k}-\boldsymbol{k}_2)|^2 - V(\boldsymbol{k}_1-\boldsymbol{k}_2)V(\boldsymbol{k}_2-\boldsymbol{k})\right]$$
$$\times\left\{f(\xi_1)f(-\xi_2)f(-\xi_3) + f(-\xi_1)f(\xi_2)f(\xi_3)\right\}\delta(\varepsilon+\xi_1-\xi_2-\xi_3).$$

Inside the braces $\{\ldots\}$, we can replace ξ_3 by $\varepsilon + \xi_1 - \xi_2$ due to the delta function. Let us first remark that, at low energy ε, the Fermi factors and the delta function impose that all momenta $\boldsymbol{k}_i$ be close to the Fermi surface: indeed the combination of Fermi factors can be rewritten as

$$\{\ldots\} = \frac{f(\xi_1)f(-\xi_1)f(\xi_2)f(-\xi_2)}{f(\varepsilon)[f(\xi_1)-f(\xi_2)] + f(-\xi_1)f(\xi_2)}.$$

In the numerator, $f(\xi_1)f(-\xi_1)f(\xi_2)f(-\xi_2)$ vanishes as soon as $\boldsymbol{k}_1$ and/or $\boldsymbol{k}_2$ deviates from the vicinity of the Fermi surface. In addition, since $\varepsilon + \xi_1 - \xi_2$ is small, the delta function also requires that ξ_3 is small. The exact evaluation of the diagram requires to perform cumbersome integrations over the angles between the vectors $\boldsymbol{k}$, $\boldsymbol{k}_1$, and $\boldsymbol{k}_2$. In order to ease the evaluation, we make two simplifications: (i) we replace the Coulomb potential $V(\boldsymbol{q})$ by a constant V_0, to be interpreted as an average of the Coulomb matrix element for wave vectors on the Fermi surface; (ii) we average the diagram on $\boldsymbol{k}$, assuming it has a weak momentum dependence. Since $\boldsymbol{k}$ only appears in ξ_3, the $\boldsymbol{k}$-average can be converted into a ξ_3 integral according to

$$\frac{1}{N}\sum_{\boldsymbol{k}}(\ldots) = \frac{\mathcal{V}}{N}\frac{1}{2}\frac{1}{\mathcal{V}}\sum_{\boldsymbol{k}\sigma}(\ldots) = \frac{1}{2}\mathcal{V}_{\mathrm{cell}}\int_{-\infty}^{\infty} d\xi_3 N_0^{\mathrm{el}}(\xi_3)(\ldots).$$

The ξ_3 integral is trivial due to the delta function and the remaining double integration on ξ_1 and ξ_2 can be done exactly for a flat DOS:

$-\mathrm{Im}\Big($ 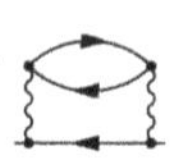 $+$ 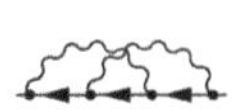 $\Big)$

$$\approx \frac{\pi}{8}\mathcal{V}_{\mathrm{cell}}V_0^2[N_0^{\mathrm{el}}(0)]^3\int_{-\infty}^{\infty} d\xi_1 d\xi_2$$
$$\times\left\{f(\xi_1)f(-\xi_2)f(-\varepsilon-\xi_1+\xi_2) + f(-\xi_1)f(\xi_2)f(\varepsilon+\xi_1-\xi_2)\right\}$$
$$= \frac{\pi}{16}\mathcal{V}_{\mathrm{cell}}V_0^2[N_0^{\mathrm{el}}(0)]^3\left[\varepsilon^2 + (\pi k_{\mathrm{B}}T)^2\right],$$

as indicated in equation (8.16). Although the two diagrams give contributions of similar amplitudes but opposite signs, they do not cancel due to a factor of two difference. The latter is due to the presence/absence of an internal spin sum.

Doc A8.5 Self-energy up to third order for the Kondo Hamiltonian

The first-order correction to the Green's function (8.28) involves the factor

$$\langle\langle T_\tau S_d^{\mu_1}(\boldsymbol{r}_1, \tau_1)\rangle_0\rangle_{\text{imp}} = \langle\langle T_\tau \sum_{\ell=1}^{N_i} s_d^{\mu_1}(\tau_1)\delta(\boldsymbol{r}_1 - \boldsymbol{R}_\ell)\rangle_0\rangle_{\text{imp}}$$

$$= \langle T_\tau s_d^{\mu_1}(\tau_1)\rangle_0 \underbrace{\left\langle \sum_{\ell=1}^{N_i} \delta(\boldsymbol{r}_1 - \boldsymbol{R}_\ell) \right\rangle_{\text{imp}}}_{n_i \int d\boldsymbol{r}\, \delta(\boldsymbol{r}) = n_i} .$$

We have proceeded like in Doc A5.6 for the impurity average. Now we use the fact that the unperturbed Hamiltonian K_0 (first term in equation (8.24)) does not depend on $s_d^{\mu_1}$, such that $[K_0, s_d^{\mu_1}] = 0$ and

$$\langle T_\tau s_d^{\mu_1}(\tau_1)\rangle_0 = \langle s_d^{\mu_1}(\tau_1)\rangle_0 = \frac{\text{Tr}e^{-\beta K_0}e^{\tau_1 K_0}s_d^{\mu_1}e^{-\tau_1 K_0}}{\text{Tr}e^{-\beta K_0}} = \frac{\text{Tr}e^{-\beta K_0}s_d^{\mu_1}}{\text{Tr}e^{-\beta K_0}} = \text{Tr}\; s_d^{\mu_1}.$$

The last equation holds because the Hilbert space is a tensor product of that of the conduction electrons and that of the impurity spin and in the absence of coupling we can write $\text{Tr}e^{-\beta K_0}s_d^{\mu_1} = \text{Tr}e^{-\beta K_0}\text{Tr}\; s_d^{\mu_1}$. We thus find, using the definition $\boldsymbol{s}_d = (\hbar/2)\boldsymbol{\tau}$ of the impurity spin as well as equation (23):

$$\langle\langle T_\tau S_d^{\mu_1}(\boldsymbol{r}_1, \tau_1)\rangle_0\rangle_{\text{imp}} = n_i \frac{\hbar}{2} \text{Tr}\; \tau^{\mu_1} = 0.$$

This result is expected: since nothing in the Hamiltonian K_0 breaks spin-rotation invariance, the average value of the impurity spin must be zero. At second order, the spin–spin correlation function does not vanish:

$$\langle\langle T_\tau S_d^{\mu_1}(\boldsymbol{r}_1, \tau_1)S_d^{\mu_2}(\boldsymbol{r}_2, \tau_2)\rangle_0\rangle_{\text{imp}} = \langle T_\tau s_d^{\mu_1}(\tau_1)s_d^{\mu_2}(\tau_2)\rangle_0 \underbrace{\left\langle \sum_{\ell=1}^{N_i} \delta(\boldsymbol{r}_1 - \boldsymbol{R}_\ell) \sum_{\ell'=1}^{N_i} \delta(\boldsymbol{r}_2 - \boldsymbol{R}_{\ell'}) \right\rangle_{\text{imp}}}_{n_i\delta(\boldsymbol{r}_1 - \boldsymbol{r}_2) + \mathcal{O}(n_i^2)} .$$

Developing the time-ordering operator and using equation (23), we find

$$\langle T_\tau s_d^{\mu_1}(\tau_1)s_d^{\mu_2}(\tau_2)\rangle_0 = \frac{\hbar^2}{4}[\theta(\tau_1 - \tau_2)\text{Tr}\; \tau^{\mu_1}\tau^{\mu_2} + \theta(\tau_2 - \tau_1)\text{Tr}\; \tau^{\mu_2}\tau^{\mu_1}] = \frac{\hbar^2}{2}\delta_{\mu_1\mu_2},$$

so that finally

$$\langle\langle T_\tau S_d^{\mu_1}(\boldsymbol{r}_1, \tau_1)S_d^{\mu_2}(\boldsymbol{r}_2, \tau_2)\rangle_0\rangle_{\text{imp}} = n_i \frac{\hbar^2}{2} \delta_{\mu_1\mu_2}\delta(\boldsymbol{r}_1 - \boldsymbol{r}_2).$$

We insert this result in equation (8.28), move from the real-space to the momentum-space electron operators, perform the spatial integrations, and finally use Wick's theorem to arrive at the following second-order contribution to the Green's function[28]:

$$\mathcal{G}^{(2)}_{\sigma\sigma'}(\boldsymbol{k},\tau) = n_i \frac{\hbar^4 J^2}{8} \int_0^\beta d\tau_1 d\tau_2 \sum_{\mu_1\sigma_2} \tau^{\mu_1}_{\sigma\sigma_2}\tau^{\mu_1}_{\sigma_2\sigma'} \frac{1}{\mathcal{V}} \sum_{\boldsymbol{q}} \mathcal{G}_0(\boldsymbol{k},\tau-\tau_1)\mathcal{G}_0(\boldsymbol{q},\tau_1-\tau_2)\mathcal{G}_0(\boldsymbol{k},\tau_2)$$

$= n_i \times$ 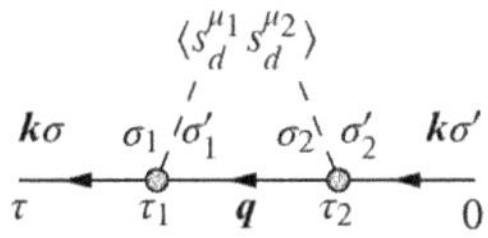

.

In the diagram, each dashed line carries a factor $\hbar J/2$, the gray circles represent the Pauli-matrix coefficients, and the constraint on spin indices ($\sigma_1 = \sigma$, $\sigma_1' = \sigma_2$, $\sigma_2' = \sigma'$) is enforced by the free propagators. Thanks to equation (24), we can evaluate $\sum_{\mu_1\sigma_2} \tau^{\mu_1}_{\sigma\sigma_2}\tau^{\mu_1}_{\sigma_2\sigma'} = (\sum_{\mu_1} \tau^{\mu_1}\tau^{\mu_1})_{\sigma\sigma'} = 3\delta_{\sigma\sigma'}$. The result is diagonal in the spin indices as imposed by the spin-rotation invariance. After a time Fourier transform we obtain:

$$\mathcal{G}^{(2)}_{\sigma\sigma'}(\boldsymbol{k}, i\omega_n) = \mathcal{G}_0(\boldsymbol{k}, i\omega_n) \left[\delta_{\sigma\sigma'} n_i \frac{3\hbar^4 J^2}{8} \frac{1}{\mathcal{V}} \sum_{\boldsymbol{q}} \mathcal{G}_0(\boldsymbol{q}, i\omega_n) \right] \mathcal{G}_0(\boldsymbol{k}, i\omega_n).$$

In this form, it is clear that what we have found is the first term of a Dyson equation, with the part in the square brackets being the self-energy (8.29). From the diagram, the self-energy $\Sigma^{(2)}_{\sigma\sigma'}(\boldsymbol{k}, \tau_1 - \tau_2)$ is obtained by removing the two external legs corresponding to the free propagators $\mathcal{G}_0(\boldsymbol{k}, \tau - \tau_1)$ and $\mathcal{G}_0(\boldsymbol{k}, \tau_2)$. The simple structure of the result without mixing of $\boldsymbol{k}$ with other wave vectors is a consequence of the factor $\delta(\boldsymbol{r}_1 - \boldsymbol{r}_2)$ from the impurity average, which itself reflects the fact that the impurity spins are local objects.

At order $n = 3$, the three-spin correlation function reads

$$\begin{aligned}&\langle\langle T_\tau S_d^{\mu_1}(\boldsymbol{r}_1, \tau_1) S_d^{\mu_2}(\boldsymbol{r}_2, \tau_2) S_d^{\mu_3}(\boldsymbol{r}_3, \tau_3)\rangle_0\rangle_{\text{imp}} \\ &= \langle T_\tau s_d^{\mu_1}(\tau_1) s_d^{\mu_2}(\tau_2) s_d^{\mu_3}(\tau_3)\rangle_0 \times n_i \delta(\boldsymbol{r}_1 - \boldsymbol{r}_2)\delta(\boldsymbol{r}_2 - \boldsymbol{r}_3) + \mathcal{O}(n_i^2)\end{aligned}$$

and also yields delta functions of the real-space coordinates. Hence the self-energy has the same simple structure as for $n = 2$ and is given by the diagram

$\Sigma^{(3)}_{\sigma\sigma'}(\boldsymbol{k}, \tau_1 - \tau_3) = n_i \times$ 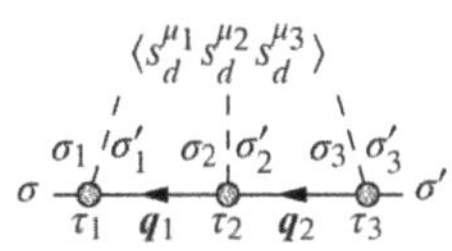

.

---

[28] The Wick theorem yields two topologically equivalent terms, only one of which must be counted.

The three-spin correlation function now has a time dependence because spin operators do not commute:

$$\begin{aligned}\langle T_\tau s_d^{\mu_1}(\tau_1)s_d^{\mu_2}(\tau_2)s_d^{\mu_3}(\tau_3)\rangle_0 = {} & \theta(\tau_1-\tau_2)\theta(\tau_2-\tau_3)\langle s_d^{\mu_1}s_d^{\mu_2}s_d^{\mu_3}\rangle_0 + \\ & \theta(\tau_1-\tau_3)\theta(\tau_3-\tau_2)\langle s_d^{\mu_1}s_d^{\mu_3}s_d^{\mu_2}\rangle_0 + \\ & \theta(\tau_2-\tau_1)\theta(\tau_1-\tau_3)\langle s_d^{\mu_2}s_d^{\mu_1}s_d^{\mu_3}\rangle_0 + \\ & \theta(\tau_2-\tau_3)\theta(\tau_3-\tau_1)\langle s_d^{\mu_2}s_d^{\mu_3}s_d^{\mu_1}\rangle_0 + \\ & \theta(\tau_3-\tau_1)\theta(\tau_1-\tau_2)\langle s_d^{\mu_3}s_d^{\mu_1}s_d^{\mu_2}\rangle_0 + \\ & \theta(\tau_3-\tau_2)\theta(\tau_2-\tau_1)\langle s_d^{\mu_3}s_d^{\mu_2}s_d^{\mu_1}\rangle_0 .\end{aligned}$$

From equation (23), we see that

$$\langle s_d^{\mu_1}s_d^{\mu_2}s_d^{\mu_3}\rangle_0 = \left(\frac{\hbar}{2}\right)^3 \mathrm{Tr}\ \tau^{\mu_1}\tau^{\mu_2}\tau^{\mu_3} = \frac{i\hbar^3}{4}\epsilon_{\mu_1\mu_2\mu_3},$$

and since $\epsilon_{\mu_1\mu_2\mu_3} = -\epsilon_{\mu_2\mu_1\mu_3}$, etc, we can rewrite

$$\begin{aligned}\langle T_\tau s_d^{\mu_1}(\tau_1)s_d^{\mu_2}(\tau_2)s_d^{\mu_3}(\tau_3)\rangle_0 = {} & \frac{i\hbar^3}{4}\epsilon_{\mu_1\mu_2\mu_3}[\theta(\tau_1-\tau_2)\theta(\tau_2-\tau_3) \\ & - \theta(\tau_1-\tau_3)\theta(\tau_3-\tau_2) - \theta(\tau_2-\tau_1)\theta(\tau_1-\tau_3) + \theta(\tau_2-\tau_3)\theta(\tau_3-\tau_1) \\ & + \theta(\tau_3-\tau_1)\theta(\tau_1-\tau_2) - \theta(\tau_3-\tau_2)\theta(\tau_2-\tau_1)].\end{aligned}$$

The value of the diagram becomes, taking into account the constraints on spin indices,

$$\begin{aligned}\Sigma^{(3)}_{\sigma\sigma'}(\boldsymbol{k}, \tau_1-\tau_3) = {} & n_i\left(\frac{\hbar J}{2}\right)^3 \frac{i\hbar^3}{4}\frac{1}{\mathscr{V}^2}\sum_{\boldsymbol{q}_1\boldsymbol{q}_2}\sum_{\mu_1\mu_2\mu_3}\sum_{\sigma_2\sigma_3}\tau^{\mu_1}_{\sigma\sigma_2}\tau^{\mu_2}_{\sigma_2\sigma_3}\tau^{\mu_3}_{\sigma_3\sigma'}\epsilon_{\mu_1\mu_2\mu_3} \\ & \times \int_0^\beta d\tau_2\, [\ldots]\mathscr{G}_0(\boldsymbol{q}_1, \tau_1-\tau_2)\mathscr{G}_0(\boldsymbol{q}_2, \tau_2-\tau_3),\end{aligned}$$

where the square brackets contain the complicated time dependence due to the time ordering. Equation (24) tells us that

$$\sum_{\mu_1\mu_2\mu_3}\sum_{\sigma_2\sigma_3}\tau^{\mu_1}_{\sigma\sigma_2}\tau^{\mu_2}_{\sigma_2\sigma_3}\tau^{\mu_3}_{\sigma_3\sigma'}\epsilon_{\mu_1\mu_2\mu_3} = \left(\sum_{\mu_1\mu_2\mu_3}\tau^{\mu_1}\tau^{\mu_2}\tau^{\mu_3}\epsilon_{\mu_1\mu_2\mu_3}\right)_{\sigma\sigma'} = 6i\delta_{\sigma\sigma'}.$$

Furthermore, since the self-energy is a function of the time difference $\tau_1 - \tau_3$, we can set $\tau_3 = 0$ and redefine $\tau_1 \equiv \tau$. In doing so, most of the theta functions in [...] either vanish or equal one and we are left with $[\ldots] = \theta(\tau - \tau_2) - \theta(\tau_2 - \tau) = \mathrm{sign}(\tau - \tau_2)$:

$$\Sigma^{(3)}_{\sigma\sigma'}(\boldsymbol{k}, \tau) = -\delta_{\sigma\sigma'}n_i\frac{3\hbar^6J^3}{16}\frac{1}{\mathscr{V}^2}\sum_{\boldsymbol{q}_1\boldsymbol{q}_2}\int_0^\beta d\tau_2\ \mathrm{sign}(\tau-\tau_2)\mathscr{G}_0(\boldsymbol{q}_1, \tau-\tau_2)\mathscr{G}_0(\boldsymbol{q}_2, \tau_2).$$

In order to perform the time integration, we need the expression of the free Green's function in imaginary time. This can be directly read from equation (17): $\mathcal{G}_0(\boldsymbol{k}, \tau) = -[\theta(\tau) - f(\xi_{\boldsymbol{k}})]e^{-\xi_{\boldsymbol{k}}\tau}$. We then have for the τ_2 integral, using the notation $\xi_i \equiv \xi_{\boldsymbol{q}_i}$,

$$\int_0^\tau d\tau_2\, [\theta(\tau - \tau_2) - f(\xi_1)]e^{-\xi_1(\tau-\tau_2)}[\theta(\tau_2) - f(\xi_2)]e^{-\xi_2\tau_2}$$

$$- \int_\tau^\beta d\tau_2\, [\theta(\tau - \tau_2) - f(\xi_1)]e^{-\xi_1(\tau-\tau_2)}[\theta(\tau_2) - f(\xi_2)]e^{-\xi_2\tau_2}$$

$$= [1 - f(\xi_1)][1 - f(\xi_2)]e^{-\xi_1\tau}\int_0^\tau d\tau_2 e^{(\xi_1-\xi_2)\tau_2} - [-f(\xi_1)][1 - f(\xi_2)]e^{-\xi_1\tau}\int_\tau^\beta d\tau_2 e^{(\xi_1-\xi_2)\tau_2}$$

$$= f(-\xi_1)f(-\xi_2)e^{-\xi_1\tau}\frac{e^{(\xi_1-\xi_2)\tau} - 1}{\xi_1 - \xi_2} + f(\xi_1)f(-\xi_2)e^{-\xi_1\tau}\frac{e^{(\xi_1-\xi_2)\beta} - e^{(\xi_1-\xi_2)\tau}}{\xi_1 - \xi_2}$$

$$= \frac{1}{\xi_1 - \xi_2}\left[f(-\xi_1)f(-\xi_2)(e^{-\xi_2\tau} - e^{-\xi_1\tau}) + \underbrace{f(\xi_1)e^{\beta\xi_1}}_{f(-\xi_1)}\underbrace{f(-\xi_2)e^{-\beta\xi_2}}_{f(\xi_2)}e^{-\xi_1\tau} - f(\xi_1)f(-\xi_2)e^{-\xi_2\tau}\right]$$

$$= \frac{1}{\xi_1 - \xi_2}\{-[1 - 2f(\xi_2)]f(-\xi_1)e^{-\xi_1\tau} + [1 - 2f(\xi_1)]f(-\xi_2)e^{-\xi_2\tau}\}$$

$$= \frac{1}{\xi_1 - \xi_2}\{[1 - 2f(\xi_2)]\mathcal{G}_0(\boldsymbol{q}_1, \tau) - [1 - 2f(\xi_1)]\mathcal{G}_0(\boldsymbol{q}_2, \tau)\}$$

$$= \frac{1 - 2f(\xi_{\boldsymbol{q}_2})}{\xi_{\boldsymbol{q}_1} - \xi_{\boldsymbol{q}_2}}\mathcal{G}_0(\boldsymbol{q}_1, \tau) + \frac{1 - 2f(\xi_{\boldsymbol{q}_1})}{\xi_{\boldsymbol{q}_2} - \xi_{\boldsymbol{q}_1}}\mathcal{G}_0(\boldsymbol{q}_2, \tau).$$

The last line shows that the two terms are equal with the roles of $\boldsymbol{q}_1$ and $\boldsymbol{q}_2$ exchanged. Since the τ dependence of $\Sigma^{(3)}_{\sigma\sigma'}(\boldsymbol{k}, \tau)$ only comes from $\mathcal{G}_0(\boldsymbol{q}_1, \tau)$, we can directly perform the time Fourier transform and write

$$\Sigma^{(3)}_{\sigma\sigma'}(\boldsymbol{k}, i\omega_n) = -\,\delta_{\sigma\sigma'}n_i\frac{3\hbar^6 J^3}{8}\frac{1}{\mathcal{V}^2}\sum_{\boldsymbol{q}_1\boldsymbol{q}_2}\frac{1 - 2f(\xi_{\boldsymbol{q}_2})}{\xi_{\boldsymbol{q}_1} - \xi_{\boldsymbol{q}_2}}\mathcal{G}_0(\boldsymbol{q}_1, i\omega_n)$$

$$= \delta_{\sigma\sigma'}n_i\frac{3\hbar^6 J^3}{8}\frac{1}{4}\int_{-\infty}^{\infty} d\xi_1 d\xi_2\, N_0^{\mathrm{el}}(\xi_1)N_0^{\mathrm{el}}(\xi_2)\frac{1 - 2f(\xi_2)}{\xi_2 - \xi_1}\frac{1}{i\omega_n - \xi_1}.$$

We finally evaluate the scattering rate given by $-\mathrm{Im}\Sigma^{(3)}$ on the real-energy axis $i\omega_n \to \varepsilon + i0^+$ using equation (10) and we thus obtain equation (8.30).

Appendix B8

Bonus B8.1 Drude formula, band mass and effective mass

Our purpose here is to generalize the expression of the Drude formula derived in section 8.2 to the case where—except for the interactions—the electrons are independent but not free. This is an occasion for illustrating the difference between the notions of 'band effective mass' and 'dynamical effective mass'. Independent electrons are described by the Hamiltonian

$$K_0 = \sum_{\alpha\sigma} \xi_\alpha \gamma^\dagger_{\alpha\sigma}\gamma_{\alpha\sigma},$$

where $\gamma^\dagger_{\alpha\sigma}$ creates an electron in the eigenstate $|\varphi_\alpha\rangle$ of K_0. It is necessary to perform a change of basis, like in Doc A3.4, in order to move from the plane-wave representation to the representation based on the eigenstates of K_0. The resulting expression for the paramagnetic current operator (2.50) is

$$\boldsymbol{j}^p(\boldsymbol{q}) = \sum_{\alpha\beta\sigma} \boldsymbol{j}_{\alpha\beta}(\boldsymbol{q})\gamma^\dagger_{\alpha\sigma}\gamma_{\beta\sigma}$$

with the matrix elements given by

$$\begin{aligned}\boldsymbol{j}_{\alpha\beta}(\boldsymbol{q}) &= \frac{\hbar}{m}\frac{1}{\mathscr{V}}\sum_{\boldsymbol{k}}\left(\boldsymbol{k}+\frac{\boldsymbol{q}}{2}\right)\varphi^*_\alpha(\boldsymbol{k})\varphi_\beta(\boldsymbol{k}+\boldsymbol{q}) = \boldsymbol{j}^*_{\beta\alpha}(-\boldsymbol{q})\\ &= \frac{-i\hbar}{2m}\int d\boldsymbol{r}\left[e^{-i\boldsymbol{q}\cdot\boldsymbol{r}}\varphi^*_\alpha(\boldsymbol{r})\nabla_{\boldsymbol{r}}\varphi_\beta(\boldsymbol{r}) + \varphi^*_\alpha(\boldsymbol{r})\nabla_{\boldsymbol{r}}\varphi_\beta(\boldsymbol{r})e^{-i\boldsymbol{q}\cdot\boldsymbol{r}}\right].\end{aligned}$$

As the translation invariance is in general broken, the conductivity tensor (8.2) depends on both $\boldsymbol{r}$ and $\boldsymbol{r}'$ independently. We must define a spatial average of the conductivity. Introducing the center-of-mass coordinate $\boldsymbol{R}$ and the relative coordinate $\boldsymbol{\rho}$, we write $\boldsymbol{r} = \boldsymbol{R} + \boldsymbol{\rho}/2$ and $\boldsymbol{r}' = \boldsymbol{R} - \boldsymbol{\rho}/2$ and we perform an average on $\boldsymbol{R}$:

$$\begin{aligned}\bar{\sigma}_{\mu\nu}(\boldsymbol{\rho}, \omega) &= \frac{1}{\mathscr{V}}\int d\boldsymbol{R}\ \sigma(\boldsymbol{R}+\boldsymbol{\rho}/2, \boldsymbol{R}-\boldsymbol{\rho}/2, \omega)\\ \bar{\sigma}_{\mu\nu}(\boldsymbol{q}, \omega) &= \int d\boldsymbol{\rho}\ \bar{\sigma}_{\mu\nu}(\boldsymbol{\rho}, \omega)e^{-i\boldsymbol{q}\cdot\boldsymbol{\rho}}.\end{aligned}$$

The Kubo formula then becomes

$$\bar{\sigma}_{\mu\nu}(\boldsymbol{q}, \omega) = \frac{ie^2}{\omega}\left[\frac{1}{\mathscr{V}}\bar{\chi}^{\mu\nu}_{jj}(\boldsymbol{q}, i\Omega_n \to \hbar\omega + i0^+) + \delta_{\mu\nu}\frac{\bar{n}}{m}\right],$$

where $\bar{n}$ is the spatial average of the electron density and

$$\begin{aligned}\bar{\chi}^{\mu\nu}_{jj}(\boldsymbol{q}, \tau) &= -\langle T_\tau j^p_\mu(\boldsymbol{q}, \tau) j^p_\nu(-\boldsymbol{q}, 0)\rangle\\ &= \sum_{\alpha\beta\alpha'\beta'\sigma}[\boldsymbol{j}_{\alpha\beta}(\boldsymbol{q})]_\mu[\boldsymbol{j}_{\alpha'\beta'}(-\boldsymbol{q})]_\nu\mathscr{G}_{\beta\alpha'}(\tau)\mathscr{G}_{\beta'\alpha}(-\tau) + \text{vertex corrections}.\end{aligned}$$

Here $\mathscr{G}_{\alpha\beta}(\tau) = -\langle T_\tau\gamma_\alpha(\tau)\gamma^\dagger_\beta(0)\rangle$ is the propagator of the independent electrons. Without interaction, $\mathscr{G}_{\alpha\beta}(\tau) = \delta_{\alpha\beta}\mathscr{G}^0_\alpha(\tau)$ because the states $|\varphi_\alpha\rangle$ are eigenstates, with

$\mathscr{G}^0_\alpha(i\omega_n) = (i\omega_n - \xi_\alpha)^{-1}$ (see section 5.2.2.1). We now take the same assumption as in section 8.2, namely that the interactions not contained in K_0 lead to a self-energy that only depends on the energy and does not induce transitions between states: $\Sigma_{\alpha\beta}(i\omega_n) = \delta_{\alpha\beta}\Sigma(i\omega_n)$. The resulting Green's function is then $\mathscr{G}_{\alpha\beta}(i\omega_n) = \delta_{\alpha\beta}[i\omega_n - \xi_\alpha - \Sigma(i\omega_n)]^{-1}$ and vertex corrections vanish. The expression of the current–current correlation function simplifies to

$$\bar{\chi}^{\mu\nu}_{jj}(\boldsymbol{q}, i\Omega_n) = \sum_{\alpha\beta\sigma}[\boldsymbol{j}_{\alpha\beta}(\boldsymbol{q})]_\mu[\boldsymbol{j}_{\beta\alpha}(-\boldsymbol{q})]_\nu \int_{-\infty}^{\infty} d\varepsilon_1 d\varepsilon_2\, A_\alpha(\varepsilon_1)A_\beta(\varepsilon_2)\frac{f(\varepsilon_1) - f(\varepsilon_2)}{i\Omega_n + \varepsilon_1 - \varepsilon_2},$$

with the spectral function $A_\alpha(\varepsilon) = \frac{-\mathrm{Im}\,\Sigma(\varepsilon)/\pi}{[\varepsilon - \xi_\alpha - \mathrm{Re}\,\Sigma(\varepsilon)]^2 + [\mathrm{Im}\,\Sigma(\varepsilon)]^2}$. The resulting expression for the dc conductivity tensor is

$$\mathrm{Re}\,\bar{\sigma}_{\mu\nu}(\boldsymbol{0}, 0) = \pi e^2\hbar \int_{-\infty}^{\infty} d\varepsilon[-f'(\varepsilon)]\frac{1}{\mathscr{V}}\sum_{\alpha\beta\sigma}[\boldsymbol{j}_{\alpha\beta}(\boldsymbol{0})]_\mu[\boldsymbol{j}^*_{\alpha\beta}(\boldsymbol{0})]_\nu A_\alpha(\varepsilon)A_\beta(\varepsilon),$$

where we have used the symmetry property of the matrix element. Following the same approach as in Doc A8.2, the dc conductivity of an isotropic ($\bar{\sigma}_{xx} = \bar{\sigma}_{yy} = \bar{\sigma}_{zz}$) system can be put in the final form

$$\sigma_{\mathrm{dc}} = \int_{-\infty}^{\infty} d\varepsilon\, [-f'(\varepsilon)][\mathrm{Im}\,\Sigma(\varepsilon)]^2$$

$$\times \frac{1}{\pi}\int_{-\infty}^{\infty} \frac{\Phi(E, E')dEdE'}{\{[\varepsilon - E - \mathrm{Re}\,\Sigma(\varepsilon)]^2 + [\mathrm{Im}\,\Sigma(\varepsilon)]^2\}\{[\varepsilon - E' - \mathrm{Re}\,\Sigma(\varepsilon)]^2 + [\mathrm{Im}\,\Sigma(\varepsilon)]^2\}}$$

$$\Phi(E, E') = \frac{e^2\hbar}{3}\frac{1}{\mathscr{V}}\sum_{\alpha\beta\sigma}|\boldsymbol{j}_{\alpha\beta}(\boldsymbol{0})|^2\delta(\xi_\alpha - E)\delta(\xi_\beta - E').$$

Bloch electrons. Let us consider the case of electrons in a periodic potential. Their Bloch wave functions are characterized by a wave vector $\boldsymbol{k}$ in the first Brillouin zone and a band index n (see section 2.4.1) and can be decomposed according to

$$\varphi_{kn}(\boldsymbol{r}) = \frac{1}{\sqrt{\mathscr{V}}}\sum_{\boldsymbol{G}} u_{kn}(\boldsymbol{G})e^{i(\boldsymbol{k}+\boldsymbol{G})\cdot\boldsymbol{r}}.$$

Using these wave functions as our basis, we can evaluate the matrix elements of the current:

$$\begin{aligned}
\boldsymbol{j}_{knk'n'}(\boldsymbol{0}) &= \frac{-i\hbar}{m}\int d\boldsymbol{r}\, \varphi^*_{kn}(\boldsymbol{r})\nabla_{\boldsymbol{r}}\varphi_{k'n'}(\boldsymbol{r}) \\
&= \frac{-i\hbar}{m}\sum_{\boldsymbol{G}\boldsymbol{G}'} u^*_{kn}(\boldsymbol{G})u_{k'n'}(\boldsymbol{G}')\underbrace{\frac{1}{\mathscr{V}}\int d\boldsymbol{r}e^{-i(\boldsymbol{k}+\boldsymbol{G})\cdot\boldsymbol{r}}\nabla_{\boldsymbol{r}}e^{i(\boldsymbol{k}'+\boldsymbol{G}')\cdot\boldsymbol{r}}}_{i(\boldsymbol{k}+\boldsymbol{G})\delta_{kk'}\delta_{\boldsymbol{G}\boldsymbol{G}'}} \\
&= \delta_{kk'}\frac{\hbar}{m}\sum_{\boldsymbol{G}}(\boldsymbol{k} + \boldsymbol{G})u^*_{kn}(\boldsymbol{G})u_{kn'}(\boldsymbol{G}).
\end{aligned}$$

The evaluation of the matrix elements seems to require the wave functions. This is actually true for inter-band transitions, but not for intra-band transitions as we shall see. Since the Drude response is due to intra-band scattering, we restrict now to $n = n'$. We first show that the matrix elements of the current do not depend on the wave functions, but can be expressed in term of the group velocity only as:

$$\boldsymbol{j}_{knk'n}(\boldsymbol{0}) = \delta_{kk'}\frac{1}{\hbar}\nabla_k\varepsilon_{kn} = \delta_{kk'}\boldsymbol{v}_{kn},$$

where ε_{kn} are the band energies and $\boldsymbol{v}_{kn}$ the group velocity. Indeed, since $\varepsilon_{kn} = \langle\varphi_{kn}|h(\boldsymbol{k})|\varphi_{kn}\rangle$ we have

$$\begin{aligned}\nabla_k\varepsilon_{kn} &= \langle\nabla_k\varphi_{kn}|h(\boldsymbol{k})\varphi_{kn}\rangle + \langle\varphi_{kn}|\nabla_k h(\boldsymbol{k})|\varphi_{kn}\rangle + \langle\varphi_{kn}h(\boldsymbol{k})|\nabla_k\varphi_{kn}\rangle \\ &= \langle\varphi_{kn}|\nabla_k h(\boldsymbol{k})|\varphi_{kn}\rangle + \varepsilon_{kn}\underbrace{[\langle\nabla_k\varphi_{kn}|\varphi_{kn}\rangle + \langle\varphi_{kn}|\nabla_k\varphi_{kn}\rangle]}_{=\nabla_k\langle\varphi_{kn}|\varphi_{kn}\rangle = 0}.\end{aligned}$$

On the other hand, the matrix elements of the Hamiltonian $h(\boldsymbol{k})$ are given by (see equation (2.52)) $h_{GG'}(\boldsymbol{k}) = \frac{\hbar^2}{2m}(\boldsymbol{k}+\boldsymbol{G})^2\delta_{GG'} + V(\boldsymbol{G}-\boldsymbol{G}')$, such that the gradient is simply $[\nabla_k h(\boldsymbol{k})]_{GG'} = \frac{\hbar^2}{m}(\boldsymbol{k}+\boldsymbol{G})\delta_{GG'}$. As a result, we can evaluate the action of $\nabla_k h(\boldsymbol{k})$ on φ_{kn} as

$$\begin{aligned}\nabla_k h(\boldsymbol{k})\varphi_{kn}(\boldsymbol{r}) &= \frac{1}{\sqrt{\mathcal{V}}}\sum_{GG'}[\nabla_k h(\boldsymbol{k})]_{GG'}u_{kn}(\boldsymbol{G}')e^{i(\boldsymbol{k}+\boldsymbol{G}')\cdot\boldsymbol{r}} \\ &= \frac{1}{\sqrt{\mathcal{V}}}\sum_{G}\frac{\hbar^2}{m}(\boldsymbol{k}+\boldsymbol{G})u_{kn}(\boldsymbol{G})e^{i(\boldsymbol{k}+\boldsymbol{G})\cdot\boldsymbol{r}}\end{aligned}$$

and the gradient of the dispersion becomes

$$\begin{aligned}\nabla_k\varepsilon_{kn} &= \frac{1}{\mathcal{V}}\int d\boldsymbol{r}\sum_{G'}u^*_{kn}(\boldsymbol{G}')e^{-i(\boldsymbol{k}+\boldsymbol{G}')\cdot\boldsymbol{r}}\sum_{G}\frac{\hbar^2}{m}(\boldsymbol{k}+\boldsymbol{G})u_{kn}(\boldsymbol{G})e^{i(\boldsymbol{k}+\boldsymbol{G})\cdot\boldsymbol{r}} \\ &= \frac{\hbar^2}{m}\sum_{G}(\boldsymbol{k}+\boldsymbol{G})u^*_{kn}(\boldsymbol{G})u_{kn}(\boldsymbol{G}) = \hbar\boldsymbol{j}_{knkn}(\boldsymbol{0}).\end{aligned}$$

This completes the proof that $\boldsymbol{j}_{knk'n}(\boldsymbol{0}) = \delta_{kk'}\boldsymbol{v}_{kn}$.

The function $\Phi(E, E')$ for intra-band scattering becomes

$$\Phi(E, E') = \delta(E-E')\Phi_{\text{intra}}(E), \qquad \Phi_{\text{intra}}(E) = \frac{e^2\hbar}{3}\frac{1}{\mathcal{V}}\sum_{kn\sigma}v^2_{kn}\delta(\xi_{kn}-E).$$

This leads to the same expression as in Doc A8.2 for the dc conductivity, except that the function $\Phi(E)$ must be replaced by $\Phi_{\text{intra}}(E)$ defined above. Neglecting the energy dependence of $\Phi_{\text{intra}}(E)$, we thus arrive at

$$\sigma_{\text{dc}} = \Phi_{\text{intra}}(0)\int_{-\infty}^{\infty}d\varepsilon\frac{[-f'(\varepsilon)]}{2|\text{Im}\,\Sigma(\varepsilon)|}.$$

As already noted in Doc A8.2, the real part of the self-energy has disappeared from this formula, together with any information regarding the *dynamical* effective mass. The dynamical effective mass m^* relates to the mass renormalization induced by the interactions *not taken into account in the band structure*. All interactions that can be treated by a mean field approximation are captured by the band structure and lead at low energy to a mass renormalization that we shall call the *band mass*: $\hbar^2k^2/(2m)\rightarrow \varepsilon_{\boldsymbol{k}} \sim \hbar^2k^2/(2m_b)$. The latter mass *is* present in the Drude formula. To see this, we can rewrite $\Phi_{\text{intra}}(0)$ by means of equation (11), which in the present context reads

$$\delta(\xi_{kn}) = \int_{S_n} dS \frac{\delta(\boldsymbol{k} - \boldsymbol{k}_{\mathrm{F}n})}{|\nabla \xi_{kn}|_{\boldsymbol{k}_{\mathrm{F}n}}} = \frac{1}{\hbar} \int_{S_n} dS \frac{\delta(\boldsymbol{k} - \boldsymbol{k}_{\mathrm{F}n})}{|\boldsymbol{v}_{\boldsymbol{k}_{\mathrm{F}n}}|}.$$

S_n is the Fermi surface corresponding to the band n and $\boldsymbol{k}_{\mathrm{F}n}$ are the corresponding Fermi wave vectors. Converting the $\boldsymbol{k}$ sum into an integral we get

$$\Phi_{\text{intra}}(0) = \frac{e^2}{3} \sum_{n\sigma} \frac{1}{(2\pi)^3} \int_{S_n} dS \, |\boldsymbol{v}_{\boldsymbol{k}_{\mathrm{F}n}}|.$$

Hence $\Phi_{\text{intra}}(0)$ is just proportional to the Fermi-surface average of the Fermi velocity. This reduces to $\Phi(0) = ne^2\hbar/m$ for free electrons but becomes $ne^2\hbar/m_b$ for a dispersion $\varepsilon_{\boldsymbol{k}} = \hbar^2k^2/(2m_b)$.

Bonus B8.2 Self-energy of the Coulomb gas in RPA

We have introduced the RPA in section 5.1.4.4, as a general approximation scheme for the polarization Π; see also equation (5.81). In section 5.1.4.6, we have seen that, in the context of the Coulomb interaction, the polarization describes the screening capability of the electrons and is related to the dielectric function. The screened Coulomb interaction is given by equation (5.90), which in RPA becomes

$$W^{\mathrm{RPA}}(\boldsymbol{q}, i\Omega_n) = \frac{V(\boldsymbol{q})}{1 - \dfrac{1}{\mathcal{V}} V(\boldsymbol{q}) \chi^0_{nn}(\boldsymbol{q}, i\Omega_n)},$$

where χ^0_{nn} is the free density–density correlation function. It is useful to rewrite this in terms of the full density–density correlation function, which is given in RPA by equation (5.82). One gets

$$W^{\mathrm{RPA}}(\boldsymbol{q}, i\Omega_n) = V(\boldsymbol{q}) + V^2(\boldsymbol{q}) \frac{1}{\mathcal{V}} \chi^{\mathrm{RPA}}_{nn}(\boldsymbol{q}, i\Omega_n).$$

Diagrammatically, this expression takes the form

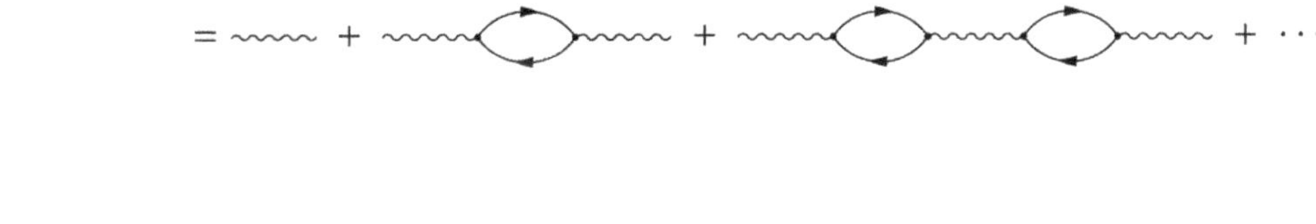

$W^{\rm RPA}(\boldsymbol{q}, i\Omega_n)$ is a retarded effective interaction, analogous to the one resulting from electron–phonon coupling, equation (5.56). In the same way as the phonon-induced electron–electron interaction is mediated by bosonic modes, the screened Coulomb interaction is mediated by bosons: these are electron–hole pairs—or density fluctuations—whose propagation is described by $\chi_{nn}^{\rm RPA}$ or more generally by χ_{nn}.

If, in the first-order exchange diagram for the self-energy (5.47), we replace the bare Coulomb interaction V by the screened interaction $W^{\rm RPA}$, an infinite series of diagrams is generated:

$\Sigma^{\rm RPA} =$ 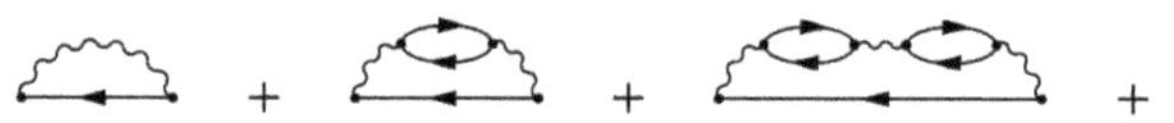 $+ \cdots$

The first diagram is the real exchange term (5.47), the second diagram is the first term of equation (5.48), and the subsequent ones contain corrections up to infinite order. The explicit self-energy formula results by applying the diagrammatic rules of Doc A5.9:

$$\Sigma^{\rm RPA}(\boldsymbol{k}, i\omega_n) = \Sigma^{\rm exch}(\boldsymbol{k}) - \frac{1}{\mathcal{V}}\sum_{\boldsymbol{q}}\frac{1}{\beta}\sum_{i\Omega_n} V^2(\boldsymbol{q})\frac{1}{\mathcal{V}}\chi_{nn}^{\rm RPA}(\boldsymbol{q}, i\Omega_n)\mathscr{G}_0(\boldsymbol{k}+\boldsymbol{q}, i\omega_n + i\Omega_n).$$

In order to evaluate the Matsubara sum, we introduce the spectral representation

$$\chi_{nn}^{\rm RPA}(\boldsymbol{q}, i\Omega_n) = \int_{-\infty}^{\infty} dE \frac{(-1/\pi)\,{\rm Im}\,\chi_{nn}^{\rm RPA}(\boldsymbol{q}, E)}{i\Omega_n - E},$$

where $\chi_{nn}^{\rm RPA}(\boldsymbol{q}, E)$ is the retarded function just above the real axis. We also use $\mathscr{G}_0(\boldsymbol{k}, i\omega_n) = 1/(i\omega_n - \xi_{\boldsymbol{k}})$ as well as equation (16). The result is, after making the analytic continuation $i\omega_n \to \varepsilon + i0^+$:[29]

$$\Sigma^{\rm RPA}(\boldsymbol{k}, \varepsilon) = \Sigma^{\rm exch}(\boldsymbol{k}) + \sum_{\boldsymbol{q}}\frac{V^2(\boldsymbol{q})}{\mathcal{V}^2}\int_{-\infty}^{\infty} dE\left(-\frac{1}{\pi}\right){\rm Im}\,\chi_{nn}^{\rm RPA}(\boldsymbol{q}, E)\frac{f(\xi_{\boldsymbol{k}+\boldsymbol{q}}) + b(E)}{\varepsilon + E - \xi_{\boldsymbol{k}+\boldsymbol{q}} + i0^+}.$$

The scattering rate follows by taking the imaginary part:

$$-\,{\rm Im}\,\Sigma^{\rm RPA}(\boldsymbol{k}, \varepsilon) = \pi\int_{-\infty}^{\infty} dE\,[f(\varepsilon + E) + b(E)] \\ \times\frac{1}{\mathcal{V}^2}\sum_{\boldsymbol{q}} V^2(\boldsymbol{q})\left(-\frac{1}{\pi}\right){\rm Im}\,\chi_{nn}^{\rm RPA}(\boldsymbol{q}, E)\delta(\varepsilon + E - \xi_{\boldsymbol{k}+\boldsymbol{q}}).$$

The quantity ${\rm Im}\,\chi_{nn}^{\rm RPA}(\boldsymbol{q}, E)$ vanishes outside the particle–hole continuum, as illustrated in figure 5.8, and is proportional to the number of particle–hole

[29] Let us point out that something was ignored here. As we saw in section 5.1.4.5, the function $\chi_{nn}^{\rm RPA}(\boldsymbol{q}, E)$ has a pole corresponding to the plasmon. When using equation (16) to perform the Matsubara sum, we have ignored the existence of this pole and thus discarded the contribution of the plasmon to the self-energy. This contribution accounts for processes in which the electron decays by emitting one or more plasmons. Since plasmons have a high energy, these processes are completely negligible for $\varepsilon \to 0$.

excitations with wave vector $\boldsymbol{q}$ and energy E. The presence of a Bose factor $b(E)$ underlines the fact that these excitations are bosons. At low energy and temperature, the combination of Fermi and Bose factors in the integrand vanishes as soon as $|E|$ exceeds a few times $k_{\mathrm{B}}T$, which means that the scattering rate is only sensitive to particle–hole excitations close to the Fermi surface as expected. Similarly, for $\boldsymbol{k}$ close to $\boldsymbol{k}_{\mathrm{F}}$, the possible $\boldsymbol{q}$ vectors extend from zero to approximately $2\boldsymbol{k}_{\mathrm{F}}$, covering all low-energy transitions close to the Fermi surface. Using equation (5.82) and the free χ^0_{nn} calculated in Doc A5.14, we can see that $\mathrm{Im}\,\chi^{\mathrm{RPA}}_{nn}(\boldsymbol{q}, E)$ is a linear function of E for $E \to 0$, with some dependence on q between 0 and $2k_{\mathrm{F}}$. Indeed we find for $q < 2k_{\mathrm{F}}$:

$$V^2(\boldsymbol{q})\left(-\frac{1}{\pi}\right)\mathrm{Im}\,\chi^{\mathrm{RPA}}_{nn}(\boldsymbol{q}, E) = \mathscr{V}\frac{\pi^2}{k_{\mathrm{F}}^3}F\left(\frac{q}{2k_{\mathrm{F}}}\right)E + \mathcal{O}(E^2),$$

where $F(x) = x/[x + 2\pi k_{\mathrm{F}}a_0x^3 + (1 - x^2)\tanh^{-1}(x)]^2$ and a_0 is the Bohr radius. We can now perform the momentum sum:

$$\begin{aligned}&\frac{1}{\mathscr{V}}\sum_{\boldsymbol{q}}F\left(\frac{q}{2k_{\mathrm{F}}}\right)\delta(\varepsilon + E - \xi_{\boldsymbol{k}+\boldsymbol{q}})\\ &= \frac{1}{(2\pi)^2}\int_0^{2k_{\mathrm{F}}}dq\, q^2F\left(\frac{q}{2k_{\mathrm{F}}}\right)\int_0^{\pi}d\vartheta\,\sin\vartheta\;\delta(\varepsilon + E - \xi_{\boldsymbol{k}+\boldsymbol{q}}),\end{aligned}$$

with ϑ the angle between $\boldsymbol{k}$ and $\boldsymbol{q}$. The angle integration is easy to perform for a parabolic dispersion:

$$\begin{aligned}&\int_0^{\pi}d\vartheta\,\sin\vartheta\;\delta\big(\varepsilon + E - \xi_{\boldsymbol{k}+\boldsymbol{q}}\big)\\ &= \frac{m}{\hbar^2kq}\begin{cases}1 & \text{if}\left(\frac{k}{k_{\mathrm{F}}} - \frac{q}{k_{\mathrm{F}}}\right)^2 < 1 + \frac{\varepsilon + E}{\varepsilon_{\mathrm{F}}} < \left(\frac{k}{k_{\mathrm{F}}} + \frac{q}{k_{\mathrm{F}}}\right)^2\\ 0 & \text{otherwise.}\end{cases}\end{aligned}$$

Since $|\varepsilon + E| \ll \varepsilon_{\mathrm{F}}$, we can neglect the energy dependence of this last expression and note that, for $k = k_{\mathrm{F}}$, the condition $(1 - q/k_{\mathrm{F}})^2 < 1 < (1 + q/k_{\mathrm{F}})^2$ is satisfied for all $q < 2k_{\mathrm{F}}$. Collecting all terms, and using the identity

$$\int_{-\infty}^{\infty}dx\;x\left(\frac{1}{e^{x+y} + 1} + \frac{1}{e^x - 1}\right) = \frac{1}{2}(\pi^2 + y^2),$$

we obtain the scattering rate on the Fermi surface as

$$-\mathrm{Im}\;\Sigma^{\mathrm{RPA}}(\boldsymbol{k}_{\mathrm{F}}, \varepsilon) = \frac{\mathcal{I}(k_{\mathrm{F}}a_0)}{\varepsilon_{\mathrm{F}}}[\varepsilon^2 + (\pi k_{\mathrm{B}}T)^2],$$

where the amplitude is

$$\mathcal{I}(k_{\mathrm{F}}a_0) = \frac{\pi}{4}\int_0^1\frac{dx\;x^2}{[x + 2\pi k_{\mathrm{F}}a_0x^3 + (1 - x^2)\tanh^{-1}x]^2} \approx \left(\frac{\pi}{16}\right)^{\frac{3}{2}}\frac{1}{\sqrt{k_{\mathrm{F}}a_0}}.$$

The last approximation is good for $k_{\mathrm{F}}a_0 \gtrsim 1$.

This calculation shows that the essential ingredient leading to $\varepsilon^2 + (\pi k_B T)^2$ behavior is that the spectrum of bosonic excitations to which the electrons are coupled increases linearly with energy. A handwaving argument goes as follows. An electron of energy ε can excite bosons of any energy E from zero to ε; the scattering rate is therefore $\Gamma \sim \int_0^\varepsilon dE\, n(E)$, where $n(E)$ is the number of bosons at energy E. Hence $\Gamma \sim \varepsilon^2$ if $n(E) \propto E$. Similarly, for electrons coupled to acoustic phonons which have a density of states $\propto E^2$ (see Doc A2.11), we expect a scattering rate $\Gamma \sim \varepsilon^3$.

Bonus B8.3 Pairing susceptibility and Thouless criterion

In the BCS theory, superconductivity arises due to the formation and condensation of Cooper pairs. The interaction responsible for the formation of these pairs is the sum of the screened Coulomb repulsion (5.90) and the effective electron–electron interaction due to the exchange of phonons (5.54), which is attractive at low energy (typically below the Debye energy $\hbar\omega_D$). If the electron–phonon coupling $g_{q\lambda}$ is large enough, the attractive part of the interaction wins at low energy resulting in a net attractive force between electrons close to the Fermi surface. In the weak-coupling BCS theory, this attractive interaction is reduced to the simplest possible model: an attractive potential $-V_0$ acting between electrons of opposite spins within a range $\hbar\omega_D$ around the Fermi surface (see figure 5.5).

The operator giving the density of Cooper pairs (see section 5.2.2.3),

$$p = \sum_{k\sigma} c_{k\sigma} c_{-k-\sigma},$$

vanishes on average in the normal (non-superconducting) state, $\langle p \rangle = 0$ for $T > T_c$, but acquires a finite value for $T < T_c$. Very much like a charge-density wave instability is signaled by a divergence of the charge susceptibility—or density–density correlation function—at zero frequency (section 5.1.4.5), the superconducting instability is signaled by a divergence of the *pairing susceptibility* or pair correlation function χ_{pp}, which describes the propagation of fluctuations in the density of Cooper pairs:

$$\chi_{pp}(\tau) = -\langle T_\tau p(\tau) p^\dagger(0) \rangle = -$$

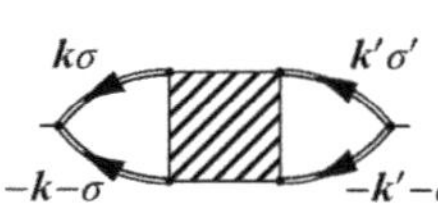

The dominant terms in χ_{pp} are likely those where two electrons repeatedly interact with each other to form a bound state (Cooper pair). These terms correspond to the ladder series in the particle–particle channel:

$$\chi_{pp}(i\Omega_n) \approx -$$

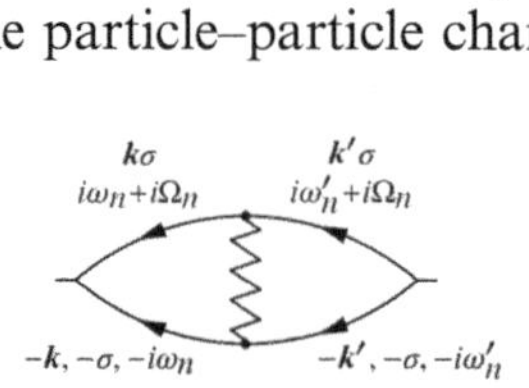

$$-$$

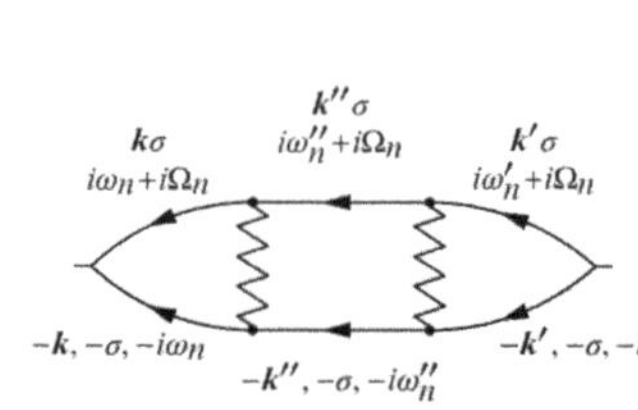

$$- \ldots$$

Here the zigzag line represents the BCS interaction, $-V_0$ for wave vectors corresponding to excitation energies within $\hbar\omega_D$ of the chemical potential and zero for other wave vectors. The first diagram gives $-2(-1)(-V_0)\zeta^2(i\Omega_n)$, where the factor 2 comes from the spin sum and

$$\zeta(i\Omega_n) = \frac{1}{\mathcal{V}} \sum_{|\xi_k|<\hbar\omega_D} \frac{1}{\beta} \sum_{i\omega_n} \mathcal{G}_0(\boldsymbol{k}, i\omega_n + i\Omega_n)\mathcal{G}_0(-\boldsymbol{k}, -i\omega_n).$$

The first minus sign is the prefactor, the second minus sign comes from the fact that the diagram is of first order and the third minus sign is the sign of the interaction. Note that there is no fermion loop in the diagram. Similarly, the second diagram gives $-2(-1)^2(-V_0)^2\zeta^3(i\Omega_n)$ and so on at higher orders, such that the complete series can be summed exactly:

$$\chi_{pp}(i\Omega_n) = -2V_0\zeta^2(i\Omega_n)\left\{1 + V_0\zeta(i\Omega_n) + [V_0\zeta(i\Omega_n)]^2 + \cdots\right\} = \frac{-2V_0\zeta^2(i\Omega_n)}{1 - V_0\zeta(i\Omega_n)}.$$

At the superconducting transition, the pair correlation function diverges at zero energy. The transition temperature is therefore found by solving $1 - V_0\zeta^R(0) = 0$. The function ζ is readily evaluated assuming $\xi_{-k} = \xi_k$ (time-reversal symmetry):

$$\zeta(i\Omega_n) = \int_{-\hbar\omega_D}^{\hbar\omega_D} d\xi\, \tilde{N}_0^{\mathrm{el}}(\xi) \underbrace{\frac{1}{\beta}\sum_{i\omega_n} \frac{1}{i\omega_n + i\Omega_n - \xi}\frac{1}{-i\omega_n - \xi}}_{-\frac{f(\xi) - f(-\xi)}{2\xi - i\Omega_n} = \frac{\tanh(\beta\xi/2)}{2\xi - i\Omega_n}}$$

$$= \int_{-\hbar\omega_D}^{\hbar\omega_D} d\xi\, \tilde{N}_0^{\mathrm{el}}(\xi)\frac{\tanh(\beta\xi/2)}{2\xi - i\Omega_n}.$$

$\tilde{N}_0^{\mathrm{el}}$ is the density of states *per spin direction*, that is, half the total DOS N_0^{el}. Usually $\tilde{N}_0^{\mathrm{el}}$ can be considered constant over the energy range $[-\hbar\omega_D, +\hbar\omega_D]$ because $\hbar\omega_D \ll \varepsilon_F$. Hence we have, using equation (20),

$$\zeta^R(0) = \tilde{N}_0^{\mathrm{el}}(0) \int_{-\hbar\omega_D}^{\hbar\omega_D} d\xi \frac{\tanh(\beta\xi/2)}{2\xi - i0^+} \approx \tilde{N}_0^{\mathrm{el}}(0)\ln\left(1.134\,\frac{\hbar\omega_D}{k_B T}\right).$$

The imaginary part of the integral vanishes because $\tanh(0) = 0$ (see equation (9)). Solving $1 - V_0\zeta^R(0) = 0$, we find the transition temperature

$$k_B T_c = 1.134\, \hbar\omega_D e^{-1/[V_0\tilde{N}_0^{\mathrm{el}}(0)]},$$

which is just the BCS result (5.144).

This calculation illustrates an alternate method for determining the superconducting transition temperature, by looking for a divergence in the pair correlation function as the temperature is lowered. This is known as the Thouless criterion for superconductivity. The Thouless criterion is more general than the method used by

BCS, since the latter relies on the gap equation which, in turn, results from a mean-field approximation. On the contrary, the Thouless criterion applies irrespective of the approximation used for the pair correlation function and is very useful, for instance, when the pairing susceptibility is calculated numerically. One should nevertheless keep in mind that, if the Thouless criterion can determine T_c, it does not provide a description of the properties *below* T_c in the ordered phase, while the mean-field approach does, at least approximately.

References

Ashcroft N W and Mermin D N 1976 *Solid State Physics* (Philadelphia, PA: Saunders)

Bruus H and Flensberg K 2004 *Many-Body Quantum Theory in Condensed Matter Physics* (Oxford: Oxford University Press)

Schrieffer J R 1964 *heory of Superconductivity* (New York: Benjamin)

Chapter 9

Electron tunneling

The ability for quantum particles to cross classically forbidden barriers is one of the many spectacular effects predicted by quantum mechanics. Very early on, this tunneling mechanism was invoked to explain the α-decay of heavy nuclei[1]. In condensed-matter physics, electron tunneling has become a very powerful tool for imaging and spectroscopy. This technique was the first to achieve the energy resolution needed for a direct observation of the gap predicted by Bardeen, Cooper, and Schrieffer (BCS) in the single-particle excitation spectrum of superconductors[2]. Thanks to its high sensitivity, this experiment revealed fine structure in the tunneling conductance which was convincingly explained as the fingerprint of a strong electron–phonon interaction[3], providing decisive support to the BCS idea of phonon-mediated superconductivity.

The invention of the scanning tunneling microscope (STM) opened new territories for electron tunneling[4]. The extraordinary spatial resolution of this instrument suddenly propelled the atom—an ancient concept that had remained an abstraction for centuries—to the front page of newspapers. In addition to being a powerful imaging technique, the STM has become one of the predominant tools for spectroscopic studies. Thanks to the Å-scale spatial resolution and sub-meV energy sensitivity, it can probe the spectroscopy of individual atoms with great detail. In superconductors, the STM was used to image the vortices and map the local density of states inside them[5], allowing one to test predictions of the BCS theory on a local

[1] Gamow G 1928 *Nature* **122** 805.
[2] Bardeen J, Cooper L N and Schrieffer J R 1957 *Phys. Rev.* **108** 1175. Giaever I 1960 *Phys. Rev. Lett.* **5** 147. Giaever I, Hart H R and Megerle K 1962 *Phys. Rev.* **126** 941.
[3] McMillan W L and Rowell J M 1965 *Phys. Rev. Lett.* **14** 108.
[4] Binnig G and Rohrer H 1982 *Helv. Phys. Acta* **55** 726; 1987 *Rev. Mod. Phys.* **59** 615.
[5] Hess H F, Robinson R B and Waszczak J V 1990 *Phys. Rev. Lett.* **64** 2711.

doi:10.1088/978-0-7503-1741-2ch9

scale[6]. One of the important playgrounds for STM nowadays is the field of high-T_c superconductivity[7], where it revealed the intrinsic inhomogeneities of the materials.

In this chapter, we first discuss the theory of electron tunneling and present in detail the approach based on the tunneling Hamiltonian. We derive expressions for the tunneling matrix element and specializations to the extreme cases of planar and STM junctions. We then briefly review the very basic ideas that relate the density of states with the electron dispersion in translation-invariant systems and finish with examples of local density of states measurements and calculations in conditions of broken translation invariance.

9.1 Electron tunneling: a phenomenon out of equilibrium

A tunneling experiment can be envisioned in several ways. A first view considers two systems that are initially isolated, each one being in contact with an infinite charge reservoir and thus maintained at a fixed chemical potential. One then brings these two systems in close proximity, preventing the electrical short-circuit by a thin insulating or vacuum layer as sketched in figure 9.1(a). The action of bringing the electrodes together amounts theoretically to lowering a tunnel barrier from infinity to some finite value (or increasing the barrier 'transparency' from zero). This method has the advantage that it is based on the properties of the two isolated electrodes, which are easier to model than the properties of the fully formed junction. The tunneling current is then evaluated as a power series in the barrier transparency. The system responding to this perturbation is not in thermal equilibrium, though, since it is made of two parts maintained at different chemical potentials. This, in principle, is outside the realm of the Matsubara formalism, which can only describe the response of systems that are in thermal equilibrium as we saw in section 6.1. Another problem for this approach is that the precise shape and properties of the tunnel barrier cannot be deduced from the knowledge of the two electrodes. Some arbitrariness necessarily results when modelling the tunnel barrier.

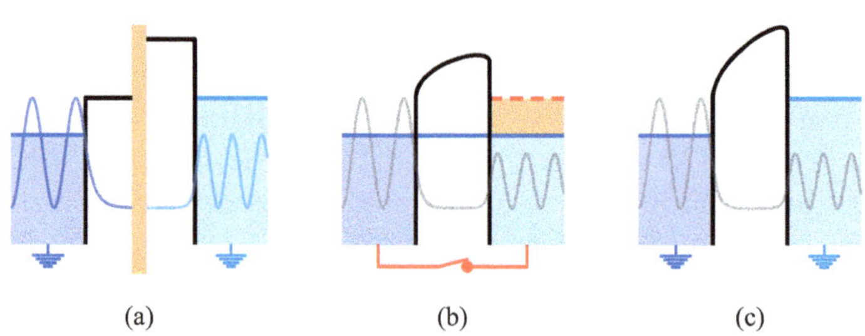

Figure 9.1. Three approaches to the problem of electron tunneling: (a) expansion in the barrier transparency; (b) expansion in the bias voltage; (c) out-of-equilibrium description.

---

[6] Caroli C, de Gennes P G and Matricon J 1964 *Phys. Lett.* **9** 307. Gygi F and Schlüter M 1991 *Phys. Rev.* B **43** 7609.

[7] For a review, see Fischer Ø, Kugler M, Maggio-Aprile I, Berthod C and Renner C 2007 *Rev. Mod. Phys.* **79** 353.

In a second view, one starts with a fully formed junction and solves it under the conditions of thermal equilibrium, i.e. with the two chemical potentials aligned and no current flowing. Although it is generally very hard, this program can be—and has been—realized using, for instance, *ab initio* methods[8]. The solution provides the shape of the tunnel barrier and the interface wave functions for the junction (figure 9.1(b)). One can then use equilibrium perturbation theory and calculate the response of this system to a difference in chemical potentials (an applied bias voltage) using the response theory and Kubo-like formulas. Beside the difficulty of solving the full junction and calculating its correlation functions, the drawback of this method is that it is perturbative in the applied voltage. In many systems of interest for tunneling spectroscopy, for instance superconductors, the current–voltage characteristic is highly non-linear, which would require pushing the response theory to high order, rendering the approach impractical.

A solution to these conceptual difficulties was given by Leonid Keldysh[9], who devised a version of many-body theory appropriate for systems that are out of equilibrium. The overall structure of the theory is similar to that in equilibrium, the price to pay for non-equilibrium being that the correlation functions become 2×2 matrices and must be calculated in real time. The 2×2 matrices have three independent components: the retarded, advanced, and Keldysh functions defined in equations (3.4), (3.5), and (3.8). The most convincing theoretical descriptions of electron tunneling so far—and the only ones in which many-body effects can be implemented in a systematic way—are based on non-equilibrium Green's functions and the Keldysh formalism[10]. These approaches synthesize the points of view presented above: the tunnel current is calculated in the fully formed junction under non-equilibrium conditions, such that both the 'transparency' and the bias voltage are treated non-perturbatively (figure 9.1(c)). The main drawback of the non-equilibrium tunneling theory is its complexity and unease of implementation. Analytical calculations based on this approach are possible only for very simple models[11].

To be able to perform calculations analytically, we need a formalism based on the isolated electrodes rather than the fully formed junction. Furthermore, in order to understand tunneling spectroscopy, we need a formalism that can reliably describe non-linear current–voltage characteristics. The best point of view for our purposes is therefore the one shown in figure 9.1(a). In the following section, we discuss the tunneling-Hamiltonian formalism, which is the simplest implementation of this approach. We will see that the problem related to non-equilibrium is 'solved' by assuming that the electron operators on both sides of the junction keep anticommuting even when the electrodes are in contact. Consequently, the system responding to

[8] See e.g. Hofer W A, Foster A S and Shluger A L 2003 *Rev. Mod. Phys.* **75** 1287, and references therein.

[9] See e.g. Rammer J and Smith H 1986 *Rev. Mod. Phys.* **58** 323.

[10] Caroli C, Combescot R, Nozières P and Saint-James D 1971 *J. Phys. C: Solid St. Phys.* **4** 916. Feuchtwang T E 1974 *Phys. Rev.* B **10** 4121. Noguera C 1990 *Phys. Rev.* B **42** 1629.

[11] See e.g. Berthod C and Giamarchi T 2011 *Phys. Rev.* B **84** 155414.

the barrier transparency is made of two disconnected parts which can legitimately be held at different chemical potentials such that the Matsubara formalism is applicable in each of them. The somewhat ill-defined problem of the barrier transparency, better known as the tunneling matrix element, will be addressed in section 9.3.

9.2 Tunneling-Hamiltonian formalism

Consider two isolated systems, say the 'left' and 'right' systems, characterized by their Hamiltonians H_L and H_R and their chemical potentials μ_L and μ_R, respectively (figure 9.2). The one-particle wave functions that we use to represent many-body states in the left and right systems are denoted $\varphi_\lambda(\boldsymbol{l})$ and $\varphi_\rho(\boldsymbol{r})$, respectively, $c_\lambda^\dagger$ and $c_\rho^\dagger$ being the corresponding creation operators. Our basic assumption is that these two systems remain independent when the tunnel junction is formed and a finite current flows. Mathematically, this implies that all fermion operators in the left system anticommute with the fermion operators in the right system: $[c_\lambda^\dagger, c_\rho]_+ = 0$. Thinking in terms of the wave functions, the assumption requires that $\varphi_\lambda(\boldsymbol{l})$ and $\varphi_\rho(\boldsymbol{r})$ be orthogonal—although, physically, the overlap of these functions in the barrier region is crucial to make the tunneling possible. The 'transparency' of the junction is implemented by means of a term H_T, the so-called *tunneling Hamiltonian* or *transfer Hamiltonian*, which describes the transfer of electrons from one sub-system to the other:

$$H_T = \sum_{\lambda\rho} T_{\lambda\rho} c_\rho^\dagger c_\lambda + \text{h.c.} \equiv \mathcal{X} + \mathcal{X}^\dagger. \tag{9.1}$$

The matrix element $T_{\lambda\rho}$ must be understood as the amplitude for an electron to jump from the state $|\varphi_\lambda\rangle$ on the left to the state $|\varphi_\rho\rangle$ on the right. As emphasized in the previous section, there is no general prescription for defining this matrix element and some kind of modeling will be necessary. This aspect of the problem is discussed in the next section. For the time being, we just assume that $T_{\lambda\rho}$ is *given* and we calculate the resulting current as a function of the applied voltage $eV = \mu_R - \mu_L$.

The calculation of the current is detailed in Doc A9.1. The main steps are the following. (i) Define the tunneling current as the rate of change of the number of electrons in the right system, say, multiplied by the electron charge:

$$I = -e\langle \dot{N}_R \rangle. \tag{9.2}$$

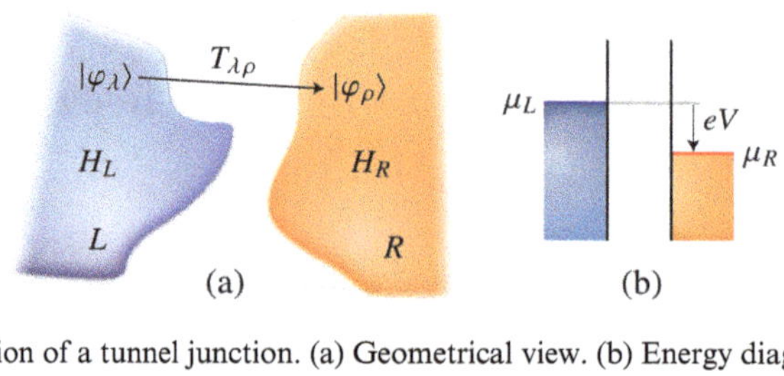

Figure 9.2. Representation of a tunnel junction. (a) Geometrical view. (b) Energy diagram; a bias voltage V is applied to the junction, resulting in a relative shift of the chemical potentials. Reprinted figure with permission from Fischer *et al* 2007 *Rev. Mod. Phys.* **79** 353. Copyright (2007) by the American Physical Society.

Our conventions are that $e = -|e|$ and that the current is positive when electrons flow from left to right. (ii) Use linear response in H_T and equation (6.3) to express $\langle \dot{N}_R \rangle$ in terms of the retarded correlation function of $\dot{N}_R$ and H_T. At this stage, we find two contributions to the current, $I = I_s + I_J(t)$ with

$$I_s = -\frac{2e}{\hbar} \operatorname{Im} C^R_{\mathcal{X}\mathcal{X}^\dagger}(eV/\hbar) \tag{9.3a}$$

$$I_J(t) = -\frac{2e}{\hbar} \operatorname{Im} e^{2ieVt/\hbar} C^R_{\mathcal{X}\mathcal{X}}(-eV/\hbar), \tag{9.3b}$$

where the operator $\mathcal{X}$ is defined in equation (9.1). The first term is the time-independent single-particle current while the second term is the Josephson current, which vanishes unless the two materials in contact are superconductors and which oscillates in time with the characteristic and universal (i.e. model-independent) frequency $2|e|V/\hbar$. (iii) Use the assumption of two independent subsystems to express the retarded correlation functions in terms of the spectral functions in the left and right systems. At this step, it is convenient to move from the abstract representation of the $c_\lambda^\dagger$ and $c_\rho^\dagger$ to the real-space representation based on the fermion operators $c_l^\dagger$ and $c_r^\dagger$. For the single-particle current this gives (see Doc A9.1)

$$\boxed{\begin{aligned} I_s = \frac{2\pi|e|}{\hbar} \int_{-\infty}^{\infty} d\varepsilon \, [f(\varepsilon - |e|V) - f(\varepsilon)] \int dl_1 dl_2 dr_1 dr_2 \\ \times \operatorname{Im}\left[T(l_1, r_1) T^*(l_2, r_2) A(l_1, l_2, \varepsilon - |e|V) \left(-\frac{1}{\pi}\right) G^R(r_2, r_1, \varepsilon) \right]. \end{aligned}} \tag{9.4}$$

In this expression, $T(\boldsymbol{l}, \boldsymbol{r})$ is the amplitude for an electron to tunnel from point $\boldsymbol{l}$ in the left system to point $\boldsymbol{r}$ in the right system, A is the spectral function of the one-electron Matsubara Green's function in real space, i.e.

$$\mathcal{G}(\boldsymbol{x}_1, \boldsymbol{x}_2, i\omega_n) = \int_{-\infty}^{\infty} d\varepsilon \frac{A(\boldsymbol{x}_1, \boldsymbol{x}_2, \varepsilon)}{i\omega_n - \varepsilon}, \quad \mathcal{G}(\boldsymbol{x}_1, \boldsymbol{x}_2, \tau) = -\langle T_\tau c_{\boldsymbol{x}_1}(\tau) c^\dagger_{\boldsymbol{x}_2}(0) \rangle, \tag{9.5}$$

and G^R is the corresponding retarded Green's function. The asymmetric treatments of the left and right systems in equation (9.4)—the former appearing with its spectral function and the latter with its Green's function—is chosen on purpose, anticipating that the left system will stand for the probing electrode and the right one for the sample to be measured. The expression of the current in the basis of the electrode's eigenstates is simpler and given below in equation (9.13). In order to illustrate the content of equation (9.4), let us consider the simplest possible situation and assume that the tunneling is forbidden except between just two particular points, say $\boldsymbol{l}_0$ on the left and $\boldsymbol{r}_0$ on the right. This can be seen as a highly idealized view of an STM junction in which electrons could only tunnel from the very end of the tip to the point of the surface immediately underneath the tip. In such a case, the tunneling matrix element would take the form $T(\boldsymbol{l}, \boldsymbol{r}) = t\delta(\boldsymbol{l} - \boldsymbol{l}_0)\delta(\boldsymbol{r} - \boldsymbol{r}_0)$ and the current reduces to

$$I_s = \frac{2\pi|e|}{\hbar}|t|^2 \int_{-\infty}^{\infty} d\varepsilon\, [f(\varepsilon - |e|V) - f(\varepsilon)]A(\boldsymbol{l}_0, \boldsymbol{l}_0, \varepsilon - |e|V)A(\boldsymbol{r}_0, \boldsymbol{r}_0, \varepsilon), \quad (9.6)$$

because $(-1/\pi)\mathrm{Im}\, G^R(\boldsymbol{x}, \boldsymbol{x}, \varepsilon) = A(\boldsymbol{x}, \boldsymbol{x}, \varepsilon)$ (equation (3.47)). As we saw in section 5.1.3.2 (equation (5.8)), the diagonal part of the spectral function is just the local density of states (LDOS): $A(\boldsymbol{x}, \boldsymbol{x}, \varepsilon) = N(\boldsymbol{x}, \varepsilon)$. Factors of two due to spins are ignored in this discussion. For simplicity, let us further assume that the left system is a probe that we have chosen for its LDOS being featureless: $N_L(\boldsymbol{l}_0, \varepsilon) \approx N_L(\boldsymbol{l}_0, 0)$. Hence

$$I_s = \frac{2\pi|e|}{\hbar}|t|^2 N_L(\boldsymbol{l}_0, 0)\int_{-\infty}^{\infty} d\varepsilon\, [f(\varepsilon - |e|V) - f(\varepsilon)]N_R(\boldsymbol{r}_0, \varepsilon). \quad (9.7)$$

Now, since the bias voltage only enters the Fermi function, we can differentiate with respect to V and deduce the differential tunneling conductance:

$$\boxed{\sigma(V) \equiv \frac{dI_s}{dV} \propto \int_{-\infty}^{\infty} d\varepsilon\, [-f'(\varepsilon - |e|V)]N_R(\boldsymbol{r}_0, \varepsilon).} \quad (9.8)$$

At $T = 0$, this gives $\sigma(V) \propto N_R(\boldsymbol{r}_0, |e|V)$ since the derivative of the Fermi function becomes a delta function. This is the main paradigm for the interpretation of local tunneling experiments: the bias dependence of the differential conductance measures the energy dependence of the LDOS near the position of the tunnel junction. At finite temperature, the tunneling spectrum is broadened (thermal smearing).

For a more realistic tunneling matrix element $T(\boldsymbol{l}, \boldsymbol{r})$, the current is influenced by the LDOS in a more extended region around the tunnel junction. Off-diagonal contributions from the probe spectral function and sample Green's function also enter and describe interference effects in the two electrodes. In the next section, we analyze two examples of tunneling matrix elements and their consequences on the differential conductance.

9.3 The tunneling matrix element

There is no unique prescription to define the tunneling matrix element $T_{\lambda\rho}$ in equation (9.1). Any model for $T_{\lambda\rho}$ will necessarily rely on a number of assumptions and will therefore not be fully satisfactory. Our goal here is to introduce the Bardeen formula for the matrix element as well as two particular applications of this formula corresponding to ideal planar and STM junctions.

9.3.1 Bardeen's formula

A formula for the matrix element to be used in equation (9.1) was given by Bardeen in a celebrated paper[12], which is somewhat difficult to read. Here we follow an

[12] Bardeen J 1961 *Phys. Rev. Lett.* **6** 57.

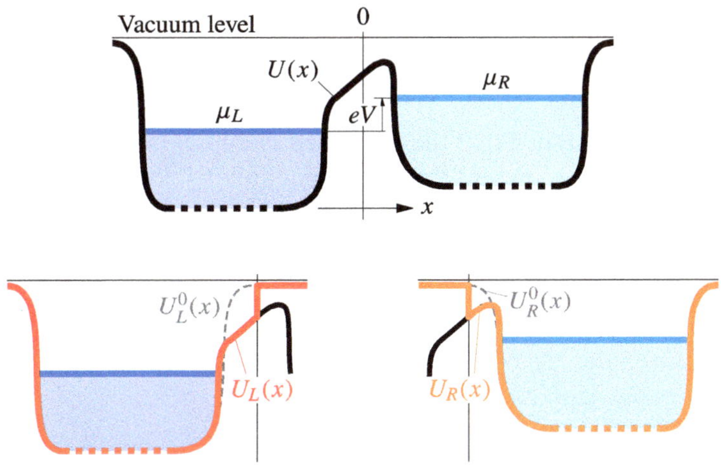

Figure 9.3. Definitions of the potentials U_L and U_R in a one-dimensional case. $U(x)$ is the exact potential of the biased junction; $U_L(x) = U(x)$ for $x < 0$ and $U_L(x) = 0$ for $x > 0$; $U_R(x) = 0$ for $x < 0$ and $U_R(x) = U(x)$ for $x > 0$. $U_L{}^0(x)$ and $U_R{}^0(x)$ are the potentials of the isolated electrodes.

alternate derivation due to Chen[13]. Let $U(\boldsymbol{x})$ be the local potential of the junction when the contact is formed and a bias is applied. Solving for the eigenfunctions in this potential is difficult. We seek a formalism allowing us to solve the left and right systems separately and then reconstitute the whole system using perturbation theory. We split the potential U in two parts, $U_L(\boldsymbol{x})$ and $U_R(\boldsymbol{x})$, in such a way that $U_L(\boldsymbol{x}) = U(\boldsymbol{x})$ in the left system, i.e. to the left of a surface S separating the two electrodes, and $U_L(\boldsymbol{x}) = 0$ to the right of S. Inversely, $U_R(\boldsymbol{x}) = U(\boldsymbol{x})$ to the right of S and $U_R(\boldsymbol{x}) = 0$ to the left of S. In this way, we have the two properties $U(\boldsymbol{x}) = U_L(\boldsymbol{x}) + U_R(\boldsymbol{x})$ and $U_L(\boldsymbol{x})U_R(\boldsymbol{x}) = 0$ everywhere in space (see figure 9.3). The potentials U_L and U_R differ from the potentials U_L^0 and U_R^0 of the *isolated* left and right systems. Therefore the wave-functions on the left and right, which we require to satisfy the Schrödinger equation

$$\left[\frac{\boldsymbol{p}^2}{2m} + U_{L,\,R}(\boldsymbol{x})\right]\varphi_{\lambda,\,\rho}(\boldsymbol{x}) = \varepsilon_{\lambda,\,\rho}\varphi_{\lambda,\,\rho}(\boldsymbol{x}), \tag{9.9}$$

are also different from the wave-functions $\varphi^0_{\lambda,\,\rho}(\boldsymbol{x})$ of the isolated electrodes. As the difference between $U_{L,\,R}$ and $U^0_{L,\,R}$ is expected to be small compared to $U_{L,\,R}$, the difference between $\varphi^0_{\lambda,\,\rho}$ and $\varphi_{\lambda,\,\rho}$ can be studied via the standard time-independent perturbation theory.

We evaluate the matrix element $T_{\lambda\rho}$ as follows. The two sub-systems are initially decoupled and the left one is governed by the Hamiltonian $\boldsymbol{p}^2/(2m) + U_L$. Upon forming the junction, the potential becomes $U = U_L + U_R$, such that the perturbation acting on the states φ_λ is simply U_R. This perturbation induces three types of change. First, it modifies the energies ε_λ due to the action of U_R on the tails of the states φ_λ in

[13] Chen C J 2007 *Introduction to Scanning Tunneling Microscopy* 2nd ed. (New York: Oxford University Press).

the region where $U_R \neq 0$. Following equation (2.43), this (small) effect, which we shall ignore, is represented by the term $\sum_{\lambda\lambda'}\langle\lambda|U_R|\lambda'\rangle c_\lambda^\dagger c_{\lambda'}$. Second, the perturbation U_R gives rise to left-right transitions through the term $\sum_{\rho\lambda}\langle\rho|U_R|\lambda\rangle c_\rho^\dagger c_\lambda$ + h.c. In writing this, we have assumed that the combined set of operators $c_\lambda^\dagger$ and $c_\rho^\dagger$ constitute an appropriate basis to represent the wave functions of the junction and we have therefore neglected the fact that the states φ_λ and φ_ρ are not orthogonal owing to their overlap in the barrier region. Third, we would have a term $\sum_{\rho\rho'}\langle\rho|U_R|\rho'\rangle c_\rho^\dagger c_{\rho'}$, but the latter is already taken into account in the definition of the functions φ_ρ calculated with the Hamiltonian $\boldsymbol{p}^2/(2m) + U_R$. Compared with equation (9.1), we see that the tunneling Hamiltonian H_T corresponds to the second effect, which allows us to define the matrix element as

$$T_{\lambda\rho} = \langle\rho|U_R|\lambda\rangle = \int_R d\boldsymbol{x}\ \varphi_\rho^*(\boldsymbol{x})U_R(\boldsymbol{x})\varphi_\lambda(\boldsymbol{x}). \tag{9.10}$$

We have indicated that the integration can be restricted to the region R, namely the region where U_R is nonzero. Under the additional assumption of elastic tunneling, i.e. $\varepsilon_\lambda = \varepsilon_\rho$, we can rewrite this as (see Doc A9.2)

$$\boxed{T_{\lambda\rho} = -\frac{\hbar^2}{2m}\int_S \left[\varphi_\rho^*(\boldsymbol{x})\nabla\varphi_\lambda(\boldsymbol{x}) - \varphi_\lambda(\boldsymbol{x})\nabla\varphi_\rho^*(\boldsymbol{x})\right]\cdot d\boldsymbol{S}.} \tag{9.11}$$

This is Bardeen's prescription for the tunneling matrix element. It is related to the single-particle current crossing a surface separating the electrodes. Due to current conservation, the detailed choice of this surface is irrelevant. In practice, the potentials U_L and U_R required to evaluate the wave functions and the matrix element are not known. One therefore has to use either the free-surface functions $\varphi_{\lambda,\rho}^0$ or approximate models for the potentials.

9.3.2 Planar junction: Harrison's cancellation

The matrix element for an ideal planar junction between electrodes hosting independent electron was calculated by Harrison[14] using the Wentzel–Kramers–Brillouin (WKB) approximation for the wave functions, namely $\psi_E(x) \propto \exp\left\{-\int^x \sqrt{2m[U(x) - E]/\hbar^2}\right\}$. We do not reproduce this calculation here, but we discuss the main features of the result and their implication for the tunneling current. Harrison's matrix element has the following structure:

$$|T_{\lambda\rho}|^2 \propto \delta_{\boldsymbol{k}_\lambda^\parallel \boldsymbol{k}_\rho^\parallel}\frac{d\xi_\lambda^\perp}{dk_\lambda^\perp}\frac{d\xi_\rho^\perp}{dk_\rho^\perp}e^{-2\kappa d}. \tag{9.12}$$

The factor $\delta_{\boldsymbol{k}_\lambda^\parallel \boldsymbol{k}_\rho^\parallel}$ expresses the conservation of momentum in the directions parallel to the junction plane. The property of exact parallel-momentum conservation is

[14] Harrison W A 1961 *Phys. Rev.* **123** 85.

sometimes called *specular transmission*. The second and third factors are the group velocities in the direction normal to the junction plane for the states φ_λ and φ_ρ, respectively. $\xi^\perp$ is the part of the energy corresponding to the motion normal to the junction, for instance $\xi^\perp = \xi_k - \hbar^2 \boldsymbol{k}_\parallel^2/(2m)$ for free electrons. These factors appear due to the gradients in equation (9.11). Finally, the last factor results from the exponential decay of the wave functions in the barrier. d is the barrier thickness and κ is the wave vector of electrons inside the barrier, namely $\hbar^2\kappa^2/(2m) = U - \xi^\perp$ with U the barrier height assumed constant.

We now want to calculate the current at a planar junction, assuming that the tunneling matrix element is given by equation (9.12). In equation (9.4), we have the expression of the current in the real space representation. In the present context, it is more convenient to use the abstract representation of the functions φ_λ and φ_ρ, which implies (see Doc A9.3)

$$I_s = \frac{2\pi|e|}{\hbar}\sum_{\lambda\rho}|T_{\lambda\rho}|^2\int_{-\infty}^{\infty} d\varepsilon\,[f(\varepsilon - |e|V) - f(\varepsilon)]A_\lambda(\varepsilon - |e|V)A_\rho(\varepsilon). \tag{9.13}$$

Substituting Harrison's matrix element, we arrive at the expression

$$I_s \propto \sum_{\boldsymbol{k}_\parallel}\sum_{k_\lambda^\perp k_\rho^\perp}\frac{d\xi_\lambda^\perp}{dk_\lambda^\perp}\frac{d\xi_\rho^\perp}{dk_\rho^\perp}e^{-2\kappa d}\int_{-\infty}^{\infty} d\varepsilon\,[f(\varepsilon - |e|V) - f(\varepsilon)]A_\lambda(\varepsilon - |e|V)A_\rho(\varepsilon). \tag{9.14}$$

For simplicity, we assume that the electrodes are simple metals that we can describe as independent-electron systems: this amounts to taking

$$A_\lambda(\varepsilon - |e|V)A_\rho(\varepsilon) = \delta(\varepsilon - |e|V - \xi_\lambda)\delta(\varepsilon - \xi_\rho). \tag{9.15}$$

It also means that each $k^\perp$ sum can be converted into a one-dimensional integral over the 'perpendicular' energy $\xi^\perp$, weighted by the corresponding density of states (DOS), which is simply $\propto (d\xi^\perp/dk^\perp)^{-1}$. One sees that these DOS factors cancel the velocities coming from the matrix element. This implies that the information about the DOS in the direction normal to the junction is removed from the tunneling current, in contrast to the simple result (9.8). This remarkable cancellation has become known as Harrison's theorem. At this step, there remains DOS factors associated with the $\boldsymbol{k}_\parallel$ sum; the latter turn out to be suppressed as well but for a different reason, as we shall see now. The current becomes:

$$I_s \propto \sum_{\boldsymbol{k}_\parallel}\int_{-\infty}^{\infty} d\xi_\lambda^\perp d\xi_\rho^\perp e^{-2\kappa d}[f(\xi_\rho - |e|V) - f(\xi_\rho)]\delta(\xi_\rho - |e|V - \xi_\lambda). \tag{9.16}$$

Writing $\xi_\rho = \xi_\rho^\perp + \varepsilon_\rho^\parallel$, $\xi_\lambda = \xi_\lambda^\perp + \varepsilon_\lambda^\parallel$, and $\kappa = \sqrt{2m(U - \xi_\rho^\perp)/\hbar^2}$, we see that the $\xi_\lambda^\perp$ integral simply drops with the delta function. In the remaining integral on $\xi_\rho^\perp$, we change variable to $\xi_\rho \equiv \xi$ and get

$$I_s \propto \int_{-\infty}^{\infty} d\xi\,[f(\xi - |e|V) - f(\xi)]\sum_{\boldsymbol{k}_\parallel} e^{-2d\sqrt{\frac{2m}{\hbar^2}(U-\xi+\varepsilon_\rho^\parallel)}}. \tag{9.17}$$

This is essentially equation (9) of Harrison's paper, specialized for a square barrier of height U. It is now clear that the $\boldsymbol{k}_{\|}$ sum is dominated by small momenta such that $\varepsilon_p^{\|}$ is small: high values of $\varepsilon_p^{\|}$ are exponentially suppressed by the matrix element. This realizes the physically intuitive idea that, at some energy ξ, the electrons that dominate the current are those with a large momentum along the direction of the current, hence a small momentum $\boldsymbol{k}_{\|}$. In order to see this more explicitly, let us take a parabolic band such that $\varepsilon_p^{\|} = \hbar^2 k_{\|}^2/(2m^*)$. The $\boldsymbol{k}_{\|}$ sum can then be performed in polar coordinates. Let us also take the limit of zero temperature for simplicity. This yields

$$\begin{aligned} I_s &\propto \frac{m^*}{m}\int_0^{|e|V} d\xi \left[1 + 2d\sqrt{\frac{2m}{\hbar^2}(U-\xi)}\right] e^{-2d\sqrt{\frac{2m}{\hbar^2}(U-\xi)}} \\ &\approx |e|\frac{m^*}{m}\left(1 + 2d\sqrt{\frac{2m}{\hbar^2}U}\right) e^{-2d\sqrt{\frac{2m}{\hbar^2}U}} \times V. \end{aligned} \tag{9.18}$$

The last line holds if $|e|V \ll U$, which is usually the case. It shows that the current–voltage characteristics of the ideal planar junction is simply and plainly ohmic, with a slope depending mostly on the properties of the barrier, U and d. This result seems to condemn tunneling spectroscopy: if the tunneling junction is ohmic, the tunneling conductance is constant instead of being proportional to the sample DOS as suggested by equation (9.8). Harrison's cancellation is not as general as it might look, however, since it relies heavily on the property of specular transmission. If the interface of the planar junction is rough, this property is likely to break down. Furthermore, as we will see in the next section, for junctions that are not ideally planar the current–voltage characteristics is no longer ohmic.

9.3.3 STM junction: local density of states

Soon after the invention of the STM, Tersoff and Hamann proposed a calculation of the Bardeen matrix element in a geometry representing an idealized STM tip[15]. The tip was modeled as a spherical potential well. The overlap of the tip wave functions with the Bloch waves of the sample surface was then evaluated by expanding the spherically-symmetric tip states as a Fourier series. The result was that $T_{\lambda p} \propto \varphi_p^*(\boldsymbol{x}_0)$, where φ_p is a sample wave function and $\boldsymbol{x}_0$ is the center of the tip potential well. One sees that the matrix element does not depend on the tip-state index λ. This result can also be obtained by a more elegant and general method elaborated by Chen[16]. Chen considers the STM junction as a microscopic atom-to-atom contact such that the tip wave function to be inserted in Bardeen's formula is a combination of atomic orbitals describing the state of the 'apex' atom. In the vacuum region, the wave functions decay

[15] Tersoff J and Hamann D R 1983 *Phys. Rev. Lett.* **50** 1998.
[16] Chen C J 1990 *Phys. Rev.* B **42** 8841.

exponentially like free electrons inside a barrier, $\boldsymbol{p}^2/(2m)\varphi = \hbar^2(i\kappa)^2/(2m)\varphi$, and we thus have

$$\left(\nabla^2 - \kappa_{\lambda,\rho}^2\right)\varphi_{\lambda,\rho}(\boldsymbol{x}) = 0 \tag{9.19}$$

with $\hbar^2\kappa_{\lambda,\rho}^2/(2m) = \phi_{L,R} - \xi_{\lambda,\rho}$, $\phi_{L,R}$ being the tip (L) and sample (R) work functions. It is easiest to solve this problem in spherical coordinates by expanding the wave functions on spherical harmonics. Let us assume that the origin is the position of the apex atom. The solutions that decay at infinity are

$$\varphi_{lm}(\boldsymbol{x}) = C_{lm}k_l(\kappa x)Y_{lm}(\vartheta, \phi), \qquad k_l(u) = (-1)^l u^l \left(\frac{d}{u\,du}\right)^l \frac{e^{-u}}{u}. \tag{9.20}$$

Chen observes that the solutions $\varphi_{lm}(\boldsymbol{x})$ can also be expressed in terms of the Green's function $g(\boldsymbol{x})$—the 'historical' Green's function, not the single-particle Green's function—defined as

$$(\nabla^2 - \kappa^2)g(\boldsymbol{x}) = -\delta(\boldsymbol{x}) \qquad \Rightarrow \qquad g(\boldsymbol{x}) = \frac{e^{-\kappa x}}{4\pi x}, \tag{9.21}$$

as can be seen with the help of equation (12). In particular, we have for an s state:

$$\varphi_{00}(\boldsymbol{x}) \equiv \varphi_s(\boldsymbol{x}) = C_{00}\,\underbrace{k_0(\kappa_s x)}_{e^{-\kappa_s x}/(\kappa_s x)}\,\underbrace{Y_{00}(\vartheta, \varphi)}_{1} = \frac{4\pi C_{00}}{\kappa_s}g(\boldsymbol{x}). \tag{9.22}$$

Thanks to this observation, the calculation of the tunneling matrix element becomes straightforward (see Doc A9.4) and one recovers the Tersoff–Hamann result

$$T_{s,\rho} \propto \varphi_\rho^*(\boldsymbol{0}), \tag{9.23}$$

where $\boldsymbol{0}$ stands for the position of the apex atom. This result provides the theoretical justification for interpreting the STM spectroscopy as a mapping of the local density of states. Indeed, the insertion of equation (9.23) into equation (9.13) leads to

$$I_s \propto \int_{-\infty}^{\infty} d\varepsilon\, [f(\varepsilon - |e|V) - f(\varepsilon)] \underbrace{\sum_\lambda A_\lambda(\varepsilon - |e|V)}_{N_L(\varepsilon - |e|V)}\, \underbrace{\sum_\rho |\varphi_\rho(\boldsymbol{0})|^2 A_\rho(\varepsilon)}_{N_R(\boldsymbol{0}, \varepsilon)}. \tag{9.24}$$

If the density of states is structureless in the tip, $N_L(\varepsilon) = N_L(0)$, we recover equation (9.8): *the differential conductance measures the sample LDOS at the position of the tip apex.*

A few remarks are in order. First, in this interpretation the LDOS is not measured inside the sample but a few Å outside its surface, where the tip is located. This explains why the tunneling conductance decreases exponentially as the tip is moved away from the surface. The LDOS is built from wave functions that all decay exponentially outside the surface: $\varphi_\rho(\boldsymbol{0}) \propto e^{-\kappa_\rho d}\varphi_\rho^{\|}(x, y)$ with d the tip-sample distance. Each sample state decays on a different length scale κ_ρ^{-1}: in a multi-band system, it is possible that different bands are probed at different distances d and thus the spectrum dI_s/dV changes with the distance. Finally, one should stress that the absence of atomic resolution on a surface does not necessarily mean that the junction quality is bad; it can also have an intrinsic explanation, namely that the LDOS has weak corrugation outside the sample.

To finish, we point out that Chen's approach can also treat cases where the relevant orbital of the apex atom is not an s state. For non-s tip states, the rule is that the matrix element is not proportional to the wave function itself, like in equation (9.23), but to one of its derivatives. For example, if the tip apex atom carries a p_x state, we have

$$T_{p_x,\,p} \propto \left.\frac{\partial}{\partial x}\varphi_p^*(\boldsymbol{x})\right|_{\boldsymbol{0}}, \tag{9.25}$$

as can be easily found by noting that, in this case, $\varphi_{p_x}(\boldsymbol{x}) \propto \partial g(\boldsymbol{x})/\partial x$. In this perspective, non-s tips may be expected to provide better spatial contrast because spatial derivatives of the LDOS are expected to vary more than the LDOS itself.

9.4 DOS and electron dispersion

Given that tunneling spectroscopy provides a way of measuring the electronic DOS, it is useful to understand the relationship between the DOS of a system and its electron dispersion. In this section, we recall basic facts about the DOS. By definition, the DOS $N(\varepsilon)$ is the number of single-electron excitations at a given energy (see section 5.1.3.2). In a translation-invariant system, the appropriate tool to describe single-electron excitations is the Green's function in momentum space $\mathscr{G}(\boldsymbol{k}, i\omega_n)$. In a paramagnetic system, we have for the DOS per unit volume

$$N(\varepsilon) = \frac{1}{\mathscr{V}}\sum_{\boldsymbol{k}\sigma} A(\boldsymbol{k}, \varepsilon),$$
$$A(\boldsymbol{k}, \varepsilon) = -\frac{1}{\pi}\text{Im}\, G^R(\boldsymbol{k}, \varepsilon) = -\frac{1}{\pi}\text{Im}\, \mathscr{G}(\boldsymbol{k}, i\omega_n \to \varepsilon + i0^+). \tag{9.26}$$

Let us first consider the simplest case of independent electrons.

9.4.1 Independent electrons

For independent electrons, the single-particle excitations are the electrons themselves and we have (see e.g. Section 3.3.1):

$$A(\boldsymbol{k}, \varepsilon) = \delta(\varepsilon - \xi_{\boldsymbol{k}}) = \sum_{\boldsymbol{k}_\varepsilon} \frac{\delta(\boldsymbol{k} - \boldsymbol{k}_\varepsilon)}{|\nabla\xi_{\boldsymbol{k}}|_{\boldsymbol{k}_\varepsilon}} \quad \text{with} \quad \xi_{\boldsymbol{k}_\varepsilon} = \varepsilon. \tag{9.27}$$

The last equality follows from equation (11). In the thermodynamic limit, the $\boldsymbol{k}_\varepsilon$ sum must in general be understood as a continuous integral along the shell of energy ε. Therefore, the DOS can be put in the form

$$N(\varepsilon) = \frac{2}{(2\pi)^d}\int d^dk \sum_{\boldsymbol{k}_\varepsilon} \frac{\delta(\boldsymbol{k} - \boldsymbol{k}_\varepsilon)}{|\nabla\xi_{\boldsymbol{k}}|_{\boldsymbol{k}_\varepsilon}} = \frac{2}{(2\pi)^d}\sum_{\boldsymbol{k}_\varepsilon} \frac{1}{|\nabla\xi_{\boldsymbol{k}}|_{\boldsymbol{k}_\varepsilon}}. \tag{9.28}$$

This expression shows that $N(\varepsilon)$ has singularities at the energies where the gradient of the dispersion vanishes. These singularities are known as *Van Hove singularities.* Whether they correspond to divergences of the DOS or weaker singularities such as discontinuities or discontinuities of the first derivative depends on the dimensionality of the system and the extension of the region in momentum space where the dispersion gradient vanishes.

9.4.1.1 One-dimensional systems

In one dimension, there are two kinds of points at which the derivative of the dispersion can vanish: extrema, either global or local, and inflection points. Accordingly there are two kinds of Van Hove singularities. Close to an extremum, the dispersion can be expanded as a power series in k and the series starts at some even power: $\xi_k \sim k^{2n}$ $(n \geqslant 1)$. The equation $\xi_{k_\varepsilon} = \varepsilon$ then implies $\varepsilon \sim k^{2n}$. On the other hand, we have $|\nabla\xi_k| \sim k^{2n-1}$. The DOS therefore varies as $1/k^{2n-1} = \varepsilon^{-\frac{2n-1}{2n}}$. The most frequent case is $n = 1$, leading to a square-root divergence of the DOS as $1/\sqrt{\varepsilon}$ at the energy of the extremum. As an example of this, consider the tight-binding chain shown in figure 9.4(a). On each site i, there are two orbitals, one for each spin, with the associated creation operators $c_{i\sigma}^\dagger$. The Hamiltonian is $H = \sum_{i\sigma} t c_{i\sigma}^\dagger c_{i+1,\,\sigma} + \text{h.c.}$, such that the dispersion is $\xi_k = 2t\cos(ka) - \mu$, which is shown in figure 9.4(b) for $t < 0$ and $\mu = 0$. The gradient of the dispersion is simply $|\nabla\xi_k| = 2a|t\sin(ka)|$ and the two solutions for $\xi_{k_\varepsilon} = \varepsilon$ are such that $\cos(k_\varepsilon a) = (\varepsilon + \mu)/(2t)$. We therefore have

$$|\nabla\xi_k|_{k_\varepsilon} = 2a|t|\sqrt{1 - \cos^2(k_\varepsilon a)} = 2a|t|\sqrt{1 - \left(\frac{\varepsilon + \mu}{2t}\right)^2}. \tag{9.29}$$

The sum over k_ε in equation (9.28) amounts to a factor 2 if $|\varepsilon + \mu| < 2|t|$, since there are two solutions, and gives zero otherwise because there is no solution if ε is outside the band. The DOS follows as:

$$N(\varepsilon) = \frac{1}{\pi a|t|}\text{Re}\left(\frac{1}{\sqrt{1 - \left(\frac{\varepsilon+\mu}{2t}\right)^2}}\right). \tag{9.30}$$

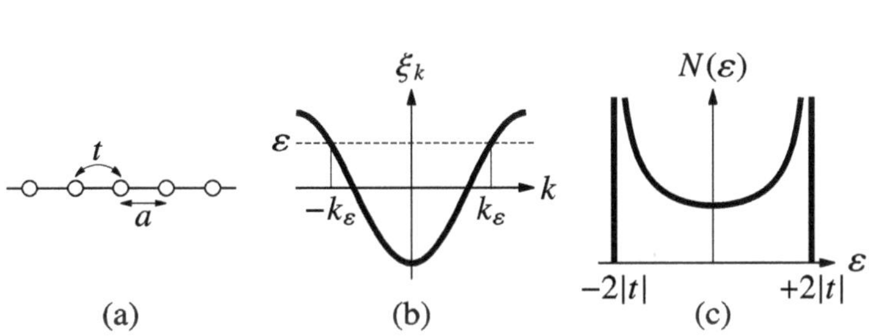

Figure 9.4. DOS of the tight-binding chain. (a) Representation of the chain with lattice parameter a and nearest-neighbor hopping energy t. (b) Band structure for $t < 0$ and $\mu = 0$; for each energy ε between $-2|t|$ and $+2|t|$, there are two solutions $\pm k_\varepsilon$ such that $\xi_{k_\varepsilon} = \varepsilon$. (c) Density of states given by equation (9.30).

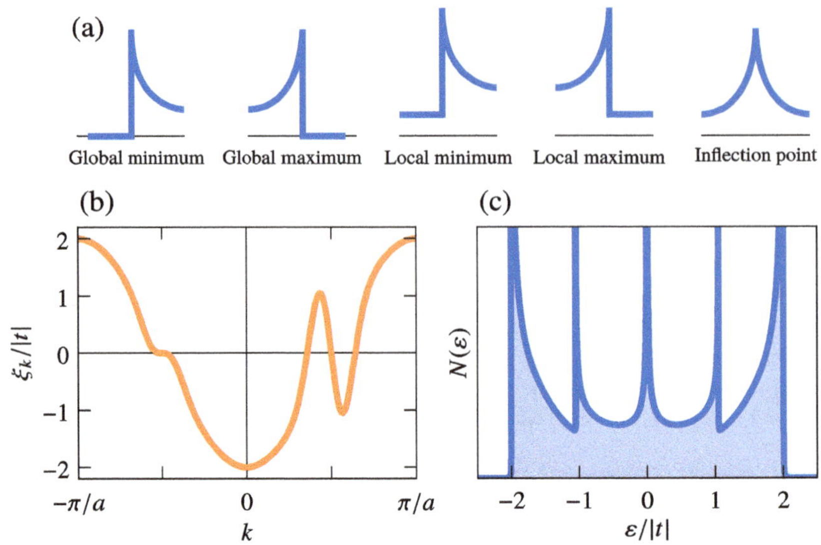

Figure 9.5. (a) Schematic representation of the five possible singularities in the DOS of one-dimensional systems. Quantitative illustration with (b) a fictitious dispersion relation and (c) the corresponding DOS featuring the five types of singularities. The dispersion is $2t\cos(ka)\{1 - \exp[-8(ka + \pi/2)^2] - 5\exp[-8(ka - \pi/2)^2]\}$.

The real part ensures that the DOS vanishes if ε is outside the band. This function is displayed in figure 9.4(c); it diverges as $1/\sqrt{\varepsilon}$ at the band edges as expected, since the dispersion varies as k^2 close to the band edges. At an inflection point, the dispersion behaves as k^{2n+1} with $n \geqslant 1$. Following the same reasoning as before, we see that the DOS diverges as $\varepsilon^{-\frac{2n}{2n+1}}$, that is $\varepsilon^{-\frac{2}{3}}$ for $n = 1$. Square-root divergences in the DOS have been observed by STM, e.g. in carbon nanotubes[17].

In the case of a global minimum or a global maximum of the dispersion, the DOS jumps abruptly from zero to infinity; in the case of a local extremum, the singularity sits on top of a background coming from the part of the band that is degenerate with the extremum but non singular. This can happen, for instance, if a next-nearest-neighbor term $2t'\cos(2ka)$ is added to the dispersion of the tight-binding chain. For an inflection point, the DOS singularity is symmetric because there are states both above and below the energy of the singularity. The five possibilities are sketched in figure 9.5(a). The other panels of the figure show a fictitious dispersion tailored to display all possible singularities together with the corresponding numerically calculated DOS.

9.4.1.2 Two- and higher-dimensional systems

In dimension $d = 1$, the dimensionality of the regions in momentum space where $\nabla\xi_k = 0$ can only be $d - 1 = 0$ (isolated points). In dimension $d > 1$, these regions can

[17] Wildöer J W G, Venema L C, Rinzler A G, Smalley R E and Dekker C 1998 *Nature* **391** 59.

have dimensions 0, …, $d-1$. Local or global extrema of the dispersion occurring at isolated points in reciprocal space have dimension 0. Clearly, the weight of these 0-dimensional singular regions in the $\boldsymbol{k}$-sum of equation (9.28) decreases as d increases and correspondingly the associated DOS singularity becomes less pronounced. Specifically, in two dimensions, an extremum of the dispersion induces a finite discontinuity of the DOS. In three dimensions, the DOS is continuous at an extremum but its *derivative* is discontinuous. This can be generalized: in a system of dimension d, an isolated extremum in the dispersion induces a discontinuity in the $(d-2)$th derivative of the DOS.

The case of the saddle points can be different. On 2D lattices, a particular type of divergence occurs due to saddle points. In such cases, although the dimensionality of the singular region is zero, a divergence rather than a discontinuity is induced in the DOS, but it is a weak divergence of logarithmic character. At a saddle point, the dispersion behaves as $\xi_{\boldsymbol{k}} = \varepsilon \sim k_x^2 - k_y^2$ such that the gradient is $|\nabla\xi_{\boldsymbol{k}}| \sim (k_x^2 + k_y^2)^{1/2} \sim (2k_x^2 - \varepsilon)^{1/2}$. The DOS behaves like the primitive of $1/|\nabla\xi_{\boldsymbol{k}}|$ at the point where the gradient vanishes, namely $N(\varepsilon) \sim \int^{k_x} (2k_x^2 - \varepsilon)^{-1/2}|_{2k_x^2\to\varepsilon} \sim -\ln|\varepsilon|$. The simplest example is the tight-binding square lattice with dispersion $\xi_{\boldsymbol{k}} = 2t[\cos(k_x a) + \cos(k_y a)] - \mu$. In addition to the band extrema at $\boldsymbol{k} = (0, 0)$ and $(\pi/a, \pi/a)$, there are saddle points at $(\pi/a, 0)$ and symmetry-equivalent points. The DOS can be shown to be $N(\varepsilon - \mu) = \theta(4|t| - |\varepsilon|)K(1 - [\varepsilon/(4t)]^2)/(\pi^2 a^2|t|)$ with $K(x)$ the complete elliptic integral of the first kind (see Doc A9.5). This function approaches $-\ln|\varepsilon/(16t)|/(\pi^2 a^2|t|)$ close to $\varepsilon = 0$ and is displayed in figure 9.6. One sees the two discontinuities at the band edges and the logarithmic divergence at the band center. A feature very similar to the logarithmic Van Hove singularity has been

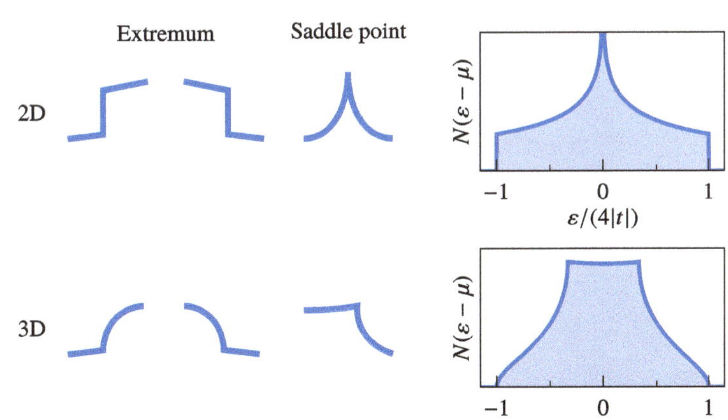

Figure 9.6. DOS singularities in 2D and 3D. Extrema in the dispersion induce discontinuities of the DOS in 2D and discontinuities of the DOS derivative in 3D. Saddle points induce logarithmic divergences of the DOS in 2D and logarithmic divergences of the DOS derivative in 3D. The illustrations show the DOS of the square (2D) and cubic (3D) nearest-neighbor tight-binding lattices.

observed by STM in the quasi-2D cuprate $Bi_2Sr_2CuO_{6+\delta}$[18]. In 3D, the saddle points induce logarithmic divergences of the DOS first derivative (see figure 9.6).

To finish, let us mention another important example of DOS singularity: BCS coherence peaks. They result from the opening of the BCS superconducting gap due to the pairing-induced change of dispersion from ξ_k to $\pm E_k = \pm(\xi_k^2 + \Delta^2)^{1/2}$ (see section 5.2.2.3). It is clear that all points such that $\xi_k = 0$, i.e. the points of the Fermi surface, lie at a minimum of E_k. Since the dimensionality of the Fermi surface is $d - 1$ for a system of dimension d, we have a singular region of dimension $d - 1$. As a result, the BCS coherence peaks are strong singularities, generally square-root singularities, in any dimension. The situation can be different for a momentum-dependent superconducting gap. For instance, a gap of $d_{x^2-y^2}$ symmetry in 2D leads to logarithmic coherence peaks because the singular regions have dimension 0 in this case (see equation (5.146)).

9.4.2 Interacting electrons

When interactions are present, the associated self-energy $\Sigma(\boldsymbol{k}, \varepsilon)$ reshapes the spectral function (see equation (7.12)). As a result, the DOS singularities generally get shifted (due to the real part of the self-energy) and broadened (due to the imaginary part) by interactions. An example of this can be found in Bonus B5.3, where figure 5.18(a) compares the DOS with and without residual interactions between Bogoliubov quasi-particles. It can also happen that the imaginary part of the self-energy vanishes in the energy range of a singularity, such that the latter is not broadened. A simple example where this happens is the Hubbard model (2.79) if the self-energy is calculated in the so-called *atomic limit* where $t/U \to 0$. In this strong-interaction regime, the kinetic energy is neglected and the problem reduces to a collection of independent one-site Hubbard 'atoms', i.e. a problem of size 4×4 that can be solved analytically. At half-filling (one electron per site), the resulting self-energy is $\Sigma(\boldsymbol{k}, i\omega_n) = U/2 + (U/2)^2/i\omega_n$ and the chemical potential is $\mu = U/2$, such that the Green's function reduces to

$$\mathcal{G}_{\rm at}(\boldsymbol{k}, i\omega_n) = \frac{1}{i\omega_n - \varepsilon_k - \dfrac{(U/2)^2}{i\omega_n}} = \frac{Z_k}{i\omega_n - E_k^+} + \frac{1 - Z_k}{i\omega_n - E_k^-} \tag{9.31}$$

with $Z_k = [1 + \varepsilon_k/(\varepsilon_k^2 + U^2)^{1/2}]/2$ and $E_k^{\pm} = [\varepsilon_k \pm (\varepsilon_k^2 + U^2)^{1/2}]/2$. The DOS can be calculated exactly and is

$$N_{\rm at}(\varepsilon) = N_0\left(\varepsilon - \frac{U^2}{4\varepsilon}\right), \tag{9.32}$$

with $N_0(\varepsilon)$ the non-interacting DOS for the dispersion ε_k. This model is very crude, but contains the essence of a Mott insulator. Electrons are immobile due to the quenching of the kinetic energy by the energetic cost of having two electrons on the

[18] Piriou A, Jenkins N, Berthod C, Maggio-Aprile I and Fischer Ø 2011 *Nature Commun.* **2** 221.

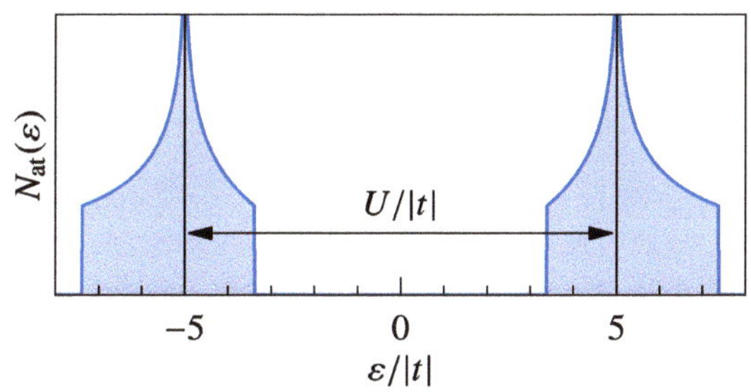

Figure 9.7. DOS of the two-dimensional Hubbard model for $U = 10|t|$ in the atomic limit for the self-energy. The Mott gap of width $\sim U$ separates the lower and upper Hubbard bands. The Van Hove singularities present in the non-interacting DOS are shifted, but not broadened in this approximation, because the imaginary part of the self-energy vanishes for $\varepsilon \neq 0$.

same site. This is seen in the DOS as a gap at zero energy—since $N_0(\pm\infty) = 0$. This gap of order U is the *Mott gap*, which separates the lower Hubbard band from the upper Hubbard band, as illustrated in figure 9.7 for the tight-binding square lattice. At half-filling, the lower Hubbard band is filled, the upper Hubbard band is empty, and the system is insulating at temperatures smaller than U. Going beyond the atomic limit and obtaining a solution of the two-dimensional Hubbard model at finite t and away from half-filling is one of the greatest challenges in theoretical condensed-matter physics.

9.5 LDOS as seen by STM

The local DOS (LDOS) $N(\boldsymbol{r}, \varepsilon)$ counts the number of single-electron excitations at some energy ε and some point $\boldsymbol{r}$ in space. In a translation-invariant system, the LDOS coincides with the DOS but in a system that breaks translation invariance, the LDOS contains much more information than the total DOS, the latter being the spatial average of the former. As we saw in section 5.1.3.2, in a paramagnetic state the LDOS is related to the diagonal elements of the real-space Green's function via

$$N(\boldsymbol{r}, \varepsilon) = -\frac{2}{\pi}\,\mathrm{Im}\,G^R(\boldsymbol{r}, \boldsymbol{r}, \varepsilon) = -\frac{2}{\pi}\,\mathrm{Im}\,\mathcal{G}(\boldsymbol{r}, \boldsymbol{r}, i\omega_n \to \varepsilon + i0^+). \tag{9.33}$$

We will consider two simple examples where the LDOS can be calculated and compared with STM measurements: local defects or impurities that scatter electrons and vortex cores in superconductors.

9.5.1 Local impurities

Consider a system that is initially invariant by translation and characterized by the real-space Green's function $\mathcal{G}_0(\boldsymbol{r} - \boldsymbol{r}', i\omega_n)$. An impurity is introduced at the origin, perturbing the system with the local potential $V(\boldsymbol{r})$. As we have seen (equations (5.17) and (7.26)), the Green's function in the presence of the potential is:

$$\mathcal{G}(\boldsymbol{r}, \boldsymbol{r}', i\omega_n) = \mathcal{G}_0(\boldsymbol{r} - \boldsymbol{r}', i\omega_n) + \int d\boldsymbol{r}''\mathcal{G}_0(\boldsymbol{r} - \boldsymbol{r}'', i\omega_n)V(\boldsymbol{r}'')\mathcal{G}(\boldsymbol{r}'', \boldsymbol{r}', i\omega_n). \tag{9.34}$$

There exists one case in which this Dyson equation can be easily solved for $\mathcal{G}$: when the impurity is so localized that it can be represented by a delta-function potential, $V(\boldsymbol{r}) = V_0\delta(\boldsymbol{r})$. In this situation, the equation becomes

$$\mathcal{G}(\boldsymbol{r}, \boldsymbol{r}', i\omega_n) = \mathcal{G}_0(\boldsymbol{r} - \boldsymbol{r}', i\omega_n) + V_0\mathcal{G}_0(\boldsymbol{r}, i\omega_n)\mathcal{G}(\boldsymbol{0}, \boldsymbol{r}', i\omega_n). \tag{9.35}$$

We can now solve for $\mathcal{G}(\boldsymbol{0}, \boldsymbol{r}', i\omega_n)$ by evaluating equation (9.35) for $\boldsymbol{r} = \boldsymbol{0}$,

$$\mathcal{G}(\boldsymbol{0}, \boldsymbol{r}', i\omega_n) = \mathcal{G}_0(-\boldsymbol{r}', i\omega_n) + V_0\mathcal{G}_0(\boldsymbol{0}, i\omega_n)\mathcal{G}(\boldsymbol{0}, \boldsymbol{r}', i\omega_n) = \frac{\mathcal{G}_0(-\boldsymbol{r}', i\omega_n)}{1 - V_0\mathcal{G}_0(\boldsymbol{0}, i\omega_n)}$$

and then reintroduce this into Dyson's equation to get[19]

$$\mathcal{G}(\boldsymbol{r}, \boldsymbol{r}, i\omega_n) = \mathcal{G}_0(\boldsymbol{0}, i\omega_n) + \frac{V_0\mathcal{G}_0(\boldsymbol{r}, i\omega_n)\mathcal{G}_0(-\boldsymbol{r}, i\omega_n)}{1 - V_0\mathcal{G}_0(\boldsymbol{0}, i\omega_n)}. \tag{9.36}$$

Therefore, the change of the LDOS induced by the impurity is given by

$$\delta N(\boldsymbol{r}, \varepsilon) = -\frac{2}{\pi}\,\mathrm{Im}\left[\frac{V_0 G_0^R(\boldsymbol{r}, \varepsilon)G_0^R(-\boldsymbol{r}, \varepsilon)}{1 - V_0 G_0^R(\boldsymbol{0}, \varepsilon)}\right] = -\frac{2V_0}{\pi}\mathrm{Im}[G_0^R(\boldsymbol{r}, \varepsilon)]^2 + \mathcal{O}(V_0^2). \tag{9.37}$$

Considering free electrons for simplicity, we have $G_0^R(\boldsymbol{r}, \varepsilon) \propto e^{ikr}/r$ with $\hbar^2k^2/(2m) = \varepsilon + \mu$ (see Doc A7.1) and we find at lowest order in V_0:

$$\delta N(\boldsymbol{r}, \varepsilon) \propto \frac{\sin[2\sqrt{(2\,m/\hbar^2)\,(\varepsilon + \mu)}\,r]}{r^2}. \tag{9.38}$$

The scattering of electrons off the impurity leads to Friedel-like spatial oscillations of the LDOS whose intensity decays as $1/r^2$ and whose wavelength varies with the energy ε. With an STM, one can measure the energy dependence of the LDOS at each point by sweeping the voltage according to $dI/dV \sim N(\boldsymbol{r}, |e|V)$. Using this method, the Friedel oscillations around impurities have been observed in many systems[20]. Such measurements offer the possibility of studying the dispersion relation ε_k by recording the wavelength of the oscillations as a function of bias voltage. For an illustration, see figure 9.8(a).

In cuprate superconductors, the LDOS oscillations around impurities have been extensively studied by STM, as they provide a way of measuring the Fermi surface and the momentum dependence of the superconducting gap[21]. The principle can be

[19] The solution of $1 - V_0G_0^R(\boldsymbol{0}, \varepsilon) = 0$, if it exists, gives the energy of an impurity bound state.

[20] Crommie M F, Lutz C P and Eigler D M 1993 *Nature* **363** 524. Kanisawa K, Butcher M J, Yamaguchi H and Hirayama Y 2001 *Phys. Rev. Lett.* **86** 3384. Lackinger M, Griessl S, Heckl W M and Hietschold M 2002 *J. Phys. Chem.* B **106** 4482. Suzuki T, Hasegawa Y, Li Z-Q, Ohno K, Kawazoe Y and Sakurai T 2001 *Phys. Rev.* B **64** 081403(R).

[21] Hoffman J E, McElroy K, Lee D-H, Lang K M, Eisaki H, Uchida S and Davis J C 2002 *Science* **297** 1148. McElroy K, Simmonds R W, Hoffman J E, Lee D-H, Orenstein J, Eisaki H, Uchida S and Davis J C 2003 *Nature* **422** 592.

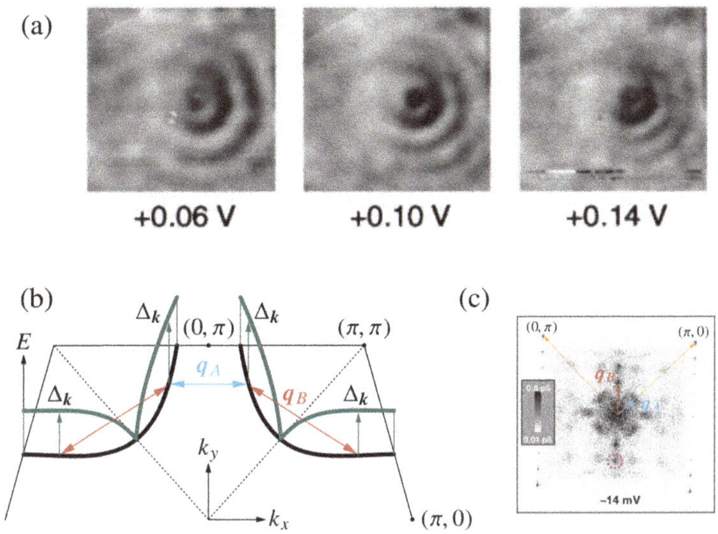

Figure 9.8. (a) STM differential conductance measured at three different biases around a point defect on the InAs(111) surface; one can see how the wavelength decreases with increasing bias. Reprinted figure with permission from Kanisawa *et al* 2001 *Phys. Rev. Lett.* **86** 3384. Copyright (2001) by the American Physical Society. (b) Representation of the Fermi surface of cuprate superconductors (black), the *d*-wave gap along the Fermi surface (green), and two vectors connecting points of the Brillouin zone with the same quasi-particle energy E_k; on the Fermi surface we have $\xi_k = 0$ and thus $E_k = |\Delta_k|$. Reprinted figure with permission from Hoffman *et al* 2002 *Science* **297** 1148. Copyright (2002) by the American Association for the Advancement of Science. (c) Fourier transform of a conductance map in $Bi_2Sr_2CaCu_2O_{8+\delta}$; the two vectors $\boldsymbol{q}_A$ and $\boldsymbol{q}_B$ are visible; the peaks associated with the atomic lattice are also visible at the corners; the peak marked by a dashed purple circle corresponds to a structural modulation. Reprinted figure with permission from McElroy *et al* 2003 *Nature* **422** 592. Copyright (2003) by Springer Nature.

understood by considering the Fourier transform of the impurity-induced LDOS equation (9.37):

$$\delta N(\boldsymbol{q}, \varepsilon) \propto \mathrm{Im} \sum_{\boldsymbol{k}} G_0^R(\boldsymbol{k}, \varepsilon) G_0^R(\boldsymbol{k} - \boldsymbol{q}, \varepsilon). \tag{9.39}$$

Now, $G_0^R(\boldsymbol{k}, \varepsilon)$ should be regarded as the Green's function of a BCS superconductor, equation (5.136), which has poles at the quasi-particle energies $\varepsilon = E_k = (\xi_k^2 + \Delta_k^2)^{1/2}$. One sees that $\delta N(\boldsymbol{q}, \varepsilon)$ has peaks when both $G_0^R(\boldsymbol{k}, \varepsilon)$ and $G_0^R(\boldsymbol{k} - \boldsymbol{q}, \varepsilon)$ have poles, namely when the condition

$$\varepsilon = E_{\boldsymbol{k}} = E_{\boldsymbol{k}-\boldsymbol{q}} \tag{9.40}$$

is fulfilled. This defines a set of ε-dependent $\boldsymbol{q}$ vectors that connect $\boldsymbol{k}$ points having the same quasi-particle energy. Two such vectors are represented in figure 9.8(b). $\boldsymbol{q}_A$ is largest at zero energy when it connects two nodal points of the Fermi surface (points where the *d*-wave superconducting gap vanishes) and it becomes shorter as ε

increases. On the contrary, $\boldsymbol{q}_B$ vanishes at zero energy and increases with ε. By computing the Fourier transform of STM differential-conductance maps, one can identify these $\boldsymbol{q}$ vectors and measure their dispersion with ε (figure 9.8(c)). From this information, simple formulas allow one to reconstruct the Fermi surface and the variation of the gap along the Fermi surface.

9.5.2 Vortices in superconductors

Superconductors immersed in an external magnetic field lower than the critical field destroying superconductivity display two types of behavior. In type-I superconductors, the magnetic field is expelled from the interior of the sample: this is the Meissner effect. In type-II superconductors, the field penetrates through discrete flux lines in the material: these are the Abrikosov vortices. Vortices form because it is energetically less expensive for the system to nucleate them than to completely expel the field. Each vortex carries a magnetic flux $\Phi_0 = h/(2e)$ and it is surrounded by a circular supercurrent of Cooper pairs. Within BCS theory[22], a vortex (or a distribution of vortices) is described by a position-dependent superconducting order parameter $\Delta(\boldsymbol{r}, \boldsymbol{r}')$. The non-locality of the order parameter stems from the possible non-locality of the pairing interaction. It is convenient to write $\Delta(\boldsymbol{r}, \boldsymbol{r}') = |\Delta(\boldsymbol{r}, \boldsymbol{r}')|e^{i\varphi(\boldsymbol{r}, \boldsymbol{r}')}$. Physically, the modulus $|\Delta|$ is related to the superfluid density (the density of Cooper pairs) and it vanishes at the center of the vortex core where superconductivity is locally destroyed. The phase φ is related to the circular supercurrent. Both quantities can be calculated by solving the BCS gap equation (5.134) in the presence of the field[23].

Once the order parameter is known, the LDOS around the vortex core can be obtained by solving the Dyson equation for the matrix $[\mathscr{G}(i\omega_n)]_{\boldsymbol{rr}'} = \mathscr{G}(\boldsymbol{r}, \boldsymbol{r}', i\omega_n)$:

$$\mathscr{G}(i\omega_n) = \mathscr{G}_0(i\omega_n)[\mathbb{1} - \Sigma(i\omega_n)\mathscr{G}_0(i\omega_n)]^{-1}, \tag{9.41}$$

where the self-energy is given by (see Doc A5.21)

$$\Sigma(i\omega_n) = -\Delta\mathscr{G}_0^T(-i\omega_n)\Delta^\dagger \tag{9.42}$$

with $\mathscr{G}_0^T$ the matrix transpose of $\mathscr{G}_0$. For an s-wave superconductor, the order parameter is local: $\Delta(\boldsymbol{r}, \boldsymbol{r}') = \Delta(\boldsymbol{r})\delta(\boldsymbol{r} - \boldsymbol{r}')$. The expression of the self-energy simplifies accordingly:

[22] In this book, 'BCS theory' is meant in a broad sense that includes pairing not mediated by phonons, non-s wave order parameters, as well as Gor'kov and Bogoliubov–de Gennes extensions to position-dependent order parameters.

[23] See e.g. Gygi F and Schlüter M 1990 *Phys. Rev.* B **41** 822. Gygi F and Schlüter M 1991 *Phys. Rev.* B **43** 7609. Han Q 2010 *J. Phys.: Condens. Matter* **22** 035702. Berthod C 2016 *Phys. Rev.* B **94** 184510.

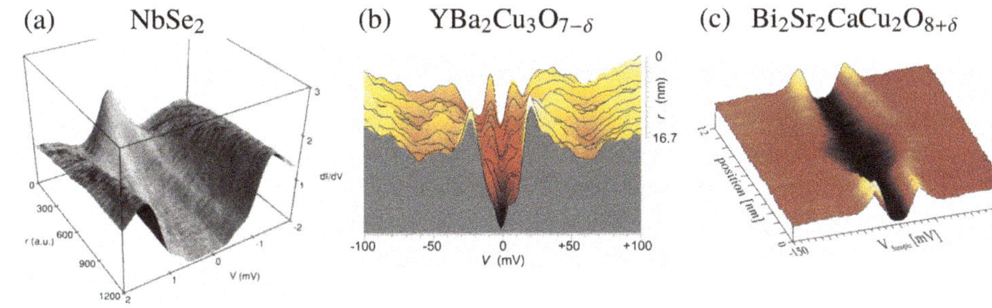

Figure 9.9. STM tunneling conductance in the core of magnetic vortices. In $NbSe_2$ (a), the LDOS shows a zero-bias peak ($r = 0$) and dispersing peaks ($r > 0$), in agreement with the BCS theory. In $YBa_2Cu_3O_{7-\delta}$ (b) and $Bi_2Sr_2CaCu_2O_{8+\delta}$ (c), there is no zero-bias peak and the spatial and energy dependencies of the LDOS do not follow the BCS predictions. In (a) and (b), the spectrum corresponding to the vortex center is at the back of the figure; in (c), the vortex core is close to the center of the figure. Reprinted figures with permission from Gygi *et al* 1991 *Phys. Rev.* B **43** 7609 and Fischer *et al* 2007 *Rev. Mod. Phys.* **79** 353. Copyright (1991, 2007) by the American Physical Society.

$$\begin{aligned}\Sigma(\boldsymbol{r}, \boldsymbol{r}', i\omega_n) &= -\int d\boldsymbol{r}_1 d\boldsymbol{r}_2\, \Delta(\boldsymbol{r}, \boldsymbol{r}_1)\mathcal{G}_0(\boldsymbol{r}_2 - \boldsymbol{r}_1, -i\omega_n)\Delta^*(\boldsymbol{r}', \boldsymbol{r}_2)\\ &= -\Delta(\boldsymbol{r})\mathcal{G}_0(\boldsymbol{r}' - \boldsymbol{r}, -i\omega_n)\Delta^*(\boldsymbol{r}').\end{aligned} \tag{9.43}$$

A practical implementation of this calculation is given in Bonus B9.1. This problem was first considered analytically in a continuum model[24] and concluded to the existence of low-energy localized states with typical inter-level spacing Δ^2/ε_F in the vortex core, leading to a zero-energy peak in the LDOS at the vortex center. The appearance of these states is due to a peculiar interference induced by the supercurrent between the electron and the hole forming the Boboliubov excitations of the superconductor[25]. Vortex-core states in close agreement with the BCS predictions were first observed by STM in $NbSe_2$ (see figure 9.9(a))[26], and later in other superconductors. The zero-energy peak can only be seen in clean materials where the mean free path is larger than the superconducting coherence length[27]. Vortex cores have also been measured in high-T_c cuprates[28], but in this case the observations disagree with the BCS theory (figure 9.9(b) and 9.9(c)): there is no zero-energy peak at the vortex center but two symmetric peaks at finite energy in $YBa_2Cu_3O_{7-\delta}$ and a broad gap in $Bi_2Sr_2CaCu_2O_{8+\delta}$. This behavior is believed to be a consequence of the anomalous normal-state properties of the cuprate superconductors[29].

---

[24] Caroli C, de Gennes P G and Matricon J 1964 *Phys. Lett.* **9** 307. Bardeen J, Kümmel R, Jacobs A E and Tewordt L 1969 *Phys. Rev.* **187** 556.
[25] Volovik G E 1993 *JETP Lett.* **57** 244. Berthod C 2005 *Phys. Rev.* B **71** 134513.
[26] Hess H F, Robinson R B, Dynes R C, Valles J M and Waszczak J V 1989 *Phys. Rev. Lett.* **62** 214.
[27] Renner C, Kent A D, Niedermann Ph, Fischer Ø and Lévy F 1991 *Phys. Rev. Lett.* **67** 1650.
[28] Maggio-Aprile I, Renner C, Erb A, Walker E and Fischer Ø 1995 *Phys. Rev. Lett.* **75** 2754.
[29] Berthod C, Maggio-Aprile I, Bruér J, Erb A and Renner C 2017 *Phys. Rev. Lett.* **119** 237001.

Appendix A9

Doc A9.1 Calculation of the tunneling current

In a tunneling experiment, electrons disappear on one side of the junction and appear on the other side by quantum mechanical tunneling. It is therefore natural to evaluate the current by counting how many electrons appear or disappear from, say, the right system per unit time: $I = -e\langle \dot{N}_R \rangle$. Here $N_R = \sum_\rho c_\rho^\dagger c_\rho$ is the total number of electrons in the right system and $\dot{N}_R$ its time derivative. Our convention is that the current is positive when electrons flow from left to right (remembering that $e = -|e|$). Since $\langle \dot{N}_R \rangle$ vanishes if the tunneling Hamiltonian H_T (equation (9.1)) is ignored, the strategy is to calculate $\langle \dot{N}_R \rangle$ using the linear-response theory for the perturbation H_T. We must tackle the difficulty that, in the absence of perturbation, we have two systems with different chemical potentials μ_L and μ_R. Therefore, the time evolution of the operators cannot be written in terms of $K = H - \mu N$ since there is no well-defined μ. We temporarily revert to the canonical description in which the time evolution is controlled by H and we use the notation $\tilde{A}(t) \equiv e^{iHt/\hbar} A e^{-iHt/\hbar}$. Later in the calculation, we shall reintroduce the grand-canonical description $A(t) = e^{iKt/\hbar} A e^{-iKt/\hbar}$ in term of a modified grand Hamiltonian $K = H_L + H_R - \mu_L N_L - \mu_R N_R$.

From the general linear-response formula (6.3), we deduce

$$\langle \dot{N}_R(t) \rangle_{H_T}^{(1)} = -\frac{i}{\hbar} \int_{-\infty}^{t} dt_1 \left\langle \left[\dot{\tilde{N}}_R(t),\, \tilde{H}_T(t_1) \right] \right\rangle$$

and, consequently, for the current at first order in H_T:

$$I(t) = \frac{e}{\hbar} \int_{-\infty}^{\infty} dt_1\, \theta(t - t_1) \left\langle \left[i\dot{\tilde{N}}_R(t),\, \tilde{H}_T(t_1) \right] \right\rangle.$$

In the canonical description, we have

$$\begin{aligned} i\hbar \dot{N}_R &= [N_R, H_L + H_R + H_T] = \underbrace{[N_R, H_L]}_{=0} + \underbrace{[N_R, H_R]}_{=0} + [N_R, H_T] \\ &= \sum_{\rho'} \sum_{\lambda\rho} T_{\lambda\rho} \underbrace{[c_{\rho'}^\dagger c_{\rho'},\, c_\rho^\dagger c_\lambda]}_{\delta_{\rho\rho'} c_\rho^\dagger c_\lambda} + \sum_{\rho'} \sum_{\lambda\rho} T_{\lambda\rho}^* \underbrace{[c_{\rho'}^\dagger c_{\rho'},\, c_\lambda^\dagger c_\rho]}_{-\delta_{\rho\rho'} c_\lambda^\dagger c_\rho} \\ &= \sum_{\lambda\rho} T_{\lambda\rho} c_\rho^\dagger c_\lambda - \sum_{\lambda\rho} T_{\lambda\rho}^* c_\lambda^\dagger c_\rho = \mathcal{X} - \mathcal{X}^\dagger. \end{aligned}$$

We have used our assumption that fermion operators in the left and right systems anticommute, the expression (9.1) of H_T, and the standard commutation rules (2.41). Note that the tildes are not required here: we use them to specify that the internal time evolution is governed by H rather than $H - \mu N$, but there is no internal time evolution involved here. Note also that the internal time evolution of the operators is governed by the full Hamiltonian H, including H_T; however, the correlation functions which give the linear response must be calculated with H_T set to zero, as usual in perturbation theory. The current becomes

$$I(t) = \frac{e}{\hbar^2} \int_{-\infty}^{\infty} dt_1\, \theta(t - t_1) \langle [\tilde{\mathcal{X}}(t) - \tilde{\mathcal{X}}^\dagger(t),\, \tilde{\mathcal{X}}(t_1) + \tilde{\mathcal{X}}^\dagger(t_1)] \rangle.$$

Using the property $-\langle[A^\dagger, B^\dagger]\rangle = \langle[A, B]\rangle^*$, this can be put in the form

$$
\begin{aligned}
I(t) &= \frac{e}{\hbar^2}\int_{-\infty}^{\infty} dt_1\, \theta(t-t_1)\Big\{\langle[\tilde{\mathcal{X}}(t), \tilde{\mathcal{X}}^\dagger(t_1)]\rangle + \langle[\tilde{\mathcal{X}}(t), \tilde{\mathcal{X}}(t_1)]\rangle + \text{c.c.}\Big\} \\
&= \frac{2e}{\hbar^2}\,\text{Re}\int_{-\infty}^{\infty} dt_1\, \theta(t-t_1)\Big\{\langle[\tilde{\mathcal{X}}(t), \tilde{\mathcal{X}}^\dagger(t_1)]\rangle + \langle[\tilde{\mathcal{X}}(t), \tilde{\mathcal{X}}(t_1)]\rangle\Big\} \\
&= -\frac{2e}{\hbar}\,\text{Im}\int_{-\infty}^{\infty} dt_1\Big\{(-i/\hbar)\theta(t-t_1)\langle[\tilde{\mathcal{X}}(t), \tilde{\mathcal{X}}^\dagger(t_1)]\rangle \\
&\quad + (-i/\hbar)\theta(t-t_1)\langle[\tilde{\mathcal{X}}(t), \tilde{\mathcal{X}}(t_1)]\rangle\Big\}.
\end{aligned}
$$

We see that the current is given by two retarded correlation functions of the form $(-i/\hbar)\theta(t)\langle[\tilde{A}(t), \tilde{B}(0)]\rangle$. The time evolution and the thermal average are governed by $H_L + H_R$ as pointed out above. We would prefer to have functions of the kind of equation (3.4), i.e. $(-i/\hbar)\theta(t)\langle[A(t), B(0)]\rangle$ with the time evolution governed by the grand Hamiltonian. To this end, we introduce $K = H_L + H_R - \mu_L N_L - \mu_R N_R$ and we note that

$$
\begin{aligned}
\tilde{\mathcal{X}}(t) &= \sum_{\lambda\rho} T_{\lambda\rho} e^{i(H_L+H_R)t/\hbar} c_\rho^\dagger c_\lambda e^{-i(H_L+H_R)t/\hbar} \\
&= \sum_{\lambda\rho} T_{\lambda\rho}\, e^{iKt/\hbar} \underbrace{e^{i(\mu_L N_L+\mu_R N_R)t/\hbar} c_\rho^\dagger c_\lambda e^{-i(\mu_L N_L+\mu_R N_R)t/\hbar}}_{e^{i\mu_R N_R t/\hbar} c_\rho^\dagger e^{-i\mu_R N_R t/\hbar} e^{i\mu_L N_L t/\hbar} c_\lambda e^{-i\mu_L N_L t/\hbar}} e^{-iKt/\hbar}.
\end{aligned}
$$

Considering equation (29), we have $e^{i\mu_R N_R t/\hbar} c_\rho^\dagger = c_\rho^\dagger e^{i\mu_R(N_R+1)t/\hbar}$ and $e^{i\mu_L N_L t/\hbar} c_\lambda = c_\lambda e^{i\mu_L(N_L-1)t/\hbar}$, which gives

$$
\tilde{\mathcal{X}}(t) = \sum_{\lambda\rho} T_{\lambda\rho}\, e^{iKt/\hbar} c_\rho^\dagger e^{i\mu_R t/\hbar} c_\lambda e^{-i\mu_L t/\hbar} e^{-iKt/\hbar} = e^{ieVt/\hbar} e^{iKt/\hbar} \mathcal{X} e^{-iKt/\hbar} = e^{ieVt/\hbar} \mathcal{X}(t),
$$

where we have used the relation $\mu_R - \mu_L = eV$ and the general definition (2.15). Hence we can rewrite the current in terms of the retarded correlation functions that we are used to manipulate:

$$
\begin{aligned}
I(t) &= -\frac{2e}{\hbar}\,\text{Im}\left\{\int_{-\infty}^{\infty} dt_1 e^{ieV(t-t_1)/\hbar} \underbrace{(-i/\hbar)\theta(t-t_1)\langle[\mathcal{X}(t), \mathcal{X}^\dagger(t_1)]\rangle}_{C^R_{\mathcal{X}\mathcal{X}^\dagger}(t-t_1)}\right. \\
&\quad \left. + \int_{-\infty}^{\infty} dt_1 e^{ieV(t+t_1)/\hbar} \underbrace{(-i/\hbar)\theta(t-t_1)\langle[\mathcal{X}(t), \mathcal{X}(t_1)]\rangle}_{C^R_{\mathcal{X}\mathcal{X}}(t-t_1)}\right\} \\
&= -\frac{2e}{\hbar}\,\text{Im}\Big\{C^R_{\mathcal{X}\mathcal{X}^\dagger}(eV/\hbar) + e^{2ieVt/\hbar} C^R_{\mathcal{X}\mathcal{X}}(-eV/\hbar)\Big\}.
\end{aligned}
$$

This is just equation (9.3). The next step of the calculation is to express the two correlation functions in terms of the Green's functions in the left and right systems. This is possible, because the correlation functions must be evaluated without H_T, i.e.

with the two systems disconnected. We deduce the retarded function from the imaginary-time equivalent:

$$\begin{aligned}
\mathscr{C}_{\mathcal{X}\mathcal{X}^\dagger}(\tau) &= -\langle T_\tau \mathcal{X}(\tau)\mathcal{X}^\dagger(0)\rangle \\
&= -\sum_{\lambda_1\rho_1}\sum_{\lambda_2\rho_2} T_{\lambda_1\rho_1}T^*_{\lambda_2\rho_2} \underbrace{\langle T_\tau c^\dagger_{\rho_1}(\tau)c_{\lambda_1}(\tau)c^\dagger_{\lambda_2}(0)c_{\rho_2}(0)\rangle_K}_{\langle T_\tau c^\dagger_{\rho_1}(\tau)c_{\rho_2}(0)\rangle_{K_R}\langle T_\tau c_{\lambda_1}(\tau)c^\dagger_{\lambda_2}(0)\rangle_{K_L}} \\
&= \sum_{\lambda_1\rho_1}\sum_{\lambda_2\rho_2} T_{\lambda_1\rho_1}T^*_{\lambda_2\rho_2}\mathscr{G}_{\lambda_1\lambda_2}(\tau)\mathscr{G}_{\rho_2\rho_1}(-\tau).
\end{aligned}$$

A tunnel junction is not invariant by translation. Hence the calculation of the tunneling current is not optimally done in the plane-wave basis or the momentum representation. Until now, we have worked in the unspecified representation of the $c^\dagger_\lambda$ and $c^\dagger_\rho$ operators introduced in section 9.2. The last step of our calculation is to move to the real-space representation in terms of the operators $c^\dagger_{\boldsymbol{l}}$ and $c^\dagger_{\boldsymbol{r}}$ (the spin is irrelevant in the tunneling problem—unless magnetic materials are involved—and thus spin indices are omitted). According to equation (2.91), the relation between $c^\dagger_{\boldsymbol{l}}$ and $c^\dagger_\lambda$ is $c^\dagger_\lambda = \int d\boldsymbol{l}\ \varphi_\lambda(\boldsymbol{l})c^\dagger_{\boldsymbol{l}}$. In other words, creating an electron in state $|\varphi_\lambda\rangle$ is like creating an electron at every point $\boldsymbol{l}$ of space with an amplitude equal to the wave function $\varphi_\lambda(\boldsymbol{l})$. The relation between the Green's function $\mathscr{G}_{\lambda_1\lambda_2}(\tau)$ and $\mathscr{G}(\boldsymbol{l}_1, \boldsymbol{l}_2, \tau)$ is thus

$$\begin{aligned}
\mathscr{G}_{\lambda_1\lambda_2}(\tau) &= -\langle T_\tau c_{\lambda_1}(\tau)c^\dagger_{\lambda_2}(0)\rangle = -\int d\boldsymbol{l}_1 d\boldsymbol{l}_2\ \varphi^*_{\lambda_1}(\boldsymbol{l}_1)\varphi_{\lambda_2}(\boldsymbol{l}_2)\langle T_\tau c_{\boldsymbol{l}_1}(\tau)c^\dagger_{\boldsymbol{l}_2}(0)\rangle \\
&= \int d\boldsymbol{l}_1 d\boldsymbol{l}_2\ \varphi^*_{\lambda_1}(\boldsymbol{l}_1)\varphi_{\lambda_2}(\boldsymbol{l}_2)\mathscr{G}(\boldsymbol{l}_1, \boldsymbol{l}_2, \tau).
\end{aligned}$$

Collecting all factors, we obtain

$$\begin{aligned}
\mathscr{C}_{\mathcal{X}\mathcal{X}^\dagger}(\tau) = \int d\boldsymbol{l}_1 d\boldsymbol{l}_2 d\boldsymbol{r}_1 d\boldsymbol{r}_2 &\underbrace{\sum_{\lambda_1\rho_1}\varphi^*_{\lambda_1}(\boldsymbol{l}_1)T_{\lambda_1\rho_1}\varphi_{\rho_1}(\boldsymbol{r}_1)}_{T(\boldsymbol{l}_1,\, \boldsymbol{r}_1)} \underbrace{\sum_{\lambda_2\rho_2}\varphi_{\lambda_2}(\boldsymbol{l}_2)T^*_{\lambda_2\rho_2}\varphi^*_{\rho_2}(\boldsymbol{r}_2)}_{T^*(\boldsymbol{l}_2,\, \boldsymbol{r}_2)} \\
&\times \mathscr{G}(\boldsymbol{l}_1, \boldsymbol{l}_2, \tau)\mathscr{G}(\boldsymbol{r}_2, \boldsymbol{r}_1, -\tau).
\end{aligned}$$

The quantity $T(\boldsymbol{l}, \boldsymbol{r})$ is the amplitude for an electron to tunnel from the point $\boldsymbol{l}$ in the left system to the point $\boldsymbol{r}$ in the right system. In order to evaluate the current, we need the Fourier transform of the retarded function at frequency $eV/\hbar$. We therefore Fourier transform from τ to $i\Omega_n$, use the spectral representation of the real-space Green's functions like in Doc A7.2, as defined in equation (9.5), evaluate the sum over Matsubara frequencies using the standard routine, and finally perform the analytic continuation. We get

$$\begin{aligned}
C^R_{\mathcal{X}\mathcal{X}^\dagger}(eV/\hbar) &= \int_0^\beta d\tau\ e^{i\Omega_n\tau}\mathscr{C}_{\mathcal{X}\mathcal{X}^\dagger}(\tau)|_{i\Omega_n\to eV+i0^+} \\
&= \int_{-\infty}^{\infty} d\varepsilon_1 d\varepsilon_2 \frac{f(\varepsilon_2) - f(\varepsilon_1)}{eV + \varepsilon_2 - \varepsilon_1 + i0^+} \\
&\times \int d\boldsymbol{l}_1 d\boldsymbol{l}_2 d\boldsymbol{r}_1 d\boldsymbol{r}_2\ T(\boldsymbol{l}_1, \boldsymbol{r}_1)T^*(\boldsymbol{l}_2, \boldsymbol{r}_2)A(\boldsymbol{l}_1, \boldsymbol{l}_2, \varepsilon_1)A(\boldsymbol{r}_2, \boldsymbol{r}_1, \varepsilon_2).
\end{aligned}$$

The spectral function $A(\boldsymbol{x}_1, \boldsymbol{x}_2, \varepsilon)$ is in general a complex quantity, if not for the diagonal part $A(\boldsymbol{x}, \boldsymbol{x}, \varepsilon)$, which is equal to the LDOS. We can relate the spectral function to the retarded Green's function by $A(\boldsymbol{x}_1, \boldsymbol{x}_2, \varepsilon) = \frac{i}{2\pi}\{G^R(\boldsymbol{x}_1, \boldsymbol{x}_2, \varepsilon) - [G^R(\boldsymbol{x}_2, \boldsymbol{x}_1, \varepsilon)]^*\}$, as can be readily checked using the spectral representation of section 3.2 or directly via equation (4.15). Using this to replace $A(\boldsymbol{r}_2, \boldsymbol{r}_1, \varepsilon_2)$ and exchanging the dummy variables $(\boldsymbol{l}_1, \boldsymbol{r}_1)$ and $(\boldsymbol{l}_2, \boldsymbol{r}_2)$ in the second term, we arrive at

$$C^R_{XX^\dagger}(eV/\hbar) = \int_{-\infty}^{\infty} d\varepsilon_1 d\varepsilon_2 \frac{f(\varepsilon_2) - f(\varepsilon_1)}{eV + \varepsilon_2 - \varepsilon_1 + i0^+}$$
$$\times \int d\boldsymbol{l}_1 d\boldsymbol{l}_2 d\boldsymbol{r}_1 d\boldsymbol{r}_2 \, \mathrm{Im}\left[T(\boldsymbol{l}_1, \boldsymbol{r}_1)T^*(\boldsymbol{l}_2, \boldsymbol{r}_2)A(\boldsymbol{l}_1, \boldsymbol{l}_2, \varepsilon_1)\left(-\frac{1}{\pi}\right)G^R(\boldsymbol{r}_2, \boldsymbol{r}_1, \varepsilon_2)\right].$$

Taking the imaginary part with equation (10) and inserting in the formula for the current, we obtain equation (9.4).

The calculation of the Josephson current is in every respect similar, with the important difference that the anomalous Green's function $\mathscr{F}^\dagger$ will replace the Green's functions $\mathscr{G}$ (see section 5.2.2.3). The final result is

$$C^R_{XX}(-eV/\hbar) = \int d\boldsymbol{l}_1 d\boldsymbol{l}_2 d\boldsymbol{r}_1 d\boldsymbol{r}_2 \; T(\boldsymbol{l}_1, \boldsymbol{r}_1)T(\boldsymbol{l}_2, \boldsymbol{r}_2)$$
$$\times \int_{-\infty}^{\infty} d\varepsilon_1 d\varepsilon_2 \; B(\boldsymbol{r}_1, \boldsymbol{r}_2, \varepsilon_1)B^*(\boldsymbol{l}_2, \boldsymbol{l}_1, \varepsilon_2)\frac{f(\varepsilon_1) - f(-\varepsilon_2)}{-eV - \varepsilon_1 - \varepsilon_2 + i0^+},$$

with B the spectral function of $\mathscr{F}^\dagger$.

Doc A9.2 Bardeen's formula for the matrix element

The matrix element (9.10) reads:

$$T_{\lambda\rho} = \int_R dx\, \varphi_\lambda(\boldsymbol{x}) \underbrace{U_R(\boldsymbol{x})\varphi_\rho^*(\boldsymbol{x})}_{\left(\varepsilon_\rho - \frac{\boldsymbol{p}^2}{2m}\right)\varphi_\rho^*(\boldsymbol{x})} = \int_R dx\left[\varphi_\lambda(\boldsymbol{x})\varepsilon_\rho\varphi_\rho^*(\boldsymbol{x}) - \varphi_\lambda(\boldsymbol{x})\frac{\boldsymbol{p}^2}{2m}\varphi_\rho^*(\boldsymbol{x})\right].$$

We assume that the tunneling process is elastic, in other words the matrix element is nonzero only for states that have the same energy: $\varepsilon_\rho = \varepsilon_\lambda$. In the integrand, the quantity $\varphi_\lambda(\boldsymbol{x})\varepsilon_\rho = \varepsilon_\lambda\varphi_\lambda(\boldsymbol{x})$ can then be replaced by $[\frac{\boldsymbol{p}^2}{2m} + U_L(\boldsymbol{x})]\varphi_\lambda(\boldsymbol{x})$. But, since $U_L(\boldsymbol{x})$ vanishes in the region of integration, this is simply $\frac{\boldsymbol{p}^2}{2m}\varphi_\lambda(\boldsymbol{x})$. We obtain

$$\begin{aligned}
T_{\lambda\rho} &= \int_R dx \left[\varphi_\rho^*(x) \frac{p^2}{2m} \varphi_\lambda(x) - \varphi_\lambda(x) \frac{p^2}{2m} \varphi_\rho^*(x) \right] \\
&= -\frac{\hbar^2}{2m} \int_R dx \left[\varphi_\rho^*(x) \nabla^2 \varphi_\lambda(x) - \varphi_\lambda(x) \nabla^2 \varphi_\rho^*(x) \right] \\
&= -\frac{\hbar^2}{2m} \int_R dx\ \nabla \cdot \left[\varphi_\rho^*(x) \nabla \varphi_\lambda(x) - \varphi_\lambda(x) \nabla \varphi_\rho^*(x) \right] \\
&= -\frac{\hbar^2}{2m} \int_S \left[\varphi_\rho^*(x) \nabla \varphi_\lambda(x) - \varphi_\lambda(x) \nabla \varphi_\rho^*(x) \right] \cdot d\boldsymbol{S}.
\end{aligned}$$

The third line uses an identity of vector analysis and the last line uses the divergence theorem, namely $\int_V (\nabla \cdot \boldsymbol{F}) dV = \int_S \boldsymbol{F} \cdot d\boldsymbol{S}$. We have thus obtained Bardeen's expression (9.11).

Doc A9.3 Single-particle current in the basis of electrode's eigenstates

Here, we rewrite equation (9.4), which is written in the real-space basis, in the abstract representation of the eigenstates of the left and right electrodes φ_λ and φ_ρ, respectively. The first step is to express the matrix element $T(\boldsymbol{l}, \boldsymbol{r})$ in terms of $T_{\lambda\rho}$:

$$T(\boldsymbol{l}, \boldsymbol{r}) = \langle \boldsymbol{r} | T | \boldsymbol{l} \rangle = \sum_{\lambda\rho} \langle \boldsymbol{r} | \rho \rangle \langle \rho | T | \lambda \rangle \langle \lambda | \boldsymbol{l} \rangle = \sum_{\lambda\rho} \varphi_\rho(\boldsymbol{r}) T_{\lambda\rho} \varphi_\lambda^*(\boldsymbol{l}).$$

We then rewrite the real-space Green's function $\mathscr{G}(\boldsymbol{l}_1, \boldsymbol{l}_2, \tau)$ using the transformation (2.91): $c_{\boldsymbol{l}}^\dagger = \sum_\lambda \varphi_\lambda^*(\boldsymbol{l}) c_\lambda^\dagger$:

$$\begin{aligned}
\mathscr{G}(\boldsymbol{l}_1, \boldsymbol{l}_2, \tau) &= -\langle T_\tau c_{\boldsymbol{l}_1}(\tau) c_{\boldsymbol{l}_2}^\dagger(0) \rangle = -\sum_{\lambda\lambda'} \varphi_\lambda(\boldsymbol{l}_1) \varphi_{\lambda'}^*(\boldsymbol{l}_2) \langle T_\tau c_\lambda(\tau) c_{\lambda'}^\dagger(0) \rangle \\
&= \sum_{\lambda\lambda'} \varphi_\lambda(\boldsymbol{l}_1) \varphi_{\lambda'}^*(\boldsymbol{l}_2) \mathscr{G}_{\lambda\lambda'}(\tau).
\end{aligned}$$

If the functions φ_λ are the ones that diagonalize the left electrode, we have $\mathscr{G}_{\lambda\lambda'}(\tau) = \delta_{\lambda\lambda'} \mathscr{G}_\lambda(\tau)$ and we therefore obtain, for the real-space spectral function,

$$A(\boldsymbol{l}_1, \boldsymbol{l}_2, \varepsilon) = \sum_\lambda \varphi_\lambda(\boldsymbol{l}_1) \varphi_\lambda^*(\boldsymbol{l}_2) A_\lambda(\varepsilon).$$

As usual, $A_\lambda(\varepsilon) = (-1/\pi) \mathrm{Im}\, \mathscr{G}_\lambda(i\omega_n \to \varepsilon + i0^+)$. With this, we have all ingredients to replace $T(\boldsymbol{l}_1, \boldsymbol{r}_1)$, $T^*(\boldsymbol{l}_2, \boldsymbol{r}_2)$, $A(\boldsymbol{l}_1, \boldsymbol{l}_2, \varepsilon - |e|V)$, and $G^R(\boldsymbol{r}_2, \boldsymbol{r}_1, \varepsilon)$ in equation (9.4). All real-space integrations are elementary, for instance

$$\int d\boldsymbol{l}_1\ \varphi_\lambda^*(\boldsymbol{l}_1) \varphi_{\lambda'}(\boldsymbol{l}_1) = \delta_{\lambda\lambda'},$$

which leads directly to equation (9.13).

Doc A9.4 Tunneling matrix element for the STM

The tunneling matrix element (9.11) can also be written in the form

$$T_{\lambda\rho} = -\frac{\hbar^2}{2m}\int_R d\boldsymbol{x}\Big[\varphi_\rho^*(\boldsymbol{x})\nabla^2\varphi_\lambda(\boldsymbol{x}) - \varphi_\lambda(\boldsymbol{x})\nabla^2\varphi_\rho^*(\boldsymbol{x})\Big]$$

using the divergence theorem (see Doc A9.2). We calculate this matrix element in the case where the tip state φ_λ is a s state. Using the correspondence between $\varphi_s(\boldsymbol{x})$ and $g(\boldsymbol{x})$, we have

$$\nabla^2\varphi_\lambda(\boldsymbol{x}) = \frac{4\pi C_{00}}{\kappa_s}\nabla^2 g(\boldsymbol{x}) = \frac{4\pi C_{00}}{\kappa_s}[\kappa_s^2 g(\boldsymbol{x}) - \delta(\boldsymbol{x})] = \kappa_s^2\varphi_s(\boldsymbol{x}) - \frac{4\pi C_{00}}{\kappa_s}\delta(\boldsymbol{x}).$$

Furthermore, we have seen (equation (9.19)) that $\nabla^2\varphi_\rho^*(\boldsymbol{x}) = \kappa_\rho^2\varphi_\rho^*(\boldsymbol{x})$. Hence

$$T_{s,\rho} = \frac{\hbar^2}{2m}\frac{4\pi C_{00}}{\kappa_s}\varphi_\rho^*(\boldsymbol{0}) - \frac{\hbar^2}{2m}\int_R d\boldsymbol{x}\left(\kappa_s^2 - \kappa_\rho^2\right)\varphi_\rho^*(\boldsymbol{x})\varphi_s(\boldsymbol{x}).$$

The second term on the right-hand side vanishes if the tunneling is elastic.

Doc A9.5 DOS of the hypercubic lattice

The hypercubic lattice is the generalization of the non-interacting one-dimensional nearest-neighbor tight-binding chain to arbitrary dimension $d \geqslant 1$. In $d = 2$, it corresponds to the square lattice and in $d = 3$ to the cubic lattice. The dispersion takes the form

$$\xi_k = 2t\sum_{i=1}^{d}\cos(k_i a) - \mu.$$

In order to evaluate the corresponding DOS, we replace the delta function by an exponential with the help of equation (3*b*):

$$N_d(\varepsilon) = \frac{2}{(2\pi)^d}\int_{-\frac{\pi}{a}}^{+\frac{\pi}{a}} d^dk\ \delta(\varepsilon - \xi_k) = \frac{2}{(2\pi)^d}\int_{-\frac{\pi}{a}}^{+\frac{\pi}{a}} d^dk\ \frac{1}{2\pi}\int_{-\infty}^{\infty} du e^{iu(\varepsilon-\xi_k)}.$$

Once the expression of ξ_k is introduced, the integrals along all directions in momentum space decouple and become equivalent:

$$N_d(\varepsilon) = \frac{2}{2\pi}\int_{-\infty}^{\infty} du e^{iu(\varepsilon+\mu)}\prod_{i=1}^{d}\int_{-\frac{\pi}{a}}^{+\frac{\pi}{a}}\frac{dk_i}{2\pi}e^{-iu2t\cos(k_i a)}.$$

The k_i integration can be performed thanks to the expansion (13):

$$\int_{-\frac{\pi}{a}}^{+\frac{\pi}{a}}\frac{dk_i}{2\pi}e^{-iu2t\cos(k_i a)} = \sum_{n=-\infty}^{+\infty} i^n J_n(-u2t)\frac{1}{a}\underbrace{\int_{-\pi}^{\pi}\frac{d\vartheta}{2\pi}e^{in\vartheta}}_{\delta_{n,0}} = \frac{1}{a}J_0(2|tu|).$$

We have used the fact that $J_0(x)$ is an even function of x. Hence

$$N_d(\varepsilon - \mu) = \frac{1}{\pi a^d}\int_{-\infty}^{\infty} du\ e^{iu\varepsilon}[J_0(2|tu|)]^d = \frac{2}{\pi a^d}\int_{0}^{\infty} du\ \cos(\varepsilon u)[J_0(2|t|u)]^d$$
$$= \frac{1}{\pi a^d|t|}\int_0^{\infty} dx\ \cos\left(\frac{\varepsilon x}{2|t|}\right)J_0^d(x).$$

For $d = 1$, the integral yields the result (9.30). We can also finish the calculation for $d = 2$: integrating by parts, we find

$$N_2(\varepsilon - \mu) = \frac{4}{\pi a^2 \varepsilon}\int_0^{\infty} dx\ \sin\left(\frac{\varepsilon x}{2|t|}\right)J_0(x)J_1(x)$$
$$= \frac{1}{\pi^2 a^2|t|}K\left(1 - [\varepsilon/(4t)]^2\right)\theta(4|t| - |\varepsilon|),$$

where $K(x)$ is the complete elliptic integral of the first kind.

Appendix B9

Bonus B9.1 Local density of states in the vortex core

The order parameter for an isolated vortex in a two-dimensional s-wave superconductor takes the form $\Delta(\boldsymbol{r}) = |\Delta(r)|e^{-i\vartheta}$, where r and ϑ are the polar coordinates, $\boldsymbol{r} = r(\cos\vartheta, \sin\vartheta)$ with the origin at the vortex center. The modulus $|\Delta(r)|$ vanishes at $r = 0$ and approaches the bulk order parameter Δ_0 over a distance of the order of the coherence length ξ. The functional form $\Delta_0\tanh(r/\xi)$ is the Ginzburg–Landau solution and provides a good approximation to the exact self-consistent profile. The phase $e^{-i\vartheta}$ winds by 2π and describes the flow of supercurrent around the vortex. Figure 9.10 shows a working code for calculating the LDOS in such a vortex core. The method is similar to the one used in Bonus B5.1. The system is a two-dimensional tight-binding lattice with dispersion $\xi_{\boldsymbol{k}} = -2t(\cos k_x + \cos k_y) - \mu$, $t = 1$, $\mu = t$, and a bulk order parameter $\Delta_0 = 0.2t$. For the modulus of the order parameter, we use $|\Delta(r)| = \Delta_0\tanh(r/\xi)$ with $\xi = 2$ in units of the lattice parameter.

```
t=1; mu=t; D0=0.2*t; N=1024; M=51;
[x y]=meshgrid(2*pi*(0:N-1)/N); xi=-2*t*(cos(x)+cos(y))-mu;
[x y]=meshgrid((1:M)-(M+1)/2); D=D0*tanh(sqrt(x.^2+y.^2)/2).*exp(-i*atan2(y,x));

w(:,1)=(-2*D0:D0/50:2*D0); for l=1:size(w)
  g0p=fft2(1./(w(l)+i/M-xi))/N^2; g0m=fft2(1./(-w(l)-i/M-xi))/N^2;
  for i1=1:M; for j1=1:M; for i2=1:M; for j2=1:M
    G0(i1+(j1-1)*M,i2+(j2-1)*M)=g0p(abs(i1-i2)+1,abs(j1-j2)+1);
    S(i1+(j1-1)*M,i2+(j2-1)*M)=-D(i1,j1)*conj(D(i2,j2))* ...
                                            g0m(abs(j1-j2)+1,abs(i1-i2)+1);
  end; end; end; end
  G=diag(G0*inv(eye(M^2)-S*G0));
  w(l,2:(M+3)/2)=-2*imag(G((M^2+1)/2:(M^2+M)/2))/pi;
end

save vortex.dat w -ascii;
```

Figure 9.10. Calculation of the local density of states (LDOS) inside a vortex core for a tight-binding model of two-dimensional s-wave superconductor, using MATLAB®.

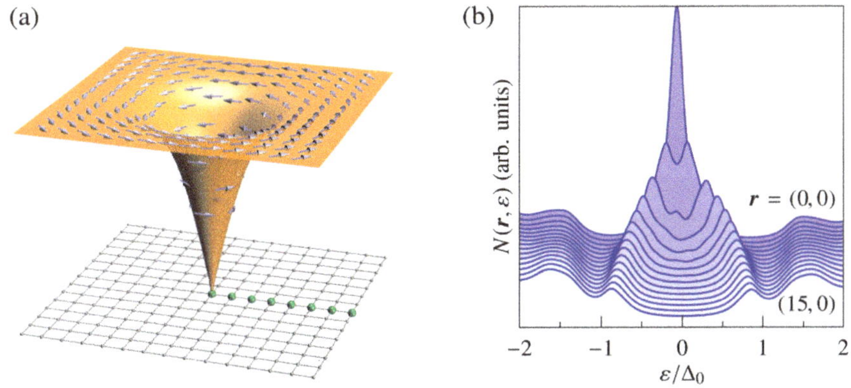

Figure 9.11. (a) Representation of a vortex. The shaded surface shows the modulus of the order parameter, while its phase $-\vartheta$ at each lattice site is represented by an arrow rotated clockwise by an angle $\pi/2 - \vartheta$. The green dots indicate some of the points where the LDOS is calculated. (b) LDOS $N(\boldsymbol{r}, \varepsilon)$ along the x axis, calculated with the code of figure 9.10. The curves are shifted vertically for clarity. Compare with figure 9.9(a).

This dispersion is built into the $N \times N$ array `xi` at line 2. For a good energy resolution, we take $N = 1024$. The complex order parameter $\Delta(\boldsymbol{r})$ is constructed at line 3. The actual real-space system considered is not of size $N \times N$—this would mean N^2 sites and much too large matrices of size $N^2 \times N^2$ to invert for the Green's function—but of a smaller size $M \times M$. With $M = 51$, the matrices of size $M^2 \times M^2$ are still manageable. The array `D` contains the order parameter at all sites (i, j) with $i, j = 1, \ldots M$ and the vortex centered in the middle of the square. The MATLAB® function `atan2(y,x)` returns the polar angle of the vector (x, y) and is thus just what we need for the phase $e^{-i\vartheta}$. Line 4 defines the energy axis extending from $-2\Delta_0$ to $+2\Delta_0$ and starts the loop over energies, which is closed at line 13. We need two free Green's functions G_0 in order to evaluate the Green's function G: $G_0(\varepsilon + i0^+)$ to be used in equation (9.41), and $G_0(-\varepsilon - i0^+)$ to be used in equation (9.43). The translation-invariant versions of these two quantities (`g0p` and `g0m`) are built at line 5 by Fourier transforming the corresponding momentum-space Green's functions using $1/M$ for 0^+. Since we do this on the $N \times N$ mesh, the G_0s are not affected by the boundaries of the $M \times M$ system. Lines 6 to 10 form a double loop on the system sites `(i1,j1)` and `(i2,j2)`, where the free matrix Green's function `G0` (line 7) and self-energy `S` (lines 8–9) are constructed using $G_0(\boldsymbol{r}_1, \boldsymbol{r}_2) = g_0(\boldsymbol{r}_1 - \boldsymbol{r}_2)$ and equation (9.43), respectively. The diagonal part of the Green's function $G(\boldsymbol{r}, \boldsymbol{r})$ is extracted at line 11 using equation (9.41). The LDOS on the $(M + 1)/2$ sites along the x axis to the right of the vortex center is stored in the array `w` at line 12. Running this code requires ~500Mb of memory and will take a couple of hours.

The result is displayed in figure 9.11(b). At the core center, the LDOS shows a sharp zero-energy peak. When moving out of the core, this peak splits in two branches and its intensity decreases. These features show the existence of localized states in the vortex core. One can infer from the behavior of the LDOS—and the fact, not shown in the figure, that the LDOS has approximate cylindrical symmetry—that the wave functions of these localized states have their maximum on a circle whose radius increases with the energy of the state.

IOP Publishing

Spectroscopic Probes of Quantum Matter

Christophe Berthod

Chapter 10

Neutron scattering

In a scattering experiment, a sample is inserted in a beam of identical particles with well-defined energy and momentum. The particles get deflected and a detector measures the distribution of energies and momenta that the particles present after going through the material (figure 10.1). By studying the energy and momentum exchanges between the particles and the sample, one learns about the modes that the particles have excited or absorbed. The chief advantage of neutrons is their neutrality: because they travel through matter without affecting the local electromagnetic environment, they are a weak perturbation and allow one to probe bulk samples in the linear regime. On the practical side, neutrons easily traverse the scientific equipment that is needed in order to keep the sample in the desired conditions of temperature, pressure, and magnetic field. The dream particle for studying condensed phases has a wavelength comparable with interatomic distances and an energy comparable with the excitations of interest: thermal neutrons offer an excellent match with a wavelength of 2 Å at an energy of 20 meV. For comparison, photons at the same wavelength are in the soft x-ray range at 6 keV. Monochromatic thermal neutrons are collected from nuclear fusion reactors or spallation sources, which use high-energy protons in order to break heavy nuclei. They have a lifetime of roughly 15 min, which is enough to perform the experiment.

The interpretation of neutron-scattering data is quite involved due to several complicating factors. First, thermal neutrons interact with matter in two different ways with comparable intensities: a contact interaction with the nuclei and a magnetic interaction with the internal magnetic field. The first process, called nuclear scattering, enables structural and lattice-dynamics studies but suffers from the fact that nuclei of a same given atomic number scatter the neutrons differently, depending on their spin and isotope mass. The magnetic interaction enables structural and dynamical studies of the magnetic texture, but there the complication is that the scattering occurs both on the unpaired electron spins and the orbital currents. In this chapter, we lay down the formalism and present the essential idea

doi:10.1088/978-0-7503-1741-2ch10

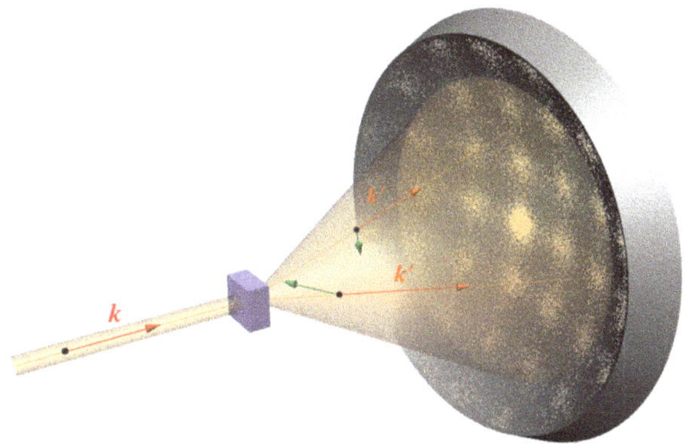

Figure 10.1. Principle of a scattering experiment. A sample is inserted in a beam of particles with wave vector $\boldsymbol{k}$, which get scattered in all directions. The green arrows show the momentum transfer $\hbar\boldsymbol{q} = \hbar(\boldsymbol{k} - \boldsymbol{k}')$ where $\boldsymbol{k}'$ is the wave vector of a scattered particle. The scattering is elastic if $k' = k$, inelastic with energy loss if $k' < k$, and inelastic with energy gain if $k' > k$. A detector counts the number of scattered particles as a function of angle and energy.

that allow one to understand the method. The reader is directed to specialized books for more in-depth discussions[1].

10.1 The differential scattering cross section

The differential cross section is to a scattering experiment what the susceptibility is to a response experiment. While the susceptibility is the proportionality coefficient between the response and the perturbation, the scattering cross section expresses the proportionality between the number of neutrons scattered in the solid angle $d\Omega$ and the number of incident neutrons crossing the surface $d\sigma$. The latter number can be written as $jd\sigma$, where j is the current density of incident neutrons. We may express it as $j = (2\pi)^{-3}\hbar k/m_n$ where $\hbar\boldsymbol{k}$ is the momentum of the monochromatic incident neutrons, m_n is the neutron mass, and $(2\pi)^{-3}$ is the density in momentum space. On the other hand, the number of neutrons scattered in the solid angle $d\Omega$ depends on the transition rate $\Gamma_{a\to b}$ defined in equation (2.27) from an initial state involving some quantum state of the target together with an incident neutron, to a final state featuring a neutron of momentum $\hbar\boldsymbol{k}'$ within $d\Omega$. As the length of $\boldsymbol{k}'$ can take any value, the number of neutrons scattered in the solid angle will be $\int_0^\infty dk'\, k'^2 d\Omega\, \Gamma_{a\to b}$. This is best written as an energy integral over the energy transfer $\varepsilon = E_{\boldsymbol{k}} - E_{\boldsymbol{k}'}$, where $E_{\boldsymbol{k}} = \hbar^2k^2/2m_n$. We find $d\sigma/d\Omega = (2\pi/\hbar)^3 m_n^2 \int_{-\infty}^{\infty} d\varepsilon\, (k'/k)\Gamma_{a\to b}$, where we could extend the upper bound of the integral from $E_{\boldsymbol{k}}$ to $+\infty$ because $\Gamma_{a\to b}$ vanishes for $\varepsilon > E_{\boldsymbol{k}}$. This allows us to identify the double-differential cross section defined by $d\sigma/d\Omega = \int_{-\infty}^{\infty} d\varepsilon\, d^2\sigma/d\Omega d\varepsilon$. We now explicitly write down the double-differential

[1] Squires G L 2012 *Introduction to the Theory of Thermal Neutron Scattering* 3rd ed. (Cambridge: Cambridge University Press). Lovesey S W 1984 *Theory of Neutron Scattering from Condensed Matter* (Oxford: Clarendon Press).

cross section using equation (2.27). The initial state can be taken as a product state $|\boldsymbol{k}\sigma;\, a\rangle = |\boldsymbol{k}\sigma\rangle|a\rangle$ of energy $E_{\boldsymbol{k}} + E_a$, where $|\boldsymbol{k}\sigma\rangle$ is a plane wave, $|a\rangle$ is an eigenstate of the target, and we should thermally average over all of them. Likewise, the final state is $|\boldsymbol{k}'\sigma'\rangle|b\rangle$ and we sum over all final states:

$$\frac{d^2\sigma}{d\Omega d\varepsilon} = \left(\frac{2\pi}{\hbar}\right)^4 m_n^2 \frac{k'}{k} \sum_{ab} \frac{e^{-\beta K_a}}{Z} |\langle b|\langle \boldsymbol{k}'\sigma'|T|\boldsymbol{k}\sigma\rangle|a\rangle|^2\, \delta(\varepsilon + K_a - K_b). \tag{10.1}$$

The t-matrix depends on the interaction potential V between the target and the neutron. For most applications, it is sufficient to work at lowest order and take $T \approx V$, which is called the Born approximation and is analogous to linear response. Neutrons interact with matter in two ways: as they are insensitive to Coulomb forces, they scatter off the nuclei via short-range forces that can be taken as purely local in a condensed-matter context, where the nuclei are treated as points. For all practical purposes, the scattering by a nucleus is spherical and can be entirely characterized by the s-wave scattering length. The effective interaction potential for a nucleus at $\boldsymbol{r}_0$ is the Fermi pseudo-potential $(2\pi\hbar^2/m_n)b\delta(\boldsymbol{r} - \boldsymbol{r}_0)$, where b is the scattering length. The scattering length can be positive or negative. It depends on the internal spin state of the nucleus and on its isotope mass. As these two properties are uncontrolled and, in principle, statistically distributed in a sample of a material, there is a distribution of possible scattering lengths for each atom type and an average over this distribution will have to be made. It is common to assume that there is no correlation between the internal state of a nucleus and its position. In particular, this requires the nuclear spins not to be ordered[2]. Neutrons have a magnetic moment, such that they also interact with the local field generated by unpaired electrons as well as with the nuclear spins. The cross section of the latter is smaller than that of the former by a factor $(\mu_{\mathrm{N}}/\mu_{\mathrm{B}})^2 \sim 10^{-7}$ and scattering on nuclear spins can be safely neglected. The nuclear spins only enter indirectly through their effect on the scattering length. We will start with nuclear scattering over a collection of nuclei and come back to magnetic scattering in section 10.3.

10.2 Nuclear scattering

Let $\boldsymbol{r}_{\mu\ell}$ be the positions of nuclei, the index μ indicating the atomic species and the index ℓ numbering the atoms of that species. The interaction potential is

$$V(\boldsymbol{r}) = \frac{2\pi\hbar^2}{m_n} \sum_{\mu\ell} b_{\mu\ell}\delta(\boldsymbol{r} - \boldsymbol{r}_{\mu\ell}), \tag{10.2}$$

where the index ℓ in $b_{\mu\ell}$ stems from the isotope-mass and nuclear-spin dependence of the scattering length. It is a peculiarity of neutron scattering that each individual atom presents a different cross section to the neutron, which complicates the interpretation of experiments. Notwithstanding, this offers an opportunity for

[2] There are systems in which the nuclear spins order at sufficiently low temperature, see Steiner M 2004 *J. Low. Temp. Phys.* **135** 545.

tuning the signal by engineering the scattering lengths via either isotope substitution or manipulation of the nuclear spins.

10.2.1 Scattering function, coherent and incoherent scattering

Substituting the potential (10.2) for the t-matrix in equation (10.1) and considering our assumptions, we find that the neutron spin is conserved in the scattering process and that the nuclear double differential cross section becomes

$$\begin{aligned}\frac{d^2\sigma_\mathrm{n}}{d\Omega d\varepsilon} &= \frac{k'}{k}\sum_{\mu\ell\nu\ell'} b_{\mu\ell}b_{\nu\ell'}\sum_{ab}\frac{e^{-\beta K_a}}{Z}\langle a|e^{-i\boldsymbol{q}\cdot\boldsymbol{r}_{\mu\ell}}|b\rangle\langle b|e^{i\boldsymbol{q}\cdot\boldsymbol{r}_{\nu\ell'}}|a\rangle\delta(\varepsilon + K_a - K_b)\\ &\equiv \frac{k'}{k}\sum_{\mu\ell\nu\ell'} b_{\mu\ell}b_{\nu\ell'}S_{\mu\ell\nu\ell'}(\boldsymbol{q},\varepsilon) \equiv \frac{k'}{k}S(\boldsymbol{q},\varepsilon).\end{aligned} \tag{10.3}$$

To evaluate the matrix element, we have treated the plane waves in a continuum limit (rather than a finite volume with periodic boundary conditions) such that, consistently with the previous argument about the neutron flux, the normalization is $1/(2\pi)^{3/2}$ rather than $1/\sqrt{\mathcal{V}}$. The quantity $S(\boldsymbol{q}, \varepsilon)$ is called the *scattering function* or *dynamic structure factor*[3]. It is an extensive property of the target, such that it is often normalized to the number of nuclei. A glimpse at equation (3.12) reveals that the scattering function has the mathematical structure of a bosonic greater spectral function, i.e. a function whose support is mostly at positive energies; see equation (3.32). This is expected, since with our convention positive energies correspond to energy loss by the neutron. Processes entailing energy gain by the neutron require that the target be excited before scattering and therefore disappear at low temperature or high (negative) energy transfer.

At this stage, we have to average over the nuclear scattering lengths. Because the latter are uncorrelated with the positions of the nuclei, we can simply replace $b_{\mu\ell}b_{\nu\ell'}$ in equation (10.3) by its average $\overline{b_{\mu\ell}b_{\nu\ell'}}$. This must be understood as an ensemble average, i.e. an average over many realizations of the same physical system with just the internal states of the nuclei changing from one realization to the next. Because of the absence of spatial correlations, $\overline{b_{\mu\ell}b_{\nu\ell'}} = \bar{b}_\mu\bar{b}_\nu$ if $\mu\ell \neq \nu\ell'$, however $\overline{b_{\mu\ell}b_{\mu\ell}} = \overline{b_\mu^2}$. The ℓ indices of course disappear as all atoms of a given species become equivalent upon averaging. Hence two different terms show up, which allow us to distinguish *coherent* from *incoherent* nuclear scattering:

[3] In the specialized literature, the name *scattering function* designated by the symbol $S(\boldsymbol{q}, \varepsilon)$ is often meant as an experimental quantity obtained by dividing the measured cross section by k'/k, and a different symbol is introduced for the correlation function to which this scattering function is equal in the Born approximation. Here we designate by $S(\boldsymbol{q}, \varepsilon)$ the correlation function itself, bearing in mind that it represents the experiment only to the extent that the Born approximation is valid.

$$\sum_{\mu\ell\nu\ell'} \overline{b_{\mu\ell} b_{\nu\ell'}} S_{\mu\ell\nu\ell'}(\boldsymbol{q}, \varepsilon) = \sum_{\mu\nu} \bar{b}_\mu \bar{b}_\nu \sum_{\ell\ell'} S_{\mu\ell\nu\ell'}(\boldsymbol{q}, \varepsilon) + \sum_{\mu}\left(\overline{b_\mu^2} - \bar{b}_\mu^2\right) \sum_{\ell} S_{\mu\ell\mu\ell}(\boldsymbol{q}, \varepsilon)$$
$$\equiv \sum_{\mu\nu} \bar{b}_\mu \bar{b}_\nu I_{\mu\nu}(\boldsymbol{q}, \varepsilon) + \sum_{\mu}\left(\overline{b_\mu^2} - \bar{b}_\mu^2\right) I_\mu^s(\boldsymbol{q}, \varepsilon). \tag{10.4}$$

The first term, called coherent, involves an undue contribution with $\mu\ell = \nu\ell'$, which is subtracted from the second term called incoherent[4]. The coherent term gives the signal one would get if all atoms were characterized by their average scattering length: all scattered spherical waves centered on the atoms then coherently add up. The actual signal is different because of the fluctuations in scattering lengths and the incoherent term is the needed correction.

Let us now consider the quantity $I_{\mu\nu}(\boldsymbol{q}, \varepsilon) = \sum_{\ell\ell'} S_{\mu\ell\nu\ell'}(\boldsymbol{q}, \varepsilon)$, commonly called *intermediate scattering function*. It involves the operator $\sum_\ell e^{-i\boldsymbol{q}\cdot\boldsymbol{r}_{\mu\ell}}$. This is nothing but the Fourier transform of the density of atoms for the species μ, $n_\mu(\boldsymbol{r}) = \sum_\ell \delta(\boldsymbol{r} - \boldsymbol{r}_{\mu\ell})$. Comparing equations (10.3) and (3.12), one sees that $I_{\mu\nu}(\boldsymbol{q}, \varepsilon)$ is the greater spectral function of the correlation function of $n_\mu(\boldsymbol{q})$ and $n_\nu(-\boldsymbol{q})$: $I_{\mu\nu}(\boldsymbol{q}, \varepsilon) = \rho^>_{n_\mu(\boldsymbol{q})n_\nu(-\boldsymbol{q})}(\varepsilon)$. Except for $\mu = \nu$, this quantity is in general not real; one should, however, obtain a real result upon summing over μ and ν. According to the general spectral representation (3.20), this is closely related to the greater correlation function $C^>_{n_\mu(\boldsymbol{q})n_\nu(-\boldsymbol{q})}(\varepsilon) = -2\pi i \rho^>_{n_\mu(\boldsymbol{q})n_\nu(-\boldsymbol{q})}(\varepsilon)$. In the neutron-scattering literature, the symbol G is usually attached to this correlation function; in order to avoid confusion with the single-particle Green's function, we will keep the superscript and use the symbol $G^>$. Thus we have to deal with three distinct but closely related quantities, the greater correlation function

$$G^>_{\mu\nu}(\boldsymbol{q}, t) = -\frac{i}{\hbar}\langle n_\mu(\boldsymbol{q}, t) n_\nu(-\boldsymbol{q}, 0)\rangle, \tag{10.5a}$$

its spectral function

$$I_{\mu\nu}(\boldsymbol{q}, \varepsilon) = \rho^>_{n_\mu(\boldsymbol{q})n_\nu(-\boldsymbol{q})}(\varepsilon) = \frac{i}{2\pi} G^>_{\mu\nu}(\boldsymbol{q}, \varepsilon), \tag{10.5b}$$

these two depending only upon atomic positions, and finally the scattering function which in addition contains information about the scattering lengths. The function $I^s_\mu(\boldsymbol{q}, \varepsilon)$ describing the incoherent scattering in equation (10.4), called *intermediate self scattering function*, has a very different character because it only encodes correlations between the positions of a given atom at different times, as opposed to coherent scattering which depends on correlations between relative positions of atom pairs. The relevant functions for the incoherent signal are

[4] Similar to, but different from, incoherent scattering is the notion of *diffuse* scattering. Diffuse scattering refers to scattering induced by chemical disorder: in that case, the variations of the scattering length at structurally equivalent sites also stem from variations in the chemical composition of that site.

$$G_\mu^{s>}(\boldsymbol{q}, t) = -\frac{i}{\hbar}\sum_\ell \langle e^{-i\boldsymbol{q}\cdot\boldsymbol{r}_{\mu\ell}(t)} e^{i\boldsymbol{q}\cdot\boldsymbol{r}_{\mu\ell}(0)}\rangle \tag{10.6a}$$

$$I_\mu^s(\boldsymbol{q}, \varepsilon) = \sum_\ell \rho^{>}_{e^{-i\boldsymbol{q}\cdot\boldsymbol{r}_{\mu\ell}} e^{i\boldsymbol{q}\cdot\boldsymbol{r}_{\mu\ell}}}(\varepsilon) = \frac{i}{2\pi} G_\mu^{s>}(\boldsymbol{q}, \varepsilon) = -\frac{1}{2\pi}\,\mathrm{Im}\, G_\mu^{s>}(\boldsymbol{q}, \varepsilon). \tag{10.6b}$$

$I_\mu^s(\boldsymbol{q}, \varepsilon)$ is real and positive-definite, being the spectral function of two adjoint operators. In term of these functions, the scattering function takes the form

$$S(\boldsymbol{q}, \varepsilon) = \sum_{\mu\nu} \bar{b}_\mu \bar{b}_\nu I_{\mu\nu}(\boldsymbol{q}, \varepsilon) + \sum_\mu \left(\overline{b_\mu^2} - \bar{b}_\mu^2\right) I_\mu^s(\boldsymbol{q}, \varepsilon). \tag{10.7}$$

Note that the detailed-balance law (3.34), together with the definitions (10.5*b*) and (10.6*b*), imply that $I_{\mu\nu}(\boldsymbol{q}, \varepsilon) = e^{\beta\varepsilon} I_{\nu\mu}(-\boldsymbol{q}, -\varepsilon)$ and $I_\mu^s(\boldsymbol{q}, \varepsilon) = e^{\beta\varepsilon} I_\mu^s(-\boldsymbol{q}, -\varepsilon)$. Accordingly, the scattering function also obeys detailed balance:

$$S(\boldsymbol{q}, \varepsilon) = e^{\beta\varepsilon} S(-\boldsymbol{q}, -\varepsilon). \tag{10.8}$$

In order to gain some insight, we specialize temporarily to a monoatomic system. In that case, we can introduce the total coherent and incoherent atomic cross sections $\sigma_{\mathrm{coh}} = 4\pi\bar{b}^2$ and $\sigma_{\mathrm{inc}} = 4\pi[\overline{b^2} - \bar{b}^2]$ and we get

$$\frac{d^2\sigma_{\mathrm{n}}}{d\Omega d\varepsilon} = \frac{k'}{k}\left[\frac{\sigma_{\mathrm{coh}}}{4\pi}\rho^{>}_{n(\boldsymbol{q})n(-\boldsymbol{q})}(\varepsilon) + \frac{\sigma_{\mathrm{inc}}}{4\pi}\sum_\ell \rho^{>}_{e^{-i\boldsymbol{q}\cdot\boldsymbol{r}_\ell} e^{i\boldsymbol{q}\cdot\boldsymbol{r}_\ell}}(\varepsilon)\right] \quad \text{(monoatomic system)}.$$

It is seen that coherent scattering measures the nuclear density–density correlations—a result that we have already encountered in section 1.1—and is closely related to the corresponding susceptibility since, using equations (3.32), (3.47), and (6.8), we get

$$\rho^{>}_{n(\boldsymbol{q})n(-\boldsymbol{q})}(\varepsilon) = -b(-\varepsilon)\rho_{n(\boldsymbol{q})n(-\boldsymbol{q})}(\varepsilon) = \frac{b(-\varepsilon)}{\pi}\,\mathrm{Im}\,\chi_{nn}(\boldsymbol{q}, \varepsilon). \tag{10.9}$$

Of course, χ_{nn} refers here to the nuclear density–density response, not to the electronic one. The poles of $\chi_{nn}(\boldsymbol{q}, \varepsilon)$ describe collective excitations of the nuclei, e.g. phonons for a crystal. Thus coherent nuclear neutron scattering provides a rather direct measure of the phonon dispersion curves in a monoatomic crystal, as we will see in more detail below. The numerical values of σ_{coh} and σ_{inc} are tabulated for all elements and vary without any systematics across the periodic table. Because the scattering length can be negative, it can happen than $\sigma_{\mathrm{coh}} \ll \sigma_{\mathrm{inc}}$ if the distribution of b values is broad and centered at $b = 0$. This is the case for vanadium, an element of choice for calibration purposes because its cross section is almost purely incoherent. To better understand the qualitative difference between coherent and incoherent scattering, we write the spectral functions in terms of the corresponding time-dependent correlation functions:

$$\rho^{>}_{n(\boldsymbol{q})n(-\boldsymbol{q})}(\varepsilon) = \frac{1}{2\pi\hbar}\int_{-\infty}^{\infty} dt\; e^{i\varepsilon t/\hbar} \sum_{\ell\ell'} \langle e^{-i\boldsymbol{q}\cdot\boldsymbol{r}_\ell(t)} e^{i\boldsymbol{q}\cdot\boldsymbol{r}_{\ell'}(0)}\rangle$$

$$\sum_\ell \rho^{>}_{e^{-i\boldsymbol{q}\cdot\boldsymbol{r}_\ell} e^{i\boldsymbol{q}\cdot\boldsymbol{r}_\ell}}(\varepsilon) = \frac{1}{2\pi\hbar}\int_{-\infty}^{\infty} dt\; e^{i\varepsilon t/\hbar} \sum_{\ell} \langle e^{-i\boldsymbol{q}\cdot\boldsymbol{r}_\ell(t)} e^{i\boldsymbol{q}\cdot\boldsymbol{r}_{\ell}(0)}\rangle .$$

$\boldsymbol{r}_\ell$ are the position operators for the nuclei. These operators commute for different nuclei, but for a given nucleus $\boldsymbol{r}_\ell(t)$ does not commute with $\boldsymbol{r}_\ell(0)$. Let us assume that the nuclei are frozen such that $\boldsymbol{r}_\ell(t) = \boldsymbol{r}_\ell(0)$. In this case, where only elastic scattering is possible ($k' = k$), the incoherent term above becomes simply $N\delta(\varepsilon)$, where N is the number of nuclei, but the coherent term gives the static structure factor $|\sum_\ell e^{-i\boldsymbol{q}\cdot\boldsymbol{r}_\ell}|^2\delta(\varepsilon)$. Here we use the fact that the operators $\boldsymbol{r}_\ell$ act on the many-body states as a multiplicative factor: $\langle a|e^{i\boldsymbol{q}\cdot\boldsymbol{r}_\ell}|a\rangle = e^{i\boldsymbol{q}\cdot\boldsymbol{r}_\ell}\langle a|a\rangle = e^{i\boldsymbol{q}\cdot\boldsymbol{r}_\ell}$, where in the last two expressions $\boldsymbol{r}_\ell$ is no longer the position operator, but the position vector. Hence, for a static monoatomic system, the differential cross section is simply

$$\frac{d\sigma_\mathrm{n}}{d\Omega} = \frac{\sigma_\mathrm{coh}}{4\pi}\left|\sum_\ell e^{-i\boldsymbol{q}\cdot\boldsymbol{r}_\ell}\right|^2 + \frac{\sigma_\mathrm{inc}}{4\pi}N \qquad \text{(static monoatomic system).}$$

Systems are never static, but this simple limiting example should make it clear that the coherent term has a strong $\boldsymbol{q}$ dependence, while the incoherent term is rather isotropic. For a crystalline lattice, the static structure factor has peaks at the vectors $\boldsymbol{q}$ coinciding with the vectors $\boldsymbol{G}$ of the reciprocal lattice.

10.2.2 Elastic, quasi-elastic, and inelastic scattering, sum rules

At very long time, the positions $\boldsymbol{r}_{\mu\ell}(t)$ and $\boldsymbol{r}_{\nu\ell'}(0)$ become uncorrelated—although this property may not hold true for too simple theoretical models lacking dissipation. Hence in the limit $t \to \infty$, the correlations functions (10.5*a*) and (10.6*a*) factorize:

$$G^{>}_{\mu\nu}(\boldsymbol{q}, t \to \infty) = -\frac{i}{\hbar}\langle n_\mu(\boldsymbol{q})\rangle\langle n_\nu(-\boldsymbol{q})\rangle$$

$$G^{s>}_{\mu}(\boldsymbol{q}, t \to \infty) = -\frac{i}{\hbar}\sum_\ell \langle e^{-i\boldsymbol{q}\cdot\boldsymbol{r}_{\mu\ell}}\rangle\langle e^{i\boldsymbol{q}\cdot\boldsymbol{r}_{\mu\ell}}\rangle = -\frac{i}{\hbar}N_\mu .$$

We can subtract from the correlation functions their long-time limit and thus define functions that vanish at long times, $\tilde{G}^{>}_{\mu\nu}(\boldsymbol{q}, t) = G^{>}_{\mu\nu}(\boldsymbol{q}, t) - G^{>}_{\mu\nu}(\boldsymbol{q}, \infty)$ and similarly for the self correlation function, $\tilde{G}^{s>}_{\mu}(\boldsymbol{q}, t) = G^{s>}_{\mu}(\boldsymbol{q}, t) + (i/\hbar)N_\mu$. Using equations (10.5*b*) and (10.6*b*), we find for the coherent and incoherent spectral functions

$$I_{\mu\nu}(\boldsymbol{q},\,\varepsilon) = \frac{i}{2\pi}\tilde{G}^{>}_{\mu\nu}(\boldsymbol{q},\,\varepsilon) + \langle n_\mu(\boldsymbol{q})\rangle\langle n_\nu(-\boldsymbol{q})\rangle\delta(\varepsilon)$$
$$I^{s}_{\mu}(\boldsymbol{q},\,\varepsilon) = \frac{i}{2\pi}\tilde{G}^{s>}_{\mu}(\boldsymbol{q},\,\varepsilon) + N_\mu\delta(\varepsilon).$$

This shows that the scattering function possesses a singular contribution at $\varepsilon = 0$, which is called the elastic peak

$$\boxed{S_{\text{el}}(\boldsymbol{q},\,\varepsilon) = \delta(\varepsilon)\left[\sum_{\mu\nu}\bar{b}_\mu\bar{b}_\nu\langle n_\mu(\boldsymbol{q})\rangle\langle n_\nu(-\boldsymbol{q})\rangle + \sum_{\mu}\left(\overline{b_\mu^2} - \bar{b}_\mu^2\right)N_\mu\right].} \tag{10.10}$$

The elastic peak appears broadened in the experiment owing to finite resolution, but its spectral weight given by the term in square brackets is independent of the resolution. Here again, the coherent contribution is modulated as a function of $\boldsymbol{q}$, becoming a single peak at $\boldsymbol{q} = 0$ for a system with homogeneous density, while the incoherent term has no reciprocal-space structure.

The elastic peak, which accounts for the correlations that survive in the infinite-time limit, does not exhaust the spectral weight at $\varepsilon = 0$ because $S(\boldsymbol{q},\,0) - S_{\text{el}}(\boldsymbol{q},\,0) \neq 0$ in general. The signal existing close to $\varepsilon = 0$ that is not contained in the elastic peak is called quasi-elastic. Subtracting the elastic peak from the scattering function, we arrive at

$$\begin{aligned} S(\boldsymbol{q},\,\varepsilon) - S_{\text{el}}(\boldsymbol{q},\,\varepsilon) = &\sum_{\mu\nu}\frac{\bar{b}_\mu\bar{b}_\nu}{2\pi\hbar}\int_{-\infty}^{\infty}dte^{i\varepsilon t/\hbar}\big[\langle n_\mu(\boldsymbol{q},\,t)n_\nu(-\boldsymbol{q},\,0)\rangle - \langle n_\mu(\boldsymbol{q})\rangle\langle n_\nu(-\boldsymbol{q})\rangle\big] \\ &+ \sum_{\mu}\left(\overline{b_\mu^2} - \bar{b}_\mu^2\right)\frac{1}{2\pi\hbar}\int_{-\infty}^{\infty}dt\;e^{i\varepsilon t/\hbar}\sum_{\ell}[\langle e^{-i\boldsymbol{q}\cdot\boldsymbol{r}_{\mu\ell}(t)}e^{i\boldsymbol{q}\cdot\boldsymbol{r}_{\mu\ell}(0)}\rangle - 1]. \end{aligned} \tag{10.11}$$

Hence, exactly at $\varepsilon = 0$, there is a signal determined by the total correlations built up over long time scales. At finite energy, the inelastic signal accounts for the correlations build up over time scales of the order of $\hbar/\varepsilon$.

With the help of equations (10.5*b*) and (10.6*b*), we obtain the sum rules $\int_{-\infty}^{\infty} d\varepsilon\; I_{\mu\nu}(\boldsymbol{q},\,\varepsilon) = i\hbar G^{>}_{\mu\nu}(\boldsymbol{q},\,t = 0) = \langle n_\mu(\boldsymbol{q})n_\nu(-\boldsymbol{q})\rangle$ and $\int_{-\infty}^{\infty} d\varepsilon\; I^{s}_{\mu}(\boldsymbol{q},\,\varepsilon) = i\hbar G^{s>}_{\mu}(\boldsymbol{q},\,t = 0) = N_\mu$. The sum rule for the scattering function (10.7) results as

$$\int_{-\infty}^{\infty} d\varepsilon\; S(\boldsymbol{q},\,\varepsilon) = \sum_{\mu\nu}\bar{b}_\mu\bar{b}_\nu\langle n_\mu(\boldsymbol{q})n_\nu(-\boldsymbol{q})\rangle + \sum_{\mu}\left(\overline{b_\mu^2} - \bar{b}_\mu^2\right)N_\mu. \tag{10.12}$$

Comparing with equation (10.10), we see that the total incoherent spectral weight is the same as the total elastic incoherent spectral weight, but it would be incorrect to conclude that all the incoherent spectral weight is elastic! Our definition of the incoherent elastic peak indeed leaves for the rest the function $(i/2\pi)\tilde{G}^{s>}_{\mu}(\boldsymbol{q},\,\varepsilon)$, which is not positive-definite and integrates to zero.

Of importance is the first moment of the scattering function, which can be interpreted as the average energy deposited in the system. In the present context, it is called the f-sum rule and reads (see Doc A10.1)

$$\boxed{\int_{-\infty}^{\infty} d\varepsilon \; \varepsilon \; S(\boldsymbol{q}, \varepsilon) = \sum_{\mu} \overline{b_{\mu}^{2}} N_{\mu} \frac{\hbar^{2} q^{2}}{2M_{\mu}}.} \tag{10.13}$$

The energy $\hbar^2 q^2/(2M_\mu)$ is called *recoil energy* of the nucleus. In a scattering event with momentum loss (for the neutron) $\hbar \boldsymbol{q}$, the nucleus must gain a momentum $\hbar\boldsymbol{q}$, such that its kinetic energy increases by $[(\boldsymbol{p} + \hbar\boldsymbol{q})^2 - \boldsymbol{p}^2]/(2M_\mu) = \hbar^2 q^2/(2M_\mu) + (\hbar/M_\mu)\boldsymbol{q} \cdot \boldsymbol{p}$. On average $\langle \boldsymbol{p} \rangle = 0$, such that the average energy increase is indeed the recoil energy.

10.2.3 Scattering by a harmonic crystal, phonon dispersion curves

In section 2.5, we have introduced the phonons in order to describe the atomic displacements in a harmonic crystal. To conform with the notation there, we rename $\boldsymbol{r}_{\mu\ell} \to \boldsymbol{r}_{\nu n} = \boldsymbol{R}_n + \boldsymbol{\tau}_\nu + \boldsymbol{u}_\nu(\boldsymbol{R}_n)$, where $\boldsymbol{R}_n$ denote the vectors of the periodic lattice and $\boldsymbol{\tau}_\nu$ is the equilibrium position of the nucleus in the unit cell. $\boldsymbol{R}_n$ and $\boldsymbol{\tau}_\nu$ are scalars, while $\boldsymbol{u}_\nu(\boldsymbol{R}_n)$ is the displacement operator given by equation (2.66). For definiteness, we rewrite the intermediate scattering functions entering equation (10.7) with the new notation:

$$I_{\nu\mu}(\boldsymbol{q}, \varepsilon) = \frac{1}{2\pi\hbar} \int_{-\infty}^{\infty} dt \; e^{i\varepsilon t/\hbar} \sum_{nm} \langle e^{-i\boldsymbol{q}\cdot\boldsymbol{r}_{\nu n}(t)} e^{i\boldsymbol{q}\cdot\boldsymbol{r}_{\mu m}(0)} \rangle \tag{10.14a}$$

$$I_{\nu}^{S}(\boldsymbol{q}, \varepsilon) = \frac{1}{2\pi\hbar} \int_{-\infty}^{\infty} dt \; e^{i\varepsilon t/\hbar} \sum_{n} \langle e^{-i\boldsymbol{q}\cdot\boldsymbol{r}_{\nu n}(t)} e^{i\boldsymbol{q}\cdot\boldsymbol{r}_{\nu n}(0)} \rangle. \tag{10.14b}$$

It is seen that the central quantity in nuclear neutron scattering is the correlator $\langle e^{-i\boldsymbol{q}\cdot\boldsymbol{r}_{\nu n}(t)} e^{i\boldsymbol{q}\cdot\boldsymbol{r}_{\mu m}(0)} \rangle$. This has the form of equation (30) and furthermore complies with the requirements for the validity of that equation: the operators $\boldsymbol{r}_{\nu n}$ are linear combinations of boson creation and annihilation operators (see equation (2.66)) and, in the harmonic approximation, there is no interaction between phonons, the latter entering the total Hamiltonian only via equation (2.65). Application of equation (30) yields

$$\begin{aligned} \langle e^{-i\boldsymbol{q}\cdot\boldsymbol{r}_{\nu n}(t)} e^{i\boldsymbol{q}\cdot\boldsymbol{r}_{\mu m}(0)} \rangle &= e^{-i\boldsymbol{q}\cdot(\boldsymbol{R}_n + \boldsymbol{\tau}_\nu - \boldsymbol{R}_m - \boldsymbol{\tau}_\mu)} \\ &\quad \times e^{-\frac{q^2}{2}\langle [u_\nu^{\parallel}(\boldsymbol{R}_n,\, t)]^2 \rangle} e^{-\frac{q^2}{2}\langle [u_\mu^{\parallel}(\boldsymbol{R}_m,\, 0)]^2 \rangle} e^{q^2 \langle u_\nu^{\parallel}(\boldsymbol{R}_n,\, t) u_\mu^{\parallel}(\boldsymbol{R}_m,\, 0) \rangle}, \end{aligned} \tag{10.15}$$

where we have introduced the symbol $u_\nu^{\parallel}$ for the projection of the displacement along the vector $\boldsymbol{q}$. We emphasize that equation (30) relies heavily on the conservation of bosons, which implies that odd powers of the creation and annihilation operators vanish on average. For harmonic atomic displacements, it is intuitively clear that odd powers of the displacements average to zero. This property is not obeyed if electron–phonon coupling is turned on, because the coupling Hamiltonian equation (2.73) does not conserve the number of phonons. Consequently, equation (10.15) is incorrect for an electron–phonon coupled system. The average quadratic displacement $\langle[u_\nu^{\parallel}(\boldsymbol{R}_n, t)]^2\rangle$ is the same in all lattice cells and at all times, so that we can simply write it as $\langle(u_\nu^{\parallel})^2\rangle$. As shown in Doc A10.2, it can be expressed in terms of the phonon frequencies and polarizations as

$$\begin{aligned}\langle(u_\nu^{\parallel})^2\rangle &= \frac{1}{\mathcal{N}}\sum_{k\lambda}\frac{\hbar}{2M_\nu\omega_{k\lambda}}\,|\,[\boldsymbol{\epsilon}_{k\lambda}^{\parallel}]_\nu|^2\,\coth\left(\frac{\hbar\omega_{k\lambda}}{2k_{\mathrm{B}}T}\right)\\ &= \frac{\hbar^2}{2M_\nu}\int_{-\infty}^{\infty}d\varepsilon\frac{1+b(\varepsilon)}{\varepsilon}N_\nu^{\mathrm{ph}}(\hat{\boldsymbol{q}},\varepsilon).\end{aligned} \tag{10.16}$$

For convenience and later use, we have defined the partial phonon density of states

$$N_\nu^{\mathrm{ph}}(\hat{\boldsymbol{q}},\varepsilon) = \frac{1}{\mathcal{N}}\sum_{k\lambda}|\,[\boldsymbol{\epsilon}_{k\lambda}^{\parallel}]_\nu|^2\,[\delta(\varepsilon-\hbar\omega_{k\lambda})+\delta(\varepsilon+\hbar\omega_{k\lambda})], \tag{10.17}$$

which counts the number of phonons of energy ε weighted by the displacement of atom ν in the direction $\hat{\boldsymbol{q}}$; this is furthermore normalized by the number of unit cells, i.e. the number of atoms ν, and symmetrized in energy thanks to the second delta function. The quantity $\exp[-(q^2/2)\langle(u_\nu^{\parallel})^2\rangle]$ in equation (10.15) is called the *Debye–Waller factor*. It expresses the suppression of positional correlations at short distance (large q) because of the uncorrelated zero-point and thermal motions of the nuclei, which however do not destroy the crystalline order at long length scales (small q).

The last factor in equation (10.15) contains the correlation function of the displacements, $\langle u_\nu^{\parallel}(\boldsymbol{R}_n, t)u_\mu^{\parallel}(\boldsymbol{R}_m, 0)\rangle$, which we can relate to the phonon Green's function (see Doc A10.2). Because this correlation function enters in an exponential, we can distinguish zero, one, two-phonon scattering processes by expanding the exponential for displacements that are small compared with $1/q$:

$$e^{q^2\langle u_\nu^{\parallel}(\boldsymbol{R}_n,t)u_\mu^{\parallel}(\boldsymbol{R}_m,0)\rangle} = 1 + q^2\langle u_\nu^{\parallel}(\boldsymbol{R}_n,t)u_\mu^{\parallel}(\boldsymbol{R}_m,0)\rangle + \cdots \tag{10.18}$$

The '1', once inserted in equation (10.15), gives a time-independent contribution such that the time integration in equation (10.14) gives $\delta(\varepsilon)$: this is the elastic scattering contribution, for which the neutron energy is conserved, $k' = k$. The corresponding double differential cross section is (see Doc A10.2 for details)

$$\left.\frac{d^2\sigma_{\mathrm{n}}}{d\Omega d\varepsilon}\right)_{\mathrm{el}} = \mathcal{N}\delta(\varepsilon)\left[|F(\boldsymbol{q})|^2\frac{(2\pi)^3}{\mathcal{V}_{\mathrm{cell}}}\sum_{\boldsymbol{G}}\delta(\boldsymbol{q}-\boldsymbol{G}) + \sum_\nu\left(\overline{b_\nu^2}-\bar{b}_\nu^2\right)e^{-q^2\langle(u_\nu^{\parallel})^2\rangle}\right]. \tag{10.19}$$

The first term is the coherent contribution, which gives Bragg peaks at the vectors $\boldsymbol{G}$ of the reciprocal lattice, weighted by the so-called *nuclear form factor of the primitive cell*, $F(\boldsymbol{q}) = \sum_\nu \bar{b}_\nu e^{-i\boldsymbol{q}\cdot\boldsymbol{\tau}_\nu}e^{-(q^2/2)\langle(u_\nu^{\parallel})^2\rangle}$. Apart from the Debye–Waller factors, this form factor is the same as one would get for a static lattice. The Debye–Waller factors account for zero-point motion and finite temperature leading to fluctuations of atomic positions around their equilibrium value, which destroys the lattice order at short distances and correspondingly the intensity of the Bragg peaks at large q. Let us emphasize that the quantum and thermal fluctuations *do not broaden* the Bragg peaks, but simply reduce their intensity. Likewise, the second term in equation (10.19), which is the incoherent contribution, is no longer isotropic as it is for a static lattice, but suppressed at large q.[5]

The coherent inelastic signal generated by one-phonon processes may be written as (see Doc A10.2)

$$\left.\frac{d^2\sigma_\text{n}}{d\Omega d\varepsilon}\right)^{\text{coh}}_{1-\text{ph}} = \mathcal{N}\frac{k'}{k}[1 + b(\varepsilon)]\sum_\lambda |F_\lambda(\boldsymbol{q})|^2\left[\delta(\varepsilon - \hbar\omega_{\check{\boldsymbol{q}}\lambda}) - \delta(\varepsilon + \hbar\omega_{\check{\boldsymbol{q}}\lambda})\right]$$
$$F_\lambda(\boldsymbol{q}) = \sum_\nu q\bar{b}_\nu\sqrt{\frac{\hbar}{2M_\nu\omega_{\check{\boldsymbol{q}}\lambda}}}\,e^{-i\boldsymbol{q}\cdot\boldsymbol{\tau}_\nu}e^{-\frac{q^2}{2}\langle(u_\nu^{\parallel})^2\rangle}[\boldsymbol{\epsilon}^{\parallel}_{\check{\boldsymbol{q}}\lambda}]_\nu. \tag{10.20}$$

A more complicated—though equivalent—expression is often reported in the literature with the conservation of momentum made explicit via delta functions. Here, we denote by $\check{\boldsymbol{q}}$ the equivalent value of $\boldsymbol{q}$ in the first Brillouin zone, in other words, $\check{\boldsymbol{q}} + \boldsymbol{G} = \boldsymbol{q}$ for some vector $\boldsymbol{G}$. The inelastic scattering at energy ε occurs through emission (first delta function) or absorption (second delta function) of a phonon of frequency $\omega_{\boldsymbol{k}\lambda}$ whose wave vector satisfies $\boldsymbol{k} = \check{\boldsymbol{q}} = \boldsymbol{q} - \boldsymbol{G}$. The emission and absorption processes look symmetric in the formula, but they are not. The emission occurs only for $\varepsilon > 0$ (energy loss by the neutron) and picks a factor $1 + b(\varepsilon)$ which approaches unity at low temperature, while the absorption at $\varepsilon < 0$ picks a factor $-1 - b(\varepsilon) = b(-\varepsilon)$ which vanishes at low T. The phonon can be emitted or absorbed in any branch λ, unless the phonon is purely transverse to $\boldsymbol{q}$ with $\boldsymbol{\epsilon}^{\parallel}_{\check{\boldsymbol{q}}\lambda} = 0$: for the neutron to lose momentum $\hbar\boldsymbol{q}$ in the scattering process, the emitted phonon must carry a momentum $\hbar\boldsymbol{q}$, but a transverse phonon carries no momentum in the direction of its propagation. Note that a phonon that is transverse to $\boldsymbol{q}$ for $\boldsymbol{q}$ in the first Brillouin zone, and therefore invisible in *Brillouin scattering* (scattering for $\boldsymbol{q}$ vectors in the first Brillouin zone), is not in general transverse to $\boldsymbol{q} + \boldsymbol{G}$ and can therefore be visible in higher-order Brillouin zones. The possible emission and absorption processes are, of course, also subject to kinematic constraints expressing the conservation of energy and momentum. For a given energy, the emission/absorption of phonons is modulated in reciprocal space by $|F_\lambda(\boldsymbol{q})|^2$, which may be considered as the *form factor of the phonon.*

[5] The higher-order terms in the expansion (10.18) only contribute to inelastic scattering, which seems to suggest that equation (10.19) exhausts the elastic scattering spectral weight. It is however expected that neutrons, like photons, experience refraction in addition to the Bragg scattering (diffraction) given by equation (10.19). This is indeed true—allowing one to build neutron waveguides—but the refraction stems from coherent multiple scattering and is therefore missed in the Born approximation.

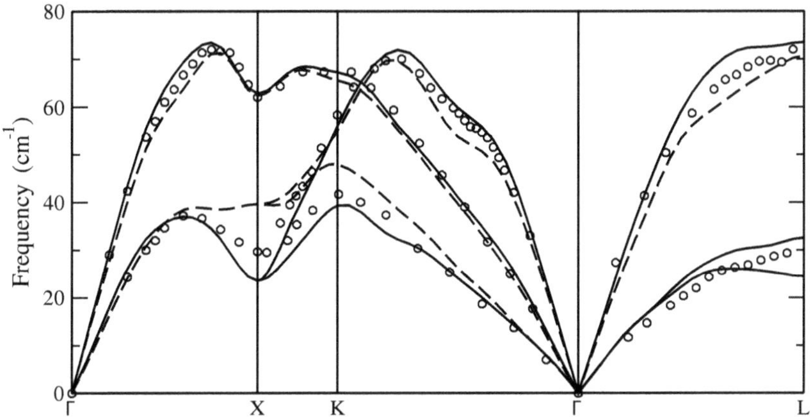

Figure 10.2. Phonon dispersion measured by inelastic neutron scattering in lead (dots). Note the three acoustic modes with linear dispersion at $\boldsymbol{k} = 0$ (Γ point). Calculations are performed with (solid) and without (dashed) spin–orbit coupling. Reprinted figure with permission from Verstraete *et al* 2008 *Phys. Rev.* B **78** 045119. Copyright (2008) by the American Physical Society.

Figure 10.2 shows the phonon dispersion measured by inelastic neutron scattering in lead[6], compared with first-principles calculations. This experimental data set was quite important as it allowed McMillan and Rowell to identify phonon signatures in the tunneling density of states of superconducting lead and thus provide a strong confirmation of the electron–phonon mechanism of superconductivity[7]. The calculations show that spin–orbit interaction has a strong influence on phonons for heavy elements like lead. Modern detectors can collect data simultaneously as a function of momentum and energy and deliver colormaps of the scattering function where the intensity and width of the phonon branches is immediately visible[8].

Finally, the incoherent one-phonon double differential cross section is (see Doc A10.2)

$$\left.\frac{d^2\sigma_{\rm n}}{d\Omega d\varepsilon}\right)_{1-{\rm ph}}^{\rm inc} = \mathcal{N}\frac{k'}{k}\frac{1+b(\varepsilon)}{\varepsilon}\sum_{\nu}\left(\overline{b_\nu^2} - \bar{b}_\nu^2\right)\frac{\hbar^2 q^2}{2M_\nu}e^{-q^2\langle(u_\nu^{\parallel})^2\rangle}N_\nu^{\rm ph}(\hat{\boldsymbol{q}}, \varepsilon). \tag{10.21}$$

We will see below in which conditions the incoherent scattering allows one to measure the phonon DOS.

With equations (10.19)–(10.21) we have accounted for the zero- and one-phonon scattering processes. These should not exhaust the f-sum rule (10.13) because of the neglect of multi-phonon processes. Indeed we find (see Doc A10.3)

$$\int_{-\infty}^{\infty} d\varepsilon\, \varepsilon\left[S_{\rm el}(\boldsymbol{q}, \varepsilon) + S_{1-{\rm ph}}(\boldsymbol{q}, \varepsilon)\right] = \sum_{\nu}\overline{b_\nu^2}\mathcal{N}\frac{\hbar^2 q^2}{2M_\nu}e^{-q^2\langle(u_\nu^{\parallel})^2\rangle}. \tag{10.22}$$

[6] Brockhouse B N, Arase T, Caglioti G, Rao K R and Woods A D B 1962 *Phys. Rev.* **128** 1099.
[7] McMillan W L and Rowell J M 1965 *Phys. Rev. Lett.* **14** 108.
[8] See, e.g. Delaire O *et al* 2011 *Nat. Mat.* **10** 614.

Because $\mathcal{N} = N_\nu$, this would reproduce the f-sum rule if there were no Debye–Waller factors. One sees that the multi-phonon processes are negligible at low q but carry most of the spectral weight at large q.

We now specialize the discussion to the case of a monoatomic cubic crystal, where particular simplifications occur. In a cubic lattice, each phonon $(\boldsymbol{k}\lambda)$ appears together with two other symmetry-equivalent modes having the same frequency, the corresponding three polarization vectors forming an orthonormal set. Restricted to these three wave vectors, the sum $\sum_k |[\epsilon_{k\lambda}^{\parallel}]|^2$ appearing in equation (10.17) gives unity. Since this is true for all $\boldsymbol{k}$ points, we can simply replace $|[\epsilon_{k\lambda}^{\parallel}]|^2$ by 1/3, which yields $N^{\rm ph}(\hat{\boldsymbol{q}}, \varepsilon) = N^{\rm ph}(|\varepsilon|)/(3\mathcal{N})$ where $N^{\rm ph}(\varepsilon)$ is the phonon DOS, an extensive quantity in our convention; see equation (2.68). The resulting parallel mean-square displacement[9] is

$$\langle (u^{\parallel})^2 \rangle = \frac{\hbar}{6M\mathcal{N}} \int_0^{\infty} d\varepsilon\, N^{\rm ph}(\varepsilon) \frac{1}{\varepsilon} \coth\left(\frac{\varepsilon}{2k_{\rm B}T}\right). \tag{10.23}$$

For a Debye spectrum, equation (2.68), this can be transformed into

$$\langle (u^{\parallel})^2 \rangle_{\rm D} = \frac{3}{k_{\rm D}^2} \int_0^{1} dx\, x \coth\left(\frac{x}{2t}\right), \tag{10.24}$$

where we have introduced a typical Debye wave vector $k_{\rm D}$ such that $\hbar^2 k_{\rm D}^2/(2M) = \hbar\omega_{\rm D}$ and the dimensionless temperature $t = T/T_{\rm D}$. For an Einstein spectrum with the same center of gravity ($\omega_{\rm E} = \frac{3}{4}\omega_{\rm D}$, see legend of figure 2.2), we get

$$\langle (u^{\parallel})^2 \rangle_{\rm E} = \frac{4}{3k_{\rm D}^2} \coth\left(\frac{3}{8t}\right). \tag{10.25}$$

Both functions are compared in figure 10.3, where experimental data for several elemental cubic metals are also shown. In contrast to the specific heat displayed in figure 2.2, where the Einstein and Debye models are very similar except at low temperatures, here they deviate significantly at high temperature. The reason is the integration kernel in equation (10.23), which peaks as $1/\varepsilon^2$ at low energy, meaning that low-frequency phonons dominate the mean-square displacement, while in equation (2.67) the kernel is smooth at $\varepsilon = 0$. Since the Einstein model has no low-frequency phonons, this effect is missed. The Debye model captures well the order of magnitude and the trend of the experimentally determined mean-square displacements.

Perhaps the most well known capability of nuclear neutron scattering is to measure the phonon DOS. Indeed, by applying the simplifications appropriate for a

[9] We must distinguish the parallel mean-square displacement $\langle (u^{\parallel})^2 \rangle$ from the mean-square displacement $\langle u^2 \rangle$, which is equal to $3\langle (u^{\parallel})^2 \rangle$ in a cubic crystal.

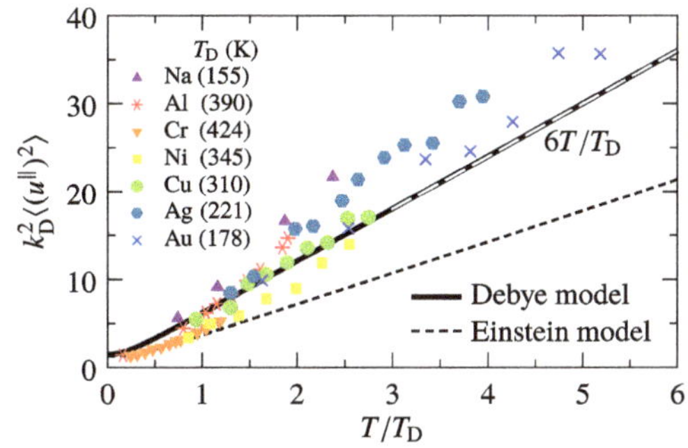

Figure 10.3. Parallel mean-square displacement for a monoatomic cubic lattice using the Debye and Einstein models for the phonon DOS. The dashed line shows the high-T behavior equal to $6T/T_D$. The experimental data[10] were rescaled using the room-temperature Debye temperature indicated next to each element. A better scaling could be achieved by taking into account the weak temperature dependence of T_D.

cubic lattice, we see that in this case the incoherent one-phonon signal is directly proportional to $N^{\mathrm{ph}}(\varepsilon)$:

$$\boxed{\left.\frac{d^2\sigma_{\mathrm{n}}}{d\Omega d\varepsilon}\right)^{\mathrm{inc}}_{1-\mathrm{ph}} = \frac{k'}{k}\frac{\hbar q^2}{6M}(\overline{b^2} - \bar{b}^2)e^{-q^2\langle(u^{\parallel})^2\rangle}\frac{1 + b(\varepsilon)}{\varepsilon}N^{\mathrm{ph}}(|\varepsilon|).} \tag{10.26}$$

It should be quite clear that the situation is more complicated for a general lattice. The pre-factors, even if not precisely known, are not important if one is interested in the energy dependence of the phonon DOS. Figure 10.4 shows the example of vanadium. As already stated, for this element $\bar{b}^2 \approx 0$ such that the coherent scattering is negligible and the one-phonon inelastic scattering function is proportional to the DOS. This material has bcc structure with lattice parameter $a = 3.03$ Å. The cubic cell contains two atoms, but the primitive cell has only one atom, so there are only acoustic phonons. The three acoustic modes give the characteristic ε^2 behavior at low energy and two peaks at ~18 and ~30 meV corresponding to a local minimum and a local maximum in the dispersion, respectively. The best fit to the ε^2 part (see Doc A2.11) gives a sound velocity of 5087 m s^{-1}, slightly larger than the measured value of 4560 m s^{-1}.

Although we have restricted our discussion to one-phonon processes, multi-phonon processes can be modeled using the same formalism: instead of expanding the exponential in equation (10.18), one must compute the correlation function first, and then exponentiate it. While one-phonon processes yield peaks at the phonon frequencies, multi-phonon processes generally yield a broad continuum due to less restrictive kinematic constraints. For one phonon, the conservation of momentum requires $\boldsymbol{k} \equiv \boldsymbol{k}' + \boldsymbol{q} = \boldsymbol{k}' + \boldsymbol{k}^{\mathrm{ph}}$ and the conservation of energy requires $E_{\boldsymbol{k}} \equiv E_{\boldsymbol{k}'} + \varepsilon = E_{\boldsymbol{k}'} + \hbar\omega^{\mathrm{ph}}$, such that the only solution is $\boldsymbol{q} = \boldsymbol{k}^{\mathrm{ph}}$ and $\varepsilon = \hbar\omega^{\mathrm{ph}}$. For two phonons, we have instead $\boldsymbol{q} = \boldsymbol{k}_1^{\mathrm{ph}} + \boldsymbol{k}_2^{\mathrm{ph}}$ and $\varepsilon = \hbar\omega_1^{\mathrm{ph}} + \hbar\omega_2^{\mathrm{ph}}$, which in general

[10] Singh N and Sharma P K 1971 *Phys. Rev.* B **3** 1141.

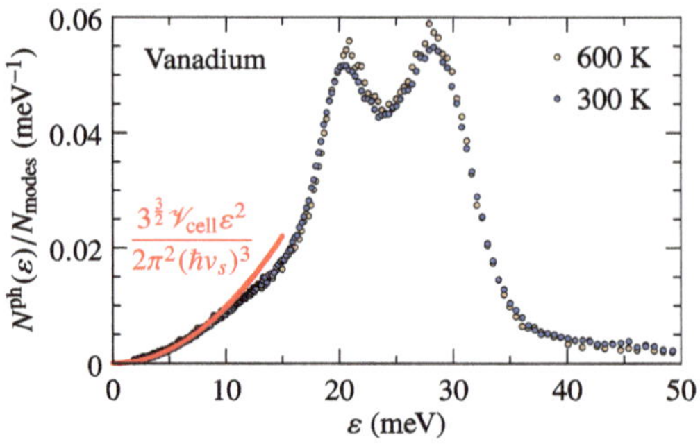

Figure 10.4. Phonon density of states of vanadium measured by inelastic neutron scattering[11]. The multi-phonon processes have been subtracted out. The very weak temperature dependence confirms the reliability of the data. The ε^2 behavior of a Debye spectrum with sound velocity $v_s = 5087$ m s^{-1} is shown in red.

has a continuum of solutions. To conclude this section, let us point out that multi-phonon processes are different from anharmonicity. The latter leads to interaction between the phonons, manifested by the appearance of a finite phonon lifetime and some renormalization of the phonon frequencies. The delta peaks, e.g. in equation (10.20), should then be replaced by the spectral function of the phonons. The width of the one-phonon peaks therefore contains, beside the unavoidable finite experimental resolution, information about the phonon lifetime.

10.3 Magnetic scattering

10.3.1 Magnetic cross section

The magnetic energy resulting from the interaction of the neutron with the electrons of a material involves two terms. The first is a Zeeman coupling between the spin density $\boldsymbol{S}(\boldsymbol{r})$ of the electrons and the field generated by the neutron magnetic moment. The second term accounts for the orbital motion of the electrons in the neutron field and relates to the current density $\boldsymbol{j}(\boldsymbol{r})$. The magnetic cross section for the case of unpolarized neutrons is derived in Doc A10.4:

$$\frac{d^2\sigma_{\mathrm{m}}}{d\Omega d\varepsilon} = \frac{k'}{k}(\gamma r_e)^2 \sum_{\mu\nu}\left(\delta_{\mu\nu} - \hat{q}_\mu \hat{q}_\nu\right) \times \sum_{ab} \frac{e^{-\beta K_a}}{Z} \langle a|\mathcal{M}_\mu(\boldsymbol{q})|b\rangle\langle b|\mathcal{M}_\nu(-\boldsymbol{q})|a\rangle \delta(\varepsilon + K_a - K_b). \tag{10.27}$$

In this expression, $\gamma = g_e g_n/4 = 1.915$, $r_e = \mu_0 e^2/(4\pi m)$ is the classical electron radius, and the operator $\mathcal{M}(\boldsymbol{q})$ is the sum of the dimensionless magnetization due to the electron spin density and an orbital contribution due to the electronic motion: $\mathcal{M}(\boldsymbol{q}) = [\boldsymbol{S}(\boldsymbol{q}) + \boldsymbol{L}(\boldsymbol{q})]/\hbar$. The orbital 'spin density' is defined as $\boldsymbol{L}(\boldsymbol{q}) = \hbar \boldsymbol{B}_L(\boldsymbol{q})/(g_e\mu_0\mu_{\mathrm{B}})$, where $\boldsymbol{B}_L$ is the orbital field obeying $\nabla \times \boldsymbol{B}_L(\boldsymbol{r}) = \mu_0 e \boldsymbol{j}(\boldsymbol{r})$. The value $\gamma r_e = 5.4$ fm is similar to

[11] Schober H 2014 *J. Neutron Res.* **17** 109.

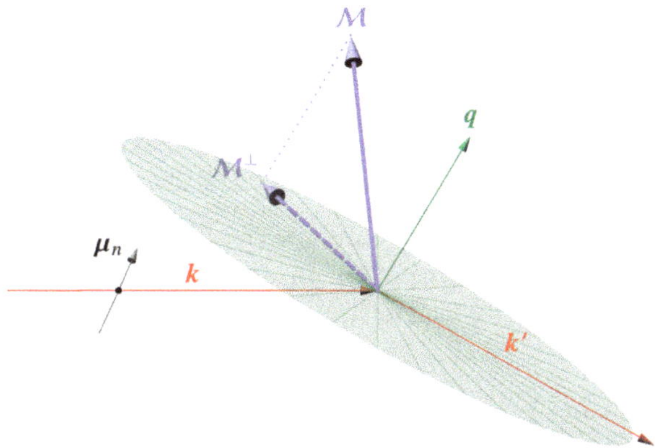

Figure 10.5. Geometrical relations for magnetic scattering of unpolarized neutrons. The neutron is scattered from $\boldsymbol{k}$ to $\boldsymbol{k}'$ by the magnetization $\boldsymbol{\mathcal{M}}$. The interaction potential goes like $-\boldsymbol{\mathcal{M}} \cdot \boldsymbol{B}_n \sim -\boldsymbol{\mathcal{M}} \cdot [\nabla \times (\boldsymbol{\mu}_n \times \boldsymbol{r})]$ $\sim \boldsymbol{\mathcal{M}} \cdot (\hat{\boldsymbol{q}} \times \boldsymbol{\mu}_n \times \hat{\boldsymbol{q}}) \sim \boldsymbol{\mathcal{M}} \cdot \boldsymbol{\mu}_n^{\perp} = \boldsymbol{\mathcal{M}}^{\perp} \cdot \boldsymbol{\mu}_n$. Consequently, the force acting on the neutron goes like $-\boldsymbol{q}(\boldsymbol{\mathcal{M}}^{\perp} \cdot \boldsymbol{\mu}_n)$. Upon averaging over the orientation of $\boldsymbol{\mu}_n$, the scattering process is only sensitive to the projection $\boldsymbol{\mathcal{M}}^{\perp}$ in the plane perpendicular to $\boldsymbol{q}$.

the typical nuclear scattering lengths and sets the scale of the magnetic cross section in the same range as the nuclear one. The factor $\delta_{\mu\nu} - \hat{q}_\mu \hat{q}_\nu$ introduces a peculiar anisotropy that is not present in nuclear scattering. Its origin lies in the anisotropy of the force exerted by a magnetic field on a magnetic dipole. The vector identity that has been used to arrive at the form (10.27), namely (see Doc A10.4)

$$(\hat{\boldsymbol{q}} \times \boldsymbol{A} \times \hat{\boldsymbol{q}}) \cdot (\hat{\boldsymbol{q}} \times \boldsymbol{B} \times \hat{\boldsymbol{q}}) \equiv \boldsymbol{A}^{\perp} \cdot \boldsymbol{B}^{\perp} = \sum_{\mu\nu} \left(\delta_{\mu\nu} - \hat{q}_\mu \hat{q}_\nu\right) A_\mu B_\nu, \tag{10.28}$$

allows one to interpret this anisotropy as a consequence of projecting the magnetization onto the plane perpendicular to the direction of the scattering vector, as illustrated in figure 10.5. The cross section measures this projection $\boldsymbol{\mathcal{M}}^{\perp} = \hat{\boldsymbol{q}} \times \boldsymbol{\mathcal{M}} \times \hat{\boldsymbol{q}}$. We recognize in equation (10.27) the greater spectral function (3.12) of the operators $\mathcal{M}_\mu(\boldsymbol{q})$ and $\mathcal{M}_\nu(-\boldsymbol{q})$. With the help of equations (3.20), (2*b*), and (3.6), we see that this spectral function is related to the Fourier transform of the greater correlation function:

$$\rho^{>}_{\mathcal{M}_\mu(\boldsymbol{q})\mathcal{M}_\nu(-\boldsymbol{q})}(\varepsilon) = \frac{1}{2\pi\hbar} \int_{-\infty}^{\infty} dt\, e^{i\varepsilon t/\hbar} \langle \mathcal{M}_\mu(\boldsymbol{q}, t) \mathcal{M}_\nu(-\boldsymbol{q}, 0) \rangle. \tag{10.29}$$

It is useful to introduce the imaginary-time susceptibility

$$\chi_{\mathcal{M}\mathcal{M}}(\boldsymbol{q}, \tau) = -\sum_{\mu\nu} \left(\delta_{\mu\nu} - \hat{q}_\mu \hat{q}_\nu\right) \langle T_\tau \mathcal{M}_\mu(\boldsymbol{q}, \tau) \mathcal{M}_\nu(-\boldsymbol{q}, 0) \rangle. \tag{10.30}$$

The spectral function of $\chi_{\mathcal{M}\mathcal{M}}$ is $(-1/\pi)\mathrm{Im}\, \chi_{\mathcal{M}\mathcal{M}}(\boldsymbol{q}, \varepsilon)$, where $\chi_{\mathcal{M}\mathcal{M}}(\boldsymbol{q}, \varepsilon)$ is the retarded function according to equation (3.47) and we omit explicitly writing the '*R*' superscript as we have been doing already for susceptibilities. The greater spectral function giving the differential cross section is obtained by multiplying by $-b(-\varepsilon) = 1 + b(\varepsilon)$, see equation (3.32); hence we arrive at:

$$\boxed{\frac{d^2\sigma_{\mathrm{m}}}{d\Omega d\varepsilon} = \frac{k'}{k}(\gamma r_e)^2[1 + b(\varepsilon)]\left(-\frac{1}{\pi}\right)\mathrm{Im}\,\chi_{\mathcal{M}\mathcal{M}}(\boldsymbol{q}, \varepsilon).} \tag{10.31}$$

Like for nuclear scattering, we can define an elastic cross section $\propto \delta(\varepsilon)$ by noting that the correlation function factorizes in the long-time limit: $\langle \mathcal{M}_\mu(\boldsymbol{q}, t \to \infty)\mathcal{M}_\nu(-\boldsymbol{q}, 0)\rangle = \langle \mathcal{M}_\mu(\boldsymbol{q})\rangle\langle \mathcal{M}_\nu(-\boldsymbol{q})\rangle$. The weight of the elastic peak can be expressed, for instance, as:

$$\left.\frac{d\sigma_{\mathrm{m}}}{d\Omega}\right)_{\mathrm{el}} = (\gamma r_e)^2|\hat{\boldsymbol{q}} \times \langle \boldsymbol{\mathcal{M}}(\boldsymbol{q})\rangle \times \hat{\boldsymbol{q}}|^2. \tag{10.32}$$

In a periodic ferromagnetic crystal, the magnetization has the same periodicity as the lattice: $\boldsymbol{\mathcal{M}}(\boldsymbol{r}) = \sum_{\boldsymbol{G}} \boldsymbol{\mathcal{M}}(\boldsymbol{G})e^{i\boldsymbol{G}\cdot\boldsymbol{r}}$. Conversely, its Fourier transform is $\boldsymbol{\mathcal{M}}(\boldsymbol{q}) = \sum_{\boldsymbol{G}} \boldsymbol{\mathcal{M}}(\boldsymbol{G})\mathcal{V}\delta_{\boldsymbol{q}\boldsymbol{G}} = (2\pi)^3\sum_{\boldsymbol{G}} \boldsymbol{\mathcal{M}}(\boldsymbol{G})\delta(\boldsymbol{q} - \boldsymbol{G})$. In analogy with equation (10.19), the measurement of the intensity at the Bragg peaks allows one to reconstruct the average magnetization distribution in real space:

$$\left.\frac{d\sigma_{\mathrm{m}}}{d\Omega}\right)_{\mathrm{el}} = (\gamma r_e)^2(2\pi)^3|\hat{\boldsymbol{q}} \times \langle \boldsymbol{\mathcal{M}}(\boldsymbol{q})\rangle \times \hat{\boldsymbol{q}}|^2 \sum_{\boldsymbol{G}} \delta(\boldsymbol{q} - \boldsymbol{G}). \tag{10.33}$$

The nuclear and magnetic signals at the Bragg peaks can be distinguished because they have different temperature dependencies. The only source of temperature dependence in the former are the Debye–Waller factors, which vary slowly with temperature (see figure 10.3). In contrast, the magnetic signal vanishes abruptly together with the disappearance of the magnetization at the Curie temperature. For an anti-ferromagnet or a system with more complex static magnetic order, the average magnetization has a different periodicity than the lattice and a different set of Bragg peaks appears that is only due to magnetism.

Two routes are typically followed in order to describe the behavior of real materials. The first route assumes 'localized magnetism'. In this view, each magnetic ion is considered as an isolated entity carrying a single spin degree of freedom which results from the combination of the spins and angular momenta of its unpaired electrons. The magnetization density of the ion is represented by a scalar cloud of magnetization shaped by the wave functions of the unpaired electrons and a single spin operator without spatial structure. In the Born–Hoppenheimer approximation, the magnetization cloud moves instantly with the ion, such that the dynamics associated with this motion is similar to the ionic dynamics studied in nuclear scattering. The magnetic contribution to the scattering cross section involve temporal correlations between the localized spins. Such a view is justified for length scales that are large compared with the ionic radius: if one looks at a magnetic ion from sufficiently far away, its magnetic field can indeed be approximated by the field of a single dipole. We discuss this approach further in section 10.3.2. The second route aims to describe 'itinerant magnetism' in systems composed of delocalized electrons. The spins and angular-momentum densities must be expressed in terms of Bloch waves. The convenient limiting case here is the homogeneous electron gas which is sketched in section 10.3.3.

10.3.2 Scattering by localized spins and magnons

As a warmup, consider the local spin density of the electrons, equation (2.47), expressed in an arbitrary basis $\varphi_\alpha(\boldsymbol{r})$ with creation operators $c^\dagger_{\alpha\sigma}$ (see also equation (2.91)):

$$\boldsymbol{S}(\boldsymbol{r}) = \frac{\hbar}{2}\sum_{\alpha\beta\sigma\sigma'} \boldsymbol{\tau}_{\sigma\sigma'}\varphi^*_\alpha(\boldsymbol{r})\varphi_\beta(\boldsymbol{r})c^\dagger_{\alpha\sigma}c_{\beta\sigma'}. \tag{10.34}$$

Now assume that each basis function is localized on a given ion sitting at position $\boldsymbol{r}_\ell$, make the replacements $\varphi_\alpha(\boldsymbol{r}) \to \varphi_\alpha(\boldsymbol{r} - \boldsymbol{r}_\ell)$ and $c^\dagger_{\alpha\sigma} \to c^\dagger_{\ell\alpha\sigma}$, and calculate the Fourier transform $\boldsymbol{S}(\boldsymbol{q})$ while neglecting the overlap of basis functions sitting on different ions:

$$\int d\boldsymbol{r}\ \varphi^*_\alpha(\boldsymbol{r} - \boldsymbol{r}_\ell)\varphi_\beta(\boldsymbol{r} - \boldsymbol{r}_{\ell'})e^{-i\boldsymbol{q}\cdot\boldsymbol{r}} \approx \delta_{\ell\ell'}e^{-i\boldsymbol{q}\cdot\boldsymbol{r}_\ell}\int d\boldsymbol{r}\ \varphi^*_\alpha(\boldsymbol{r})\varphi_\beta(\boldsymbol{r})e^{-i\boldsymbol{q}\cdot\boldsymbol{r}}.$$

The result is

$$\begin{aligned}\boldsymbol{S}(\boldsymbol{q}) &\approx \sum_\ell e^{-i\boldsymbol{q}\cdot\boldsymbol{r}_\ell}\int d\boldsymbol{r}\ e^{-i\boldsymbol{q}\cdot\boldsymbol{r}}\underbrace{\sum_{\alpha\beta}\varphi^*_\alpha(\boldsymbol{r})\varphi_\beta(\boldsymbol{r})\frac{\hbar}{2}\sum_{\sigma\sigma'}\boldsymbol{\tau}_{\sigma\sigma'}c^\dagger_{\ell\alpha\sigma}c_{\ell\beta\sigma'}}_{\approx|\varphi_\ell(\boldsymbol{r})|^2\boldsymbol{S}_\ell} \\ &\approx \sum_\ell \mathscr{F}_\ell(\boldsymbol{q})e^{-i\boldsymbol{q}\cdot\boldsymbol{r}_\ell}\boldsymbol{S}_\ell,\end{aligned} \tag{10.35}$$

where $\mathscr{F}_\ell(\boldsymbol{q}) = \int d\boldsymbol{r}\ |\varphi_\ell(\boldsymbol{r})|^2e^{-i\boldsymbol{q}\cdot\boldsymbol{r}}$ and $|\gamma r_e\mathscr{F}_\ell(\boldsymbol{q})|^2$ is the *magnetic form factor of the ion.* The possibility of replacing the spin density operator on a given ion by the product of a scalar magnetization cloud $|\varphi_\ell(\boldsymbol{r})|^2$ and a spin operator $\boldsymbol{S}_\ell$ free of any spatial structure is the essence of localized magnetism. It can be justified on length scales large compared with the ion, i.e. when $1/q$ is much larger than the spatial extent of the wave functions occupied by unpaired electrons. In this regime, the ion behaves as a single magnetic dipole while multipole corrections only become visible at short distances within the ion.

Equation (10.35) is not complete because it only gives the electronic spin, while the neutrons scatter off the *total* magnetic moment of the ion. In order to represent the operator $\boldsymbol{\mathcal{M}}(\boldsymbol{q})$, the spin $\boldsymbol{S}_\ell$ in equation (10.35) must be replaced by some effective angular momentum whose size depends on the total spin and angular momenta. Which size to use depends on a balance between different effects: Coulomb repulsion leading to Hund's coupling, spin–orbit coupling, crystal field. For isolated atoms, there is no crystal field and the total angular momentum $\boldsymbol{J} = \boldsymbol{S} + \boldsymbol{L}$ is conserved. The total magnetic moment, however, is not conserved:

$\boldsymbol{m}_J = g_e\mu_\mathrm{B}\boldsymbol{S} - \mu_\mathrm{B}\boldsymbol{L} = -\mu_\mathrm{B}[\boldsymbol{J} - (g_e + 1)\boldsymbol{S}]$. This means that the moment $\boldsymbol{m}_J$ precesses around the conserved vector $\boldsymbol{J}$ such that, on average, the relevant effective moment is the projection of $\boldsymbol{m}_J$ along $\boldsymbol{J}$:

$$\frac{(\boldsymbol{m}_J \cdot \boldsymbol{J})\boldsymbol{J}}{\boldsymbol{J} \cdot \boldsymbol{J}} \equiv g_J\mu_\mathrm{B}\boldsymbol{J},$$

$$g_J = -1 + \frac{g_e + 1}{2}\frac{J(J + 1) + S(S + 1) - L(L + 1)}{J(J + 1)}.$$

To obtain this expression, we have written $2\boldsymbol{S} \cdot \boldsymbol{J}$ as $\boldsymbol{J} \cdot \boldsymbol{J} + \boldsymbol{S} \cdot \boldsymbol{S} - (\boldsymbol{J} - \boldsymbol{S})^2 = J(J + 1) + S(S + 1) - L(L + 1)$. In a crystal, one has to compare the crystal field with the spin–orbit interaction. If the crystal field is larger, like in $3d$ transition metals, the angular-momentum degeneracies are lifted by the breaking of spherical symmetry and the ground state is usually non degenerate with angular momentum $L = 0$. In this case, the ion magnetic moment is dominated by the spin, with a small orbital correction due to spin–orbit. The value of the spin is fixed by the atomic Hund's rules if the Hund's coupling is stronger than the crystal field. In the opposite regime of a crystal field stronger than Hund's coupling, Hund's rules must be applied in the crystal-field multiplet rather than the atomic multiplet. If the spin–orbit interaction is larger than the crystal field, like for rare-earth ions, one must start with the atomic multiplet of total angular momentum $\boldsymbol{J}$ and degeneracy $2J+1$ and study how the crystal field lifts this degeneracy. For the rest of this section, we will assume that equation (10.35) is valid with $\boldsymbol{S}_\ell$ understood as an effective spin multiplied by an effective g-factor, such that the operator probed by the neutrons is assumed to take the form

$$\mathcal{M}(\boldsymbol{q}) = \frac{1}{\hbar}\sum_\ell g_\ell \mathscr{F}_\ell(\boldsymbol{q})e^{-i\boldsymbol{q}\cdot\boldsymbol{r}_\ell}\boldsymbol{S}_\ell. \tag{10.36}$$

Upon evaluating the susceptibility (10.30), we will furthermore rewrite the ionic positions as $\boldsymbol{r}_\ell = \boldsymbol{R}_n + \boldsymbol{\tau}_\nu + \boldsymbol{u}_\nu(\boldsymbol{R}_n)$ like in section 10.2.3, use equation (10.15), and neglect any possible correlation between the displacement of an ion and its spin:

$$\begin{aligned}\chi_{\mathcal{M}\mathcal{M}}(\boldsymbol{q}, \tau) = -\frac{1}{\hbar^2}\sum_{\nu\mu}&\left[g_\nu\mathscr{F}_\nu(\boldsymbol{q})e^{-i\boldsymbol{q}\cdot\boldsymbol{\tau}_\nu}e^{-\frac{q^2}{2}\langle(u_\nu^\parallel)^2\rangle}\right]\left[g_\mu\mathscr{F}_\mu(\boldsymbol{q})e^{-i\boldsymbol{q}\cdot\boldsymbol{\tau}_\mu}e^{-\frac{q^2}{2}\langle(u_\mu^\parallel)^2\rangle}\right]^*\\ &\times\sum_{nm}e^{-i\boldsymbol{q}\cdot(\boldsymbol{R}_n-\boldsymbol{R}_m)}e^{q^2\langle T_\tau u_\nu^\parallel(\boldsymbol{R}_n,\,\tau)u_\mu^\parallel(\boldsymbol{R}_m,\,0)\rangle}\langle T_\tau\boldsymbol{S}_{n\nu}^\perp(\tau)\cdot\boldsymbol{S}_{m\mu}^\perp(0)\rangle.\end{aligned} \tag{10.37}$$

Note that the indices ν and μ in this expression refer to different magnetic ions in the unit cell, unlike in equation (10.30) where the same indices refer to the Cartesian coordinates. Based on equation (10.37), we can distinguish four types of processes, depending on whether the scattering is elastic or inelastic in the phonon and spin sectors, respectively.

The purely elastic signal is obtained by replacing the displacement correlation function by its zero-phonon value, $e^{q^2\langle T_\tau u_\nu^{\|}(\boldsymbol{R}_n,\,\tau)u_\mu^{\|}(\boldsymbol{R}_m,\,0)\rangle} \to 1$—see equation (10.18)—and the spin–spin correlation function by its long-time uncorrelated value, $\langle T_\tau \boldsymbol{S}_{n\nu}^{\perp}(\tau)\cdot\boldsymbol{S}_{m\mu}^{\perp}(0)\rangle \to \langle \boldsymbol{S}_{n\nu}^{\perp}\rangle \cdot \langle \boldsymbol{S}_{m\mu}^{\perp}\rangle$[12]. The result is a cross section of the form (10.32) with

$$\langle \mathcal{M}(\boldsymbol{q})\rangle = \frac{1}{\hbar}\sum_{\nu} g_\nu \mathscr{F}_\nu(\boldsymbol{q})\langle \boldsymbol{S}_\nu^{\perp}(\boldsymbol{q})\rangle e^{-i\boldsymbol{q}\cdot\boldsymbol{\tau}_\nu} e^{-\frac{q^2}{2}\langle (u_\nu^{\|})^2\rangle}. \tag{10.38}$$

If the spins are ordered and present the same periodicity as the lattice, $\langle \boldsymbol{S}_{n\nu}^{\perp}\rangle \equiv \langle \boldsymbol{S}_{\nu}^{\perp}\rangle$, this scattering is similar to the nuclear coherent elastic scattering given by the first term in equation (10.19), with the quantity $\gamma g_\nu r_e \mathscr{F}_\nu(\boldsymbol{q})\langle \boldsymbol{S}_\nu^{\perp}\rangle/\hbar$ playing the role of the average scattering length $\bar{b}_\nu$. The former length depends on $\boldsymbol{q}$ because the magnetic interaction is long ranged, while the latter does not because the Fermi pseudo-potential (10.2) is short ranged. The two contributions can be distinguished by their different temperature and field dependencies.

The so-called magneto-vibrational signal, which is inelastic in the phonon sector and elastic in the spin sector, obtains by the replacement $\langle T_\tau \boldsymbol{S}_{n\nu}^{\perp}(\tau)\cdot\boldsymbol{S}_{m\mu}^{\perp}(0)\rangle \to \langle \boldsymbol{S}_{\nu}^{\perp}\rangle \cdot \langle \boldsymbol{S}_{\mu}^{\perp}\rangle$ and by taking the second and subsequent terms in equation (10.18). Here we have assumed for simplicity that the spins have the same periodicity as the lattice. The resulting one-phonon cross section again has the same form as the nuclear one, equation (10.20), with $\bar{b}_\nu$ replaced by $\gamma g_\nu r_e \mathscr{F}_\nu(\boldsymbol{q})\langle \boldsymbol{S}_\nu^{\perp}\rangle/\hbar$. Thus magnetic scattering contributes to the phonon emission and absorption peaks, however with a specific intensity that can be tuned with a magnetic field and vanishes above the Curie temperature, where $\langle \boldsymbol{S}_\nu\rangle = 0$. For instance, in a monoatomic ferromagnet fully polarized in the z direction by an applied field, we have $\langle S^x\rangle = \langle S^y\rangle = 0$, such that the magnetic cross section is given by equation (10.20) with the replacement

$$\begin{aligned}|\bar{b}|^2 &\to |\gamma g r_e \mathscr{F}(\boldsymbol{q})|^2 |\hat{\boldsymbol{q}} \times \langle \boldsymbol{S}(\boldsymbol{q})\rangle \times \hat{\boldsymbol{q}}|^2/\hbar^2\\ &= |\gamma g r_e \mathscr{F}(\boldsymbol{q})|^2 (1 - \hat{q}_z^2)\langle S_z\rangle^2/\hbar^2.\end{aligned}$$

This scattering channel can therefore be closed by polarizing the ferromagnet along the direction of $\boldsymbol{q}$ such that $\hat{q}_z = 1$.

A slightly different situation of magneto-vibrational scattering occurs in a paramagnet. In a paramagnet, the spin operators commute with the Hamiltonian; they are constants of the motion and thus independent of time. The spin correlation function in equation (10.37) is also independent of time and therefore leads to purely elastic scattering in the spin sector. Furthermore, the different spin components are uncorrelated and one gets

[12] Note that the latter replacement is incorrect for a paramagnet, where there is no correlation between spins on different ions. We return to the case of a paramagnet below.

$$\langle T_\tau \boldsymbol{S}_{n\nu}^{\perp}(\tau)\cdot \boldsymbol{S}_{m\mu}^{\perp}(0)\rangle/\hbar^2 = \sum_{ij}\left(\delta_{ij}-\hat{q}_i\hat{q}_j\right)\langle S_{n\nu}^i S_{m\mu}^j\rangle/\hbar^2$$
$$= \delta_{nm}\delta_{\nu\mu}\sum_i\left(1-\hat{q}_i^2\right)\underbrace{\langle S_{n\nu}^i S_{n\nu}^i\rangle}_{\frac{1}{3}\langle S_\nu^2\rangle}/\hbar^2 = \delta_{nm}\delta_{\nu\mu}\frac{2}{3}S_\nu(S_\nu+1).$$

The absence of magnetic correlations between different ions in a paramagnet leads to purely incoherent magnetic scattering in the phonon sector. There is an elastic term equivalent to the second term in equation (10.19) where $\overline{b_\nu^2}-\bar{b}_\nu^2$ is replaced by $|\gamma g_\nu r_e \mathscr{F}_\nu(\boldsymbol{q})|^2\frac{2}{3}S_\nu(S_\nu+1)$. There is also an incoherent inelastic term equivalent to equation (10.21) with the same replacement. Under an applied field, long-range correlations between different ions appear, because $\langle S_{n\nu}^z S_{m\nu}^z\rangle = \langle S_\nu^z\rangle^2$, and consequently there appears coherent elastic scattering at the Bragg peaks and inelastic scattering at the phonon peaks.

The most interesting signal in magnetic systems is the one that is elastic in the phonon sector and inelastic in the spin sector. Apart from the Debye–Waller and magnetic form factors, this signal allows one to probe the genuine spin–spin correlations according to:

$$\boxed{\begin{aligned}\chi_{\mathcal{MM}}(\boldsymbol{q},\tau) = \sum_{\nu\mu}&\left[g_\nu\mathscr{F}_\nu(\boldsymbol{q})e^{-i\boldsymbol{q}\cdot\boldsymbol{\tau}_\nu}e^{-\frac{q^2}{2}\langle(u_\nu^{\parallel})^2\rangle}\right]\left[g_\mu\mathscr{F}_\mu(\boldsymbol{q})e^{-i\boldsymbol{q}\cdot\boldsymbol{\tau}_\mu}e^{-\frac{q^2}{2}\langle(u_\mu^{\parallel})^2\rangle}\right]^*\\ &\times\sum_{ij}\left(\delta_{ij}-\hat{q}_i\hat{q}_j\right)\left(-\frac{1}{\hbar^2}\right)\langle T_\tau S_\nu^i(\boldsymbol{q},\tau)S_\mu^j(-\boldsymbol{q},0)\rangle.\end{aligned}} \tag{10.39}$$

In section 2.6.2, we have recalled the theory of spin waves for localized ferromagnets. Based on this theory, it is possible to calculate the cross section for inelastic scattering of the neutrons on these spin waves. All what we have to do is to express the operator $\boldsymbol{S}(\boldsymbol{q})$ in terms of the Holstein–Primakoff bosons, and then evaluate the susceptibility (10.39), considering that the bosons are independent and that, therefore, the only non-vanishing correlator is $-\langle T_\tau b_{\boldsymbol{q}}(\tau)b_{\boldsymbol{q}}^\dagger(0)\rangle = \mathscr{C}_{b_{\boldsymbol{q}}b_{\boldsymbol{q}}^\dagger}(\tau)$. Restricting to a monoatomic system for simplicity, we find the inelastic differential cross section (see Doc A10.5):

$$\left.\frac{d^2\sigma_{\mathrm{m}}}{d\Omega d\varepsilon}\right)_{1-\mathrm{ma}}^{\mathrm{FM,\ inel}} = \mathscr{N}\frac{k'}{k}|\gamma g r_e \mathscr{F}(\boldsymbol{q})|^2 e^{-q^2\langle(u^{\parallel})^2\rangle}\frac{S}{2}\left(1+\hat{q}_z^2\right)$$
$$\times[1+b(\varepsilon)]\left[\delta(\varepsilon-\hbar\omega_{\boldsymbol{q}})-\delta(\varepsilon+\hbar\omega_{-\boldsymbol{q}})\right]. \tag{10.40}$$

This is a 'one-magnon' cross section, whose meaning is analogous to the coherent nuclear one-phonon cross section in equation (10.20). The linear spin-wave theory cannot account for multi-magnon processes. Note that the minus sign in front of the second delta function (magnon absorption and energy gain by the neutron) is compensated by the negative sign of $1+b(\varepsilon)$ at negative energies. Equation (10.40) applies to a mono-domain ferromagnet with magnetization along the z axis, taken as

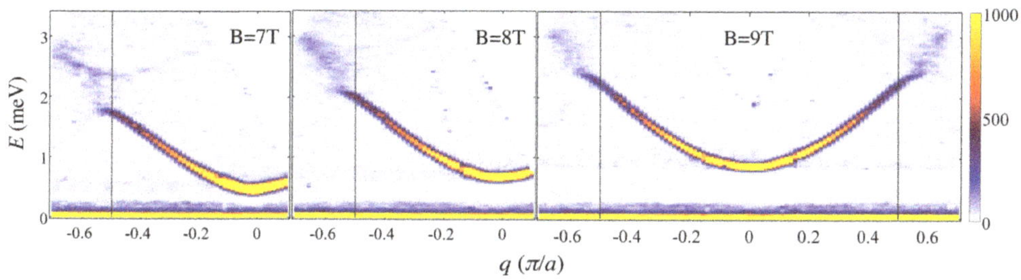

Figure 10.6. High-resolution neutron scattering measurement showing the field dependence of the magnon mode in $CoNb_2O_6$. The magnon disappears near 2 meV as it interacts with the two-magnon continuum. Reprinted figure with permission from Robinson *et al* 2014 *Phys. Rev.* **B 90** 174406. Copyright (2014) by the American Physical Society.

the spin quantization axis. Thus q_z is to be understood as the projection of $\boldsymbol{q}$ along the direction of the average magnetization. This can be used to disentangle the magnetic scattering from the nuclear one: if a magnetic field is applied, which is sufficiently strong to polarize all domains in the direction parallel or normal to $\boldsymbol{q}$, then $\hat{q}_z = 1$ or $\hat{q}_z = 0$, respectively. Equation (10.40) shows that there is a factor two difference in the magnetic scattering between the two field orientations, while the nuclear scattering is not sensitive to the field and remains unchanged.

Figure 10.6 shows high-resolution neutron scattering data for the quasi-one dimensional ferromagnet $CoNb_2O_6$. At sufficiently high applied field, this system is a transverse-field Ising chain, i.e. the leading term in the Hamiltonian is $\sum_\ell (JS_\ell^z S_{\ell+1}^z + h_x S_\ell^x)$, $J < 0$, with a magnon dispersion given by $[J^2/4 + h_x^2 + h_x J\cos(qa)]^{1/2}$. The data show this magnon and its variation with increasing field. Remarkably, the magnon broadens dramatically around 2 meV and reappears at the zone boundary ($q = \pm\pi/a$, outside the figure). As this broadening occurs away from $q = \pi/(2a)$, it cannot be ascribed to a doubling of the unit cell due for instance to dimerization. The effect was ascribed to magnon decay into the two-magnon continuum, based on calculations involving interaction terms beyond the linear theory.

In localized anti-ferromagnets, the situation is slightly more complicated because the bosonic modes that diagonalize the Hamiltonian mix spin-lowering excitations in one sublattice with spin-raising excitations in the other sublattice. In this case, we find the inelastic one-magnon cross section (see Doc A10.5):

$$\left.\frac{d^2\sigma_\mathrm{m}}{d\Omega d\varepsilon}\right)_{1-\mathrm{ma}}^{\mathrm{AF,\ inel}} = \mathcal{N}\frac{k'}{k}\,|\gamma g r_e \mathscr{F}(\boldsymbol{q})|^2 e^{-q^2\langle(u^{\parallel})^2\rangle}\frac{S}{2}\left(1 + \hat{q}_z^2\right)\Big[(u_q + v_q)^2 + (u_q - v_q)^2 b^2\Big][1 + b(\varepsilon)]\Big[\delta(\varepsilon - \hbar\omega_q) - \delta(\varepsilon + \hbar\omega_{-q})\Big], \quad (10.41)$$

where $b = \sin\vartheta$ is proportional to the applied field (equation (2.85)). The intensity of the magnon peaks is now modulated by the amplitudes u_q and v_q of the modes, which are given in Doc A2.18. Using the expressions and notations of Doc A2.18, we find

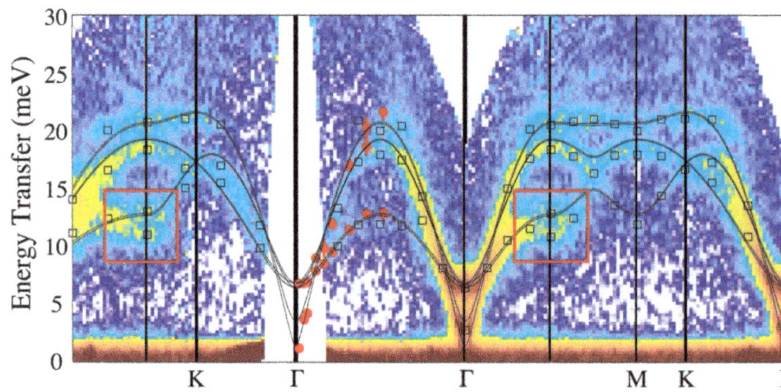

Figure 10.7. Neutron scattering data for the two-dimensional anti-ferromagnet $LuMnO_3$. The symbols show peak positions and the lines are a fit based on linear spin-wave theory. The peaks are broad in the red squares, indicating magnon decay. Reprinted figure with permission from Oh *et al* 2013 *Phys. Rev. Lett.* **111** 257202. Copyright (2013) by the American Physical Society.

$$(u_q \pm v_q)^2 = \frac{J_1(\mathbf{0})S}{\hbar\omega_q}\Big[\big|1 + b^2 j_1(\boldsymbol{q}) + j_2(\boldsymbol{q})\big| \mp (1 - b^2) j_1(\boldsymbol{q})\Big].$$

It follows that, in the absence of an applied field, the peak intensity takes the explicit form

$$(u_q + v_q)^2 = \frac{J_1(\mathbf{0})S}{\hbar\omega_q}\left\{|1 + j_2(\boldsymbol{q})| - \frac{\sum_n J(\boldsymbol{R}_n + 2\boldsymbol{\tau})\cos[\boldsymbol{q}\cdot(\boldsymbol{R}_n + 2\boldsymbol{\tau})]}{\sum_n J(\boldsymbol{R}_n + 2\boldsymbol{\tau})}\right\}.$$

In a bipartite lattice, the vector $2\boldsymbol{\tau}$ is equal to one of the lattice vectors $\boldsymbol{R}_n$. On the points $\boldsymbol{G}$ of the nuclear reciprocal lattice, we therefore have $\boldsymbol{G}\cdot(\boldsymbol{R}_n + 2\boldsymbol{\tau}) = 2\pi$ and the intensity is minimal, even exactly zero if $j_2(\boldsymbol{G}) = 0$. On the contrary, on the points $\boldsymbol{G}'$ of the anti-ferromagnet superlattice, we have $\boldsymbol{G}'\cdot(\boldsymbol{R}_n + 2\boldsymbol{\tau}) = \pi$ and the intensity is maximal.

Figure 10.7 shows inelastic neutron scattering data for the triangular $S = 2$ anti-ferromagnet $LuMnO_3$. This material has six spins in the magnetic unit cell with noncollinear 120 degree arrangement, hence six magnon modes. The linear spin-wave theory gives a good account of the observed excitation spectrum, except near some momenta where the observed magnon peaks are broad. This is taken as evidence for magnon decay, a phenomenon that can be predicted by extending the spin-wave theory beyond linear order[13].

10.3.3 Scattering by itinerant electrons

For insulators and metals with weakly-dispersing bands, it appears meaningful to approximate the spin and orbital angular-momentum densities by localized effective spins, but the same approach is questionable for wide-band metals with delocalized

[13] Zhitomirsky M E and Chernyshev A L 2013 *Rev. Mod. Phys.* **85** 219.

electrons. For those, a better starting point is the homogeneous electron gas, which we consider here.

In the plane-wave basis, the spin-density operator is given by equation (2.47) and—in the absence of an applied magnetic field—the current operator has the form (2.50). We can deduce the expression of the operator $\boldsymbol{\mathcal{M}}^{\perp}(\boldsymbol{q})$ (see Doc A10.4):

$$\begin{aligned}\boldsymbol{\mathcal{M}}^{\perp}(\boldsymbol{q}) &= \frac{1}{\hbar}\left[\hat{\boldsymbol{q}} \times \boldsymbol{S}(\boldsymbol{q}) \times \hat{\boldsymbol{q}} + \frac{ie\hbar}{g_e \mu_B q}\hat{\boldsymbol{q}} \times \boldsymbol{j}(\boldsymbol{q})\right] \\ &= \sum_{\boldsymbol{k}\sigma_1\sigma_2}\left(\frac{1}{2}\hat{\boldsymbol{q}} \times \boldsymbol{\tau}_{\sigma_1\sigma_2} \times \hat{\boldsymbol{q}} + \delta_{\sigma_1\sigma_2}\frac{i}{q}\,\hat{\boldsymbol{q}} \times \boldsymbol{k}\right)c^{\dagger}_{\boldsymbol{k}\sigma_1}c_{\boldsymbol{k}+\boldsymbol{q}\sigma_2} \\ &\equiv \sum_{\boldsymbol{k}\sigma_1\sigma_2}\boldsymbol{m}_{\sigma_1\sigma_2}(\boldsymbol{k},\, \boldsymbol{q})c^{\dagger}_{\boldsymbol{k}\sigma_1}c_{\boldsymbol{k}+\boldsymbol{q}\sigma_2}.\end{aligned} \tag{10.42}$$

At the second line, we have used the approximation $g_e \approx -2$ for simplicity. This formula is expressed in the plane-wave basis, but is nevertheless general. A change of basis, like in Doc A2.7, Doc A3.4, or Bonus B8.1, allows one to retrieve the expression of the operator in all situations, in particular for Bloch electrons. The imaginary-time susceptibility giving the cross section in equation (10.31) may be represented by a bubble diagram as

$$\chi_{\mathcal{M}\mathcal{M}}(\boldsymbol{q}, \tau) = -\langle T_\tau \boldsymbol{\mathcal{M}}^{\perp}(\boldsymbol{q}, \tau) \cdot \boldsymbol{\mathcal{M}}^{\perp}(-\boldsymbol{q}, 0)\rangle = - \tag{10.43}$$

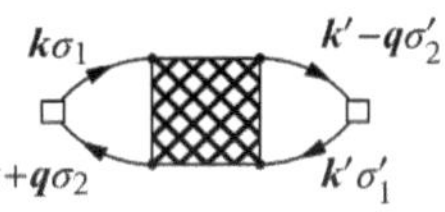

The left square in the diagram represents $\boldsymbol{m}_{\sigma_1\sigma_2}(\boldsymbol{k},\, \boldsymbol{q})$ as defined in equation (10.42), while the right square stands for $\boldsymbol{m}_{\sigma_1'\sigma_2'}(\boldsymbol{k}',\, -\boldsymbol{q})$ and a scalar product of these two vectors is implied. Apart from these vertices, the structure of the correlator is the same as that of χ_{nn} in equation (5.75) and χ_{jj} in equation (8.5). In the electron gas, all interactions preserve the spin such that the particle–hole pair created with spins σ_1 and σ_2 can only be annihilated with the same spins. In this case, the part of the diagram in-between the vertices is proportional to $\delta_{\sigma_1'\sigma_2}\delta_{\sigma_2'\sigma_1}$ and otherwise does not depend on the spins: as a result, the spin sum involves only the vertices and simplifies to:

$$\begin{aligned}\sum_{\sigma_1\sigma_2}\boldsymbol{m}_{\sigma_1\sigma_2}(\boldsymbol{k},\, \boldsymbol{q}) \cdot \boldsymbol{m}_{\sigma_2\sigma_1}(\boldsymbol{k}',\, -\boldsymbol{q}) &= \frac{1}{4}\sum_{\sigma_1\sigma_2}\sum_{\mu\nu}\left(\delta_{\mu\nu} - \hat{q}_\mu\hat{q}_\nu\right)\tau^{\mu}_{\sigma_1\sigma_2}\tau^{\nu}_{\sigma_2\sigma_1} \\ &\quad + \frac{1}{q^2}\sum_{\sigma_1\sigma_2}\delta_{\sigma_1\sigma_2}(\hat{\boldsymbol{q}} \times \boldsymbol{k}) \cdot (\hat{\boldsymbol{q}} \times \boldsymbol{k}') \\ &= 1 + \frac{2}{q^2}(\hat{\boldsymbol{q}} \times \boldsymbol{k}) \cdot (\hat{\boldsymbol{q}} \times \boldsymbol{k}').\end{aligned}$$

We have used equations (10.28) and (23). The '1' corresponds to the spin-only scattering and the second term is the orbital contribution. The cross terms vanish because they are proportional to the trace of a Pauli matrix. We see that the spin-only contribution to $\chi_{\mathcal{MM}}$ is just half the electronic density–density correlation function χ_{nn} defined in equation (5.75). Hence

$$\chi_{\mathcal{MM}}(\boldsymbol{q}, \tau) = \frac{1}{2}\chi_{nn}(\boldsymbol{q}, \tau) - \frac{2}{q^2}\sum_{kk'}(\hat{\boldsymbol{q}} \times \boldsymbol{k}) \cdot (\hat{\boldsymbol{q}} \times \boldsymbol{k}') \times \tag{10.44}$$

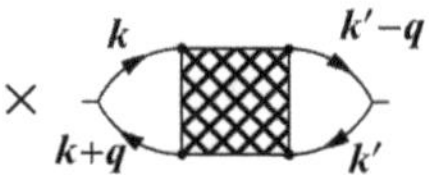

Equation (10.44) is the starting point for studying the scattering cross section of the electron gas, in particular how it is influenced by the Coulomb interaction and by the plasmon. As for the density–density correlation function, much of the physics is already present in the free-electron limit. For free electrons, the diagram reduces to a free bubble, $\boldsymbol{k}' = \boldsymbol{k} + \boldsymbol{q}$, and we simply get at $T = 0$ (see Doc A10.6):

$$\frac{d^2\sigma_{\mathrm{m}}^0}{d\Omega d\varepsilon} = \frac{k'}{k}(\gamma r_e)^2 N_0^{\mathrm{el}}(0)\frac{\theta(\varepsilon)}{16x}\left\{\theta(1 - a_-^2)\left[1 - a_-^2 + \frac{(1 - a_-^2)^2}{4x^2}\right]\right.$$
$$\left. - \theta(1 - a_+^2)\left[1 - a_+^2 + \frac{(1 - a_+^2)^2}{4x^2}\right]\right\}, \tag{10.45}$$

where $x = q/(2k_{\mathrm{F}})$, $y = \varepsilon/(4\varepsilon_{\mathrm{F}})$, and $a_\pm = (y \pm x^2)/x$. The superscript '0' reminds that this is a free-electron result at $T = 0$. There is no elastic scattering, because at $\varepsilon = 0$ we have $a_+^2 = a_-^2 = x^2$ and the cross section vanishes. Otherwise, the cross section is nonzero in a finite domain of momenta given by $k_{\mathrm{F}}(\sqrt{1 + \varepsilon/\varepsilon_{\mathrm{F}}} - 1) < q < k_{\mathrm{F}}(\sqrt{1 + \varepsilon/\varepsilon_{\mathrm{F}}} + 1)$, i.e. the limits of the particle–hole continuum discussed in section 5.1.4.5; see figure 10.8. For $\varepsilon < \varepsilon_{\mathrm{F}}$, the spin-only scattering diverges as $2y/x$ before being cut at the edge of the particle–hole continuum, while in the same regime the orbital currents dominate with a contribution of order $y(1/x^3 - 1/x - y^2/x^5)$. The situation is reversed at higher energy, where the orbital contribution dies as $y^{-3/2}$ while the spin-only term vanishes only as $y^{-1/2}$.

Appendix A10

Doc A10.1 f-Sum rule for the nuclear scattering function

With the help of equations (10.5*b*), (10.5*a*), and (3*b*), we can write

$$\int_{-\infty}^{\infty} d\varepsilon\, e^{-i\varepsilon t/\hbar} I_{\mu\nu}(\boldsymbol{q}, \varepsilon) = \langle n_\mu(\boldsymbol{q}, t) n_\nu(-\boldsymbol{q}, 0)\rangle.$$

At $t = 0$, this gives the sum rule already indicated in the text. Differentiating with respect to t and setting t to zero gives

$$\int_{-\infty}^{\infty} d\varepsilon\, \varepsilon\, I_{\mu\nu}(\boldsymbol{q}, \varepsilon) = \langle [n_\mu(\boldsymbol{q}), K] n_\nu(-\boldsymbol{q})\rangle,$$

where we have used $i\hbar\partial_t n_\mu(\boldsymbol{q}, t) = [n_\mu(\boldsymbol{q}, t), K]$. With this form of the sum rule, the right-hand side is not quite so easy to evaluate, so we use a detour. In section 3.4.5, we have derived a general expression for the first moment of the total spectral function. Applying equation (3.46), we can immediately write

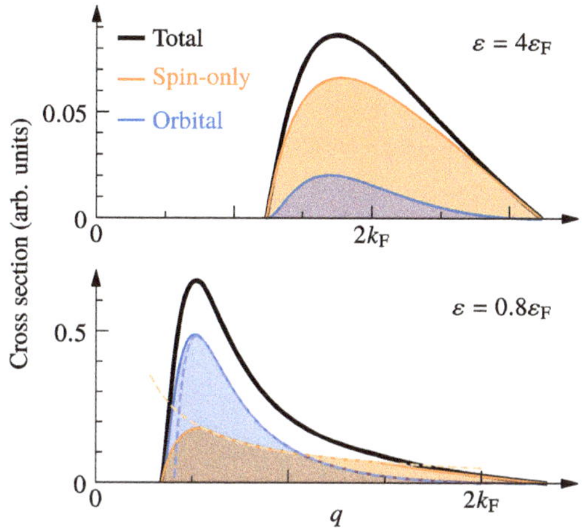

Figure 10.8. Magnetic neutron scattering cross section for the free-electron gas at low (bottom) and high (top) energy relative to the Fermi energy. The q interval where there is scattering corresponds to the particle–hole continuum. The orbital contribution in blue dominates the spin-only contribution in orange at low energy, while the situation is reversed at high energy. The dashed lines show the analytical dependencies given in the text.

$$\int_{-\infty}^{\infty} d\varepsilon\ \varepsilon\rho_{n_\mu(\boldsymbol{q})n_\nu(-\boldsymbol{q})}(\varepsilon) = \langle[[n_\mu(\boldsymbol{q}),\ K],\ n_\nu(-\boldsymbol{q})]\rangle.$$

Taking into account equation (3.32) as well as the detailed balance (3.34), this can be manipulated as follows:

$$\begin{aligned}
\langle[[n_\mu(\boldsymbol{q}),\ K],\ n_\nu(-\boldsymbol{q})]\rangle &= -\int_{-\infty}^{\infty} d\varepsilon\frac{\varepsilon}{b(-\varepsilon)}\rho^{>}_{n_\mu(\boldsymbol{q})n_\nu(-\boldsymbol{q})}(\varepsilon)\\
&= \int_{-\infty}^{\infty} d\varepsilon\ \varepsilon(1 - e^{-\beta\varepsilon})\rho^{>}_{n_\mu(\boldsymbol{q})n_\nu(-\boldsymbol{q})}(\varepsilon)\\
&= \int_{-\infty}^{\infty} d\varepsilon\ \varepsilon\ I_{\mu\nu}(\boldsymbol{q},\ \varepsilon) - \int_{-\infty}^{\infty} d\varepsilon\ \varepsilon\ I_{\nu\mu}(-\boldsymbol{q},\ -\varepsilon)\\
&= \int_{-\infty}^{\infty} d\varepsilon\ \varepsilon\ I_{\mu\nu}(\boldsymbol{q},\ \varepsilon) + \int_{-\infty}^{\infty} d\varepsilon\ \varepsilon\ I^{*}_{\mu\nu}(\boldsymbol{q},\ \varepsilon)\\
&= 2\int_{-\infty}^{\infty} d\varepsilon\ \varepsilon\ \mathrm{Re}\ I_{\mu\nu}(\boldsymbol{q},\ \varepsilon),
\end{aligned}$$

where we have used $I_{\nu\mu}(-\boldsymbol{q},\ \varepsilon) = I^{*}_{\mu\nu}(\boldsymbol{q},\ \varepsilon)$, which follows from the general expression of the greater spectral function, equation (3.12). In the alternate form

$$\mathrm{Re}\int_{-\infty}^{\infty} d\varepsilon\ \varepsilon\ I_{\mu\nu}(\boldsymbol{q},\ \varepsilon) = \frac{1}{2}\langle[[n_\mu(\boldsymbol{q}),\ K],\ n_\nu(-\boldsymbol{q})]\rangle,$$

the right-hand side is very easily evaluated because the commutator is a scalar. Usually, the interactions between nuclei as well as between nuclei and electrons do not depend on the nucleus velocity. As a result, the only term in K that does not commute with $\boldsymbol{r}_{\mu\ell}$ is the nucleus kinetic energy $\boldsymbol{p}_{\mu\ell}^2/(2M_\mu)$. Using the fact that, for any function $f(\boldsymbol{r})$, the commutator with the corresponding momentum is $[f(\boldsymbol{r}), \boldsymbol{p}] = i\hbar\nabla f(\boldsymbol{r})$, we obtain

$$[n_\mu(\boldsymbol{q}), K] = -\frac{\hbar^2 q^2}{2M_\mu} n_\mu(\boldsymbol{q}) + \frac{\hbar \boldsymbol{q}}{M_\mu} \sum_\ell e^{-i\boldsymbol{q}\cdot\boldsymbol{r}_{\mu\ell}} \boldsymbol{p}_{\mu\ell}.$$

The first term on the right commutes with $n_\nu(-\boldsymbol{q})$ and thus disappears from the sum rule. The second also commutes if $\nu \neq \mu$, but contributes a term if $\nu = \mu$, whence finally

$$[[n_\mu(\boldsymbol{q}), K], n_\nu(-\boldsymbol{q})] = \delta_{\mu\nu} \frac{\hbar^2 q^2}{M_\mu} N_\mu.$$

The exact same procedure with $n_\mu(\boldsymbol{q})$ and $n_\nu(-\boldsymbol{q})$ replaced by $e^{-i\boldsymbol{q}\cdot\boldsymbol{r}_{\mu\ell}}$ and $e^{i\boldsymbol{q}\cdot\boldsymbol{r}_{\mu\ell}}$, respectively, leads to the same result for the self intermediate scattering function:

$$\mathrm{Re} \int_{-\infty}^{\infty} d\varepsilon\, \varepsilon\, I_\mu^s(\boldsymbol{q}, \varepsilon) = \frac{1}{2} \sum_\ell \langle [[e^{-i\boldsymbol{q}\cdot\boldsymbol{r}_{\mu\ell}}, K], e^{i\boldsymbol{q}\cdot\boldsymbol{r}_{\mu\ell}}] \rangle = \frac{\hbar^2 q^2}{2M_\mu} N_\mu.$$

Upon summing over atomic species in equation (10.7), the final result is real and gives equation (10.13).

Doc A10.2 Debye–Waller factor and one-phonon processes

The expression of the harmonic displacements, equation (2.66), translates into the following:

$$\begin{aligned}\langle u_\nu^{\parallel}(\boldsymbol{R}_n, t) u_\mu^{\parallel}(\boldsymbol{R}_m, t')] \rangle = & \frac{1}{\mathcal{N}} \sum_{\boldsymbol{k}\lambda} \sum_{\boldsymbol{k}'\lambda'} \sqrt{\frac{\hbar}{2M_\nu \omega_{\boldsymbol{k}\lambda}}} \sqrt{\frac{\hbar}{2M_\mu \omega_{\boldsymbol{k}'\lambda'}}} [\boldsymbol{\epsilon}_{\boldsymbol{k}\lambda}^{\parallel}]_\nu [\boldsymbol{\epsilon}_{\boldsymbol{k}'\lambda'}^{\parallel}]_\mu \\ & \times \left(\langle b_{\boldsymbol{k}\lambda}(t) b_{-\boldsymbol{k}'\lambda'}^\dagger(t') \rangle + \langle b_{-\boldsymbol{k}\lambda}^\dagger(t) b_{\boldsymbol{k}'\lambda'}(t') \rangle \right) e^{i\boldsymbol{k}\cdot\boldsymbol{R}_n} e^{i\boldsymbol{k}'\cdot\boldsymbol{R}_m},\end{aligned}$$

where we have discarded terms of the kind $\langle bb \rangle$ and $\langle b^\dagger b^\dagger \rangle$, which vanish because of phonon-number conservation. Comparing with equations (3.6) and (3.7), we recognize the greater and lesser correlation functions of the single phonon operators, which are diagonal because the phonons are free; see Doc A5.10 and equation (5.51):

$$\begin{aligned}\langle b_{\boldsymbol{k}\lambda}(t) b_{-\boldsymbol{k}'\lambda'}^\dagger(t') \rangle &= \delta_{\boldsymbol{k}, -\boldsymbol{k}'} \delta_{\lambda\lambda'} i\hbar C_{b_{\boldsymbol{k}\lambda} b_{\boldsymbol{k}\lambda}^\dagger}^{>}(t - t') \\ \langle b_{-\boldsymbol{k}\lambda}^\dagger(t) b_{\boldsymbol{k}'\lambda'}(t') \rangle &= \delta_{\boldsymbol{k}, -\boldsymbol{k}'} \delta_{\lambda\lambda'} i\hbar C_{b_{\boldsymbol{k}\lambda} b_{\boldsymbol{k}\lambda}^\dagger}^{<}(t' - t).\end{aligned}$$

Successive use of equations (2*a*), (3.20) and (3.21), and (3.32) leads to:

$$i\hbar C^{>}_{b_{k\lambda}b^{\dagger}_{k\lambda}}(t-t') + i\hbar C^{<}_{b_{k\lambda}b^{\dagger}_{k\lambda}}(t'-t) = \int_{-\infty}^{\infty} d\varepsilon\, \rho_{b_{k\lambda}b^{\dagger}_{k\lambda}}(\varepsilon)[b(\varepsilon)e^{i\varepsilon(t-t')/\hbar} - b(-\varepsilon)e^{i\varepsilon(t'-t)/\hbar}]$$
$$= b(\hbar\omega_{k\lambda})e^{i\omega_{k\lambda}(t-t')} - b(-\hbar\omega_{k\lambda})e^{i\omega_{k\lambda}(t'-t)},$$

where the last line holds for free phonons with $\rho_{b_{k\lambda}b^{\dagger}_{k\lambda}}(\varepsilon) = \delta(\varepsilon - \hbar\omega_{k\lambda})$. Setting $t' = t$, $\mu = \nu$, and $m = n$, we get equation (10.16) for the mean square displacement:

$$\langle[u^{\parallel}_{\nu}(\boldsymbol{R}_n, t)]^2\rangle \equiv \langle(u^{\parallel}_{\nu})^2\rangle = \frac{1}{\mathcal{N}}\sum_{k\lambda}\frac{\hbar}{2M_{\nu}\omega_{k\lambda}} \mid [\boldsymbol{\epsilon}^{\parallel}_{k\lambda}]_{\nu}\mid^2 \coth\left(\frac{\hbar\omega_{k\lambda}}{2k_{\mathrm{B}}T}\right).$$

Setting $t' = 0$ yields the correlator of the positions for different atoms, which enters in the expression of the intermediate scattering function (10.14*a*):

$$\langle u^{\parallel}_{\nu}(\boldsymbol{R}_n, t)u^{\parallel}_{\mu}(\boldsymbol{R}_m, 0)]\rangle = \frac{\hbar}{2\sqrt{M_{\nu}M_{\mu}}}\frac{1}{\mathcal{N}}\sum_{k\lambda}[\boldsymbol{\epsilon}^{\parallel}_{k\lambda}]_{\nu}[\boldsymbol{\epsilon}^{\parallel *}_{k\lambda}]_{\mu}$$
$$\times\left[\frac{b(\omega_{k\lambda})}{\omega_{k\lambda}}e^{i\omega_{k\lambda}t} - \frac{b(-\omega_{k\lambda})}{\omega_{k\lambda}}e^{-i\omega_{k\lambda}t}\right]e^{ik\cdot(\boldsymbol{R}_n-\boldsymbol{R}_m)},$$

while setting $t' = 0$, $\mu = \nu$, and $m = n$ yields the self-correlator entering the intermediate self scattering function (10.14*b*):

$$\langle u^{\parallel}_{\nu}(\boldsymbol{R}_n, t)u^{\parallel}_{\nu}(\boldsymbol{R}_n, 0)]\rangle = \frac{\hbar}{2M_{\nu}}\frac{1}{\mathcal{N}}\sum_{k\lambda}\mid[\boldsymbol{\epsilon}^{\parallel}_{k\lambda}]_{\nu}\mid^2\left[\frac{b(\omega_{k\lambda})}{\omega_{k\lambda}}e^{i\omega_{k\lambda}t} - \frac{b(-\omega_{k\lambda})}{\omega_{k\lambda}}e^{-i\omega_{k\lambda}t}\right].$$

The elastic coherent and incoherent scatterings do not depend on the above two time-dependent correlation functions, but only on the mean-square displacements. The coherent term involves the sum $\sum_{nm}e^{-i\boldsymbol{q}\cdot(\boldsymbol{R}_n-\boldsymbol{R}_m)} = \mathcal{N}^2\sum_{\boldsymbol{G}}\delta_{\boldsymbol{q}\boldsymbol{G}}$, which we rewrite using a continuous notation for $\boldsymbol{q}$ in line with the convention in this chapter: $\delta_{\boldsymbol{q}\boldsymbol{G}} = \delta(\boldsymbol{q}-\boldsymbol{G})\times(2\pi)^3/\mathcal{V}$. This leads to the first term in the square brackets in equation (10.19). The incoherent elastic contribution follows without difficulty and gives the second term in that equation. The one-phonon intermediate scattering function results by replacing $e^{q^2\langle u^{\parallel}_{\nu}(\boldsymbol{R}_n, t)u^{\parallel}_{\mu}(\boldsymbol{R}_m, 0)\rangle}$ in equation (10.15) by $q^2\langle u^{\parallel}_{\nu}(\boldsymbol{R}_n, t)u^{\parallel}_{\mu}(\boldsymbol{R}_m, 0)]\rangle$ and inserting in equation (10.14*a*). The time integrations yield delta functions at $\varepsilon = \pm\hbar\omega_{k\lambda}$ and the phonon wave vector $\boldsymbol{k}$ (not to be confused with the incident neutron wave vector also denoted $\boldsymbol{k}$) is constrained to be equal to $\boldsymbol{q}-\boldsymbol{G}$: this is the consequence of the sum $\sum_n e^{-i(\boldsymbol{q}-\boldsymbol{k})\cdot\boldsymbol{R}_n} = \mathcal{N}\sum_{\boldsymbol{G}}\delta_{\boldsymbol{q}-\boldsymbol{k},\,\boldsymbol{G}}$. Recall that $\boldsymbol{k}$ is in the first Brillouin zone, such that the constraint $\boldsymbol{k} = \boldsymbol{q}-\boldsymbol{G}$ selects a unique $\boldsymbol{G}$ vector (see remark before equation (2.52)); the $\boldsymbol{G}$ sum is therefore of order 1 and not of order $\mathcal{N}$. In order to avoid possible misunderstandings, we introduce the notation $\check{\boldsymbol{q}}$ for the irreducible value of $\boldsymbol{q}$ in the first Brillouin zone. Thus $\sum_n e^{-i(\boldsymbol{q}-\boldsymbol{k})\cdot\boldsymbol{R}_n} = \mathcal{N}\delta_{\boldsymbol{k}\check{\boldsymbol{q}}}$. Because $(\delta_{\boldsymbol{k}\check{\boldsymbol{q}}})^2 = \delta_{\boldsymbol{k}\check{\boldsymbol{q}}}$, the m sum simply contributes another factor $\mathcal{N}$. We get

$$I_{\nu\mu}^{1-\mathrm{ph}}(\boldsymbol{q},\,\varepsilon) = \mathcal{N}\sum_{\lambda}\left\{q\sqrt{\frac{\hbar}{2M_\nu}}e^{-i\boldsymbol{q}\cdot\boldsymbol{\tau}_\nu}e^{-\frac{q^2}{2}\langle(u_\nu^{\|})^2\rangle}[\boldsymbol{\epsilon}_{\check{\boldsymbol{q}}\lambda}^{\|}]_\nu\right\}$$
$$\times\left\{q\sqrt{\frac{\hbar}{2M_\mu}}e^{-i\boldsymbol{q}\cdot\boldsymbol{\tau}_\mu}e^{-\frac{q^2}{2}\langle(u_\mu^{\|})^2\rangle}[\boldsymbol{\epsilon}_{\check{\boldsymbol{q}}\lambda}^{\|}]_\mu\right\}^*\frac{1+b(\varepsilon)}{\omega_{\check{\boldsymbol{q}}\lambda}}\left[\delta(\varepsilon-\hbar\omega_{\check{\boldsymbol{q}}\lambda})-\delta(\varepsilon+\hbar\omega_{\check{\boldsymbol{q}}\lambda})\right],$$

where we have used the delta functions in order to replace $b(\pm\hbar\omega_{\check{\boldsymbol{q}}\lambda})$ by $b(-\varepsilon)$. Upon multiplying by $\bar{b}_\nu\bar{b}_\mu$ and summing over ν and μ as in equation (10.7), we get the coherent inelastic cross section given in equation (10.20). For the one-phonon intermediate self-scattering function, the n sum simply contributes a factor $\mathcal{N}$ without constraining momenta, yielding

$$I_\nu^{s,\,1-\mathrm{ph}}(\boldsymbol{q},\,\varepsilon) = \frac{\hbar^2q^2}{2M_\nu}e^{-q^2\langle(u_\nu^{\|})^2\rangle}\sum_{\boldsymbol{k}\lambda}|\,[\boldsymbol{\epsilon}_{\boldsymbol{k}\lambda}^{\|}]_\nu|^2\,\frac{1+b(\varepsilon)}{\varepsilon}[\delta(\varepsilon-\hbar\omega_{\boldsymbol{k}\lambda})+\delta(\varepsilon+\hbar\omega_{\boldsymbol{k}\lambda})].$$

We recognize the partial phonon DOS defined in equation (10.17), and thus obtain equation (10.21).

Doc A10.3 f-Sum rule for the one-phonon processes

The first moment of the scattering function for the one-phonon processes (the elastic contribution has no first moment) is easily calculated by noting that, for any ε_0,

$$\int_{-\infty}^{\infty}d\varepsilon\,\varepsilon\frac{1+b(\varepsilon)}{\varepsilon}[\delta(\varepsilon-\varepsilon_0)+\delta(\varepsilon+\varepsilon_0)] = 1.$$

Using this, we get

$$\int_{-\infty}^{\infty}d\varepsilon\,\varepsilon S_{1-\mathrm{ph}}^{\mathrm{coh}}(\boldsymbol{q},\,\varepsilon) = \mathcal{N}\sum_{\lambda}|F_\lambda(\boldsymbol{q})|^2\hbar\omega_{\check{\boldsymbol{q}}\lambda} = \mathcal{N}\sum_{\nu\mu}\frac{\hbar^2q^2\bar{b}_\nu\bar{b}_\mu}{2\sqrt{M_\nu M_\mu}}$$
$$\times\,e^{-i\boldsymbol{q}\cdot(\boldsymbol{\tau}_\nu-\boldsymbol{\tau}_\mu)}e^{-\frac{q^2}{2}[\langle(u_\nu^{\|})^2\rangle+\langle(u_\mu^{\|})^2\rangle]}\sum_{\lambda}[\boldsymbol{\epsilon}_{\check{\boldsymbol{q}}\lambda}^{\|}]_\nu[\boldsymbol{\epsilon}_{\check{\boldsymbol{q}}\lambda}^{\|*}]_\mu$$

$$\int_{-\infty}^{\infty}d\varepsilon\,\varepsilon S_{1-\mathrm{ph}}^{\mathrm{inc}}(\boldsymbol{q},\,\varepsilon) = \mathcal{N}\sum_{\nu}\left(\overline{b_\nu^2}-\bar{b}_\nu^2\right)\frac{\hbar^2q^2}{2M_\nu}e^{-q^2\langle(u_\nu^{\|})^2\rangle}\frac{1}{\mathcal{N}}\sum_{\boldsymbol{k}\lambda}|\,[\boldsymbol{\epsilon}_{\boldsymbol{k}\lambda}^{\|}]_\nu\,|^2.$$

The phonon polarization vectors form, for each wave vector $\boldsymbol{k}$, a complete orthonormal set in the dN_a-dimensional space of the vibrational modes, where d is the dimensionality and N_a is the number of atoms in the unit cell (see section 2.5.1). The orthonormality conditions have already been used in Doc A2.10 and read

$$\sum_{\nu i}[\boldsymbol{\epsilon}_{\boldsymbol{k}\lambda}^*]_{\nu i}[\boldsymbol{\epsilon}_{\boldsymbol{k}\lambda'}]_{\nu i} = \delta_{\lambda\lambda'},\qquad \sum_{\lambda}[\boldsymbol{\epsilon}_{\boldsymbol{k}\lambda}^*]_{\nu i}[\boldsymbol{\epsilon}_{\boldsymbol{k}\lambda}]_{\mu j} = \delta_{\nu\mu}\delta_{ij}.$$

With the help of these conditions, we can evaluate the expressions entering the first moment:

$$\sum_{\lambda}[\epsilon^{\parallel}_{\check{q}\lambda}]_{\nu}[\epsilon^{\parallel *}_{\check{q}\lambda}]_{\mu} = \sum_{\lambda}\sum_{ij}\hat{q}_i\hat{q}_j[\epsilon_{\check{q}\lambda}]_{\nu i}[\epsilon^{*}_{\check{q}\lambda}]_{\mu j} = \sum_{ij}\hat{q}_i\hat{q}_j\delta_{\nu\mu}\delta_{ij}$$
$$= \delta_{\nu\mu}\sum_{i}\hat{q}_i^2 = \delta_{\nu\mu}$$
$$\frac{1}{\mathcal{N}}\sum_{k\lambda}|[\epsilon^{\parallel}_{k\lambda}]_{\nu}|^2 = \frac{1}{\mathcal{N}}\sum_{k\lambda}[\epsilon^{\parallel *}_{k\lambda}]_{\nu}[\epsilon^{\parallel}_{k\lambda}]_{\nu} = \frac{1}{\mathcal{N}}\sum_{k\lambda}\sum_{ij}\hat{q}_i\hat{q}_j[\epsilon^{*}_{k\lambda}]_{\nu i}[\epsilon_{k\lambda}]_{\nu j}$$
$$= \frac{1}{\mathcal{N}}\sum_{k}\sum_{ij}\hat{q}_i\hat{q}_j\delta_{ij} = \frac{1}{\mathcal{N}}\sum_{k} = 1.$$

This leads us immediately to equation (10.22).

Doc A10.4 Matrix element for magnetic neutron scattering

The operator representing the neutron magnetic moment, whose strength is $\mu_n = -0.966 \times 10^{-26}\ \mathrm{JT}^{-1}$, is $\boldsymbol{\mu}_n = (1/2)g_n\mu_{\mathrm{N}}\boldsymbol{\tau}$, where $g_n = -3.83$ is the neutron Landé gyromagnetic factor, μ_{N} is the nuclear magneton, and $\boldsymbol{\tau}$ is the vector of Pauli matrices. This dipole moment produces a field $\boldsymbol{B}_n(\boldsymbol{r}) = \nabla \times \boldsymbol{A}_n(\boldsymbol{r})$, where

$$\boldsymbol{A}_n(\boldsymbol{r}) = \frac{\mu_0}{4\pi}\frac{\boldsymbol{\mu}_n \times \boldsymbol{r}}{r^3}$$

if the neutron sits at the origin of coordinates. It is a simple exercise to calculate the Fourier transform of this vector potential using spherical coordinates and to deduce the Fourier transform of the field:

$$\boldsymbol{A}_n(\boldsymbol{k}) = -i\mu_0\frac{\boldsymbol{\mu}_n \times \boldsymbol{k}}{k^2}, \qquad \boldsymbol{B}_n(\boldsymbol{k}) = i\boldsymbol{k} \times \boldsymbol{A}_n(\boldsymbol{k}) = \frac{g_n\mu_0\mu_{\mathrm{N}}}{2}\,\hat{\boldsymbol{k}} \times \boldsymbol{\tau} \times \hat{\boldsymbol{k}}.$$

The field $\boldsymbol{B}_n$ acts on the electron spins as an external magnetic field $\boldsymbol{H} = \boldsymbol{B}_n/\mu_0$. The associated perturbation is given by the Zeeman energy (6.13), where $\boldsymbol{M}(\boldsymbol{r}) = g_e\mu_{\mathrm{B}}\boldsymbol{S}(\boldsymbol{r})/\hbar$ is the magnetization with $g_e = -2.0023$ the electron gyromagnetic factor, μ_{B} the Bohr magneton, and $\boldsymbol{S}(\boldsymbol{r})$ the spin density operator given by equation (2.47). The neutron vector potential furthermore shifts the electron momenta by $-e\boldsymbol{A}_n$ and thus triggers an orbital response. Let us first consider the Zeeman term:

$$V_S = -\int d\boldsymbol{r}\ \boldsymbol{M}(\boldsymbol{r}) \cdot \boldsymbol{B}_n(\boldsymbol{r} - \boldsymbol{r}_n)$$
$$= -\frac{g_e\mu_{\mathrm{B}}g_n\mu_0\mu_{\mathrm{N}}}{2\hbar}\int d\boldsymbol{r}\ \boldsymbol{S}(\boldsymbol{r}) \cdot \frac{1}{\mathcal{V}}\sum_{\boldsymbol{k}}(\hat{\boldsymbol{k}} \times \boldsymbol{\tau} \times \hat{\boldsymbol{k}})e^{i\boldsymbol{k}\cdot(\boldsymbol{r}-\boldsymbol{r}_n)}$$
$$= -\frac{\pi g_e g_n}{2}\frac{\hbar r_e}{m_p}\frac{1}{\mathcal{V}}\sum_{\boldsymbol{k}''}\int d\boldsymbol{r}\ \boldsymbol{\tau} \cdot [\hat{\boldsymbol{k}}'' \times \boldsymbol{S}(\boldsymbol{r}) \times \hat{\boldsymbol{k}}'']e^{i\boldsymbol{k}''\cdot(\boldsymbol{r}-\boldsymbol{r}_n)}.$$

We have introduced the neutron position $\boldsymbol{r}_n$ and used the classical electron radius

$$r_e = \frac{\mu_0 e^2}{4\pi m} = \frac{m_p \mu_0 \mu_B \mu_N}{\pi \hbar^2}$$

in order to eliminate the Bohr and nuclear magnetons in favor of the proton mass m_p. At the last line, we have made use of the vector identity $\boldsymbol{S} \cdot (\hat{\boldsymbol{k}} \times \boldsymbol{\tau} \times \hat{\boldsymbol{k}}) = \boldsymbol{\tau} \cdot (\hat{\boldsymbol{k}} \times \boldsymbol{S} \times \hat{\boldsymbol{k}})$ that we can apply without concern because the neutron and electron spin operators commute. We have also renamed $\boldsymbol{k} \to \boldsymbol{k}''$ in order to reserve the symbols $\boldsymbol{k}$ and $\boldsymbol{k}'$ for the neutron incident and scattered wave vectors. The matrix element entering the cross section (10.1) in the Born approximation is then easily obtained:

$$\langle \boldsymbol{k}'\sigma' | V_S | \boldsymbol{k}\sigma \rangle = -\frac{\pi g_e g_n}{2(2\pi)^3} \frac{\hbar r_e}{m_p} \boldsymbol{\tau}_{\sigma'\sigma} \cdot [\hat{\boldsymbol{q}} \times \boldsymbol{S}(-\boldsymbol{q}) \times \hat{\boldsymbol{q}}].$$

We recall that, in the present context, the plane wave describing the neutron is normalized to the impulsion and reads $(2\pi)^{-3/2} e^{i\boldsymbol{k}\cdot\boldsymbol{r}_n}$. The integration over $\boldsymbol{r}_n$ then yields $\mathscr{V}\delta_{\boldsymbol{k}-\boldsymbol{k}', \boldsymbol{k}''} = \mathscr{V}\delta_{\boldsymbol{q}\boldsymbol{k}''}$, where $\boldsymbol{q}$ is the scattering vector defined in equation (1.2). The quantity $\boldsymbol{\tau}_{\sigma'\sigma} = \langle \sigma' | \boldsymbol{\tau} | \sigma \rangle$ relates the interaction matrix element to the spin states of the incident and scattered neutron. For an unpolarized neutron beam and a spin-insensitive detector, we will have to average the cross section over the two possible incident spin states with equal probability 1/2, and to sum over the two possible final states. The occurrence of $r_e \propto \mu_B$ in the matrix element explains why the scattering on nuclear spins is negligible: the latter has μ_N instead of μ_B and consequently the cross section is smaller by a factor $(\mu_N/\mu_B)^2 = (m/m_p)^2 \sim 10^{-7}$.

We now turn to the orbital term. The neutron vector potential couples to the electron motion in the same way as the photon field: the interaction energy is given by equation (6.17):

$$V_L = -e \int d\boldsymbol{r}\, \boldsymbol{A}_n(\boldsymbol{r} - \boldsymbol{r}_n) \cdot \boldsymbol{j}(\boldsymbol{r}).$$

The matrix element follows by introducing the Fourier transform of the vector potential and performing the integration over the neutron coordinate:

$$\begin{aligned}
\langle \boldsymbol{k}'\sigma' | V_L | \boldsymbol{k}\sigma \rangle &= \frac{ie\mu_0 g_n \mu_N}{2(2\pi)^3} \frac{\boldsymbol{\tau}_{\sigma'\sigma} \times \boldsymbol{q}}{q^2} \cdot \boldsymbol{j}(-\boldsymbol{q}) \\
&= \frac{ie\mu_0 g_n \mu_N}{2(2\pi)^3} \frac{1}{q} \boldsymbol{\tau}_{\sigma'\sigma} \cdot [\hat{\boldsymbol{q}} \times \boldsymbol{j}(-\boldsymbol{q})] \\
&= -\frac{\pi g_e g_n}{2(2\pi)^3} \frac{\hbar r_e}{m_p} \boldsymbol{\tau}_{\sigma'\sigma} \cdot [\hat{\boldsymbol{q}} \times \boldsymbol{L}(-\boldsymbol{q}) \times \hat{\boldsymbol{q}}].
\end{aligned}$$

On the second line, we have used the vector identity $(\boldsymbol{\tau} \times \boldsymbol{q}) \cdot \boldsymbol{j} = \boldsymbol{\tau} \cdot (\boldsymbol{q} \times \boldsymbol{j})$. The manipulation on the third line is just a rewriting of the matrix element in order to recover the same form as the Zeeman term. For this, we must introduce an angular-momentum density, or 'orbital spin density' $\boldsymbol{L}(\boldsymbol{q})$, which satisfies the relation:

$$-\boldsymbol{L}(-\boldsymbol{q}) \times \hat{\boldsymbol{q}} = \frac{ie\hbar}{g_e\mu_\mathrm{B}}\frac{1}{q}\boldsymbol{j}(-\boldsymbol{q}) \qquad \text{or} \qquad \boldsymbol{L}(\boldsymbol{q}) \times \hat{\boldsymbol{q}} = \frac{ie\hbar}{g_e\mu_\mathrm{B}}\frac{1}{q}\boldsymbol{j}(\boldsymbol{q}).$$

The interpretation goes as follows. Let $\boldsymbol{B}_L$ be the orbital magnetic field generated by the current $\boldsymbol{j}$: then $\nabla \times \boldsymbol{B}_L(\boldsymbol{r}) = \mu_0 e\boldsymbol{j}(\boldsymbol{r})$, which in momentum space reads $i\boldsymbol{q} \times \boldsymbol{B}_L(\boldsymbol{q}) = \mu_0 e\boldsymbol{j}(\boldsymbol{q})$. Multiplying both sides by i/q, we get $\boldsymbol{B}_L(\boldsymbol{q}) \times \hat{\boldsymbol{q}} = (ie\mu_0/q)\boldsymbol{j}(\boldsymbol{q})$. The magnetization corresponding to $\boldsymbol{B}_L$ is $\boldsymbol{B}_L/\mu_0$; it can be parametrized by an effective electronic 'spin' according to $\boldsymbol{B}_L/\mu_0 = g_e\mu_\mathrm{B}\boldsymbol{L}/\hbar$. This leads us to the above expression.

The quantity $\hat{\boldsymbol{q}} \times \boldsymbol{S} \times \hat{\boldsymbol{q}} = \boldsymbol{S} - (\boldsymbol{S} \cdot \hat{\boldsymbol{q}})\hat{\boldsymbol{q}}$ gives the component of $\boldsymbol{S}$ in the plane normal to the direction of $\boldsymbol{q}$. Let us therefore define the dimensionless operator

$$\boldsymbol{\mathcal{M}}(\boldsymbol{q}) = \frac{1}{\hbar}\,[\boldsymbol{S}(\boldsymbol{q}) + \boldsymbol{L}(\boldsymbol{q})]$$

as well as its projection $\boldsymbol{\mathcal{M}}^{\perp}(\boldsymbol{q}) = \hat{\boldsymbol{q}} \times \boldsymbol{\mathcal{M}}(\boldsymbol{q}) \times \hat{\boldsymbol{q}}$ and thus express the matrix element for magnetic neutron scattering as

$$\langle \boldsymbol{k}'\sigma' | V_S + V_L | \boldsymbol{k}\sigma \rangle = -\left(\frac{\hbar}{2\pi}\right)^2 \frac{\gamma r_e}{m_p}\, \boldsymbol{\tau}_{\sigma'\sigma} \cdot \boldsymbol{\mathcal{M}}^{\perp}(-\boldsymbol{q})$$

with $\gamma = g_e g_n/4 = 1.915$. Inserting this into equation (10.1) leads us to the expression:

$$\begin{aligned}
&\left(\frac{2\pi}{\hbar}\right)^4 m_n^2\, |\langle b|\langle \boldsymbol{k}'\sigma' | V_S + V_L | \boldsymbol{k}\sigma\rangle |a\rangle|^2 \\
&= \left(\frac{m_n}{m_p}\right)^2 (\gamma r_e)^2 \langle a|\boldsymbol{\tau}_{\sigma\sigma'} \cdot \boldsymbol{\mathcal{M}}^{\perp}(\boldsymbol{q})|b\rangle\langle b|\boldsymbol{\tau}_{\sigma'\sigma} \cdot \boldsymbol{\mathcal{M}}^{\perp}(-\boldsymbol{q})|a\rangle.
\end{aligned}$$

From here on, we shall replace $(m_n/m_p)^2 = 1.003$ by unity. We can now average over the neutron spin coordinates. To this end, we make use of equation (23):

$$\sum_\sigma \frac{1}{2} \sum_{\sigma'} \tau^{\mu_1}_{\sigma\sigma'}\tau^{\mu_2}_{\sigma'\sigma} = \frac{1}{2}\,\mathrm{Tr}\ \tau^{\mu_1}\tau^{\mu_2} = \delta_{\mu_1\mu_2}.$$

It follows that

$$\begin{aligned}
&\sum_\sigma \frac{1}{2} \sum_{\sigma'} \langle a|\boldsymbol{\tau}_{\sigma\sigma'} \cdot \boldsymbol{\mathcal{M}}^{\perp}(\boldsymbol{q})|b\rangle\langle b|\boldsymbol{\tau}_{\sigma'\sigma} \cdot \boldsymbol{\mathcal{M}}^{\perp}(-\boldsymbol{q})|a\rangle \\
&= \sum_\sigma \frac{1}{2} \sum_{\sigma'} \sum_{\mu_1\mu_2} \langle a|\tau^{\mu_1}_{\sigma\sigma'}\mathcal{M}^{\perp}_{\mu_1}(\boldsymbol{q})|b\rangle\langle b|\tau^{\mu_2}_{\sigma'\sigma}\mathcal{M}^{\perp}_{\mu_2}(-\boldsymbol{q})|a\rangle \\
&= \sum_\mu \langle a|\mathcal{M}^{\perp}_{\mu}(\boldsymbol{q})|b\rangle\langle b|\mathcal{M}^{\perp}_{\mu}(-\boldsymbol{q})|a\rangle \\
&= \sum_{\mu\nu} \left(\delta_{\mu\nu} - \hat{q}_\mu \hat{q}_\nu\right)\langle a|\mathcal{M}_\mu(\boldsymbol{q})|b\rangle\langle b|\mathcal{M}_\nu(-\boldsymbol{q})|a\rangle.
\end{aligned}$$

For the last line, we have used $\boldsymbol{\mathcal{M}}^{\perp} = \boldsymbol{\mathcal{M}} - (\boldsymbol{\mathcal{M}} \cdot \hat{\boldsymbol{q}})\hat{\boldsymbol{q}}$. Collecting the results and using equation (10.1) yields the cross section given in equation (10.27).

Specific operators can be probed with polarized neutrons. For instance, if the incident neutrons are fully polarized with $|\sigma\rangle = |\uparrow\rangle$ and one counts only neutrons with $|\sigma'\rangle = |\uparrow\rangle$, then the relevant operator is $\boldsymbol{\tau}_{\uparrow\uparrow}\cdot\boldsymbol{\mathcal{M}}^{\perp} = \mathcal{M}_z^{\perp} = \mathcal{M}_z - (\boldsymbol{\mathcal{M}}\cdot\hat{\boldsymbol{q}})\hat{q}_z$. If, instead, $|\sigma\rangle = |\uparrow\rangle$ and $|\sigma'\rangle = |\downarrow\rangle$, the probed operator is the spin-lowering one, $\mathcal{M}_-^{\perp} = \mathcal{M}_x^{\perp} - i\mathcal{M}_y^{\perp}$, etc.

Doc A10.5 Scattering of neutrons by spin waves

In a monoatomic ferromagnet, the bosonization (see Doc A2.15) gives the following representation of the spin operator to linear order:

$$S^x(\boldsymbol{q}) = \hbar\sqrt{\frac{\mathcal{N}S}{2}}\left(b_{\boldsymbol{q}} + b_{-\boldsymbol{q}}^{\dagger}\right)$$

$$S^y(\boldsymbol{q}) = -i\hbar\sqrt{\frac{\mathcal{N}S}{2}}\left(b_{\boldsymbol{q}} - b_{-\boldsymbol{q}}^{\dagger}\right)$$

$$S^z(\boldsymbol{q}) = \mathcal{N}\hbar S\delta_{\boldsymbol{q}0} - \hbar\sum_{\boldsymbol{k}} b_{\boldsymbol{k}}^{\dagger}b_{\boldsymbol{k}+\boldsymbol{q}}.$$

It is easy to check that the nonzero correlations functions are

$$\begin{aligned}
-\langle T_\tau S^x(\boldsymbol{q},\tau)S^x(-\boldsymbol{q},0)\rangle &= -\langle T_\tau S^y(\boldsymbol{q},\tau)S^y(-\boldsymbol{q},0)\rangle \\
&= \mathcal{N}\hbar^2\frac{S}{2}\left[\mathscr{C}_{b_{\boldsymbol{q}}b_{\boldsymbol{q}}^{\dagger}}(\tau) + \mathscr{C}_{b_{-\boldsymbol{q}}b_{-\boldsymbol{q}}^{\dagger}}(-\tau)\right] \\
-\langle T_\tau S^x(\boldsymbol{q},\tau)S^y(-\boldsymbol{q},0)\rangle &= +\langle T_\tau S^y(\boldsymbol{q},\tau)S^x(-\boldsymbol{q},0)\rangle \\
&= \mathcal{N}\hbar^2\frac{S}{2}i\left[\mathscr{C}_{b_{\boldsymbol{q}}b_{\boldsymbol{q}}^{\dagger}}(\tau) - \mathscr{C}_{b_{-\boldsymbol{q}}b_{-\boldsymbol{q}}^{\dagger}}(-\tau)\right] \\
-\langle T_\tau S^z(\boldsymbol{q},\tau)S^z(-\boldsymbol{q},0)\rangle &= \mathcal{N}\hbar^2 S\delta_{\boldsymbol{q}0}\left(-\mathcal{N}S + 2\sum_{\boldsymbol{k}}\langle b_{\boldsymbol{k}}^{\dagger}b_{\boldsymbol{k}}\rangle\right).
\end{aligned}$$

In the $\langle S^zS^z\rangle$ correlation function, we have discarded a four-boson term in order to remain consistent within the linearized theory. The crossed terms $\langle S^xS^y\rangle$ and $\langle S^yS^x\rangle$ cancel in the sum (10.39) and the $\langle S^zS^z\rangle$ term is time-independent, hence it contributes to the elastic scattering. Collecting all inelastic terms leads to

$$\begin{aligned}
\chi_{\mathcal{M}\mathcal{M}}^{\text{inel}}(\boldsymbol{q},\tau) &= \mathcal{N}g^2|\mathcal{F}(\boldsymbol{q})|^2 e^{-q^2\langle(u^{\parallel})^2\rangle}\left(2 - \hat{q}_x^2 - \hat{q}_y^2\right)\frac{S}{2}\left[\mathscr{C}_{b_{\boldsymbol{q}}b_{\boldsymbol{q}}^{\dagger}}(\tau) + \mathscr{C}_{b_{-\boldsymbol{q}}b_{-\boldsymbol{q}}^{\dagger}}(-\tau)\right] \\
&= \mathcal{N}g^2|\mathcal{F}(\boldsymbol{q})|^2 e^{-q^2\langle(u^{\parallel})^2\rangle}\left(1 + \hat{q}_z^2\right)\frac{S}{2}\left[\mathscr{C}_{b_{\boldsymbol{q}}b_{\boldsymbol{q}}^{\dagger}}(\tau) + \mathscr{C}_{b_{-\boldsymbol{q}}b_{-\boldsymbol{q}}^{\dagger}}(-\tau)\right].
\end{aligned}$$

Upon transforming to bosonic Matsubara frequencies $i\Omega_n$, making the analytic continuation $i\Omega_n \to \varepsilon + i0^+$, and taking the imaginary part as in equation (10.31), $\mathscr{C}_{b_{\boldsymbol{q}}b_{\boldsymbol{q}}^{\dagger}}(\tau)$ becomes $\rho_{b_{\boldsymbol{q}}b_{\boldsymbol{q}}^{\dagger}}(\varepsilon) = \delta(\varepsilon - \hbar\omega_{\boldsymbol{q}})$ and $\mathscr{C}_{b_{-\boldsymbol{q}}b_{-\boldsymbol{q}}^{\dagger}}(-\tau)$ becomes $-\delta(\varepsilon + \hbar\omega_{-\boldsymbol{q}})$ (see Doc A3.6.3 for the spectral function of independent bosons). We thus arrive at equation (10.40).

For an anti-ferromagnet on a bipartite lattice, the relations between the spin-wave modes and the original spins are given at the end of Doc A2.18. There, we read

$$S^x(\boldsymbol{q}) = \hbar \sin \vartheta \sqrt{\mathcal{N} S}\ (u_q - v_q)\left(\beta_q + \beta_{-q}^\dagger\right)$$
$$S^y(\boldsymbol{q}) = i\hbar \sqrt{\mathcal{N} S}\ (u_q + v_q)\left(\beta_q - \beta_{-q}^\dagger\right),$$

with $\sin\vartheta$ given by equation (2.85) and proportional to the magnetic field applied along z. Like for the ferromagnet, the S^z component has a term at $\boldsymbol{q} = \boldsymbol{0}$ and a term quadratic in the spin-wave operators and does not contribute to the inelastic scattering. The $\langle S^x S^x \rangle$ and $\langle S^y S^y \rangle$ correlation functions are given by

$$-\langle T_\tau S^x(\boldsymbol{q}, \tau) S^x(-\boldsymbol{q}, 0)\rangle = \mathcal{N}\hbar^2 S \sin^2\vartheta (u_q - v_q)^2 \left[\mathscr{C}_{\beta_q \beta_q^\dagger}(\tau) + \mathscr{C}_{\beta_{-q}\beta_{-q}^\dagger}(-\tau)\right]$$
$$-\langle T_\tau S^y(\boldsymbol{q}, \tau) S^y(-\boldsymbol{q}, 0)\rangle = \mathcal{N}\hbar^2 S (u_q + v_q)^2 \left[\mathscr{C}_{\beta_q \beta_q^\dagger}(\tau) + \mathscr{C}_{\beta_{-q}\beta_{-q}^\dagger}(-\tau)\right],$$

while the $\langle S^x S^y \rangle$ and $\langle S^y S^x \rangle$ are opposite and cancel. We deduce the quantity entering the spin susceptibility $\chi_{\mathcal{M}\mathcal{M}}$:

$$\sum_{ij}\left(\delta_{ij} - \hat{q}_i \hat{q}_j\right)\left(-\frac{1}{\hbar^2}\right)\langle T_\tau S^i(\boldsymbol{q}, \tau) S^j(-\boldsymbol{q}, 0)\rangle = \mathcal{N} S$$
$$\times\left[(1 - \hat{q}_x^2)(u_q - v_q)^2 \sin^2\vartheta + (1 - \hat{q}_y^2)(u_q + v_q)^2\right]\left[\mathscr{C}_{\beta_q \beta_q^\dagger}(\tau) + \mathscr{C}_{\beta_{-q}\beta_{-q}^\dagger}(-\tau)\right].$$

In the calculation of section 2.6.3 and Doc A2.18, the spins live in the (x, z) plane. In reality, all planes parallel to the z axis are equivalent by symmetry. Our choice breaks the symmetry between x and y in the formula for $\chi_{\mathcal{M}\mathcal{M}}$. The symmetry can be restored by averaging the above expression with the one where x and y are exchanged. We are thus led to the one-magnon cross section given in equation (10.41).

Doc A10.6 Scattering of neutrons by free electrons

For free electrons, equation (10.44) and the standard diagrammatic rules give

$$\left(-\frac{1}{\pi}\right)\operatorname{Im}\chi^0_{\mathcal{M}\mathcal{M}}(\boldsymbol{q}, \varepsilon) = \sum_{\boldsymbol{k}}\left(1 + \frac{2}{q^2}|\hat{\boldsymbol{q}} \times \boldsymbol{k}|^2\right) f(\xi_{\boldsymbol{k}})$$
$$\times\left[\delta(\varepsilon + \xi_{\boldsymbol{k}} - \xi_{\boldsymbol{k}+\boldsymbol{q}}) - \delta(\varepsilon + \xi_{\boldsymbol{k}-\boldsymbol{q}} - \xi_{\boldsymbol{k}})\right].$$

The calculation at $T = 0$ is analogous to the one presented in Doc A5.14. One converts the $\boldsymbol{k}$ sum to an integral in spherical coordinates chosen such that $\hat{\boldsymbol{k}}_z$ is parallel to $\boldsymbol{q}$, hence $|\hat{\boldsymbol{q}} \times \boldsymbol{k}| = k \sin\vartheta$ and $\boldsymbol{q} \cdot \boldsymbol{k} = qk\cos\vartheta$, one introduces the dimensionless variables

$$x = \frac{q}{2k_{\mathrm{F}}}, \qquad y = \frac{\varepsilon}{4\varepsilon_{\mathrm{F}}}, \qquad u = \frac{k}{k_{\mathrm{F}}}$$

as well as the Fermi level DOS $N_0^{\mathrm{el}}(0) = mk_{\mathrm{F}}\mathcal{V}/(\pi^2\hbar^2)$, and one expresses the delta function as, e.g.

$$\delta(\varepsilon + \xi_{\boldsymbol{k}} - \xi_{\boldsymbol{k}+\boldsymbol{q}}) = \frac{1}{4\varepsilon_{\mathrm{F}}xu}\delta\left(\frac{y - x^2}{xu} - \cos\vartheta\right).$$

The result of these substitutions is

$$\left(-\frac{1}{\pi}\right)\mathrm{Im}\,\chi^0_{\mathcal{M}\mathcal{M}}(\boldsymbol{q}, \varepsilon) = \frac{N_0^{\mathrm{el}}(0)}{8x}\int_0^1 du \int_0^\pi d\vartheta\, u \sin\vartheta\left(1 + \frac{u^2\sin^2\vartheta}{2x^2}\right)$$
$$\times\left[\delta\left(\frac{y - x^2}{xu} - \cos\vartheta\right) - \delta\left(\frac{y + x^2}{xu} - \cos\vartheta\right)\right].$$

The first delta function contributes only if $u > |(y - x^2)/x|$ and the second only if $u > |(y + x^2)/x|$, and their contribution, according to equation (11), is $1/\sin\vartheta$ because $\sin\vartheta$ is positive in the domain of integration. The remaining integrals can be deduced from the identity

$$\int_0^1 du\, u\left[1 + \frac{1}{2x^2}(u^2 - a^2)\right]\theta(u - |a|) = \frac{1}{2}\theta(1 - a^2)\left[1 - a^2 + \frac{(1 - a^2)^2}{4x^2}\right].$$

This leads to equation (10.45).

www.ingramcontent.com/pod-product-compliance
Ingram Content Group UK Ltd.
Pitfield, Milton Keynes, MK11 3LW, UK
UKHW051906100926
8556IPUK00005B/779

9 780750 317399